PRIMARY IMMUNODEFICIENCY DISORDERS

PRIMARY IMMUNODEFICIENCY DISORDERS

A HISTORIC AND SCIENTIFIC PERSPECTIVE

Edited by

AMOS ETZIONI

HANS D. OCHS

ELSEVIER

AMSTERDAM • BOSTON • HEIDELBERG • LONDON • NEW YORK • OXFORD • PARIS
SAN DIEGO • SAN FRANCISCO • SINGAPORE • SYDNEY • TOKYO

Academic Press is an Imprint of Elsevier

Academic Press is an imprint of Elsevier
The Boulevard, Langford Lane, Kidlington, Oxford OX5 1GB, UK
225 Wyman Street, Waltham, MA 02451, USA

First edition 2014

Notice
No responsibility is assumed by the publisher for any injury and/or damage to persons or property as a matter of products liability, negligence or otherwise, or from any use or operation of any methods, products, instructions or ideas contained in the material herein. Because of rapid advances in the medical sciences, in particular, independent verification of diagnoses and drug dosages should be made

British Library Cataloguing in Publication Data
A catalogue record for this book is available from the British Library

Library of Congress Cataloging-in-Publication Data
A catalog record for this book is availabe from the Library of Congress

ISBN: 978-0-12-407179-7

For information on all Academic Press publications
visit our website at http://store.elsevier.com/

Typeset by Thomson Digital

14 15 16 17 18   10 9 8 7 6 5 4 3 2 1

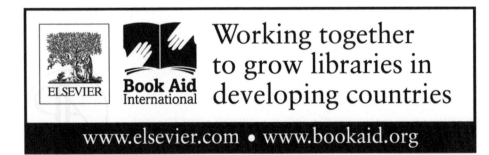

Contents

Contributors

Bernd H. Belohradsky, MD Professor of Pediatrics, Haunerschen Kinderspital, Ludwig Maximilians University, München, Germany

Melvin Berger, MD, PhD CSL Behring LLC, King of Prussia, PA, and Adjunct Professor of Pediatrics and Pathology, Case Western Reserve University, Cleveland, OH, USA

Aziz A. Bousfiha Clinical Immunology Unit, IbnRoshd University Hospital. King Hassan II University, Casablanca, Morocco

Jean-Laurent Casanova, MD, PhD Laboratory of Human Genetics of Infectious Diseases, Necker Branch, INSERM U980, Necker Medical School, Imagine Institute, Paris Descartes University, Paris, France; Pediatric Hematology-Immunology Unit, Necker Hospital for Sick Children, Paris, France; St Giles Laboratory of Human Genetics of Infectious Diseases, Rockefeller Branch, The Rockefeller University, New York, NY, USA

Marina Cavazzana-Calvo, MD, PhD INSERM U1163, Paris; Sorbonne Paris Cité, Université Paris Descartes, Imagine Institute, Paris; Biotherapy Clinical Investigation Center, Groupe Hospitalier Universitaire Ouest, Assistance Publique–Hôpitaux de Paris, INSERM, Paris, France

Helen M. Chapel, MA MD, FRCP Nuffield Department of Medicine, University of Oxford, Oxford University Hospitals, Oxford, UK

Antonio Condino-Neto Department of Immunology, Institute of Biomedical Sciences, University of São Paulo, Brazil

Max D. Cooper, MDM Professor of Pathology and Laboratory Medicine, Emory University School of Medicine, Atlanta, GA

Charlotte Cunningham-Rundles, MD, PhD Mount Sinai School of Medicine, New York, NY, USA

Robert Currier, PhD Genetic Disease Screening Program, California Department of Public Health, Richmond, CA, USA

Geneviève de Saint Basile Inserm U1163, Paris; CEDI, Assistance Publique-Hôpitaux de Paris, Paris; Univ. Paris Descartes, Sorbonne Paris Cité, Imagine Institute, Paris, France

Carol Ann Demaret The David Center, Texas Children's Hospital, Baylor College of Medicine, Houston, Texas; The Department of Pediatrics, Immunology, Allergy, and Rheumatology, Baylor College of Medicine, Houston, Texas

Anne Durandy Institut National de la Santé et de la Recherche Médicale U768, Hôpital Necker Enfants Malades, Paris; Faculté de Médecine, Descartes-Sorbonne Paris Cité University of Paris, *Imagine* Institute, Paris; Centre d'Étude des Déficits Immunitaires, Hôpital Necker Enfants Malades, Paris, France

Karin R. Engelhardt Centre for Chronic Immunodeficiency, University Medical Centre, Freiburg, Germany

Amos Etzioni, MD Meyer Children Hospital, Rambam Medical Campus, Rappaport Faculty of Medicine, Technion, Haifa, Israel

Alain Fischer, MD, PhD INSERM U1163, Paris; Sorbonne Paris Cité, Université Paris Descartes, *Imagine* Institute, Paris; Immunology and Pediatric Hematology Department, Necker Children's Hospital, Assistance Publique–Hôpitaux de Paris, INSERM, Paris; Collège de France, Paris, France

Thomas A. Fleisher, M.D. Chief, Department of Laboratory Medicine, NIH Clinical Center, National Institutes of Health, Bethesda, MD, USA

Michael M. Frank, MD Duke University Medical Center, Durham, NC, USA

Richard A. Gatti, MD Departments of Pathology and Laboratory Medicine, and Human Genetics, David Geffen School of Medicine at UCLA, Los Angeles, CA, USA

Raif S. Geha, MD James L. Gamble Professor of Pediatrics, Harvard Medical School, Boston, MA, USA; Chief, Division Immunology, Boston Children's Hospital, Boston, MA, USA

Bodo Grimbacher Centre for Chronic Immunodeficiency, University Medical Centre, Freiburg, Germany

Salima Hacein-Bey-Abina, PharmD, PhD INSERM U1163, Paris; Sorbonne Paris Cité, Université Paris Descartes, Imagine Institute, Paris; Biotherapy Clinical Investigation Center, Groupe Hospitalier Universitaire Ouest, Assistance Publique–Hôpitaux de Paris, INSERM, Paris, France

Michael S. Hershfield Professor of Medicine and Biochemistry, Duke University School of Medicine, Durham, NC, USA

Rochelle Hirschhorn Professor Emerita of Medicine, Cell Biology and Pediatrics, Research Professor of Medicine, New York University Medical Center, NY, USA

Steven M. Holland, MD Laboratory of Clinical Infectious Diseases, National Institute of Allergy and Infectious Diseases, National Institutes of Health, Bethesda, Maryland, USA

Leila Jeddane Clinical Immunology Unit, IbnRoshd University Hospital. King Hassan II University, Casablanca, Morocco

Sven Kracker Institut National de la Santé et de la Recherche Médicale U768, Hôpital Necker Enfants Malades, Paris; Faculté de Médecine, Descartes-Sorbonne Paris Cité University of Paris, *Imagine* Institute, France

Warren J. Leonard Laboratory of Molecular Immunology and The Immunology Center, National Heart, Lung, and Blood Institute, National Institutes of Health, Bethesda, MD, USA

Deborah McCurdy, MD Department of Pediatrics, David Geffen School of Medicine at UCLA, Los Angeles, California, USA

Donna M. McDonald-McGinn The Division of Human Genetics, The Children's Hospital of Philadelphia, Philadelphia, PA, USA

Hilaire J. Meuwissen Department of Pediatrics, Albany Medical Center, Albany, NY, USA

Fred M. Modell Jeffrey Modell Foundation, NY, USA

Vicki M. Modell Jeffrey Modell Foundation, NY, USA

Luigi Daniele Notarangelo Professor of Pediatrics, Children's Hospital, Boston, MA, USA

Hans D. Ochs, MD Professor of Pediatrics, University of Washington School of Medicine, Seattle Children's Research Institute, Seattle, WA, USA

Capucine Picard, MD, PhD Laboratory of Human Genetics of Infectious Diseases, Necker Branch, INSERM U980, Necker Medical School, *Imagine* Institute, Paris Descartes University, Paris, France; Study Center for Primary Immunodeficiencies, Assistance Publique-Hôpitaux de Paris, Necker Hospital for Sick Children, Paris, France; St Giles Laboratory of Human Genetics of Infectious Diseases, Rockefeller Branch, The Rockefeller University, New York, NY, USA

Jennifer M. Puck, MD Division of Allergy, Immunology and Blood and Bone Marrow Transplantation, Department of Pediatrics, University of California San Francisco and UCSF Benioff Children's Hospital, San Francisco, CA, USA

William T. Shearer, MD, PhD The David Center, Texas Children's Hospital, and the The Department of Pediatrics, Immunology, Allergy, and Rheumatology, Baylor College of Medicine, Houston, Texas

C.I. Edvard Smith, MD, PhD Professor of Molecular Genetics, Clinical Research Center, Department of Laboratory Medicine, Karolinska Institutet, Stockholm/Huddinge, Sweden

E. Richard Stiehm, MD Department of Pediatrics, David Geffen School of Medicine at UCLA, Los Angeles CA, USA

Rainer Storb, MD Fred Hutchinson Cancer Research Center, University of Washington, School of Medicine, Seattle, WA, USA

Kathleen E. Sullivan The Division of Allergy Immunology, The Children's Hospital of Philadelphia, Philadelphia, PA, USA

Karl Welte Department of Molecular Hematopoiesis, Medical School Hannover, Hannover, Germany

Jerry A. Winkelstein, MD Emeritus Professor of Pediatrics, Medicine and Pathology, Johns Hopkins University School of Medicine, Baltimore, Maryland, USA

Foreword

Hans D. Ochs and Amos Etzioni

THE GIANTS OF OUR FIELD

The recognition, study and quest for effective therapy of primary immunodeficiency diseases (PIDD) have been around for barely 60 years. In this historical review, we commemorate the discovery of X-linked agammaglobulinemia (XLA) and the treatment of this condition with immunoglobulin. With most of those who have witnessed the early days of our field getting on in age, if still living, memories turn hazy and details are being forgotten. Therefore, this memorial to the giants in our field, many of whom were our mentors and teachers, is both a worthy and pleasurable exercise.

To become a giant in any field, one has to be at the right place at the right time. One of the outstanding scientists of the 19th century, the microbiologist Robert Koch (see Chapter 1) recognized and acknowledged this when, at the end of his career over 100 years ago, he mused: "If my efforts have led to greater success than usual, this is due, I believe, to the fact that during my wanderings in the field of medicine, I have strayed onto paths where the gold was still lying by the wayside. It takes a little luck to distinguish gold from dross, but that is all". There is a handful of giants in our specialty, who, for a short moment of their lives, were truly at the right place at the right time, but who then left the field to pursue other careers. Nevertheless, they entered PIDD history because of one important trait: they knew – or sensed – that they had struck gold; they became fascinated – but not consumed – by what they had observed.

As Bruton describes his discovery of XLA: "No gammaglobulin, no antibody, ergo infections". This was his moment of glory; then he returned to pursue a career as a general pediatrician. Familial thrombocytopenia affecting boys was recognized by Alfred Wiskott as odd, not being idiopathic thrombocytopenia (ITP), and something never seen before, and years later caught the attention of Bob Aldrich, who recognized the X-linked inheritance of this syndrome; both went on to work in other areas of medicine (see Chapter 9). Siblings with ataxia associated with ocular telangiectasia enticed Syllaba and Henner, two Czech neurologists, in 1926, to write a paper – in French; then forgot about it until the syndrome was rediscovered by Boder and Sedgwick in 1957 (see Chapter 8). While these keen observers published their unique findings, the rest of the world was not ready and nobody noticed until years later, when the medical establishment caught up to their vanguard. However, as described in this book, many of those early giants went beyond scientific "one night stands", and pursued a career in this new clinical field of immunology and primary immunodeficiency.*

THE RIGHT TIME

It was not until post-World War II, when medical science was ready to look for and recognize PIDD. Infant mortality in the developed world had plummeted, related directly to safe water, improved nutrition, progress in microbiology,

---

*Many of the black and white figures shown in the Foreword and the individual chapters are reproduced in color picture section located in the insert at the end of the book.

the discovery of antibiotics, a well-organized public health system, enforced vaccination and publicly funded medical research. These prerequisites were in place when Ogden Bruton, a general Army Pediatrician, discovered XLA in 1952.

THE RIGHT PLACE

As illustrated in the first chapter of this book, the action in biology and medicine during the 19th and early 20th century was Berlin, Paris, London and Vienna, the capital cities of Europe's 19th century super powers. However, the two world wars changed the landscape and new cities emerged where biomedical research began to flourish and to which the young and bright students of medical science were drawn. One such city was Boston. As part of the war effort during World War II, the US Army requested the development of safe plasma collection and fractionation in order to obtain the large quantity of albumin needed for the treatment of hypovolemic shock on the battlefield. To facilitate this request, the National Research Council charged and funded Edward Cohn, a biochemist working in the Department of Physical Chemistry at Harvard Medical School, with the isolation of stable protein fractions derived from human blood. In collaboration with his colleague Jay Oncley, Cohn succeeded in the large scale fractionation of human plasma. In a single volume of the *Journal of Clinical Investigation* (Vol 23, Feb 17, 1944), a series of reports were published that described the technique of fractionating plasma, characterized the protein fractions and, importantly, suggested that the gammaglobulin containing fraction II also included specific antibodies, and that concentrated gammaglobulin derived from normal human serum was highly effective in preventing or attenuating measles. One of the authors who had contributed to this series of papers was pediatrician Charles Janeway (Fig. F.1), who went on to explore the therapeutic value of gammaglobulin in humans

FIGURE F.1 Charles Janeway: Introduced gammaglobulin as source of antibodies.

and animals. He communicated with Bruton following the discovery of XLA, and identified several agammaglobulinemic patients of his own. As Chairman of the Department of Pediatrics at Harvard, Janeway attracted a cadre of young physician-scientists to Boston to work with him in this new field. David Gitlin focused on the characterization of gammaglobulin and its use in the treatment of patients with agammaglobulinemia. Ralph Wedgwood (Fig. F.2) started his career in immunology/rheumatology at Boston Children's Hospital by exploring the role of complement in collagen diseases. He

FIGURE F.2 Ralph Wedgwood: Connected immunodeficiency with autoimmunity.

FIGURE F.3 Fred Rosen: Instrumental in creating the field of PIDD as a specialty in medicine.

FIGURE F.4 Robert Good: Considered the Father of PIDD in North America.

then moved to Seattle, and together with Hans Ochs, a young physician newly emigrated from Germany, established the largest PIDD center on the West Coast. Fred Rosen (Fig. F.3) began to investigate the clinical and laboratory characteristics of agammaglobulinemia, Wiskott–Aldrich syndrome (WAS) and "dysgammaglobulinemia", which he subsequently defined clinically and, in 1961, named hyper IgM syndrome. Boston rapidly became a center of PIDD research, a legacy that continues with a new generation of Bostonians spearheading cutting edge research including Raif Geha, Gigi Notarangelo, Talal Chatila, and Francisco Bonilla.

At the time when Boston's reputation as an Immune Deficiency Center was in full swing, a young, MD/PhD student from Minnesota began his own stellar career in PIDD. Robert Good (Fig. F.4), born in 1922, had just recovered from a polio-like illness that temporarily placed him in a wheelchair and caused a permanent limp, when he entered medical school at age 22. After completing his clinical training in Pediatrics at the University of Minnesota, he began a one-year fellowship at the Rockefeller University in New York, where he was fired up by immunologist and rheumatologist Henry Kunkel. Bob Good returned to the University of Minnesota, where he pursued this new interest, focusing on adaptive

immunity and inherited immune defects. Like Janeway, he attracted a group of brilliant and enthusiastic young investigators, many of whom became giants in the field in their own right. Max Cooper (Fig. F.5), while a student and collaborator with Good, discovered the dual roles of the thymus- and bursa-dependent cell lineages, now known as T and B cells (see Chapter 2). While a member of Good's team in 1968, Cooper demonstrated that WAS represented an immunodeficiency disorder (see Chapter 9). Another "Good Guy", R. Peterson, recognized the immune defect in ataxia telangiectasia (see Chapter 8), and

FIGURE F.5 Max Cooper: Established the concept of B and T lymphocytes and adaptive immunity.

Paul Quie discovered that neutrophils from patients with chronic granulomatous disease (CGD) failed to kill certain ingested bacteria. With his students, Richard Gatti (who reported the first successful bone marrow transplantation in a patient with severe combined immunodeficiency (SCID); see Chapter 24), Richard Hong, Hillaire Meuwissen and Richard O'Reilly, Good explored the technique of bone marrow transplantation for patients with cellular immune deficiency. Meuwissen and Ben Polara, both trained in Good's laboratory, recognized a new SCID phenotype, which was molecularly identified by Elo Giblett (Fig. F.6) as adenosine deaminase (ADA) deficiency (see Chapter 20). Mary Ann South, while in Good's laboratory, studied IgA deficiency and later, while in Houston, Texas, diagnosed a newly born boy as a SCID patient and placed him in gnotobiotic isolation, hoping, in vain, that "bubble boy" David's immune defect would eventually mature (see Chapter 25). Charlotte Cunningham-Rundles had joined Bob Good as a fellow while he was President of the Sloan Kettering Institute in NY. As an internist, she was drawn to common variable immune deficiency (CVID) and together with Helen Chapel from Oxford, England, devoted her medical career to CVID.

At NIH, Tom Waldmann studied gammaglobulin metabolism in patients with PIDD, demonstrating hyper-katabolism of these proteins in WAS patients and, in 1968, together with Mike Blaese, defined the cellular and humoral immune defects in WAS (see Chapter 9). Together with Steve Polmar, Waldmann carried out measurements of serum IgE in normal and antibody-deficient patients. Warren Leonard, also at NIH, discovered that in X-linked SCID the molecular defect resided in the common γ-chain, required for the interleukin 2 receptor (IL-2R) and a number of other lymphokine receptors, and that in autosomal recessive SCID with a similar clinical phenotype, the defect was in JAK3 (see Chapter 15). John Gallin, Harry Malech and Steve Holland, with a group of young investigators at NIH, have investigated CGD over the last twenty years (see Chapter 13). More recently, Holland studied the molecular basis of atypical mycobacterial disease, co-discovered the hyper-IgE syndromes (see Chapter 19) and investigated other molecularly defined diseases affecting innate immunity, with the support of Tom Fleisher, an expert in immunological laboratory diagnosis (see Chapter 5).

Another giant in the field is Rebecca Buckley (Fig. F.7), who trained at Duke University as a fellow in allergy and stayed on (in loyalty to

FIGURE F.6 Elo Giblett: Connected PIDD with genetics by discovering the first gene causing PIDD, adenosine deaminase.

FIGURE F.7 Rebecca Buckley: Pioneered early bone marrow transplantation for SCID.

Duke's basketball team) to build one of the most successful bone marrow transplant units for the treatment of SCID, with special interest in early transplantation using haploidentical donors. Mary Ellen Conley, a trainee of Max Cooper, investigated autosomal recessive agammaglobulinemia and identified at least six genes that, if defective, result in lack of B cells and autosomal recessive agammaglobulinemia.

On the other side of the Atlantic, contemporary and equally accomplished giants were making great strides in the field of primary immunodeficiency – there universally known as PID. In France, Maxime Seligmann (Fig. F.8) developed a profound interest in immunology and PID. After finishing medical school, he started his career at the Pasteur Institute, where he was surrounded by basic immunologists. In 1957, he became head of clinical immunology at Saint-Louis Hospital in Paris. He was the first to recognize anti-DNA antibodies in patients with systemic lupus erythematosus (SLE). When investigating complement, he became aware of PID. In 1968, he proposed to generate a classification of primary immunodeficiency diseases with the statement: "We wish to underline that in spite of increased knowledge, the nature of the basic defect remains unknown in most PID syndromes and most patients' diagnoses resemble

a wastebasket". He joined other pioneers in a special WHO committee which was charged to classify the various forms of PID known in those early days. One of Seligmann's most successful fellows was Claude Griscelli, born in Morocco, who joined Seligmann's lab in Paris before spending time at the laboratory of Nobel Laureate Baruj Benacerraf at New York University. Griscelli returned to Paris to start the first clinical unit at Hopital Necker-Enfants Malades for the care and study of children with PID. There he established a laboratory research unit of hereditary immunodeficiency to understand better the immune defects of those patients. He founded the club "The Young Lymphomaniaques" which, over time, transformed into one of the most successful groups in the PID field. Many unique syndromes were described by these "maniaques", including bare lymphocyte syndromes, Griscelli syndrome and a list of hyper IgM syndromes. One of Griscelli's students, Alain Fischer, succeeded Griscelli who, in 1996, went on to become directeur general de l'institute national de la santé et de la recherché medicale (INSERM). Fischer further expanded the Immunodeficiency Unit at Necker, and led his group to discover new molecularly defined PIDs, to refine bone marrow transplantation for PID patients and, eventually, to introduce, for the first time, gene therapy for the treatment of X-linked SCID (see Chapter 26). Over the years, the primary immunodeficiency center in Paris, started by Maxime Seligmann, expanded by Claude Griscelli and transformed into a superb clinical and research entity by Alain Fischer, has become the premier location for innovative research in the field, addressing both basic science and its transfer to clinical care. Not surprisingly, the second generation of immunologists from this institution (now known as Imagine Institute) has, over the years, attained "giant" status in their own rights: Anne Durandy (see Chapter 16), Geneviève de Saint Basile (see Chapter 12), Marina Cavazzana-Calvo (see Chapter 26) and Jean-Laurent

FIGURE F.8   Maxime Seligman: Early pioneer of PID in Europe.

FIGURE F.9   Walter Hitzig: Discovered severe combined immunodeficiency.

FIGURE F.10   Silvio Barandun: His study of gamma-globulin as treatment for antibody deficiency led to the first IVIG preparation.

Casanova (see Chapter 3) who has defined the molecular basis of innate immunity.

In Switzerland, Walter Hitzig (Fig. F.9), born in Mexico to Swiss parents, entered medical school in Zurich, and studied pediatrics at the local Children's Hospital. There he observed several infants with a lymphocytosis (described earlier by Swiss pathologists as "Essentielle Lymphozytophtise" without recognizing the immune deficiency) and realized that these patients were different from the recently described patients with agammaglobulinemia. As both boys and girls were affected, he concluded autosomal recessive inheritance, a form of SCID later described as "Swiss type agammaglobulinemia". Hitzig and his colleagues published their observations in 1958, pointing out that these patients were more severely affected, since treatment with gammaglobulin did not improve the clinical course, which was inevitably lethal. In 1970, a WHO committee designated the syndrome "severe combined immunodeficiency". Hitzig was fascinated by this new specialty and decided to visit the place where the action was in the 1950s. He headed for Boston to become a post-doc in Charles Janeway's laboratory, where he spent a year studying congenital and acquired agammaglobulinemia. Upon his return

to Switzerland, he was rewarded by his chairman, Prof. Guido Fanconi, who created the first chair in "Pediatric Immunology", and named Walter Hitzig the first occupant. Hitzig attracted young colleagues to work with him on the study of PID and the development of immunoglobulin preparations that eventually resulted in the first intravenous immunoglobulin (IVIG) preparation: Silvio Barandun (Fig. F.10), who experimented extensively with immunoglobulin preparations; Alfred Hässig, who founded the Swiss Red Cross blood bank system and started the movement promoting unpaid blood donations in Switzerland and other countries, and developed Sandoglobulin; Paul Imbach who discovered the effect of IVIG on idiopathic thrombocytopenia (ITP); and Rainhart Seger, who devoted his career to the study and treatment of chronic granulomatous disease.

In the UK, John Soothill (Fig. F.11) started to create interest in PID. He graduated from medical school in Cambridge, and subsequently became the first Hugh Greenwood Professor of Immunology at Great Ormond Street Hospital. Soothill began his clinical career by studying immune complex diseases of the kidney; he subsequently changed his focus of research to PID. He was one of the original members of

FIGURE F.11 John Soothill: Started the field of PID in England.

the WHO committee for PID and, in 1970, suggested the term severe combined immunodeficiency. He was the first to recognize the syndrome of leukocyte adhesion deficiency. One of his students was Roland Levinski, who joined Soothill while working on his thesis on immune complex diseases and collaborated with him for many years on investigating SCID and other PIDs. He eventually succeeded Soothill as Head of Pediatrics at Great Ormond Street and co-founded the European Group for Primary Immunodeficiency (EGID) which later became ESID. He was one of the pioneers of bone marrow transplantation in the UK. He died of a tragic accident in 2007.

Other post-war European pioneers in the PID field: Jaak Vossen, together with Leonard Dooren and Dirk van Bekkum from the Netherlands, performed the first successful bone marrow transplantation in a SCID patient in Europe. Germany had a late start, recovering from World War II and being politically divided. One German pioneer was Karl Welte (see Chapter 10), who studied medicine in Tübingen and Berlin, followed by a fellowship and young faculty position at Sloan-Kettering Memorial Cancer Center in New York, where he worked for six years on the purification and characterization of cytokines. After his return to Hannover Medical School in Germany, he founded the International Chronic Neutropenia Registry and introduced granulocyte colony-stimulating factor (G-CSF) treatment for neutropenia. Together with Christoph Klein, now chairman of Pediatrics at the University of Munich, Welte made Hannover the place to be for exploring the molecular basis of inherited neutropenias. A few years later, a second German center devoted to PID was created at the University of Freiburg where Hans-Hartmut Peter, Klaus Warnatz and Bodo Grimbacher (see Chapter 19) are heading a team of investigators that focuses on the genetic basis of CVID, hyper-IgE syndromes, and chronic mucocutaneous candidiasis.

Further east, young physicians caught on quickly after being exposed to PIDD during post-doctoral training in the USA. Izzet Berkel, whose family originated in Rhodos, spent time during the late 1980s in Oklahoma before returning to Hacettepe University, Ankara, Turkey. His main interest has been ataxia telangiectasia (A-T) and he participated in the discovery of the *ATM* gene. Amos Etzioni studied pediatric immunology in Philadelphia and subsequently returned to Meyer Children's Hospital in Haifa, where he focused on leukocyte adhesion defects, discovering the molecular basis of LAD2 and LAD3 (see Chapter 21). Yoshi Shiloh, while at the University of Tel Aviv, discovered *ATM*, the gene responsible for A-T (see Chapter 8).

In the tradition of the giants of the 19th century, who introduced modern science into medicine, the pioneers in our field discovered and described the clinical features, solved the molecular and genetic basis, and designed effective or even curative treatments for PID patients. Those following these giants find fertile ground to reap rich harvests.

Introduction

Raif S. Geha

James L. Gamble, Professor of Pediatrics, Harvard Medical School
Chief, Division Immunology, Boston Children's Hospital, Boston, MA, USA

Primary immunodeficiency diseases (PIDDs), as we know them, were born six decades ago. Their saga started with the discovery by Ogden Bruton at Walter Reed Hospital and Charles Janeway and David Gitlin at Children's Hospital in Boston that serum immunoglobulins are absent in agammaglobulinemia. This discovery was made possible by a new technology, protein-electrophoresis. In the intervening years, particularly in the past decade, the field has grown tremendously thanks to advances in flow cytometry, biochemistry and genetics. The study of PIDDs has garnered the interest of basic immunologists who have come to the realization that PIDDs are unique experiments of nature that inform us about the workings of the human immune system in the natural environment we live in. In this book, the 60-year saga of PIDDs is recounted by many of the pioneers in the field.

Early in the days of PIDD research, Robert Good and Max Cooper realized that PIDDs bisect the microbial word into two groups of organisms. The first group consists mostly of encapsulated bacteria that infect patients who lack serum immunoglobulins, become sick after the first six months of life, but generally grow well. The second group consists of microbes that include fungi and viruses that chronically infect patients who manifest the disease early in their life and tend to grow poorly and to die quickly of their disease. These observations led to the view that there are two components in the immune system that deal with distinct types of microbes. This theory was verified by a series of observations. These include the discovery by Bruce

Glick and Max Cooper of the bursa of Fabricius as a B-cell factory in chickens, the subsequent discovery of T-cells and the role of the thymus in their generation, and the earlier finding that the thymus is rudimentary and lacks lymphoid cells in "Swiss type" agammaglobulinemia, a severe combined immunodeficiency (SCID) intensely studied by Walter Hitzig in Zurich, and later identified to be autosomal recessive due to mutations in JAK3 or RAG1/RAG2 or X-linked due to mutations in the IL-2R gamma chain. No one conveyed the excitement about PIDD in the late 1960s, 1970s and 1980s better than Bob Good, an inspirational leader, a frequent traveler and the only immunologist to have made the cover of TIME magazine. Many of the contributors of this book remember Bob, Fred Rosen, Max Cooper, Walter Hitzig from Zurich, Maxime Seligman from Paris, John Suthill from London and Ralph Wedgwood from Seattle organizing the bi-yearly meetings of the WHO (later IUIS) committee for PIDD. These meetings were held for two days in charming and isolated places, often with rather primitive accommodation, which forced memorable interactions between the giants of the PIDD world and young trainees. Young investigators were coaxed to take the podium and share their science and their plans for the future and were offered the unqualified and enthusiastic support of the senior scientists. I am among the many who owe the early giants in PIDD a great debt.

Specific treatment for PIDDs followed very quickly after their initial discovery. Many investigators were involved, but the initial advances came from the Boston and Minnesota groups, who maintained an ongoing amicable competition. Janeway and Rosen in Boston were instrumental in the use of gammaglobulin replacement therapy for antibody deficiency, while Bob Good and Max Cooper in Minnesota pioneered matched allogeneic bone marrow transplantation (BMT) for SCID, shortly followed by Boston and other groups. The early days of BMT conveyed to all involved a feeling of excitement. Everyone was treading into the unknown, and life and death decisions were made with pretty thin data and little precedent. Soon after the initial problems were solved, the challenge became to cross the histocompatibility barrier. A new technology came to the rescue. Several groups took advantage of the recently generated anti-T-cell monoclonal antibodies, or plant lectins and sheep red blood cells that selectively bound to T-cells, to deplete successfully the BM of T-cells and overcome the barrier. Once genes that cause SCID were identified, the next challenge was gene therapy. It was met in the 1990s by Alain Fischer and his group at Hôpital Necker, again building on a newly available technology. Every success carries with it risks and challenges. BMT carried the risk of severe graft-versus-host disease. Gene therapy resulted in some cases in malignant transformation and T-cell leukemia caused by random integration of the retrovirus, resulting in dysregulated expression of oncogenes. In each case, the battle will be eventually won.

Defects in innate immunity were initially slow to be identified, with the exception of neutropenia syndromes and chronic granulomatous disease. Basic immunology came to the rescue with the discovery of Toll-like receptors by Jules Hoffman and Charles Janeway, Jr. In a short period of time, a number of discoveries, many pioneered by Jean-Laurent Casanova, put defects in innate immunity on the PID map.

Since its discovery, PID has been suspected to be associated with autoimmunity. Over the past 20 years, mutations in an increasing number of genes have been associated with autoimmunity and shown to play an important role in central tolerance, peripheral tolerance or apoptosis. In particular, the discovery of FOXP3 deficiency by Hans Ochs and Talal Chatila has been instrumental in highlighting the role of T-regulatory cells in tolerance, and has opened an ever expanding field of investigation, with important implications for a variety of common autoimmune diseases and the mechanisms by which PIDs can result in autoimmunity.

The past five years have witnessed another revolution in PID caused by two technological breakthroughs: the increasing ability to sequence the entire exome, or whole genome, at relatively low cost, and advances in bioinformatics that allow the analysis of enormous amounts of data. This revolution is resulting in the discovery of novel genes at a furious pace, and in the realization that hypomorphic mutations of known genes can give rise to unexpected clinical phenotypes. Integral to the success of this revolution is the ease of global electronic communications and the establishment of collaborative networks such as the one supported by the Jeffrey Modell Foundation. These have facilitated collaborations between immunologists in areas of the world where PIDs are prevalent due to consanguinity and their colleagues in research centers who have access to cutting edge facilities and technologies. There is little doubt that mutations in the approximately 10 000 genes expressed in immune cells will be identified in the coming decade. Efficient methodology for the derivation of induced pluripotent stem cells (iPSCs) derived from PID patients, and for repairing defects in one or more genes in iPSCs already exist. They are being refined by Luigi Notarangelo and others and are expected to be applicable to PIDs in the next decade.

There are, however, clouds on the horizon. In a global economy in recession, the resources available for research in general and PID in particular are becoming increasingly limited. Despite ample evidence, it has been hard to convince granting agencies that research in PID is of ultimate importance for common diseases such as autoimmunity, allergy and cancer. A more pressing issue is the need for young blood in the field. PID is the field par excellence for physician scientists. Yet, not many medical school curriculums pay attention to PID. The authors of the chapters in this book and colleagues of their generation have the tremendous responsibility of following in the steps of the early giants in the field, who have inspired and relentlessly encouraged young scientists interested in PID to stay the course. This book is a welcome and excellent step in this direction.

Immunity: From Serendipitous Observations to Science-Based Specialty

Hans D. Ochs

Professor of Pediatrics, University of Washington School of Medicine,
Seattle Children's Research Institute, Seattle, WA, USA

OUTLINE

SERENDIPITOUS OBSERVATIONS

There is little evidence that the concept of microbes, infections and immunity was considered by healers and physicians in the ancient world. The idea of disease caused by microorganisms did not fit into a world of superstition, where sickness was believed to be God's punishment for the sins committed by men, and was not an option until the 19th century when, based on the new philosophy of rationalism, medicine turned to alchemy and then to science. There were, however, a few scattered reports in ancient history that considered protection from pestilence following an initial exposure. When recording the plague that hit Athens in 430 BCE, the historian Thucydides records: "Those who recovered from the disease were never attacked twice, at least not fatally". Rhazes, the famous Arab physician of the ninth century, remarked that survival from smallpox infection guaranteed protection from subsequent exposures to the disease. However, the systematic study of acquired immunity from smallpox did not occur until the 18th century, when isolated reports, almost unnoticed by the medical establishment, were circulated by physicians traveling to China and the Ottoman Empire, suggesting that the deliberate exposure protected from smallpox, a much feared epidemic disease that threatened disfigurement and possible death. As recounted by Dixon (1), the introduction of variolation as an effective protection from smallpox was introduced to the Western world by Lady Mary Wortley Montagu (Fig. 1.1), the wife of the British ambassador to Istanbul, the center of the Ottoman Empire. In a letter to a friend,

FIGURE 1.1 Lady Mary Wortley Montagu, wife of the British Ambassador to Istanbul, capital of the Ottoman Empire. She introduced variolation to the West. (*Source: Wikipedia, Public Domain.*) This figure is reproduced in color in the color section.

written in 1717, she described the local custom of people coming together for "smallpox parties", where "usually an old woman would collect in a nutshell of the matter of the best sort of smallpox" and inoculate into a vein "as much matter as can lay upon the head of her needle". Lady Montagu had her son inoculated in 1718 while still in Turkey; in 1721, her daughter became the first person to be variolated in England, without any complications. Possibly influenced by Lady Montagu's courageous action, another smallpox immunization trial, known as the "Royal Experiment" was carried out in England at about the same time. As recounted by Silverstein (2), during the smallpox epidemic of 1721, the daughter of King George I, Caroline, initiated experiments based on rumors that suggested that subcutaneous inoculation of a small amount of material from a human smallpox lesion would protect against the disease. Apparently, Princess Caroline, after consulting the King's advisors, requested that a consortium of royal physicians conduct safety and efficacy tests on six prisoners and five orphan children, including smallpox challenge of the inoculated prisoners. Satisfied with the experiment, she allowed the variolation of her 11-year-old daughter Amelia and 3-year-old daughter Mary, supposedly with no detrimental results. Following these experiments, variolation became a popular procedure in England, the American colonies and eventually central Europe, in spite of occasional complications and the transmission of other infectious diseases via this procedure.

The concept of a safer and scientifically proven immunization strategy was introduced by Edward Jenner (Fig. 1.2) in 1796, following experiments inoculating material from harmless cowpox lesions (vaccination) into the arms of several teenage boys (3). One of them, James Phipps, was subsequently exposed to smallpox and reported to be fully protected.

FIGURE 1.2 Edward Jenner (1749–1823), performed the first smallpox vaccination, using material from cowpox lesions. (*Source: Wikipedia, Public Domain.*)

A CELL DERIVES FROM A CELL, A MICROORGANISM FROM A MICROBE

During the second half of the 19th century, a handful of scientists utilized the principles of chemistry, pathology, microbiology and epidemiology to transform medicine into a science-based discipline. Using rigorous animal and human experiments and a series of groundbreaking observations at top European Universities, the field of infectious disease was created, which finally spawned a new specialty, immunology.

One of these early 19th century giants was Rudolf Virchow (Fig. 1.3), a pathologist, activist, politician and archaeologist, born in 1821 in eastern Prussia, now part of Poland, to a working-class family of butchers. A brilliant but rebellious student while attending the local gymnasium, Virchow was given the opportunity to study medicine at The Berlin Military Academy (Friedrich Wilhelm Institute), where gifted students from poor families received free education with the expectation that, after graduation, they would serve in the Prussian Army for 10 years. After completing medical school in 1843, Virchow became an assistant at the Charité, a large teaching

FIGURE 1.3 Rudolf Carl Virchow (1821–1902), scientist, statesman, activist. Founder of modern pathology and the concept of cell-based anatomy ("omni cellula e cellula"). *(Source: Wikipedia, Public Domain.)*

hospital in Berlin where he studied microscopy and, in 1847, was appointed Privat Dozent. There he founded the "Archiv für Patologische Anatomie und für Clinische Medizin", a journal still published as "Virchow's Archiv". Because of his stellar performance as a young scientist, Virchow was eventually released from military obligation. However, because of his political activism and radical views (he participated in the uprising of 1848 by helping in the construction of barricades in Berlin), he was fired from his position at the Charité. Interestingly, as an agnostic Protestant and anti-Catholic, Virchow accepted a position as chair of pathology/anatomy at the University of Würzburg in conservative and Catholic Bavaria, where he developed his idea of "Zellular Pathologie" which he summed up as "omni cellula e cellula". In 1856, he was allowed to return to Berlin, where he accepted the chair in Pathology at the Charité, which he kept until his death in 1902.

Virchow was a non-conformist, both scientifically and politically. By introducing scientific principles into medicine through the development of cellular pathology, he directly challenged the old philosophy of Galenic humoralism. On the other hand, he was not convinced that infection based on the germ theory had merit, opposed Koch's principles and Ignatz Semmelweis' advocacy of hand washing and attacked Darwin's theory of evolution. Politically, he was an outsider and a liberal who founded his own progressive party (Deutsche Fortschritts Partei) and was an elected member of the German Reichstag (Parliament). There he verbally attacked von Bismarck, who challenged Virchow to a duel. Two versions of this 1865 episode exist. The first had Virchow declining the challenge because he considered dueling an uncivilized method of solving conflict. The second version went "viral" and was well documented in the scientific literature of the time: Virchow, being challenged, was entitled to select the weapon; he chose two pork sausages, one normal, one loaded with *Trichinella* larvae (plausible, Virchow being the son of a butcher, but disputable as an opponent of germ theory). As the story goes, his challenger declined the proposition, either as being too risky or undignified. Virchow died at the age of 80 of complications from a fractured leg he sustained while jumping off a streetcar in Berlin.

Louis Pasteur (1822–1895) (Fig. 1.4) is considered by many to be the father of Immunology, for his seminal contributions to the field of active immunization. He recognized that the principle of vaccination, introduced almost a century earlier by Edward Jenner for smallpox, could be applied to any microbe-related disease. By introducing the concept of "attenuated" – weakened – microbial organisms, and the technique of multiple exposures to the same infectious agent to increase protection, and by propagating the idea of post-infection prophylaxis as in the case of rabies, Pasteur envisioned that infectious diseases could not only be prevented but also treated by vaccination.

FIGURE 1.4 Louis Pasteur (1822–1895), developed germ theory of diseases ("omne vivum ex vivo") and developed the concept of active immunization. Founder of Pasteur Institute in Paris. *(Source: Wikipedia, Public Domain.)*

Pasteur started his career as a chemist and physicist but, in the 1860s, while studying fermentation, turned his attention to microbes. He designed experiments demonstrating that fermentation can be prevented if the air used in these experiments was filtered through cotton. Based on the observation that microbes cause fermentation and putrefaction, he introduced the concept of biogenesis: *"omne vivum ex vivo"* or "all life from life". Pasteur's experiments convincingly refuted the centuries-old belief in spontaneous generation. For this work, Louis Pasteur is also considered "the father of germ theory". His procedure of low grade heating of milk to eliminate microbes is known as Pasteurization and led to the techniques used by Joseph Lister for sterilizing surgical instruments and wounds, and implementing sterile obstetrical procedures. Based on his understanding of microbes, Pasteur developed vaccines to protect livestock – and humans – from anthrax, and demonstrated the safety and effectiveness of a vaccine against rabies, which he prepared from dried spinal cords collected from lethally infected rabbits. In July 1885, such a tissue emulsion was repeatedly injected subcutaneously into Joseph Meister, Pasteur's most famous patient, who had received multiple bites from a rabid dog. The boy survived and the fame of Pasteur was established. While Pasteur's and Koch's discoveries had set the stage for understanding infectious diseases and designing vaccines, the field of Immunology was in its infancy, and based more on theory than facts. Pasteur, having observed that attenuated bacteria depended on certain nutritional requirements, postulated that the attenuated and weakened bacteria might deplete the "immunized" host of trace substances, thus rendering the host no longer suitable to support the growth of the virulent form of the microbe he had safely used for immunization. The trick, Pasteur thought, was to attenuate (using a stressful environment) each specific strain of microbe, either *in vitro* (as he did for anthrax), or *in vivo* (for his vaccine against rabies he had made in rabbits). This theory placed Pasteur in direct conflict with many of his colleagues, who were able successfully to immunize animals with killed bacteria, such as M. Toussaint, a French veterinarian who demonstrated protection by immunizing dogs and sheep with killed anthrax bacillus, and Behring, who showed that inactivated toxins produced by bacteria, if injected into mice and rabbits, induced immunity. Pasteur, nevertheless, stuck to his theory.

Pasteur's accomplishments as a chemist, microbiologist and immunologist propelled him to being the most prestigious scientist in France. He was elected as a member of the French Academy of Sciences (1862), the Academie National du Medecine (1873) and the Academie Française (1881). Based on his reputation, Pasteur was able to fund and establish the Pasteur Institute in Paris, which supported his research and attracted top scientists from various specialties to combine basic research with clinical application. In subsequent years, Pasteur Institutes were established in many parts of the world. Pasteur died of a stroke in 1895, was given a state funeral in Paris, buried in the Cathedral of Notre Dame and later interred in the Pasteur Institute.

FIGURE 1.5   Robert Koch (1843–1910) discovered the anthrax and the tubercle bacilli by developing modern methods of bacteriology. *(Source: Wikipedia, Public Domain.)*

SERUM THERAPY, ACTIVE IMMUNIZATION AND THE CONCEPT OF ANTIBODIES

Robert Koch (Fig. 1.5), influenced by Virchow and Pasteur, is known as the father of medical microbiology. He graduated from medical School in Göttingen and, after a short stint in Berlin with Virchow, worked as a practicing general physician, participated as a volunteer physician during the Franco-German War of 1870–1871 and finally settled as a "district physician" in Wollstein, a small town in Prussia. Using his own home as a laboratory, he began experimental studies of anthrax using novel techniques that allowed the isolation of anthrax bacillus from blood, and culturing the organisms in liquid and solid media. By using agar in the Petri dish, which was invented by his assistant, Julius Petri, he was able to isolate pure cultures of anthrax. He realized that the anthrax bacillus could not survive for long outside a host, but that anthrax formed endospores that could survive for a long time in their natural environment. Following his appointment in 1880 as public health officer in Berlin, Koch focused on

tuberculosis (TB) and two years later published the discovery of the tubercle bacillus (4) in a classic paper that outlined his principles/postulates. He subsequently isolated and characterized the agent causing cholera, *Vibrio cholerae*, and speculated that the bacterium generated a toxin that is responsible for the clinical symptoms. Based on the fact that organisms were isolated from drinking water, he introduced the technique of filtering water to stop the spread of the disease.

After the discovery of the tubercle bacilli, Koch became obsessed with finding a cure for this highly contagious and often lethal disease that affected a large proportion of the population. He used a glycerine extract from killed tubercle bacilli, which he called tuberculin, to treat patients affected with tuberculosis. Koch's "remedy" created an enormous expectation, but after several thousand patients were treated, tuberculin was found to cause fever, muscle ache and nausea, but not to cure tuberculosis. Tuberculin, however, was found to be a most useful diagnostic tool to identify tuberculosis-infected individuals. Koch never accepted the failure of his treatment for curing TB.

Koch's contributions to the field of microbiology, public health and infectious diseases made him a national and international celebrity. Following his seminal experiments for identifying the etiology of "Milzbrand" (anthrax), he was asked in 1882 to head the Imperial Health Office in Berlin, and in 1885 he was nominated to be Professor of Hygiene at the Friedrich Wilhelm University of Berlin. He was instrumental in obtaining funding for the construction of The Institute for Infectious Diseases and became its first director in 1892. "Koch's Institute", as it was known from the beginning, had several sections that focused on research and patient care, both at the national and international level. In 1912, the Institute was officially named the "Königlich Preussische Institut für Infektionskrankheiten, Robert Koch" and is now an independent federal entity, the "Robert Koch Institute" in Berlin. For his contributions to microbiology and medicine, he was elected member of the Prussian, the Austrian and the Royal Swedish

FIGURE 1.6 Emil Adolf von Behring (1854–1917). He developed passive immunization using serum from immunized animals to protect individuals from tetanus and diphtheria (serum therapy). *(Source: Wikipedia, Public Domain.)*

Academies of Sciences, Foreign Member of the Royal Society (London) and, in 1905, received the Nobel Prize for Physiology and Medicine.

Like Virchow, Emil Behring (Fig. 1.6), born in 1854, came from East Prussia, and like Virchow's, Behring's career was facilitated by acceptance into the Friedrich Wilhelm Institute in Berlin, where future military physicians received free medical education. During his payback service in the Prussian Army, he had the opportunity to spend an internship at the Charité in Berlin. While working as a military surgeon in Prussia, he studied disinfection of wounds and later, as a trainee at the Pharmacological Institute in Bonn, he learned how to use animal experimentation in toxicological research. Thus, Behring was well prepared to join Robert Koch at the Institut für Hygiene at the Friedrich Wilhelm University of Berlin, where, between 1889 and 1895, he developed his pioneering concepts of passive (serum) therapy and active immunization. While serving in the Prussian Army and studying the effect of iodoform and other compounds in curing infections such

as anthrax (an approach that, as he found out through animal experimentation, was too toxic to be used in humans), he discovered that previously exposed rats became "immune" to anthrax and that their serum was able to kill anthrax bacteria (5). When Behring joined Koch's Institute, he began to collaborate with Shibasaburo Kitasato, who had arrived there three years earlier. Kitasato had been sent to Berlin by the Japanese Government as part of Japan's attempt to quickly adopt Western science, following the end of seclusion from the outside world ("sakoku") that was made possible after the Meiji Restoration in 1868. While working in Koch's Laboratory, Kitasato, born in 1853 in the Kumamoto prefecture, developed a method to cultivate tetanus bacillus in pure culture (6). This allowed the carefully planned and executed animal experiments that led to the landmark publication by Behring and Kitasato in 1890 entitled "The mechanism of immunity in animals to diphtheria and tetanus". The two page article had not a single literature reference, as there had been no previous work to be listed (7).

Behring and Kitasato concluded the following from their experiments with rabbits and mice, using various immunization schemes to induce immunity to the tetanus toxin:

1. The blood of animals that were rendered immune to tetanus had the ability to neutralize or destroy the tetanus toxin
2. This property could also be demonstrated in extravascular blood samples and cell-free serum from tetanus-immune rabbits
3. This property was very stable and was even effective in the bodies of other animals, so that it was possible, through whole blood or serum transfer, to achieve outstanding therapeutic effects
4. The property that destroys tetanus toxin does not exist in the blood of non-immune animals.

Behring and Kitasato ended their report by warning against the increasing trend for replacing blood transfusions with normal saline by citing Wolfgang von Goethe's Faust, where

Mephistopheles exclaimed: *"Blut ist ein ganz besonderer Saft"* (blood is quite a special juice) when asking Dr Faust to sign his pact in blood – not quite in line with the message Behring and Kitasato propagated in their 1890 article.

In a second paper, published one week later in the same journal, Behring extended the work with tetanus toxin to the more clinically pressing disease, diphtheria (8). In collaboration with Ehrlich and others (9), he subsequently succeeded in producing standardized "diphtheria antitoxin" in cows, and to commercialize the process; first by contracting with Farbwerke Höchst AG, located near Frankfurt, then by creating the Behringwerke in Marburg where he had accepted a position as Professor of Hygiene. In 1894, a total of 75 225 vials of diphtheria anti-serum were sold and successfully used in Germany, France, England and the USA. For this accomplishment, Emil von Behring received the first Nobel Prize for Medicine in 1901.

Elie Metchnikoff (Fig. 1.7), born in 1845 in the Ukraine, studied natural sciences in Russia and marine biology at the Universities of Giessen, Göttingen and Munich. At the age of 22, he started his academic career at the University of Odessa where, in 1870, he was appointed Professor of Zoology and Comparative Anatomy.

Influenced by Darwin, Metchnikoff tried to understand phylogenetic relationships through the study of mesodermic digestive cells, using invertebrate models. While working in a small private laboratory at Messina, Italy in 1882, he experimented by introducing a rose thorn into starfish larvae. Using a microscope, he observed amoeboid phagocytes surrounding the tip of the thorn and, based on this discovery, developed his theory of cellular host defense. Rather than deleterious to the host, Metchnikoff considered inflammation associated with the influx of phagocytic cells and the killing of microorganisms as an important defense against bacteria and parasites (10). Metchnikoff's theory of a cellular component of the immune system was immediately accepted by the "cellular" pathologist Virchow, who visited him in Messina, but vehemently rejected by most pathologists, who considered inflammation to be a serious condition. When Behring, Kitasato and Ehrlich published their pioneering work establishing the effectiveness of "serum-therapy", Metchnikoff, who had joined the Pasteur Institute in 1899, was forced onto the defensive. The controversy, which seemed to position French and German scientists in two separate camps, was finally settled when it was recognized that phagocytes can ingest bacteria only if opsonized with either heat-labile complement components or heat-stabile antibodies, the latter providing specificity.

The Nobel committee's decision to select both Metchnikoff and Ehrlich to share the 1908 Nobel Prize for Medicine was instrumental in bridging the gap between these two camps and emphasizing the importance of both immune systems (but "could not prevent World War I").

Paul Ehrlich (Fig. 1.8) was born in 1854 in lower Silesia, at that time part of the Prussian Kingdom. Coming from a wealthy family, he was able to enroll at the Medical School in Breslau and subsequently attended the Universities of Strasbourg, Freiburg and Leipzig, where he obtained a doctorate with a dissertation entitled "Contribution to the theory and practice of histological staining". Ehrlich used recently synthesized aniline pigments to differentiate histologically distinct tissue structures and blood cell types. In his doctoral thesis, delivered at the age of 28,

FIGURE 1.7 Elie Metchnikoff (1845–1916), a Russian zoologist, introduced "cellular" immunology by discovering phagocytic cells. He joined the Pasteur Institute in 1888. *(Source: Wikipedia, Public Domain.)*

FIGURE 1.8 Paul Ehrlich (1854–1915), collaborated with Behring to standardize diphtheria anti-toxin. He discovered mast cells through modern staining techniques he had developed. He is considered to be the first to propose the existence of cell surface receptors as part of his "side-chain" theory, and is considered as the father of chemotherapy by developing Salvarsan to treat syphilis. *(Source: Public Domain; Library of Congress Prints and Photographs Division Washington, DC.)*

Ehrlich described the entire spectrum of staining techniques and the chemistry of the pigments employed. While working on his dissertation, he discovered a new cell type, which he identified by the presence of large granules. He was able to visualize them with the help of an alkaline dye. He thought the granules represented good nourishment and named them "mast" cells, using the German word "mast", meaning animal fattening. After having obtained his medical degree, he joined the medical staff at the Charité in Berlin in 1886, where he worked on methods to improve the staining of bacteria and to differentiate white blood cells into lymphocytes, monocytes, polynuclear leukocytes and mast cells. This work led to his habilitation.

His academic career was interrupted in 1888, when he left for Egypt to find a cure for the tuberculosis he had contracted during his work in the laboratory. Upon his return to Berlin, he established a private medical practice and laboratory. Three years later, he was invited by Koch to join his Institute, where he collaborated with Behring, developing methods to standardize diphtheria anti-toxin sera and test them for safety and efficacy (9). To facilitate these efforts, Ehrlich established the Institut für Serumforschung und Serumprüfung, which he moved to Frankfurt in 1899, renaming it Institut für Experimentelle Therapie. There he developed his famous side-chain theory (11, 12), which postulates that "toxins" (antigens) are recognized and fixed by cell surface molecules (complementary chemical structures or "side-chains", equivalent to B-cell receptors) protruding from the cell wall. Ehrlich postulated that these "toxins" combine like a "lock and key" with the side-chains, and thus stimulate the cell to produce more side-chains, which would finally become so abundant that some of them (as antibodies) would leave the cell surface and circulate freely in the blood (Fig. 1.9). With this theory, Ehrlich introduced

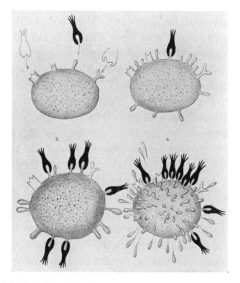

FIGURE 1.9 Ehrlich's illustration of his side-chain theory, which he used in a lecture to the Royal Society of London in 1900. *(Source: Ehrlich, Paul. "On immunity with special reference to cell life." Croonian Lecture, Proceedings of the Royal Society, 1900. Courtesy of the Wellcome Library, London.)*

the concept of specific antigen receptors and antibody secretion by cellular elements, a process providing active protection. Following Buchner's discovery of what he called "Anexin", a bacteriocidal substance which he had found in the blood of non-immune animals (13), Ehrlich recognized the complexity of this system, referring to it as "complement" (14). For his contributions to the field of immunity, specifically his side-chain theory, Ehrlich received the Nobel Prize for Medicine in 1908.

In search of a "magic bullet" to cure syphilis, Ehrlich's group, which included Sahachiro Hata from Japan, screened hundreds of newly synthesized organic arsenic compounds for anti-spirochetal activity. This search resulted in the first chemotherapeutic agent (preparation #606) with strong inhibitory action toward spirochetes. It was introduced in 1910 as Salvarsan (15). These revolutionizing ideas paved the way for the discovery, in 1935, of the sulfonamide Prontosil, an industrially produced red dye, by Gerhard Domagk (16), while working at the IG Farben Institute of Pathology and Bacteriology. The first antibacterial compound was now available for clinical use, followed by the serendipitously discovered antibiotic penicillin by Alexander Fleming on September 28, 1928 (17). Each received the Nobel Prize for their discoveries, Domagk in 1939 [which he was forced by the then German government to decline, an interesting story in itself; he was finally able to accept the award in 1947 (without the elapsed monetary portion)], and Sir Fleming in 1945.

The term "protein" (derived from the Greek word "proteos", meaning "the first") was introduced in 1839 in a description of the properties of fibrin isolated from blood (18). As the interest of physician scientists in serum proteins increased, methods to separate and identify these proteins were needed. A number of novel ideas, explored in Sweden, included electrophoresis, in which serum proteins were separated according to their electric charge (19), and ultracentrifugation,

which separated proteins according to their weight (20). The introduction of paper electrophoresis (21, 22) allowed routine protein identification. The push to develop techniques for plasma fractionation on a large scale came from the US Armed Forces during World War II in an effort to control hemodynamic shock in severely wounded soldiers. On the basis of this request, the National Research Council charged Edwin Cohn with isolating stable protein fractions from human blood. This effort was successful and, in 1943, Cohn's team reported that alloantibodies are sequestered in the gamma-globulin fraction of human serum.

The importance of the bone marrow as the source of all blood cells was recognized as early as 1868 (23), resulting in the systematic analysis of bone marrow for the description and diagnosis of blood diseases (24). Finally, the discovery of human blood groups provided the basis for a safe and effective blood transfusion system (25).

The astounding advances in biochemistry/physiological chemistry (split from traditional organic chemistry in the 1870s by Felix Hoppe-Seyler) made it possible for Albert Kossel, working in the laboratory of Hope-Seyler at the University of Strasbourg, to discover the chemical composition of nucleic acid (26) for which he received the Nobel Prize in 1910.

EPICRISIS

The pioneering research accomplished in the 19th and first half of the 20th century firmly established the basic tenets of cellular pathology; the microbial basis of infectious diseases; adaptive and innate, humoral and cellular immunity; the principle of active and passive immunization; antibacterial therapy; protein chemistry and the composition of DNA. The stage was set to recognize, clinically define, molecularly dissect and effectively treat primary immune deficiency diseases during the decades following World War II.

References

1. Dixon CW. *Smallpox*. London: J and A Churchill, Ltd; 1962.

2. Silverstein AM. *A history of immunology*. 2nd edn. Elsevier; 2009.

3. Jenner E. *An inquiry into the causes and effects of the variolae vaccine*. London: Sampson-Low; 1798.

4. Koch R. Die Aetiologie der Tuberculose. *Berl klin Wochenschr* 1882;**19**:221–30.

5. Behring E. Ueber die Ursachen der Immunität von Ratten gegen Milzbrand. *Zbl klin Med* 1889;**9**:681–3.

6. Kitasato S. Ueber den Tetanuserreger. *Dtsch med Wochenschr* 1899;**15**:635–6.

7. Behring E, Kitasato S. Ueber das Zustandekommen der Diphtherie-Immunität und der Tetanus-Immunität bei Thieren. *Dtsch med Wochenschr* 1890;**16**:1113–4.

8. Behring E. Untersuchungen über das Zustandekommen der Diphtherie-Immunität bei Thieren. *Dtsch med Wochenschr* 1890;**16**:1145–8.

9. Behring E, Erlich P. Zur Diphtherie immunisirungs- und Heilungsfrage. *Dtsch med Wochenschr* 1894;**20**:437–8.

10. Metchnikoff C. Ueber eine Sprosspilzkrankheit der Daphnien. Beitrag zur Lehre über den Kampf der Pagocyten gegen Krankeitserreger. *Arch Pathol Amat (Virchow's Archiv)* 1884;**96**:177–95.

11. Ehrlich P. On immunity with special reference to cell life. Croonian lecture. *Proc R Soc* 1900;**66**:424–48.

12. Heymann B. Zur Geschichte der Seitenketten theorie Paul Ehrlichs. *Klin Wochenschr* 1928;**7**:1257–60 305-309.

13. Buchner H. Ueber die nähere Natur der bacterien tödtenden Substanz im Blutserum. *Zentralbl Bakteriol* 1889;**16**:561–5.

14. Ehrlich P, Sachs H. Ueber die Vielheit der Complemente des Serums. In: Erlich P, editor. Gesammelte Arbriten Zur Immunitätsforschung, Velag von August Hirschwald, Berlin, 1904; 282–302.

15. Ehrlich P. Die Behandlung der Syphilis mit dem Ehrlichschen Präparat 606. *Dtsch med Wochenschr* 1910;**41**:1893–6.

16. Domagk G. Ein Beitrag zur Chemotherapie der bakteriellen Infektionen. *Dtsch med Wochenschr* 1935;**61**:250–3.

17. Fleming A. On the antibacterial action of cultures of a penicillium, with special reference to their use in the isolation of B. Influenzae. *Br J Exp Pathol* 1929;**10**:226–36.

18. Mulder GJ. Zusammensetzung von Fibrin, Albumin, Leimzucker, Leucin u.s.w. *Ann Pharm* 1839;**28**:73–82.

19. Tiselius A, Kabat EA. An electrophoretic study of immune sera and purified antibody preparations. *J Exp Med* 1939;**69**:119–31.

20. Svedberg T, Pedersen KO. *The ultracentrifuge*. Oxford: Clarendon Press; 1940.

21. Dettker A, Anduren H. A new time saving apparatus for paper electrophoresis. *Scand J Clin Lab Invest* 1954;**6**:74–5.

22. Schneider G, Wunderly C. Die Papierelektrophorese als Schnellmethode des klinisch-chemischen Laboratoriums. Möglichkeiten und Grenzen ihrer Anwendung. *Schweiz med Wochenschr* 1952;**82**:445–9.

23. Neumann E. Ueber die Bedeutung des Knochenmarkes für die Blutbildung. *Zentralbl Med Wissensch* 1868;**6**:689.

24. Naegeli O. Ueber rothes Knochenmark und Myeloblasten (Apropos red bone marrow and myeoloblasts). *Dtsch med Wochenschr* 1900;**26**:285–90.

25. Landsteiner K. Ueber Agglutinationserscheinungen normalen menschlichen Blutes. *Wien klin Wochenschr* 1901;**14**:1132–4.

26. Kossel A. Weitere Beiträge zur Chemie des Zellkerns. *Zschr physiol Chem* 1886;**10**:248.

2

Discovery of the T- and B-Cell Compartments

Max D. Cooper

Professor of Pathology and Laboratory Medicine, Emory University School of Medicine, Atlanta, GA

O U T L I N E

INTRODUCTION

Since Ogden Bruton's 1952 discovery of agammaglobulinemia as the underlying cause of repeated bacterial infections in a young boy (1), studies of patients with primary immunodeficiency diseases have taught us much about the complexities of the way that our immune system develops and functions. Bruton's discovery was made possible by the use of Arne Tiselius' electrophoresis method for separating blood proteins (2) to demonstrate his patient's absence of gamma globulins, the serum fraction shown to contain antibodies by Tiselius and Elvin Kabat (3). This finding of agammaglobulinemia led Bruton to initiate the successful treatment of his patient with

Primary Immunodeficiency Disorders: A Historic and Scientific Perspective

monthly injections of gamma globulin derived from healthy donors. Other physicians confronted with patients having unusual susceptibilities to infections were drawn to the field of immunology in search of the underlying immune system defects and better treatment options. One of the most successful of these physician-scientists was Robert Good, who often referred to the heritable immunodeficiencies as "experiments of nature".

DISCOVERY OF THE IMPORTANCE OF THE THYMUS IN IMMUNE SYSTEM DEVELOPMENT

In examining the lymph nodes from agammaglobulinemic boys, Good noted the absence of plasma cells (4), which Astrid Fagraeus (5) and he considered as the cellular source of antibodies, an association that Albert Coons verified using his immunofluorescence technique (6). Intrigued by finding agammaglobulinemia and plasma cell deficiency in an adult with thymoma, Good postulated a functional link between the thymus and antibody-producing plasma cells. Disappointingly, however, no effect on antibody production was seen when the thymus was removed from rabbits. Good and his coworkers would later revisit this issue after learning about studies in chickens that had gone unnoticed by immunologists who focused on studying mammalian models. In the journal *Poultry Science*, Bruce Glick and his coworkers reported that removal of a hindgut lymphoid organ, the bursa of Fabricius, in newly hatched chicks, severely compromised their subsequent ability to produce antibodies (7). Harold Wolfe and his colleagues found that testosterone treatment of chick embryos also inhibited bursal development and the ability to produce antibodies (8). Ben Papermaster, Good's associate, learned of these findings during a visit to the Wolfe lab, and realized that it would be important to remove the thymus early in life to test its role in immune system development. Indeed, the Good group then found that thymectomy in

neonatal rabbits and mice inhibited lymphocyte development and antibody production (9). Meanwhile, Jacques Miller in England was conducting experiments in mice to determine why removal of the thymus prevented the development of virus-induced lymphomas, when he discovered that neonatal thymectomy resulted in a fatal runting disease characterized by severe lymphopenia and immunological incompetence (10). One year later, the group of Byron Waksman reported that neonatal thymectomy in rats profoundly inhibits development of the immune system (11). Needless to say, the discovery that the thymus is essential for normal immune system development stimulated great interest in how the thymus functions, and an international conference on this topic was convened by Good in 1962 (12).

The key to thymus function was thought to lie in its capacity to produce lymphocytes. Robert Auerbach found that the embryonic thymus, which appeared to be composed entirely of epithelial cells in 12-day-old mouse embryos, was capable of generating lymphocytes *ex vivo*. His results implied that lymphocytes were derived from the epithelial cells in the thymus (13). The idea that lymphocytes produced in the thymus were seeded into the peripheral lymphoid tissues was later supported by the immunological reconstitution of thymectomized mice with thymus grafts or thymocyte infusions (14, 15), and then by tracing the migration of radiolabeled thymocytes (16). A hormonal influence of the thymus was also suggested by the beneficial effects of thymus implants in thymectomized recipients, even when the implants were placed within filter chambers that were considered to be cell impenetrable. James Gowans used radiolabeled lymphocytes to trace their circular migratory route out of the blood vessels into the lymphoid tissues and then back into the blood stream via connecting lymphatic channels like the thoracic duct (17). Gowans also showed that radiolabeled lymphocytes could become plasma cells (18). These findings collectively supported the contemporaneous model schematically illustrated in Fig. 2.1.

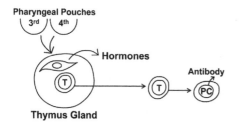

FIGURE 2.1   The role of the thymus in immune system development as viewed in the early 1960s. T: T-cells; PC: plasma cells.

This was an exciting time in immunology. The paired heavy and light chain composition of antibodies had been elucidated in biochemical studies by Rodney Porter and Gerald Edelman (19, 20), and the clonal selection theory of Macfarlane Burnet (21) provided a guiding principle for immunologists eager to understand the molecular basis of antibody diversity. Nevertheless, this period is sometimes referred to as "the dark age of immunology" because the knowledge base and available research tools were so primitive by comparison with present day standards. Experiments at that time were conducted largely in whole animal models, by passive transfer of immunity using cells or plasma, and by analysis of serum proteins. Methods for maintaining cells in culture had not been refined, recombinant DNA technology was not yet on the horizon, and the dawn of molecular biology was just beginning to break. Studies of patients with inherited immunodeficiency diseases were just beginning to be appreciated as a rich source of information about the functional components of our complex immune system.

EN ROUTE TO BOB GOOD'S RESEARCH GROUP IN MINNESOTA

Let me digress briefly to recount background events that led me to join Good's laboratory. I trained in Pediatrics at Tulane University and then worked as a house officer at the Hospital for Sick Children in London before moving to the University of California at San Francisco (UCSF) for training in allergy and immunology. En route to San Francisco, I took a detour to learn the immunofluorescence technique in John Holborow's laboratory, ostensibly for the purpose of working with future UCSF colleagues to examine the role of cellular immunity in phlyctenular keratoconjunctivitis. To my considerable embarrassment, Holborow informed me that immunofluorescence techniques were then useful for studies of antibodies, but not cell-mediated immunity. Merrill Chase and Karl Landsteiner had shown earlier that, whereas humoral immunity could be passively transferred by plasma infusion, lymphocytes were required for transfer of contact hypersensitivity and delayed-type hypersensitivity (22, 23). This revelation led me to the laboratory library where I found Burnet's book entitled "The Clonal Selection Theory of Acquired Immunity", the reading of which was an epiphany for me. I vowed then, if nothing else, I would learn the difference between cellular and humoral immunity.

My interest in immunology grew during fellowship training at UCSF, where Linus Pauling, Edelman and Burnet all came to give lectures on their views of the field. Returning to Tulane as an Instructor in Pediatrics, I participated in general pediatrics and allergy/immunology training and inpatient care. My research activities were limited to clinical investigations, the most successful of which was the demonstration of immunological tolerance in blood group O infants who had received exchange transfusions with maternal A- or B-type blood for treatment of erythroblastosis fetalis (24). These studies were derivative of Ray Owen's demonstration of lasting blood cell chimerism conferred by intrauterine exchange of hematopoietic cells between calf twins with different fathers (25), and Billingham, Brent and Medawar's experimental demonstration of tolerance induction in newborn mice (26). It was unmistakably clear that, in order to have any chance to make an original contribution, I needed training in how to conduct laboratory research.

THE ATMOSPHERE IN BOB GOOD'S LABORATORY

The atmosphere in Bob Good's laboratory was electric when I joined his group in 1963. He had assembled a team of energetic and talented individuals with diverse clinical and basic science backgrounds, whose research themes reflected his broad interests in phylogeny and ontogeny of the immune system, immune deficiency diseases, autoimmune diseases, lymphoid and non-lymphoid malignancies, the complement cascade, and organ transplantation, in addition to the role of the thymus in immune system development. The most energetic person in the group was Bob Good himself, who slept about five hours a night and was in constant motion for the rest of the day. Good was curious, charismatic, eager to learn and exceptionally quick in assimilating new information. He was also a dynamic and entertaining speaker, who adhered to the philosophy that researchers in academic institutions were obligated to inform practicing physicians and the public about what they learned.

In conducting his heavy schedule of speaking engagements and other professional activities, Bob Good was an early frequent flyer. Through these activities, his publications and outstanding reputation as a physician, Bob attracted a continuous flow of patients, who were referred because of illnesses that might reflect immunological dysfunction. Virtually everyone in his research group participated in the analysis of these patients and their diseases. Each patient's problems initiated lengthy discussion and debate about the diagnosis and current treatment options before moving on to questions about disease pathogenesis, which we needed to answer in order to beneficially modify the disease course. The constant to and fro from patients'

diseases and laboratory experiments by the same group of people ensured firm linkage between these activities. This stimulating environment was infectious. On any day of the week and virtually any time of the day, members of Good's research group could be found in the laboratory or in the hallway debating issues with which we were concerned.

Good's mentoring philosophy was to provide his trainees with a stimulating environment, encouragement and the freedom to choose their own research projects. He advised me to begin with studies to explore the ideas that I had recounted in applying for a position in his research group. As one example, I had found that infant recipients of exchange transfusions were unable to manifest delayed-type hypersensitivity reactions that should have been conferred by the lymphocytes they received from their sensitized adult blood donors. My pursuit of an explanation for this paradox and a bevy of other experimental enquiries were informative, but failed to provide satisfactory solutions.

To my good fortune, Raymond Peterson, a well-established member of the research team took me under his wing and invited me to join him in his studies on the role of the bursa in the pathogenesis of an avian lymphoma. Ray and Ben Burmester, the virologist-oncologist director of the US Regional Poultry Laboratory in Michigan, had shown that avian leukosis virus-induced lymphomas in chickens could be prevented by bursectomy, whereas thymectomy had no effect (27). This finding contrasted with the fact that mouse lymphomas were preventable by thymectomy, as mentioned earlier. The virus-induced lymphomas in chickens appeared 5–9 months after virus inoculation at the time of hatching. Our follow-up studies showed that removal of the bursa any time before 5 months of age prevented lymphoma development, but exactly how the bursa controlled lymphoma development was unclear (28). The possibility of hormonal control by the bursa was favored because of the widespread distribution of the tumors in the spleen, liver and other visceral organs, whereas the bursa appeared to be unaffected. Later, it would become obvious how to readdress this puzzle.

CONNECTING STUDIES IN CHICKS AND IMMUNODEFICIENCY DISEASES

In examining the immunological effects of *in ovo* treatment with testosterone, Noel Warner, Alex Szenberg and Burnet observed that the cortical region of the thymus was atrophic in some of the treated chicks. Delayed rejection of skin allografts was observed in this subgroup, and also in chicks subjected to surgical thymectomy after hatching. Their analysis of the composite testosterone effects on immune system development led these investigators to propose a tripartite model in which the thymus governed skin graft rejection, the bursa controlled antibody production and delayed-type hypersensitivity, and the bone marrow controlled graft-versus-host reactions (29). This view of the avian immune system was difficult to integrate with results of the contemporaneous studies of mammalian immune systems. Good interpreted these results to be indicative of complementary roles for the thymus and the bursa. With David Sutherland and Olga Archer, he had found that early removal of either the rabbit appendix or thymus impaired antibody production, and the combination of appendectomy and thymectomy had a synergistic inhibitory effect (30). When Ed Yunis and I re-examined the effects of thymectomy and bursectomy in newly hatched chicks, we could easily confirm the inhibitory effect of bursectomy on antibody production, but could find no effect of thymectomy on either antibody production or skin graft rejection. After verifying these disappointing results, we abandoned this experimental approach as a blind alley.

Observations in individuals with the Wiskott–Aldrich syndrome (WAS) would later bring us back to the avian model. The median life span of boys with the identifying features of WAS (thrombocytopenia, eczema and recurrent infections) was then about three and a half years (see Chapter 9). In search of clues regarding their undue susceptibility to a wide variety of infectious agents, including fatal herpes virus infections, a review of the clinical findings and hospital records of 18 WAS patients revealed a progressive depletion of lymphocytes in the circulation, spleen and lymph nodes. In examining sections of stored tissues from deceased patients, I identified what appeared to be a very abnormal thymus, which my colleagues agreed was essentially devoid of lymphocytes. When I presented these findings to Bob Good, he also became very excited and announced to everyone "Cooper has discovered another thymus disease". The joy of my discovery proved short-lived, however. On further review of the evidence, we watched as Good's broad smile faded into a more serious countenance and then a fierce frown as he stepped back from the microscope to bellow: "You bastards, this is not the thymus! It's the pancreas"!

When I later succeeded in obtaining sections of the thymus from deceased WAS patients, they were indeed very deficient in lymphocytes, although this could have been a secondary consequence of the stressful terminal illness. The most puzzling finding was the contrast between the general paucity of lymphocytes in the face of an abundance of plasma cells and their immunoglobulin products (31). This paradox was not easily explained by the single lymphocyte lineage model, but could fit well with the possibility that plasma cells were not the progeny of thymus-derived lymphocytes.

THE KEY EXPERIMENTS IN IRRADIATED BIRDS

The avian immune system appeared to be the most suitable model for testing the idea of an alternative pathway of plasma cell development, although our prior experience indicated that a different experimental strategy was needed. By then, it had become clear that the effects of thymectomy depended upon the immunological maturity of the subject when the thymus was removed. This suggested that ablation of the thymus or the bursa in chick embryos would be an ideal experiment. However, surgical excision was then the only precise way to eliminate these organs, and the necessary *in ovo* surgery presented a seemingly insurmountable hurdle, especially in view of the difficulty in removing the seven pairs of thymus lobes that extend along the neck blood vessels into the thoracic region of birds. An alternative experimental design would be to roll back the clock of immune system development by destroying lymphocytes or lymphocyte precursors that had been produced *in ovo*. If this could be done, the removal of the thymus or bursa after hatching would theoretically afford a better view of their roles in immune system development. From a limited list of available lymphotoxic agents, I elected to combine whole body X-irradiation with thymectomy or bursectomy of newly hatched chicks. Pilot experiments to determine the dose of irradiation that chicks could tolerate indicated that approximately half of the newly hatched chicks could survive exposure to 700 rads of whole body irradiation. This led to my decision to use 600 rads, in combination with surgical removal of the thymus, bursa, both or neither.

When the immunological competence of the different experimental groups of birds was evaluated after recovery from the effects of irradiation six weeks later, the results were remarkably clear cut (Fig. 2.2). Every irradiated and bursectomized animal was agammaglobulinemic (32). None of these birds had germinal centers, plasma cells, or the ability to make antibodies, although they had normal thymus development, an abundance of lymphocytes elsewhere in the body and intact cell-mediated immune responses. Conversely, the irradiated and thymectomized birds had severe lymphocyte deficiency, and this was associated with impairment of delayed-type

Impaired Cellular Immunity

1. **Lymphocyte depletion**

2. **Impaired allograft rejection, delayed type hypersensitivity and GVH potential**

3. **Many plasma cells, but impaired antibody responses**

Impaired Humoral Immunity

1. **No germinal centers or plasma cells**

2. **No antibodies**

3. **Normal cellular immunity**

FIGURE 2.2   The immune system effects of thymectomy or bursectomy combined with near-lethal irradiation. GVH: graft versus host. This figure is reproduced in color in the color section.

hypersensitivity, allograft rejection ability and graft-versus-host reactivity. The birds that were irradiated, thymectomized and bursectomized had severely impaired cellular and humoral immunity, but were able to clear particulate antigens from their blood stream as evidence for the integrity of their reticular endothelial system. These results (32, 33) provided an unmistakably clear view of two lymphocyte lineages, one of which was thymus-dependent and the other bursa-dependent. Furthermore, infusion of autologous bursal lymphocytes restored the development of germinal centers and plasma cells in irradiated and bursectomized birds (34).

THE INITIAL T- AND B-CELL MODEL AND ITS CLINICAL IMPLICATIONS

The new model provided by these results (Fig. 2.3) indicated that the thymus and the bursa were central lymphoid tissues, which served as source organs for two separate lymphocyte

THYMUS SYSTEM DEVELOPMENT

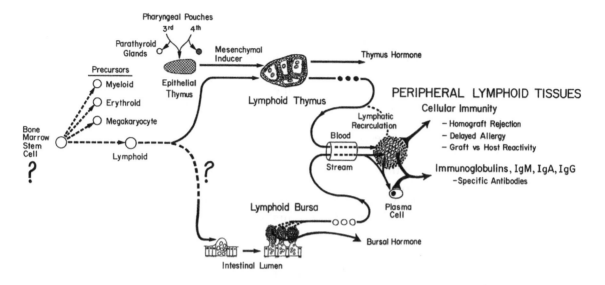

BURSAL SYSTEM DEVELOPMENT

FIGURE 2.3 1965 model of thymus-dependent and bursa-dependent lymphocyte lineages (35). The two types of lymphocytes were later given the names T-cells and B-cells in a review by Ivan Roitt and colleagues (36).

lineages. The thymus-derived lineage of lymphocytes was primarily responsible for cell-mediated immune responses, whereas the bursa-derived lineage was responsible for the production of antibodies and humoral immunity. The two lineages of cells were physically juxtaposed in the peripheral lymphoid tissues, and their functional interaction was implied by the impairment of antibody responses in thymectomized and irradiated birds, just as had been shown in thymectomized mice and rabbits. The precursors for both lymphoid lineages were postulated to be descendants of the bone marrow stem cells, whose multilineage differentiation potential had been demonstrated in mice by the Till and McCullough group (37) and whose lymphocyte differentiation potential had been shown by Charles Ford and Spedding Micklem (38). The postulated hematopoietic stem cell origin for both the thymus-dependent (T) and

bursa-dependent (B) lymphocyte lineages was later verified experimentally by Malcolm Moore and John Owen (39). The major tenets of this model have withstood the test of time, although the postulated hormonal roles for the thymus and bursa could not be substantiated (40).

The two lymphocyte lineage model profoundly altered our view of the development and function of the immune system, and it is difficult to describe the level of our excitement as we began to explore its implications. The new model could easily accommodate many of the experimental results that had been obtained in mammalian models and in studies of patients. Importantly, it could also be used to begin to classify immunodeficiency diseases according to those which resulted from defects in the thymus-dependent pathway (41), those, like Bruton's agammaglobulinemia, which reflected defects in the antibody-producing lineage of cells, and those with familial deficiencies in both cellular and humoral immunity that had been described by Eduard Glanzmann, Walter Hitzig and their colleagues (42, 43) (Fig. 2.4). The best example that we could find of a developmental defect that primarily affected the thymus-derived lineage was Christian Nezelof's description of thymic hypoplasia and lymphopenia in siblings who had an abundance of plasma cells and immunoglobulins (44). A more compelling example came with Angelo DiGeorge's dramatic announcement of his findings in patients with congenital absence of the thymus; he had observed that these congenitally athymic infants had severe lymphopenia and impaired cellular immunity in the face of a paradoxical abundance

of plasma cells and immunoglobulins (41, 45). As Angelo noted during a brisk debate about the validity of our model in humans, the findings in his patients were not consistent with the prevailing view of a single thymus-derived lineage of cells, but they were easily explained by our two lymphocyte lineage model. It was also clear that the model provided a strategic road map toward the repair of immunodeficiency diseases. Bob Good was especially excited by the possibility of transplanting healthy counterparts of the defective cell types to replace the missing components in patients with immunodeficiency diseases, an endeavor he termed "cellular engineering". He and his colleagues would achieve the first successful bone marrow transplant in 1968,

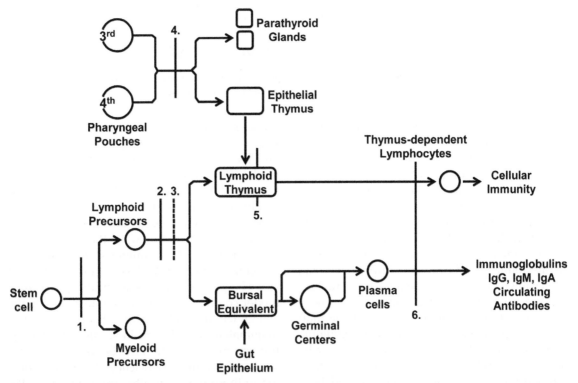

FIGURE 2.4 Early (1968) schematic view of lymphocyte differentiation defects in patients with immunodeficiency diseases (72). (1) Reticular dysgenesis; (2) Swiss-type agammaglobulinemia/SCID; (3) sex-linked recessive lymphopenic hypogammaglobulinemia; (4) aplasia of thymus and parathyroids; (5) thymus dysplasia of the Nezelof type with normal immunoglobulin levels; and (6) congenital intestinal lymphangiectasia.

permanently curing a boy who was born with a severe combined immunodeficiency (46).

The new model also implied that lymphoid malignancies in humans could be viewed according to whether they involved T-lineage or B-lineage cells (47). Indeed, the derivation of the model was significantly influenced by insight gained from the studies of lymphoma pathogenesis in mice and chickens mentioned earlier. It seemed logical to conclude that leukemia associated with thymic lymphomas probably originated in the thymus, whereas follicular lymphomas, plasmacytomas and multiple myelomas were B-lineage malignancies representative of later stages in this differentiation pathway. Although this seems perfectly obvious now, at that time, lymphomas were generally considered to be derived by *in situ* transformation of "lymphoreticular cells". Pathologist colleagues were not immediately accepting of the idea that lymphoid malignancies evolved during the continual differentiation of bone marrow precursors along the two lymphocyte differentiation pathways. Ray Peterson and I made an appointment with a highly respected pathologist, who had been trained in lymphoma classification at the Armed Forces Institute of Pathology, to discuss our idea of a new classification scheme for lymphoid malignancies and to seek his help in developing this new classification. As we began to describe the basis for our reclassification scheme, he almost had an apoplectic fit and we made a hasty retreat to avoid being thrown out of his office.

BASIC QUESTIONS POSED BY THE SEPARATE T- AND B-LINEAGE MODEL

The compartmentalized lymphocyte differentiation model also raised fundamental questions that would not be answered easily or quickly, despite their importance. (1) How do T-cells recognize antigens, given that they do not make immunoglobulins? (2) How do T-cells cooperate with B-cells to facilitate antibody responses? (3) Where is the bursa-equivalent in mammals? Over the next several decades, misleading theories, inadequate experimental tools and misinterpreted results would lead into many blind alleys before valid answers to these questions could be obtained.

As one way of addressing the first question, we made antisera against T-cells from agammaglobulinemic chicks. An avian MHC class I protein was identified in these experiments (48), but none of our anti-T-cell antisera recognized an obvious T-cell receptor candidate. Not surprisingly, neither did they contain antibodies to immunoglobulins. Nonetheless, the most widely held theories about the nature of the T-cell receptors for antigens were that (1) T-cells make their own IgM, and (2) T-cells make soluble antigen-specific factors as hybrid proteins composed of a constant region combined with variable regions like those of antibodies. In a test of the endogenous T-cell IgM hypothesis, Sue Webb and I confirmed prior findings in mice by showing that T-cells from normal pre-immunized chickens could bind the sheep erythrocytes that had been used for immunization (49). However, T-cells from similarly immunized agammaglobulinemic birds were unable to do so, unless they were pre-incubated with serum from immunized controls. The antibodies responsible for the passive antigen binding proved to be of IgM isotype, suggesting that T-cells have receptors for IgM. Perhaps the most popular view then was that T-cells make helper and suppressor factors that employ variable regions for their antigen specificity (reviewed in references 50, 51). This view gained support from studies showing that T-helper and suppressor factors could be inhibited by antisera made against the Ig V region. However, when Hiromi Kubagawa later made monoclonal anti-VH antibodies, these failed to react with T-cells (52). The basic solution to the question of how T-cells recognize antigen would come later, with

Rolf Zinkernagel and Peter Doherty's discovery of the MHC restriction of cytotoxic T-cell recognition of virus-infected cells (53), and Tak Mak and Mark Davis' cloning of T-cell receptor genes in humans (54) and mice (55) almost 10 years after Susumu Tonegawa and colleagues cloned the genes for antibodies (56).

As a first attempt to address the second question about cooperation between the T- and B-cell lineages, Ray Peterson initiated studies using our irradiation, thymectomy and bursectomy model system. This avian model offered the potential for the transfer of T-cells from B-less donors or B-cells from T-deficient donors into recipients which were deficient in either one or the other or both lymphocyte lineages. However, it quickly became apparent that our inbred chickens, which had been inbred for virus susceptibility studies, were not sufficiently inbred for these cell transfer studies. Inbred mice proved much better models for elucidating the basic principles of cooperative interactions between T- and B-cells in pioneering studies conducted by Henry Clamen, Graham Mitchell, Jacque Miller and Avrion Mitchison (57–59), and for the ongoing exploration of the complex interactions between the numerous subpopulations of lymphocytes that have since been recognized.

THE LONG SEARCH FOR THE MAMMALIAN BURSA-EQUIVALENT

The identification of the mammalian bursa equivalent was important for several reasons. This information was a necessary starting point for definition of early events in the B-cell differentiation pathway in mammalian species. Identification of the mammalian bursa-equivalent would also allow us to determine when, where and how B-cell differentiation might go awry in B-cell deficient patients. Furthermore, the identification of an extra-thymic origin for

mammalian B-cells would offer convincing support for the general validity of our dual lymphocyte differentiation model. Elsewhere I have summarized our decade-long search for a mammalian bursa equivalent (60), which began with the assumption that clonally diverse B lymphocytes would be generated in a follicular lymphoepithelial tissue analogous to the bursa of Fabricius. However, we could find no evidence of a hindgut lymphoid organ in mammals that was similar to the avian bursa. My attention was directed next to the pharyngeal tonsils, but removal of the entire ring of pharyngeal lymphoid tissue in newborn rabbits had no effect on their ability to produce antibodies. Daniel Perry and I then began an exploration of the hypothesis that the gut-associated lymphoepithelial tissues (GALT), appendix and Peyer's patches were the source of B-cells. In testing this theory, which became known as the GALT hypothesis, we found that neonatal appendectomy in rabbits combined with subsequent removal of the Peyer's patches and whole body irradiation significantly impaired antibody responsiveness, but had no effect on skin allograft rejection (61). These results implied that the mammalian GALT serves as the bursa-equivalent. Experiments which my colleagues and I conducted over the next few years at the University of Alabama at Birmingham were consistent with this hypothesis, but failed to provide conclusive evidence for its validity (62).

Compelling evidence against the GALT hypothesis would be obtained later, in two experimental approaches conducted during sabbatical studies in Av Mitchison's department at University College London. We postulated that, if the mammalian GALT were the true equivalent of the avian bursa, removal of the intestines early in embryonic life would prevent the development of B-cells elsewhere in the body. Geoffrey Dawes offered the opportunity to do this experiment in lambs with the help of his fetal physiology research group in Oxford. Bill Gathings and I made antibodies against sheep immunoglobulin, which we used to show that B-cells normally appear in fetal lambs at 65–70 days of gestation. Surgical removal of the entire intestines at 60 days of gestation proved to have absolutely no effect on the emergence of B-cells. This result, which was never published, provided devastatingly clear evidence against our long-standing hypothesis that the GALT is the primary source of mammalian B-cells.

Experiments conducted simultaneously in London with John Owen and Martin Raff would instead support their hypothesis of a hematopoietic origin of B-cells in mice. John had developed an organ culture method that allowed us to test the idea that B-cells could be generated in the hematopoietic fetal liver of a mouse. After showing that Ig-positive cells appear on the 17th or 18th day of mouse embryonic life, we placed 14-day liver fragments in culture and monitored them for the appearance of Ig-bearing B-cells. To our great jubilation, B-cells appeared 4–5 days later in the fetal liver cultures (63). In contemporaneous experiments, Dennis Osmond, Gus Nossal, Jean Ryser and Pierre Vassalli showed that Ig-negative cells in bone marrow could give rise to Ig-positive cells (64, 65), and we found that B-cells were generated *ex vivo* within the marrow of femoral bones from mouse embryos (66). These composite findings clearly indicated that the B-cells in mice are generated in their hematopoietic tissues.

THE IDENTIFICATION OF B-LINEAGE PROGENITORS

In earlier experiments summarized elsewhere (67), Paul Kincade, Sandy Lawton, Richard Asofsky and I had shown that the development of immature IgM-bearing B-cells can be inhibited in chickens (68) and mice (69) by IgM-specific antibodies, and that IgM-bearing B-cells undergo isotype switching to give rise to the cells that produce IgG and IgA. This led us to test the inhibitory effects of anti-IgM antibodies on B-cell generation in the fetal liver cultures. This treatment effectively eliminated development of all Ig-bearing B-cells as anticipated, but John Owen insisted that he could still find cells with lymphocyte morphology in these cultures. Much to my surprise, when these cells were fixed and permeabilized, their cytoplasm was stained by the anti-IgM antibodies (70). In this way, the precursors of B-cells were first identified and, by combining cell surface and cytoplasmic staining with radiolabeling of the dividing cells, we could trace the sequential development of large Ig-negative progenitor B-cells into large cytoplasmic Ig-positive precursor B-cells and thence to non-dividing small B-cells in mouse hematopoietic tissues (Fig. 2.5) (71). Using

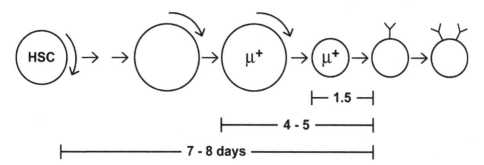

FIGURE 2.5 Model of B lineage differentiation events in hematopoietic tissues of mice. This model, which is modified from the original (73), incorporates the discovery of the sequential onset of heavy chain and light chain expression (74) and the underlying asynchronous V-D-J and V-J gene rearrangements (88). Y: IgM; YY: IgM & IgD.

the insight and methods gained from these experiments, we could show that B-lineage cells follow the same differentiation pattern during their development in the hematopoietic tissues of rabbits (75) and humans (76). These findings contributed a refined definition of the stages in B-cell differentiation, which could also be used to determine where B-cell development goes awry in patients with antibody deficiency diseases and B-cell malignancies. This information allowed us to show that B-lineage differentiation was stalled at the pre-B-cell stage in boys with X-linked agammaglobulinemia (77, 78), at the B-cell stage in patients with acquired agammaglobulinemia (CVID) (79), specifically at the IgM B-cell stage in patients with hyper-IgM (80), and at the IgA B-lymphocyte stage in patients with isolated IgA deficiency. The acute lymphocytic leukemias of childhood could be defined as malignancies of pro-B and pre-B-cells (81, 82). The identification of cellular differentiation blocks using these simple models was helpful, but clearly only an early prelude to the current genetic and pathophysiological definition of immunodeficiency diseases and lymphocytic malignancies.

AN ANCIENT BLUEPRINT FOR THE T- AND B-CELL LINEAGES

We have known about the developmentally separate T- and B-cell lineages and their functional intertwinement for almost 50 years now, a relatively long period of time measured in terms of recent history. More recently, we have learned that this basic pattern of lymphocyte development and differentiation was a very ancient evolutionary invention. Recent studies have revealed that alternative adaptive immune systems are present in all of the surviving representatives of the two sister groups of

vertebrates (reviewed in references 83–85). In a remarkable example of convergent evolution, all of the vertebrates with jaws (gnathostomes) use immunoglobulin-based T-cell receptors (TCRαβ and TCRγδ) and B-cell receptors (BCR) for antigen recognition, whereas the extant jawless vertebrates (cyclostomes, lampreys and hagfish) instead use leucine-rich-repeat proteins to construct equally diverse repertoires of antigen receptors. The cyclostomes have been shown to have three types of variable lymphocyte receptors, VLRA, VLRB and VLRC, each of which is expressed by a separate lymphocyte lineage (Fig. 2.6) (86, 87). Like our B-cells, the VLRB-bearing lymphocytes are generated in hematopoietic tissues, and they respond to antigenic stimulation with proliferation and differentiation into plasma cells which secrete VLRB antibodies. On the other hand, lamprey VLRA$^+$ and VLRC$^+$ lymphocytes express orthologs of genes that αβ and γδ T-cells use for their differentiation, undergo *VLRA* and *VLRC* gene assembly and repertoire diversification in a thymus-equivalent region of the gills, and express their VLRs solely as cell surface receptors. Our composite findings thus indicate that the genetic programs for two primordial T-cell lineages and a prototypic B-cell lineage were already present in the last common vertebrate ancestor ≈500 million years ago (87). Hence, the compartmentalized development of T and B-cells constitutes a fundamental organizing principle for all of the lymphocyte-based adaptive immune systems that have evolved in vertebrates.

Many complex systems of immunity, both innate and adaptive, have evolved in unicellular and multicellular organisms during the on-going struggle for survival on our planet. Within this evolutionary context, it is also clear that studies of the underlying defects in patients with immunodeficiency diseases will continue to be a rich source of information in the quest to understand our complex immune system.

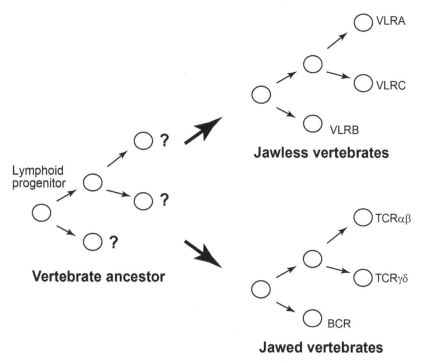

FIGURE 2.6 Evolutionary model depicting the ancient origin of T- and B-cell pathways. *(Modified from the original Supplemental Figure 6 in reference 87).*

Acknowledgments

I gratefully acknowledge the guidance, support, help and friendship of my invaluable mentors, colleagues and students, and the research support by the National Institutes of Health, Howard Hughes Medical Institute, and Georgia Research Alliance. Finally, I wish to express my immense gratitude for the unequivocal love and support of my wife Rosalie.

References

1. Bruton OC. Agammaglobulinemia. *Pediatrics* 1952;**9**:672–8.
2. Tiselius A. A new apparatus for electrophoretic analysis of colloidal mixtures. *Trans Faraday Soc* 1937;**33**:524–31.
3. Tiselius A, Kabat E. An electrophoretic study of immune serum and purification antibody preparations. *J Exp Med* 1939;**69**:119–31.
4. Good RA, Varco RL. Clinical and experimental study of agammaglobulinemia. *J Lancet* 1955;**6**:245–71.
5. Fagraeus A. Antibody production in relation to the development of plasma cells. *Acta Med Scand* 1948;**130**: 1–11.
6. Coons AH, Leduc EH, Connolly JM. Studies on antibody production. A method for the histochemical demonstration of specific antibody and its application to a study of the hyperimmune rabbit. *J Exp Med* 1955;**102**:49–60.
7. Glick B, Chang TS, Jaap RG. The bursa of Fabricius and antibody production. *Poult Sci* 1956;**35**:224–5.
8. Mueller AP, Wolfe HR, Meyer RK. Precipitation production in chickens. XXI Antibody production in bursetomized chickens and in chickens treated with 19 non-testosterone on the fifth day of incubation. *J Immunol* 1960;**85**:172–81.
9. Good RA, Dalmasso AP, Martinez C, et al. The role of the thymus in development of immunologic capacity in rabbits and mice. *J Exp Med* 1962;**116**:773–6.
10. Miller JFAP. The immunological function of the thymus. *Lancet* 1961;**2**:748–9.
11. Jankovic BD, Waksman BH, Arnason BG. Role of the thymus in immune reactions in rats. *J Exp Med* 1962;**116**: 159–76.
12. Good RA, Gabrielsen A, editors. *The Thymus in Immunobiology.* New York: Hoeber-Harper; 1964.
13. Auerbach R. Experimental analysis of the origin of cell types in the development of the mouse thymus. *Develop Biol* 1961;**3**:336–43.

14. Miller JFAP. Effect of thymic ablation and replacement. In: Good RA, Gabrielsen A, editors. *The Thymus in Immunobiology*. New York: Hoeber-Harper; 1964.

15. Hilgard R, Yunis EJ, Sjodin K, et al. Reversal of wasting in thymectomized mice by the injection of syngeneic spleen or thymus cell suspensions. *Nature* 1964;**202**:668–70.

16. Weissman IL. Thymus cell migration. *J Exp Med* 1967;**126**: 291–304.

17. Gowans JL, Knight EJ. The route of recirculation of lymphocytes in the rat. *Proc R Soc London Ser B Biol Sci* 1964;**159**:257–82.

18. Gowans JL, McGregor DD. The origin of antibody-forming cells. In: Grabar P, Miescher PA, editors. *Immunopathology, IIIrd International Symposium*. Basel: Schwabe; 1963. p. 89.

19. Porter RR. The hydrolysis of rabbit γ-globulin and antibodies with crystalline papain. *Biochem J* 1959;**73**: 119–27.

20. Edleman GM. Dissociation of γ-globulin. *J Am Chem Soc* 1959;**81**:3155–6.

21. Burnet FM. *The Clonal Selection Theory of Acquired Immunity*. Nashville, TN: Vanderbilt University Press; 1959.

22. Landsteiner K, Chase MW. Experiments on transfer of cutaneous sensitivity to simple compounds. *Proc Soc Exp Biol Med* 1942;**49**:688–90.

23. Chase MW. The cellular transfer of cutaneous hypersensitivity to tuberculin. *Proc Soc Exp Biol Med* 1945;**59**: 134–5.

24. Cooper MD, Lusher J. Immunologic tolerance of ABO incompatible erythrocytes in human neonates. *J Pediatr* 1964;**65**:831–8.

25. Owen RD. Immunogenetic consequences of vascular anastomoses between bovine twins. *Science* 1945;**102**:400–3.

26. Billingham RE, Brent L, Medawar PB. 'Actively acquired tolerance' of foreign cells. *Nature* 1953;**172**:603–7.

27. Peterson RDA, Burmester BR, Fredrickson TN, et al. Effect of bursectomy and thymectomy on the development of visceral lymphomatosis in the chicken. *J Natl Cancer Inst* 1964;**32**:1343–54.

28. Peterson RDA, Purchase MG, Burmester BR, et al. Relationships among visceral lymphomatosis, bursa of Fabricius, and bursa-dependent lymphoid tissue of the chicken. *J Natl Cancer Inst* 1966;**36**:585–98.

29. Warner NL, Szenberg A, Burnet FM. The immunological role of different lymphoid organs in the chicken. I. Dissociation of immunological responsiveness. *Aust J Exp Biol Med Sci* 1962;**40**:373–437.

30. Sutherland DER, Archer OK, Good RA. Role of the appendix in development of immunologic capacity. *Exp Biol Med* (Maywood) 1964;**115**:673.

31. Cooper MD, Chase HP, Lowman JT, et al. Wiskott-Aldrich syndrome. An immunologic deficiency disease involving the afferent limb of immunity. *Am J Med* 1968;**44**:499–513.

32. Cooper MD, Peterson RDA, Good RA. Delineation of the thymic and bursal lymphoid systems in the chicken. *Nature* 1965;**205**:143–6.

33. Cooper MD, Peterson RDA, South MA, et al. The functions of the thymus system and the bursa system in the chicken. *J Exp Med* 1966;**23**:75–102.

34. Cooper MD, Schwartz MM, Good RA. Restoration of gamma globulin production in agammaglobulinemic chickens. *Science* 1966;**151**:471–3.

35. Mayerson HS. *Lymph and Lymphatic System: Proceedings of the Conference on Lymph and the Lymphatic System*. Springfield, IL: Charles C Thomas; 1968.

36. Roitt IM, Greaves MF, Torrigiani G, et al. Cellular basis of immunological responses. *Lancet* 1969;**294**:367–71.

37. Becker AJ, McCulloch EA, Till JE. Cytological demonstration of the clonal nature of spleen colonies derived from transplanted mouse marrow cells. *Nature* 1963;**197**:452–4.

38. Ford CE, Micklem HS. The thymus and lymph nodes in radiation chimaeras. *Lancet* 1963;**281**:359–62.

39. Moore MAS, Owen JJT. Stem cell migration in developing myeloid and lymphoid systems. *Lancet* 1967;**290**: 658–9.

40. Dent PB, Perey DYE, Cooper MD, et al. Non-specific stimulation of antibody production in surgically bursectomized chickens by bursa–containing diffusion chambers. *J Immunol* 1968;**101**:799–805.

41. Cooper MD, Peterson RDA, Good RA. A new concept of the cellular basis of immunity (discussion). *J Pediatr* 1965;**67**:907–8.

42. Glanzmann E, Riniker P. Essentielle lymphocytophthise: Ein neues krankheitsbild ans der Sauglingspathologie. *Ann Pédiat* 1950;**175**:1–32.

43. Hitzig WH, Biro Z, Bosch H, et al. Agammaglobulinämie und Alymphocytose mit Schwund des lymphatischen Gewebes. *Helv Paediat Acta* 1958;**13**:551–85.

44. Nezelof C, Jammet ML, Lertholary P, et al. L'hypoplasie héréditaire du thymus: sa place et sa responsabilité dans une observation d'aplasie lymphocytaire normoplasmocytaire et normoglobulinémique du nourrison. *Arch Franc Pediatr* 1964;**21**:897.

45. DiGeorge AM. Congenital absence of the thymus and its immunologic consequences: Concurrence with congenital hypoparathyroidism. *Birth Defects, Original Article Series, The National Foundation March of Dimes* 1968;116–21.

46. Gatti RA, Meuwissen HJ, Allen HD, et al. Immunological reconstitution of sex-linked lymphopenic immunological deficiency. *Lancet* 1968;**292**:1366–9.

47. Cooper MD, Peterson RDA, Gabrielsen AE, et al. Lymphoid malignancy and development, differentiation and function of the lymphoreticular system. *Cancer Res* 1966;**26**:1165–9.

48. Ewert DL, Cooper MD. Ia-like alloantigens in the chicken: Serologic characterization and ontogeny of cellular expression. *Immunogenetics* 1978;**7**:521–35.

49. Webb CS, Cooper MD. T cells can bind antigen via cytophilic IgM antibody made by B cells. *J Immunol* 1973;**111**:275–7.

50. Eichmann K. Expression and function of idistypes on lymphocytes. *Adv Immunol* 1978;**26**:195–254.

51. Tada T, Okumura K. The role of antigen-specific T cell factors in the immune response. *Adv Immunol* 1979;**28**:1–87.

52. Kubagawa H, Mayumi M, Kearney JF, et al. Immunoglobulin V$_H$ determinants defined by monoclonal antibodies. *J Exp Med* 1982;**156**:1010–24.

53. Zinkernagel RM, Doherty PC. Cytotoxic thymus-derived lymphocytes in cerebrospinal fluid of mice with lymphocytic choriomeningitis. *J Exp Med* 1973;**138**:1266–9.

54. Yanagi Y, Yoshikai Y, Leggett SP, et al. A human T cell specific cDNA clone encodes a protein having extensive homology to immunoglobulin chains. *Nature* 1984;**308**:145–9.

55. Hedrick SM, Cohen DI, Nielsen EA, et al. Isolation of cDNA clones encoding T cell-specific membrane-associated proteins. *Nature* 1984;**308**:149–53.

56. Tonegawa S. Somatic generation of antibody diversity. *Nature* 1983;**302**:571–81.

57. Clamen HN, Chaperon EA, Triplett RF. Thymus-marrow cell combinations. Synergism in antibody production. *Proc Soc Exp Biol Med* 1966;**122**:1167–71.

58. Mitchell GF, Miller JFAP. Cell to cell interaction in the immune response. II. The source of homolysin-forming cells in irradiated mice given bone marrow and thymus or thoracic duct lymphocytes. *J Exp Med* 1968;**128**:126–35.

59. Mitchison NA. The carrier effect in the secondary response to hapten-protein conjugates. II. Cellular cooperation. *Eur J Immunol* 1971;**1**:18–27.

60. Cooper MD. A life of adventure in immunobiology. *Annu Rev Immunol* 2010;**28**:1–19.

61. Cooper MD, Perry DY, McKneally MF, et al. A mammalian equivalent of the avian bursa of Fabricius. *Lancet* 1966;**287**:1388–91.

62. Cooper MD, Lawton AR. The mammalian "bursa-equivalent": Does lymphoid differentiation along plasma cell lines begin in the gut-associated lymphoepithelial tissues (GALT) of mammals? In: Hanna MG, editor. *Contemporary Topics in Immunobiology*. New York: Plenum; 1972. p. 49–68.

63. Owen JJT, Cooper MD, Raff MC. In vitro generation of B lymphocytes in mouse fetal liver, a mammalian "bursa equivalent". *Nature* 1974;**249**:361–3.

64. Osmond DG, Nossal GJV. Differentiation of lymphocytes in mouse bone marrow. II. Kinetics of maturation and renewal of antiglobulin-binding cells studied by double labeling. *Cell Immunol* 1974;**13**:132–45.

65. Ryser JE, Vassalli P. Mouse bone marrow lymphocytes and their differentiation. *J Immunol* 1974;**113**:719–28.

66. Owen JJT, Raff MC, Cooper MD. Studies on the generation of B lymphocytes in the mouse embryo. *Eur J Immunol* 1975;**5**:468–73.

67. Cooper MD, Lawton AR. The development of the immune system. *Sci Am* 1974;**231**:58–72.

68. Kincade PW, Lawton AR, Bockman DE, et al. Suppression of immunoglobulin G synthesis as a result of antibody-mediated suppression of immunoglobulin M synthesis in chickens. *Proc Natl Acad Sci USA* 1970;**67**:1918–25.

69. Lawton AR, Asofsky R, Hylton MB, et al. Suppression of immunoglobulin class synthesis in mice: I. Effects of treatment with antibody to μ chain. *J Exp Med* 1972;**35**:277–97.

70. Raff MC, Megson M, Owen JJT, Cooper MD. Early production of intracellular IgM by B lymphocyte precursors in mouse. *Nature* 1976;**259**:224–6.

71. Owen JJT, Raff MC, Cooper MD. Studies on the generation of B lymphocytes in the mouse embryo. *Eur J Immunol* 1975;**5**:468–73.

72. Hoyer JR, Cooper MD, Gabrielsen AE, et al. Lymphopenic forms of congenital immunologic deficiency: Clinical and pathologic patterns. In: Bergsma MDC, Good RA, editors. *Immunologic Deficiency Diseases in Man*. New York: The National Foundation; 1968. p. 91–103.

73. Cooper MD. B cell development in birds and mammals. In: Clnader B, Miller RG, editors. *Progress in Immunology VI*. Academic Press Inc; 1986. p. 18–31.

74. Burrows PD, LeJeune M, Kearney JF. Evidence that murine pre-B cells synthesise μ heavy chains but no light chains. *Nature* 1979;**280**:838–40.

75. Hayward AR, Simons MA, Lawton AR, et al. Pre-B and B cells in rabbits: Ontogeny and allelic exclusion of kappa light chain genes. *J Exp Med* 1978;**148**:1367–77.

76. Gathings WE, Lawton AR, Cooper MD. Immunofluorescent studies of the development of pre–B cells, B lymphocytes and immunoglobulin isotype diversity in humans. *Eur J Immunol* 1977;**7**:804–10.

77. Vogler LB, Pearl ER, Gathings WE, et al. B–lymphocyte precursors in the bone marrow of patients with immunoglobulin deficiency diseases. *Lancet* 1976;376 ii.

78. Pearl ER, Vogler LB, Okos AJ, et al. B lymphocyte precursors in bone marrow. An analysis of normal individuals and patients with antibody deficiency states. *J Immunol* 1978;**120**:1169–75.

79. Cooper MD, Lawton AR, Bockman DE. Agammaglobulinemia with B lymphocytes: A specific defect of plasma cell differentiation. *Lancet* 1971;791–5 ii.

80. Levitt D, Haber P, Rich K, et al. Hyper IgM immunodeficiency: A primary dysfunction of B lymphocyte isotype switching. *J Clin Invest* 1983;**72**:1650–7.

81. Vogler LB, Crist WM, Bockman DE, et al. Pre–B cell leukemia: A new phenotype of childhood lymphoblastic leukemia. *N Engl J Med* 1978;**298**:872–8.

82. Vogler LB, Preud'homme JL, Seligmann M, et al. Diversity of immunoglobulin expression in leukemic cells resembling B lymphocyte precursors. *Nature* 1981;**290**:339–41.

83. Flajnik MF, Kasahara M. Origin and evolution of the adaptive immune system: genetic events and selective pressures. *Nat Rev Genet* 2010;**11**:47–59.

84. Litman GW, Rast JP, Fugmann SD. The origins of vertebrate adaptive immunity. *Nat Rev Immunol* 2010;**10**:543–53.

85. Hirano M, Das S, Guo P, Cooper MD. The evolution of adaptive immunity in vertebrates. *Adv Immunol* 2011;**109**:125–57.

86. Guo P, et al. Dual nature of the adaptive immune system in lampreys. *Nature* 2009;**459**:796–801.

87. Hirano M, Guo P, McCurley N, et al. Evolutionary implications of a third lymphocyte lineage in lampreys. *Nature* 2013;**501**:435-38.

88. Alt FW, Rosenberg N, Lewis S, et al. Organization and reorganization of immunoglobulin genes in A-MuLV-transformed cells: Rearrangement of heavy but not light chain genes. *Cell* 1981;**27**:391–400.

Evolution of the Definition of Primary Immunodeficiencies

Capucine Picard[1,2,3],
Jean-Laurent Casanova[1,3,4,5]

[1]Laboratory of Human Genetics of Infectious Diseases,
Necker Branch, INSERM UMR 1163, Necker Medical School, Imagine Institute,
Paris Descartes University, Paris, France
[2]Study Center for Primary Immunodeficiencies, Assistance Publique-Hôpitaux de Paris,
Necker Hospital for Sick Children, Paris, France
[3]St Giles Laboratory of Human Genetics of Infectious Diseases, Rockefeller Branch,
The Rockefeller University, New York, NY, USA
[4]Pediatric Hematology-Immunology Unit, Necker Hospital for Sick Children, Paris, France
[5]Howard Hughes Medical Institute, New York, NY, USA

INTRODUCTION

Single-gene inborn errors of immunity are commonly referred to as primary immunodeficiencies (PIDs). PIDs were first defined in the early

1950s and now include over 250 known clinical entities, at least 240 of which have well defined molecular genetic bases (1–3). Patients with PIDs typically present with multiple infectious diseases. Before the introduction of antibiotics in the 1940s, children with PIDs died during the first infectious episode, preventing the expression and recognition of this phenotype. When antibiotics became widely available in the 1950s, the attention of a small group of physician-scientists was drawn to rare children who displayed recurrent infectious diseases, such as recurrent pyogenic bacterial diseases in particular (4–12). These children were soon found to display multiple infections and to have not only an unusual clinical presentation, but also unusual hematoimmunological features, such as neutropenia or agammaglobulinemia (13, 14). They also displayed opportunistic infections not seen in children without hematological and immunological abnormalities. These syndromes were inherited as Mendelian traits, some of which were autosomal recessive (AR) (e.g. neutropenia) (6), whereas others were X-linked recessive (XR) traits (e.g. agammaglobulinemia) (14). These discoveries were the first step towards development of the concept of PID and, as we discuss here, of "classical" PIDs. These diseases have since been described as rare, penetrant, Mendelian recessive traits predisposing patients to multiple and recurrent infections, including infections caused by weakly virulent (opportunist) microorganisms, occurring as the consequence of a detectable immunological phenotype (1, 15, 16). However, the concept of PID was rapidly extended beyond susceptibility to infections, to become associated with other clinical symptoms, such as autoimmunity, allergy and malignancy (17). This concept, for example, was extended to patients with Wiskott–Aldrich syndrome (WAS), first described in 1937 as a hematological disorder different from idiopathic thrombocytopenia (18, 19). Many other diverse phenotypes (autoinflammation, angioedema, hemophagocytosis, microangiopathies, and others) were progressively shown to result from inborn errors of immunity (3, 20). This led to a fascinating diversification of the field over the last two decades, with the discovery of a wide array of noninfectious phenotypes caused by inborn errors of immunity, which we will not review here due to space constraints.

The definition of PID also evolved within the field of infectious diseases itself, with the discovery that severe infections in otherwise healthy individuals, including common infections not seen in patients with classical PIDs, may result from inborn errors of immunity. Indeed, infectious diseases due to any specific microbe rarely recur, and typically strike otherwise healthy individuals with no detectable immunological phenotype. The occurrence of familial forms, parental consanguinity, or both, suggested that some of these "idiopathic" infections also reflected a Mendelian predisposition, particularly in cases of recurrent infection. This observation paved the way for genetic dissection of the etiologies of these conditions. This intriguing group of "nonconventional" PIDs (21), characterized by a narrow spectrum of specific infections, limited to one microbial genus, or even a single species, has expanded over the last two decades (21–24). Several of these disorders were actually described before the first classical PIDs were recognized in the 1950s but, for various reasons, they made no major contribution to the development of the definition of PID until recently. These conditions include, in particular, epidermodysplasia verruciformis (EV), described as an autosomal recessive predisposition to oncogenic human papilloma viruses in 1946 (25, 26), X-linked lymphoproliferative (XLP) disorder, described as an XR predisposition to Epstein–Barr virus (EBV) in 1974 (27), and defects of the terminal components of complement and properdin, described in the 1980s (28, 29) as a cause of invasive *Neisseria* disease. Had these disorders been historically associated with the initial definition of PID, our current view of the field and its development might have been very different. In this review, we focus on the historic events that illustrate the evolution of PIDs from the standpoint of infection. As we will see, the definition of PID has shifted from fully penetrant Mendelian traits that confer a predisposition to multiple, opportunistic infections in individuals with hematological abnormalities to single-gene inborn errors of immunity, not necessarily displaying complete penetrance or affecting only leukocytes, and underlying severe infections with a single group of microbes in otherwise healthy individuals. We will illustrate the history of the field by focusing on four conditions, each marking a landmark in the evolution of the PID definition.

THE BIRTH OF CLASSICAL PIDs, ILLUSTRATED BY THE DISCOVERY OF X-LINKED AGAMMAGLOBULINEMIA

Various constitutive features of PIDs were identified as important from the 1950s onwards, but the idea that infections in patients with PID could be "recurrent" was arguably the most important seminal feature leading to the description and definition of PID. The best example of recurrent disease is perhaps that provided by Bruton's first patient with agammaglobulinemia, reported in 1952: a boy who had suffered from 19 episodes of invasive pneumococcal disease (4). This extraordinary clinical phenotype, rendered visible by the advent of antibiotics, led to a clinical investigation that discovered an absence of serum gamma globulins (see Chapter 13). It was also subsequently shown that patients with the X-linked form of agammaglobulinemia (XLA) have very few or no circulating B-cells (30–33) and most have low pre-B-cell numbers in bone marrow and few immature B-cells in the bloodstream (34, 35). It took more than 40 years to elucidate the genetic etiology of XLA, with the identification in 1993 of mutations in the gene encoding Bruton's tyrosine kinase (BTK), a cytoplasmic tyrosine kinase specific to most cells of the hematopoietic lineage (36, 37). BTK is expressed throughout B-cell differentiation, from early precursor B-cell stages to mature B-cell stages (38). BTK protein is also expressed in platelets and phagocytes, but not in T-cells and plasma cells (39). *BTK* mutations account for 85% of agammaglobulinemia

cases (40) and most of these mutations result in an absence of detectable BTK protein (41, 42). Milder immunological and clinical phenotypes have been observed in some patients with *BTK* mutations that do not affect protein stability (41, 43, 44). There is currently no functional assay able to distinguish easily between hypomorphic and loss-of-function mutations, precluding studies of correlations between genotype and immunological or infectious phenotypes.

Several years after the clinical and immunological description of XLA patients, AR forms of agammaglobulinemia (ARA) were identified (45). Patients with ARA carry mutations in genes encoding components of the pre-B-cell receptor (BCR) and BCR, including the μ heavy chain (46–49), the surrogate light chain protein λ5 (50), the signal transduction molecules Igα and Igβ (46, 51, 52), and *BLNK* (53). Mutations have also recently been identified in the genes encoding PI3 kinase (PI3K) p85α (54) and the E47 isoform of E2A (55), the latter defining the first AD form of agammaglobulinemia. Patients with XLA or ARA are prone to recurrent, multiple, pyogenic, bacterial infections, which is clearly the principal infectious phenotype of these patients (56, 57). They may also develop enteroviral infections (56–60), mycoplasma and ureaplasma infections (61) and *Giardia* gut infections (62). Intriguingly, only a few of these infections are clearly opportunistic, as illustrated by disseminated enteroviral disease, a presentation almost never seen in otherwise healthy children. Significant infectious complications and early death are still sometimes reported, but most patients now survive into adulthood, thanks to IgG substitution (57, 63). By studying this experiment of Nature, it was possible to identify microbial infections for which humoral immunity was critical. Some of the infections (enterovirus, mycoplasma, *Giardia*) seen in these patients remain intriguing, as does the apparent absence of certain other infections. The spectrum of infections in patients with XLA or ARA is relatively broad but quite specific, as for most of the classical PIDs described in the

1950s and 1960s, including severe combined immunodeficiency (SCID), which was recognized in 1958 (64, 65), chronic granulomatous disease (CGD), first described in 1957 (66), and the hyper-IgM syndrome, first described in 1960 (67). Many PIDs corresponding to this definition have been discovered since these early days and are still being identified today (1, 3).

THE CONCEPT OF PIDs WITH NARROW PHENOTYPE, ILLUSTRATED BY X-LINKED LYMPHOPROLIFERATIVE DISEASE

PIDs were originally defined by the identification of immunological phenotypes, as illustrated by agammaglobulinemia and neutropenia. Given the limited techniques available in the 1950s and 1960s, only major abnormalities could be detected, and PIDs were therefore, understandably, associated with multiple, recurrent, and often opportunistic infections. The first exception to this rule, or at least the first exception to be accepted within the field of PIDs itself, was provided by the description of X-linked lymphoproliferative (XLP) disorder, which manifests as various EBV-driven clinical phenotypes in previously healthy individuals (27, 68). Strikingly, these patients displayed no immunological marker before exposure to EBV occurred. This situation is reminiscent of that for EV (25), which was not accepted by the PID community until 2002 (69), but contrasts with that of defects involving terminal complement components, which are characterized by selective vulnerability to *Neisseria* and were delineated on the basis of their complement phenotype. The four clinical features of XLP, hemophagocytic lymphohistiocytosis (HLH), lymphoproliferative disease (LPD), hypogammaglobulinemia and B-cell lymphoma, were all described in 1974 and were attributed to the XR inheritance of susceptibility to EBV in one large kindred (27, 68). This clinical and laboratory phenotype was confirmed by studies of additional kindreds

(70). The first XLP-causing gene, *SH2D1A*, was identified in 1998 (two years after the first MSMD-causing gene was found, see below) and was found to be specifically expressed in natural killer (NK) and T-cells (71, 72). *SH2D1A* encodes a small adaptor molecule, SLAM-associated protein (SAP) (73, 74). The pathogenesis of XLP in SAP-deficient patients was recently clarified by two very elegant studies of female carriers and somatic mutants, which genetically established the role of CD8$^+$ cytoxic T lymphocytes (CTLs) (75, 76). CD8$^+$ CTLs seem to kill EBV-infected B-cells through SAP (73, 77). In the absence of SAP, these activating pathways are inhibited and EBV infection remains uncontrolled (74). The persistence of the virus within B lymphocytes triggers the continuous activation of virus-specific CD8$^+$ CTLs, resulting in LPD or B-cell lymphoma and the release of large amounts of interferon γ (IFN-γ), underlying HLH (78). SAP is also important for the function of follicular helper T-cells, which promote memory B-cell development and plasmablast differentiation (79). Consistently, most SAP-deficient patients develop hypogammaglobulinemia, with a lack of memory B-cells (80). SAP is also important for the development of invariant natural killer T (iNKT) lymphocytes (81) and it has recently been shown that these cells might contribute to the early innate control of B-cell EBV infection (82).

SAP deficiency is a serious disease, as most untreated patients die from HLH or B-cell lymphoma, and survivors often display hypogammaglobulinemia (78, 83). Hematopoietic stem cell transplantation (HSCT) is the only curative treatment available, and should ideally be attempted before primary EBV infection occurs in genetically affected individuals (83). This disease displays complete clinical penetrance. A second, related XR disorder associated with EBV-driven LPD and HLH was found to be due to hemizygous mutations of *BIRC4*, which encodes the X-linked inhibitor of apoptosis (XIAP) (84). Unlike SAP deficiency, XIAP deficiency is not associated with lymphoma and most of the affected subjects present with HLH. Patients with

XIAP deficiency may even develop HLH triggered by another herpes virus in the absence of EBV infection (78). XIAP-deficient patients may also develop chronic hemorrhagic colitis, mimicking Crohn's disease (78).

XIAP-deficient patients also display a quantitative deficiency of iNKT-cells, but to a lesser extent than SAP-deficient patients (84, 85). The cellular basis of disease in XIAP-deficient patients remains largely elusive. HSCT is the only curative treatment for XIAP-deficiency, but the outcome is poor (86). The loss of XIAP anti-apoptotic functions may contribute to the high incidence of toxicity observed when busulfan-containing myeloablative conditioning regimens are used (86). Three new T-cell deficiencies have recently been identified in patients prone to severe and chronic EBV infection: XR MAGT1 (87), AR IKT (88) and AR CD27 deficiency (89). The discovery of MAGT1 deficiency (XMEN) revealed a particularly elegant mechanism, as the molecular defect prevents Mg influx in T-cells and can be rescued by treatment with magnesium *in vitro* and *in vivo* (87, 90). These studies confirmed the essential role of T-cells in the control of EBV. More surprisingly, as for SAP deficiency, it could be shown that some T-cell molecules are essential for immunity against EBV but are otherwise redundant. In any case, studies of the genetic disorders selectively causing EBV-driven diseases have clearly demonstrated the occurrence of Mendelian defects associated with specific vulnerability to a single infectious agent. Incidentally, they have also contributed to the demonstration that PIDs are not limited to a typical infectious phenotype, as some of these patients develop HLH (SAP, XIAP and CD27 deficiencies), neoplasia (SAP, MAGT1, ITK and CD27 deficiencies) or inflammatory bowel features (XIAP deficiency). It is interesting, from a historical point of view, that XLP was immediately considered a PID when it was first identified in 1974, years before the identification of the first genetic etiology in 1998, whereas EV, which had clearly been shown to involve Mendelian predisposition to virus-induced benign and malignant lesions in the 1930s and 1940s, was not considered to be a PID until 2002 (25, 91).

THE DEVELOPMENT OF PIDs WITH NARROW PHENOTYPE, ILLUSTRATED BY MENDELIAN SUSCEPTIBILITY TO MYCOBACTERIAL DISEASE

The clinical description of XLP was also paradoxically made two decades after that of Mendelian susceptibility to mycobacterial disease (MSMD), another syndrome conferring predisposition to a single infectious agent or a small group of pathogens, which was probably first observed and described in Algeria in 1951 (92). The idea that PIDs might underlie infectious diseases in patients normally resistant to most other microbes, with no overt immunological abnormality, gained ground from the mid-1990s onwards. MSMD, originally described in the 1950s as "idiopathic" infections caused by the Bacille Calmette–Guérin (BCG) (92) vaccine, was not recognized as a PID until 1995–1996 (93), shortly before the identification of AR IFN-γR1 deficiency as its first genetic etiology (94, 95). MSMD has since been shown to be genetically heterogeneous, with at least 15 genetic etiologies identified by the year 2013. Morbid mutations in seven autosomal genes (*IFNGR1, IFNGR2, IL12B, IL12RB1, STAT1, IRF8, ISG15*) and two X-linked genes (*NEMO, CYBB*) have been discovered (94–101). There is also considerable allelic heterogeneity, with recessive and dominant traits, and complete and partial defects associated with mutations at several loci. The gene most recently implicated in MSMD, *ISG15*, was identified through a combination of whole-exome sequencing (WES) and genome-wide linkage (GWL) analysis (96). All MSMD genetic defects result in an impairment of IFN-γ immunity, affecting the production of or response to this cytokine. IFN-γR1, IFN-γR2 and STAT1 deficiencies impair cellular responses to IFN-γ, particularly in macrophages (101). *ISG15* mutations disrupt a circuit involving principally granulocytes and NK cells, thereby impairing IFN-γ production (96). IL-12p40 and IL-12Rβ1-deficiencies impair the IL-12-dependent induction of IFN-γ production in NK and T-cells (101–103). These two defects also impair the IL-23-dependent induction of IL-17, accounting for the chronic mucocutaneous candidiasis (CMC) documented in some patients (102–104). AD IRF8 deficiency prevents the development of IL-12-producing CD1c⁺ CD11c⁺ dendritic cells (98). MSMD-causing mutations of *CYBB* impair the IFN-γ-dependent respiratory burst in macrophages (100). MSMD-causing mutations of *NEMO* impair the T-cell- and CD40L-dependent induction of IL-12 by dendritic cells (97). In these last two conditions, patients carry particular mutations in genes for which null alleles are known to underlie more complex PIDs: CGD and anhidrotic ectodermal dysplasia (EDA)-immunodeficiency (ID), respectively (105). Known genetic etiologies account for only about half the observed cases of MSMD. We therefore expect new genetic etiologies of MSMD to be discovered in the future.

Patients with MSMD display clinical disease caused by weakly virulent mycobacteria, such as BCG vaccines and environmental mycobacteria, but they are also highly prone to *Mycobacterium tuberculosis* infection (102, 103). MSMD patients, particularly those with IL-12p40 and IL-12Rβ1 deficiencies, often suffer from non-typhoidal, extra-intestinal salmonellosis (102, 103). Poor producers of IFN-γ, due to IL-12Rβ1 deficiency, and patients with reduced IFN-γ response due to partial deficiencies of IFN-γR1, IFN-γR2 and STAT1, may benefit from curative treatment with recombinant IFN-γ (102, 103, 106). Conversely, patients whose cells do not respond to IFN-γ, such as those with complete IFN-γR1, IFN-γR2 and STAT1 deficiencies may benefit from HSCT (107). Studies of these genetic defects have shown that IFN-γ is essential for immunity against mycobacteria and a few other intracellular bacteria, fungi and parasites that infect macrophages. Surprisingly, however, these patients are not prone to

infections with other intracellular agents, including most viruses. The rare viral illnesses occurring in these patients may have been favored by mycobacterium-induced   immunosuppression (102, 103, 106, 108). Based on these observations, it has been suggested that mycobacterial diseases in other medical settings, including tuberculosis, may result from impaired IFN-γ immunity. Finally, these studies lend weight to the idea that otherwise healthy children with isolated infectious diseases may suffer from single-gene inborn errors of immunity (109, 110). The study of MSMD has proved fruitful, as it led to the identification of AR IL-12Rβ1 deficiency as the first genetically defined severe and isolated form of pediatric tuberculosis (111–113). Some patients with genetic etiologies of MSMD may never develop MSMD because of incomplete clinical penetrance but, nevertheless, may be vulnerable to the more virulent causal agent of tuberculosis. This is perhaps the most exciting and rewarding development of this line of research, as MSMD was initially studied with a view to dissect the genetic basis of human tuberculosis (109, 111). These studies therefore suggest that relatively common infections in the general population, such as tuberculosis, which is fundamentally different from EBV-driven HLH and BCG disease, may also, surprisingly, result from single-gene inborn errors of immunity (15, 17, 23, 101, 110).

PIDs WITH NARROW PHENOTYPES UNDERLYING COMMON INFECTIONS, ILLUSTRATED BY HERPES SIMPLEX VIRUS ENCEPHALITIS OF CHILDHOOD

Herpes simplex virus 1 (HSV1) encephalitis (HSE) is the first example of an isolated, severe and relatively common childhood infection, not seen in any patients with conventional PIDs or acquired immunodeficiencies, that was eventually shown to result from single-gene inborn errors of immunity (114). HSE is the most common sporadic viral encephalitis in Western countries. HSV1 is ubiquitous and typically innocuous. In patients with HSE, the virus reaches the central nervous system (CNS) via cranial nerves. Intriguingly, the virus is restricted to the CNS in the course of HSE. The disease is sporadic in the vast majority of cases, with only four multiplex kindreds reported in 60 years (114). The sporadic nature of this disease argued against a genetic origin, or at least against a genetic etiology with complete penetrance. However, parental consanguinity was found in 12% of HSE patients recruited in a French survey, suggesting that HSE might be due to single-gene inborn errors of immunity (114). Patients with the most severe PIDs, including children without T-cells, display no particular susceptibility to HSE. By contrast, patients with complete STAT1 deficiency and impaired responses to both IFN-γ and IFN-α/β were found to be prone to multiple viral diseases, including HSE (115, 116). Moreover, patients with *NEMO* mutations are broadly susceptible to viral infections, including HSE, reflecting the impairment of antiviral IFN-α/β production (115–117). The first genetic etiology of HSE was identified as AR UNC-93B deficiency, through a combination of GWL and a candidate gene approach. UNC-93B deficiency results in an impairment of cellular responses to the four intracellular TLRs (TLR3, 7–9) (118). A morbid role of the TLR3 pathway was suspected, because IRAK-4- and MyD88-deficient patients, whose cells do not respond to TLR7-9, are not prone to HSE (119, 120). TLR3 was formally implicated in the disease when AD and AR TLR3 deficiencies were discovered in other children with HSE (121). The subsequent identification of children with AR and AD TRIF deficiency confirmed the role of the TLR3 signaling pathway and further suggested that childhood HSE might result from a collection of highly diverse but immunologically related single-gene lesions (122).

HSE-causing heterozygous mutations of *TRAF3* further highlighted the potential morbid role of subtle mutations in pleiotropic genes, even for narrow clinical phenotypes (123). In AD TRAF3 deficiency, the mutation is dominant-negative, and impaired TLR3 responses account for HSE, whereas the other cellular phenotypes, such as impaired responses to members of the TNF receptor superfamily, are clinically silent (123). In AD TBK1 deficiency, a more recently defined etiology of HSE, there is a narrow cellular phenotype, apparently restricted to the TLR3 pathway, although TBK1, like TRAF3, is involved in multiple signaling pathways (124). While genetic etiologies of isolated HSE display complete penetrance at the cellular level (in fibroblasts), the clinical penetrance is incomplete, accounting for the sporadic nature of HSE. Interestingly, the defects predisposing subjects to childhood HSE during primary infection with HSV1 do not seem to impair immunity to other viruses. However, one patient with coxsackievirus myocarditis has been reported to carry a dominant-negative *TLR3* mutation that has also been found in HSE patients (125). Immunity to latent HSV1 also seems to be maintained, although several TLR3-deficient patients have presented relapses of HSE (121, 126). The TLR3 pathway has been shown to be largely redundant in keratinocytes and leukocytes, accounting for the lack of disseminated disease in the course of HSE. By contrast, the TLR3 pathway is essential for cellular responses to polyI:C (a dsRNA mimic) and for immunity to HSV1 in both fibroblasts and induced pluripotent stem cell (iPSC)-derived CNS cells (127). Indeed, anti-HSV1 immunity has been shown to be critically dependent on the TLR3-dependent production of IFN-α/β in neurons and oligodendrocytes (127). These findings suggest that HSE results from a collection of single-gene inborn errors of TLR3 intrinsic immunity operating in neurons and oligodendrocytes, in particular. This is another leap in defining PIDs which were initially restricted to defects in leukocytes or plasma proteins produced elsewhere (e.g. complement produced in the liver). The genetic dissection of HSE has shed new light on host defenses in general, revealing that the TLR3 pathway is responsible for ensuring CNS-intrinsic protective immunity against HSV1 during primary, but possibly not latent infection. Future studies will search for new genetic etiologies of HSE and will determine whether HSE is a valid, general model for the genetic architecture of sporadic, isolated, life-threatening, childhood infectious diseases that are not opportunistic (109). They will also explore the molecular and cellular basis of the incomplete penetrance of the genetic etiologies of HSE.

CONCLUSIONS AND PERSPECTIVES

The last 15 years have witnessed the emergence of a new group of PIDs in otherwise healthy patients with isolated and "idiopathic" infections, such as XLP, MSMD, and HSE (128). Other examples include invasive dermatophytic disease (129), CMC (117, 130, 131), Kaposi sarcoma (132, 133) and invasive pneumococcal disease (120, 134–137). There is much hope and little doubt that next-generation sequencing (NGS) will allow the identification of new PIDs in patients with various infectious diseases. Moreover, iPSC-based approaches can be used to study almost any human cell type from any patient, including, in particular, non-hematopoietic cells that cannot otherwise be studied in humans (127). These new NGS and iPSC technologies are ushering in a new era in which they will facilitate the discovery and characterization of new PIDs in patients with infectious and other phenotypes. The definition of PID in 2014 is clearly much broader than that in 1952. Exciting times lie ahead.

Acknowledgments

We thank Anne Puel, Shen-Ying Zhang, Emmanuelle Jouanguy, Stéphanie Boisson-Dupuis, Jacinta Bustamante, Guillaume Vogt and Laurent Abel for helpful discussions and advice.

References

1. Al-Herz W, Bousfiha A, Casanova JL, et al. Primary immunodeficiency diseases: an update on the classification from the international union of immunological societies expert committee for primary immunodeficiency. *Front Immunol* 2011;**2**:54.
2. Good RA. Historical aspects of immunologic deficiency diseases. In Kagen BM, Stiehm ER (eds.) Immunologic Incompetence: Year Book Medical Publishing, 1971, pp. 149–177.
3. Conley ME, Notarangelo LD, Casanova JL. Definition of primary immunodeficiency in 2011: a "trialogue" among friends. *Ann NY Acad Sci* 2011;**1238**:1–6.
4. Bruton OC. Agammaglobulinemia. *Pediatrics* 1952;**9**:722–8.
5. Good RA, Zak SJ. Disturbances in gamma globulin synthesis as experiments of nature. *Pediatrics* 1956;**18**:109–49.
6. Kostmann R. Infantile genetic agranulocytosis; agranulocytosis infantilis hereditaria. *Acta Paediatr Suppl* 1956;**45**(Suppl 105):1–78.
7. Bruton OC, Apt L, Gitlin D, Janeway CA. Absence of serum gamma globulins. *Am J Dis Child* 1952;**84**:632–6.
8. Good RA. Absence of plasma cells from bone marrow and lymph nodes following antigenic stimulation in patients with a gamma globulinemia. *Rev Hematol* 1954;**9**:502–3.
9. Gitlin D, Hitzig WH, Janeway CA. Multiple serum protein deficiencies in congenital and acquired agammaglobulinemia. *J Clin Invest* 1956;**35**:1199–204.
10. Bruton OC. A decade with agammaglobulinemia. *J Pediatr* 1962;**60**:672–6.
11. Good RA. Experiments of nature in immunobiology. *N Engl J Med* 1968;**279**:1344–5.
12. Gitlin D. Low resistance to infection: relationship to abnormalities in gamma globulin. *Bull NY Acad Med* 1955;**31**:359–65.
13. Boztug K, Klein C. Novel genetic etiologies of severe congenital neutropenia. *Curr Opin Immunol* 2009;**21**:472–80.
14. Conley ME, Dobbs AK, Farmer DM, et al. Primary B cell immunodeficiencies: comparisons and contrasts. *Annu Rev Immunol* 2009;**27**:199–227.
15. Casanova JL, Abel L. Primary immunodeficiencies: a field in its infancy. *Science* 2007;**317**:617–9.
16. Notarangelo LD, Casanova JL. Primary immunodeficiencies: increasing market share. *Curr Opin Immunol* 2009;**21**:461–5.
17. Casanova JL, Abel L. The genetic theory of infectious diseases: a brief history and selected illustrations. *Annu Rev Genomics Hum Genet* 2013;**14**:215–43.
18. Wiskott A. Familiärer, angeborener Morbus Werlhofii? *Monatsschr Kinderheilkd* 1937;**68**:212–6.

19. Sullivan KE, Mullen CA, Blaese RM, Winkelstein JA. A multiinstitutional survey of the Wiskott–Aldrich syndrome. *J Pediatr* 1994;**125**:876–85.

20. Bousfiha AA, Jeddane L, Ailal F, et al. A [henotypic approach for IUIS PID classification and diagnosis: guidelines for clinicians at the bedside. *J Clin Immunol* 2013;**33**:1078–87.

21. Casanova JL, Fieschi C, Bustamante J, et al. From idiopathic infectious diseases to novel primary immunodeficiencies. *J Allergy Clin Immunol* 2005;**116**:426–30.

22. Casanova JL, Abel L. Inborn errors of immunity to infection: the rule rather than the exception. *J Exp Med* 2005;**202**:197–201.

23. Casanova JL, Abel L. The human model: a genetic dissection of immunity to infection in natural conditions. *Nat Rev Immunol* 2004;**4**:55–66.

24. Casanova JL, Schurr E, Abel L, Skamene E. Forward genetics of infectious diseases: immunological impact. *Trends Immunol* 2002;**23**:469–72.

25. Cockayne EA. Epidermodysplasia verruciformis. In Press O, editor. *Inherited abnormalities of the skin and its appendages*. London: Oxford University Press; 1933. p. 156.

26. Lutz W. A propos de l'epidermodysplasie verruciforme. *Dermatologica* 1946;**92**:30–43.

27. Purtilo DT, Cassel C, Yang JP. Letter: Fatal infectious mononucleosis in familial lymphohistiocytosis. *N Engl J Med* 1974;**291**:736.

28. Ross SC, Densen P. Complement deficiency states and infection: epidemiology, pathogenesis and consequences of neisserial and other infections in an immune deficiency. *Medicine* 1984;**63**:243–73.

29. Sjoholm AG, Braconier JH, Soderstrom C. Properdin deficiency in a family with fulminant meningococcal infections. *Clin Exp Immunol* 1982;**50**:291–7.

30. Geha RS, Rosen FS, Merler E. Identification and characterization of subpopulations of lymphocytes in human peripheral blood after fractionation on discontinuous gradients of albumin. The cellular defect in X-linked agammaglobulinemia. *J Clin Invest* 1973;**52**:1726–34.

31. Cooper MD, Lawton AR, Bockman DE. Agammaglobulinaemia with B lymphocytes. Specific defect of plasma-cell differentiation. *Lancet* 1971;**2**:791–4.

32. Preud'Homme JL, Griscelli C, Seligmann M. Immunoglobulins on the surface of lymphocytes in fifty patients with primary immunodeficiency diseases. *Clin Immunol Immunopathol* 1973;**1**:241–56.

33. Siegal FP, Pernis B, Kunkel HG. Lymphocytes in human immunodeficiency states: a study of membrane-associated immunoglobulins. *Eur J Immunol* 1971;**1**:482–6.

34. Conley ME. B cells in patients with X-linked agammaglobulinemia. *J Immunol* 1985;**134**:3070–4.

35. Nunez C, Nishimoto N, Gartland GL, et al. B cells are generated throughout life in humans. *J Immunol* 1996;**156**:866–72.

36. Tsukada S, Saffran DC, Rawlings DJ, et al. Deficient expression of a B cell cytoplasmic tyrosine kinase in human X-linked agammaglobulinemia. *Cell* 1993;**72**:279–90.

37. Vetrie D, Vorechovsky I, Sideras P, et al. The gene involved in X-linked agammaglobulinaemia is a member of the src family of protein-tyrosine kinases. *Nature* 1993;**361**:226–33.

38. de Weers M, Verschuren MC, Kraakman ME, et al. The Bruton's tyrosine kinase gene is expressed throughout B cell differentiation, from early precursor B cell stages preceding immunoglobulin gene rearrangement up to mature B cell stages. *Eur J Immunol* 1993;**23**:3109–14.

39. Smith CI, Baskin B, Humire-Greiff P, et al. Expression of Bruton's agammaglobulinemia tyrosine kinase gene, BTK, is selectively down-regulated in T lymphocytes and plasma cells. *J Immunol* 1994;**152**:557–65.

40. Conley ME, Mathias D, Treadaway J, Minegishi Y, Rohrer J. Mutations *in btk* in patients with presumed X-linked agammaglobulinemia. *Am J Hum Genet* 1998;**62**:1034–43.

41. Futatani T, Miyawaki T, Tsukada S, et al. Deficient expression of Bruton's tyrosine kinase in monocytes from X-linked agammaglobulinemia as evaluated by a flow cytometric analysis and its clinical application to carrier detection. *Blood* 1998;**91**:595–602.

42. Gaspar HB, Lester T, Levinsky RJ, Kinnon C. Bruton's tyrosine kinase expression and activity in X-linked agammaglobulinaemia (XLA): the use of protein analysis as a diagnostic indicator of XLA. *Clin Exp Immunol* 1998;**111**:334–8.

43. Broides A, Yang W, Conley ME. Genotype/phenotype correlations in X-linked agammaglobulinemia. *Clin Immunol* 2006;**118**:195–200.

44. Plebani A, Soresina A, Rondelli R, et al. Clinical, immunological, and molecular analysis in a large cohort of patients with X-linked agammaglobulinemia: an Italian multicenter study. *Clin Immunol* 2002;**104**:221–30.

45. Conley ME. Genetics of hypogammaglobulinemia: what do we really know? *Curr Opin Immunol* 2009;**21**:466–71.

46. Ferrari S, Zuntini R, Lougaris V, et al. Molecular analysis of the pre-BCR complex in a large cohort of patients affected by autosomal-recessive agammaglobulinemia. *Genes Immun* 2007;**8**:325–33.

47. Lopez Granados E, Porpiglia AS, Hogan MB, et al. Clinical and molecular analysis of patients with defects in micro heavy chain gene. *J Clin Invest* 2002;**110**:1029–35.

48. van Zelm MC, Geertsema C, Nieuwenhuis N, et al. Gross deletions involving IGHM, BTK, or Artemis: a model for genomic lesions mediated by transposable elements. *Am J Hum Genet* 2008;**82**:320–32.

49. Yel L, Minegishi Y, Coustan-Smith E, et al. Mutations in the mu heavy-chain gene in patients with agammaglobulinemia. *N Engl J Med* 1996;**335**:1486–93.

50. Minegishi Y, Coustan-Smith E, Wang YH, Cooper MD, Campana D, Conley ME. Mutations in the human lambda5/14.1 gene result in B cell deficiency and agammaglobulinemia. *The Journal of experimental medicine* 1998;**187**(1):71–7.

51. Dobbs AK, Yang T, Farmer D, Kager L, Parolini O, Conley ME. Cutting edge: a hypomorphic mutation in Igbeta (CD79b) in a patient with immunodeficiency and a leaky defect in B cell development. *J Immunol* 2007;**179**:2055–9.

52. Minegishi Y, Coustan-Smith E, Rapalus L, Ersoy F, Campana D, Conley ME. Mutations in Igalpha (CD79a) result in a complete block in B-cell development. *J Clin Invest* 1999;**104**:1115–21.

53. Minegishi Y, Rohrer J, Coustan-Smith E, et al. An essential role for BLNK in human B cell development. *Science* 1999;**286**:1954–7.

54. Conley ME, Dobbs AK, Quintana AM, et al. Agammaglobulinemia and absent B lineage cells in a patient lacking the p85alpha subunit of PI3K. *J Exp Med* 2012;**209**:463–70.

55. Boisson B, Wang Y-D, Bosompem A, et al. EA recurrent dominant negative E47 mutation causes agammaglobulinemia and BCR⁻ B cells. *J Clin Invest* 2013;**123**:47418–85.

56. Conley ME, Howard V. Clinical findings leading to the diagnosis of X-linked agammaglobulinemia. *J Pediatr* 2002;**141**:566–71.

57. Winkelstein JA, Marino MC, Lederman HM, et al. X-linked agammaglobulinemia: report on a United States registry of 201 patients. *Medicine* 2006;**85**:193–202.

58. Davis LE, Bodian D, Price D, Butler IJ, Vickers JH. Chronic progressive poliomyelitis secondary to vaccination of an immunodeficient child. *N Engl J Med* 1977;**297**:241–5.

59. McKinney Jr RE, Katz SL, Wilfert CM. Chronic enteroviral meningoencephalitis in agammaglobulinemic patients. *Rev Infect Dis* 1987;**9**:334–56.

60. Quartier P, Foray S, Casanova JL, Hau-Rainsard I, Blanche S, Fischer A. Enteroviral meningoencephalitis in X-linked agammaglobulinemia: intensive immunoglobulin therapy and sequential viral detection in cerebrospinal fluid by polymerase chain reaction. *Pediatr Infect Dis J* 2000;**19**:1106–8.

61. Furr PM, Taylor-Robinson D, Webster AD. Mycoplasmas and ureaplasmas in patients with hypogammaglobulinaemia and their role in arthritis: microbiological observations over twenty years. *Ann Rheum Dis* 1994;**53**:183–7.

62. LoGalbo PR, Sampson HA, Buckley RH. Symptomatic giardiasis in three patients with X-linked agammaglobulinemia. *J Pediatr* 1982;**101**:78–80.

63. Howard V, Greene JM, Pahwa S, et al. The health status and quality of life of adults with X-linked agammaglobulinemia. *Clin Immunol* 2006;**118**:201–8.

64. Hitzig WH, Biro Z, Bosch H, Huser HJ. [Agammaglobulinemia & alymphocytosis with atrophy of lymphatic tissue]. *Helvet Paediatr Acta* 1958;**13**:551–85.

65. Tobler R, Cottier H. [Familial lymphopenia with agammaglobulinemia & severe moniliasis: the essential lymphocytophthisis as a special form of early childhood agammaglobulinemia]. *Helvet Paediatr Acta* 1958;**13**:313–38.

66. Berendes H, Bridges RA, Good RA. A fatal granulomatosus of childhood: the clinical study of a new syndrome. *Minnesota Med* 1957;**40**:309–12.

67. Israel-Asselain R, Burtin P, Chebat J. [A new biological disorder: agammaglobulinemia with beta2-macroglobulinemia (a case)]. *Bull Mem Soc Med Hop Paris* 1960;**76**:519–23.

68. Purtilo DT, DeFlorio Jr D, Hutt LM, et al. Variable phenotypic expression of an X-linked recessive lymphoproliferative syndrome. *N Engl J Med* 1977;**297**:1077–80.

69. Ramoz N, Rueda LA, Bouadjar B, Montoya LS, Orth G, Favre M. Mutations in two adjacent novel genes are associated with epidermodysplasia verruciformis. *Nat Genet* 2002;**32**:579–81.

70. Hamilton JK, Paquin LA, Sullivan JL, et al. X-linked lymphoproliferative syndrome registry report. *J Pediatr* 1980;**96**:669–73.

71. Sayos J, Wu C, Morra M, et al. The X-linked lymphoproliferative-disease gene product SAP regulates signals induced through the co-receptor SLAM. *Nature* 1998;**395**:462–9.

72. Coffey AJ, Brooksbank RA, Brandau O, et al. Host response to EBV infection in X-linked lymphoproliferative disease results from mutations in an SH2-domain encoding gene. *Nat Genet* 1998;**20**:129–35.

73. Cannons JL, Tangye SG, Schwartzberg PL. SLAM family receptors and SAP adaptors in immunity. *Annu Rev Immunol* 2011;**29**:665–705.

74. Latour S, Veillette A. The SAP family of adaptors in immune regulation. *Semin Immunol* 2004;**16**:409–19.

75. Palendira U, Low C, Chan A, et al. Molecular pathogenesis of EBV susceptibility in XLP as revealed by analysis of female carriers with heterozygous expression of SAP. *PLoS Biol* 2011;**9**:e1001187.

76. Palendira U, Low C, Bell AI, et al. Expansion of somatically reverted memory CD8+ T cells in patients with X-linked lymphoproliferative disease caused by selective pressure from Epstein–Barr virus. *J Exp Med* 2012;**209**:913–24.

77. Dong Z, Cruz-Munoz ME, Zhong MC, Chen R, Latour S, Veillette A. Essential function for SAP family adaptors in the surveillance of hematopoietic cells by natural killer cells. *Nat Immunol* 2009;**10**:973–80.

78. Pachlopnik Schmid J, Canioni D, Moshous D, et al. Clinical similarities and differences of patients with X-linked lymphoproliferative syndrome type 1 (XLP-1/SAP deficiency) versus type 2 (XLP-2/XIAP deficiency). *Blood* 2011;**117**:1522–9.

79. Qi H, Cannons JL, Klauschen F, Schwartzberg PL, Germain RN. SAP-controlled T-B cell interactions underlie germinal centre formation. *Nature* 2008;**455**:764–9.

80. Ma CS, Hare NJ, Nichols KE, et al. Impaired humoral immunity in X-linked lymphoproliferative disease is associated with defective IL-10 production by CD4+ T cells. *J Clin Invest* 2005;**115**:1049–59.

81. Nichols KE, Hom J, Gong SY, et al. Regulation of NKT cell development by SAP, the protein defective in XLP. *Nat Med* 2005;**11**:340–5.

82. Chung BK, Tsai K, Allan LL, et al. Innate immune control of EBV-infected B cells by invariant natural killer T cells. *Blood* 2013;.

83. Booth C, Gilmour KC, Veys P, et al. X-linked lymphoproliferative disease due to SAP/SH2D1A deficiency: a multicenter study on the manifestations, management and outcome of the disease. *Blood* 2011;**117**:53–62.

84. Rigaud S, Fondaneche MC, Lambert N, et al. XIAP deficiency in humans causes an X-linked lymphoproliferative syndrome. *Nature* 2006;**444**:110–4.

85. Gerart S, Siberil S, Martin E, et al. Human iNKT and MAIT cells exhibit a PLZF-dependent proapoptotic propensity that is counterbalanced by XIAP. *Blood* 2013;**121**:614–23.

86. Marsh RA, Rao K, Satwani P, et al. Allogeneic hematopoietic cell transplantation for XIAP deficiency: an international survey reveals poor outcomes. *Blood* 2013;**121**:877–83.

87. Li FY, Lenardo MJ, Chaigne-Delalande B. Loss of MAGT1 abrogates the Mg2+ flux required for T cell signaling and leads to a novel human primary immunodeficiency. *Magnesium Res* 2011;**24**:S109–114.

88. Huck K, Feyen O, Niehues T, et al. Girls homozygous for an IL-2-inducible T cell kinase mutation that leads to protein deficiency develop fatal EBV-associated lymphoproliferation. *J Clin Invest* 2009;**119**:1350–8.

89. van Montfrans JM, Hoepelman AI, Otto S, et al. CD27 deficiency is associated with combined immunodeficiency and persistent symptomatic EBV viremia. *J Allergy Clin Immunol* 2012;**129**:787–93.

90. Chaigne-Delalande B, Li FY, O'Connor GM, et al. Mg2+ regulates cytotoxic functions of NK and CD8 T cells in chronic EBV infection through NKG2D. *Science* 2013;**341**:186–91.

91. Orth G. Host defenses against human papillomaviruses: lessons from epidermodysplasia verruciformis. *Curr Topics Microbiol Immunol* 2008;**321**:59–83.

92. Mimouni J. [Our experiences in three years of BCG vaccination at the center of the O.P.H.S. at Constantine; study of observed cases (25 cases of complications from BCG vaccination)]. *Algerie Med* 1951;**55**:1138–47.

93. Casanova JL, Blanche S, Emile JF, et al. Idiopathic disseminated bacillus Calmette-Guerin infection: a French national retrospective study. *Pediatrics* 1996;**98**:774–8.

94. Jouanguy E, Altare F, Lamhamedi S, et al. Interferon-gamma-receptor deficiency in an infant with fatal bacille Calmette-Guerin infection. *N Engl J Med* 1996;**335**:1956–61.

95. Newport MJ, Huxley CM, Huston S, et al. A mutation in the interferon-gamma-receptor gene and susceptibility to mycobacterial infection. *N Engl J Med* 1996;**335**:1941–9.

96. Bogunovic D, Byun M, Durfee LA, et al. Mycobacterial disease and impaired IFN-gamma immunity in humans with inherited ISG15 deficiency. *Science* 2012;**337**:1684–8.

97. Filipe-Santos O, Bustamante J, Haverkamp MH, et al. X-linked susceptibility to mycobacteria is caused by mutations in NEMO impairing CD40-dependent IL-12 production. *J Exp Med* 2006;**203**:1745–59.

98. Hambleton S, Salem S, Bustamante J, et al. IRF8 mutations and human dendritic-cell immunodeficiency. *N Engl J Med* 2011;**365**:127–38.

99. Kong XF, Vogt G, Itan Y, et al. Haploinsufficiency at the human IFNGR2 locus contributes to mycobacterial disease. *Hum Mol Genet* 2013;**22**:769–81.

100. Bustamante J, Arias AA, Vogt G, et al. Germline CYBB mutations that selectively affect macrophages in kindreds with X-linked predisposition to tuberculous mycobacterial disease. *Nat Immunol* 2011;**12**:213–21.

101. Casanova JL, Abel L. Genetic dissection of immunity to mycobacteria: the human model. *Annu Rev Immunol* 2002;**20**:581–620.

102. de Beaucoudrey L, Samarina A, Bustamante J, et al. Revisiting human IL-12Rbeta1 deficiency: a survey of 141 patients from 30 countries. *Medicine* 2010;**89**:381–402.

103. Prando C, Samarina A, Bustamante J, et al. Inherited IL-12p40 deficiency: genetic, immunologic, and clinical features of 49 patients from 30 kindreds. *Medicine* 2013;**92**:109–22.

104. de Beaucoudrey L, Puel A, Filipe-Santos O, et al. Mutations in STAT3 and IL12RB1 impair the development of human IL-17-producing T cells. *J Exp Med* 2008;**205**:1543–50.

105. Bustamante J, Picard C, Boisson-Dupuis S, Abel L, Casanova JL. Genetic lessons learned from X-linked Mendelian susceptibility to mycobacterial diseases. *Ann NY Acad Sci* 2011;**1246**:92–101.

106. Dorman SE, Picard C, Lammas D, et al. Clinical features of dominant and recessive interferon gamma receptor 1 deficiencies. *Lancet* 2004;**364**:2113–21.

107. Roesler J, Horwitz ME, Picard C, et al. Hematopoietic stem cell transplantation for complete IFN-gamma receptor 1 deficiency: a multi-institutional survey. *J Pediatr* 2004;**145**:806–12.

108. Camcioglu Y, Picard C, Lacoste V, et al. HHV-8-associated Kaposi sarcoma in a child with IFNgammaR1 deficiency. *J Pediatr* 2004;**144**:519–23.

109. Alcais A, Quintana-Murci L, Thaler DS, Schurr E, Abel L, Casanova JL. Life-threatening infectious diseases of childhood: single-gene inborn errors of immunity? *Ann NY Acad Sci* 2010;**1214**:18–33.

110. Casanova JL, Conley ME, Notarangelo L. Inborn errors of immunity. In: Paul W, editor. *Fundamental Immunology*. 7th edn. Philadelphia: Lippincott Williams & Wilkins; 2012. p. 1235–66.

111. Alcais A, Fieschi C, Abel L, Casanova JL. Tuberculosis in children and adults: two distinct genetic diseases. *J Exp Med* 2005;**202**:1617–21.

112. Altare F, Ensser A, Breiman A, et al. Interleukin-12 receptor beta1 deficiency in a patient with abdominal tuberculosis. *J Infect Dis* 2001;**184**:231–6.

113. Tabarsi P, Marjani M, Mansouri N, et al. Lethal tuberculosis in a previously healthy adult with IL-12 receptor deficiency. *J Clin Immunol* 2011;**31**:537–9.

114. Abel L, Plancoulaine S, Jouanguy E, et al. Age-dependent Mendelian predisposition to herpes simplex virus type 1 encephalitis in childhood. *J Pediatr* 2010;**157**:623–9.

115. Dupuis S, Jouanguy E, Al-Hajjar S, et al. Impaired response to interferon-alpha/beta and lethal viral disease in human STAT1 deficiency. *Nat Genet* 2003;**33**:388–91.

116. Chapgier A, Wynn RF, Jouanguy E, et al. Human complete Stat-1 deficiency is associated with defective type I and II IFN responses *in vitro* but immunity to some low virulence viruses *in vivo*. *J Immunol* 2006;**176**: 5078–83.

117. Audry M, Ciancanelli M, Yang K, et al. NEMO is a key component of NF-kappaB- and IRF-3-dependent TLR3-mediated immunity to herpes simplex virus. *J Allergy Clin Immunol* 2011;**128**:610–7.

118. Casrouge A, Zhang SY, Eidenschenk C, et al. Herpes simplex virus encephalitis in human UNC-93B deficiency. *Science* 2006;**314**:308–12.

119. Yang K, Puel A, Zhang S, et al. Human TLR-7-, -8-, and -9-mediated induction of IFN-alpha/beta and -lambda Is IRAK-4 dependent and redundant for protective immunity to viruses. *Immunity* 2005;**23**:465–78.

120. Picard C, von Bernuth H, Ghandil P, Chrabieh M, Levy O, Arkwright PD, et al. Clinical features and outcome of patients with IRAK-4 and MyD88 deficiency. *Medicine* 2010;**89**(6):403–25.

121. Zhang SY, Jouanguy E, Ugolini S, et al. TLR3 deficiency in patients with herpes simplex encephalitis. *Science* 2007;**317**:1522–7.

122. Sancho-Shimizu V, Perez de Diego R, Lorenzo L, et al. Herpes simplex encephalitis in children with autosomal recessive and dominant TRIF deficiency. *J Clin Invest* 2011;**121**:4889–902.

123. Perez de Diego R, Sancho-Shimizu V, Lorenzo L, et al. Human TRAF3 adaptor molecule deficiency leads to impaired Toll-like receptor 3 response and susceptibility to herpes simplex encephalitis. *Immunity* 2010;**33**:400–11.

124. Herman M, Ciancanelli M, Ou YH, et al. Heterozygous TBK1 mutations impair TLR3 immunity and underlie herpes simplex encephalitis of childhood. *J Exp Med* 2012;**209**:1567–82.

125. Gorbea C, Makar KA, Pauschinger M, et al. A role for Toll-like receptor 3 variants in host susceptibility to enteroviral myocarditis and dilated cardiomyopathy. *J Biol Chem* 2010;**285**:23208–23.

126. Lim HK, Zhang SY, Casanova JL. Novel forms of TLR3 deficiency in childhood onset recurrent herpes simplex virus encephalitis: molecular, cellular, and clinical features. in preparation. 2013.

127. Lafaille FG, Pessach IM, Zhang SY, et al. Impaired intrinsic immunity to HSV-1 in human iPSC-derived TLR3-deficient CNS cells. *Nature* 2012;**491**:769–73.

128. Bustamante J, Boisson-Dupuis S, Jouanguy E, et al. Novel primary immunodeficiencies revealed by the investigation of paediatric infectious diseases. *Curr Opin Immunol* 2008;**20**:39–48.

129. Lanternier F, Pathan S, Abel L, et al. Human deep dermatophytosis is caused by inborn errors of CARD9. *N Engl J Med* 2013;**369**(18):1704-14.

130. Liu L, Okada S, Kong XF, et al. Gain-of-function human STAT1 mutations impair IL-17 immunity and underlie chronic mucocutaneous candidiasis. *J Exp Med* 2011;**208**:1635–48.

131. Puel A, Cypowyj S, Marodi L, Abel L, Picard C, Casanova JL. Inborn errors of human IL-17 immunity underlie chronic mucocutaneous candidiasis. *Curr Opin Allergy Clin Immunol* 2012;**12**:616–22.

132. Byun M, Abhyankar A, Lelarge V, et al. Whole-exome sequencing-based discovery of STIM1 deficiency in a child with fatal classic Kaposi sarcoma. *J Exp Med* 2010;**207**:2307–12.

133. Byun M, Ma CS, Akcay A, et al. Inherited human OX40 deficiency underlying classic Kaposi sarcoma of childhood. *J Exp Med* 2013;.

134. Picard C, Puel A, Bonnet M, et al. Pyogenic bacterial infections in humans with IRAK-4 deficiency. *Science* 2003;**299**:2076–9.

135. von Bernuth H, Picard C, Jin Z, et al. Pyogenic bacterial infections in humans with MyD88 deficiency. *Science* 2008;**321**:691–6.

136. Picard C, Casanova JL, Puel A. Infectious diseases in patients with IRAK-4, MyD88, NEMO, or IkappaBalpha deficiency. *Clin Microbioly Rev* 2011;**24**:490–7.

137. Boisson B, Laplantine E, Prando C, et al. Immunodeficiency, autoinflammation and amylopectinosis in humans with inherited HOIL-1 and LUBAC deficiency. *Nat Immunol* 2012;**13**:1178–86.

4

From Immunodeficiency to Autoimmunity

Luigi Daniele Notarangelo

Children's Hospital, Boston, MA, USA

OUTLINE

A HISTORICAL JOURNEY INTO RECOGNITION OF IMMUNODEFICIENCY AND AUTOIMMUNITY

Inborn errors of immunity, or primary immunodeficiency diseases (PIDs), were recognized as a distinct group of disorders in the 1950s, following the identification of infants and children who experienced severe and/or recurrent infections because of the intrinsic inability to produce antibodies (1) or to generate lymphocytes (2). The number of such conditions rapidly grew to include also defects of innate immunity, such as chronic granulomatous disease (3), congenital neutropenia (4) and complement deficiency (5).

Unique susceptibility to infections was rapidly identified as the clinical hallmark of PID, consistent with the notion of an intrinsic defect in immune function. Importantly, identification of these disorders was often instrumental to better understanding (or even identifying) basic mechanisms that govern immune system development and function.

By contrast, speculation about autoimmunity dates back to the first years of the twentieth century, with Paul Ehrlich's theory of "horror autotoxicus" (according to which animals fail to react when immunized with self-antigens), and, a few years later, the demonstration of cold autohemolysins by Donath and Landsteiner (reviewed in 6). Remarkably, the concept of autoimmunity as a mechanism of human disease was almost completely neglected for decades, probably due to attention being given mostly to investigation of antibody responses to foreign antigens as the result of natural infection or of active immunization. In 1946, use of the Coombs' test allowed the identification of congenital and acquired forms of autoimmune hemolytic anemia (7). This discovery was soon followed by recognition of the so-called lupus erythematosus (LE) cell effect in the bone marrow of patients with systemic lupus erythermatosus (SLE), and the possibility of inducing LE cell formation by incubating a marrow sample from a normal subject with plasma from SLE patients (8). These advances led to the rapid identification of a number of human conditions characterized by the aberrant production of autoantibodies and/or the presence of self-reactive T-cells. The discovery of naturally occurring mouse models of SLE (9) brought further evidence in support of autoimmunity. Thus, by 1960–1970, both PID and autoimmune diseases were clearly recognized as disorders of altered immune function.

AUTOIMMUNITY: AN OFTEN NEGLECTED FEATURE OF IMMUNE DEFICIENCY

Following their identification as disorders of the immune system, congenital immune deficiencies and autoimmune diseases were taken as opposite, mutually exclusive examples of impaired ability to mount immune responses (PIDs), and of aberrant generation of immune reaction to self-antigens (autoimmune diseases), respectively. For many decades, this led to the misconception that autoimmunity may not be a cardinal feature of PIDs. Yet, many clinical observations argued against this. For example, bi-allelic deficiency of early components of the classical pathway of complement was identified as a significant disease susceptibility trait for SLE (10), autoimmune cytopenias and

granulomatous disease were known to be frequent in common variable immunodeficiency (11) and autoimmunity was known to be an important feature of the Wiskott–Aldrich syndrome (12). Indeed, there are many mechanisms whereby abnormal immune function in patients with PID may favor development of autoimmune manifestations (Table 4.1). It is also possible that the monogenic nature of most forms of PIDs, which contrasts with the polygenic susceptibility that is most commonly seen in autoimmune diseases, contributed to the general attitude to keep immune dysregulation manifestations of PIDs in the dark, while highlighting infections as the key feature of the disease. This view was particularly strong when considering the most severe form of PID, namely severe combined immune deficiency (SCID). Infants with SCID lack adaptive immunity, with an absence of circulating T-cells. B lymphocytes may be absent or, when present, their function is profoundly impaired due to lack of helper T-cells or because of intrinsic signaling defects (13). This very same severe impairment of adaptive immunity was viewed as the key pathophysiological mechanism to account for the increased frequency of life-threatening infections soon after birth, while making it impossible to generate self-reactive

TABLE 4.1 Mechanisms of immune dysregulation in immunodeficiency diseases

Defective negative selection of autoreactive T-cells

Impaired generation/function of regulatory T-cells

Defective apoptosis of autoreactive lymphocytes

Homeostatic lymphocyte proliferation

Defective receptor editing (impaired purging of bone marrow self-reactive B-cells)

Increased BAFF serum levels (rescue of peripheral self-reactive B lymphocytes)

Increased load or decreased clearance of apoptotic cells

Defective clearance of immune complexes

Persistent infection/inflammation (leading to increased production of inflammatory cytokines)

manifestations. However, Nature often deceives us. The heterogeneity of mutations that may occur in the same SCID-causing gene is the culprit that may explain the coexistence of immune deficiency and autoimmunity even in the most severe forms of inborn errors of immunity.

OMENN SYNDROME: INFLAMMATION AND AUTOIMMUNITY IN IMMUNE DEFICIENCY

My personal interest in immune dysregulation associated with immune deficiency was generated by the clinical observation of several infants who were referred for diagnosis and treatment to the Department of Pediatrics at the University of Brescia. Starting in the late 1980s, a team of clinicians that included Alberto Ugazio, Fulvio Porta, Evelina Mazzolari (whom I married a few years later), myself and a group of excellent nurses, developed a program for hematopoietic cell transplantation (HCT) in severe forms of PID. This made Brescia rapidly become one of the leading centers in the field in Italy and Europe. With a background in genetics and molecular and cellular immunology that I had developed at the NIH, my role was mostly of a physician scientist who aimed to characterize the pathophysiology of the underlying disease in affected children who were referred to our institution. In particular, we had the opportunity to treat several patients with Omenn syndrome (OS), a rare and mysterious disease.

Reported for the first time in 1965 in *The New England Journal of Medicine* by Gilbert Omenn (at that time still a medical student!), OS is a combined immunodeficiency characterized by severe, early-onset infections associated with systemic tissue damage (14). The latter reflects infiltration by oligoclonal, poorly functioning, autologous T-cells, which cause generalized erythroderma, lymphadenopathy (obviously an unusual finding in SCID!) and liver and/or spleen enlargement.

Cytopenias and graft-versus-host-like manifestations in the gastrointestinal tract are also common. Eosinophilia and elevated serum IgE (in spite of profound hypogammaglobulinemia and the frequent absence of circulating B-cells) are other typical features of the disease. Thus, OS is a perfect example of a severe congenital immunodeficiency also characterized by prominent immune dysregulation (15).

The pathophysiology of OS had remained undefined for more than two decades until, in 1991, Genevieve de Saint Basile reported the simultaneous occurrence of OS and of SCID with the absence of both T and B lymphocytes (T$^-$ B$^-$ SCID) in two siblings (16). The observation of two very rare disorders in the same sibship led to the hypothesis that OS and T$^-$ B$^-$ SCID could in fact be allelic disorders. The molecular mechanisms involved in the generation of adaptive immunity, and of T and B lymphocytes in particular, were just beginning to be unraveled. In 1983, Tonegawa had identified V(D)J recombination as the process that allows somatic generation of antibody diversity (17), and the V(D)J recombination activating gene 1 (*RAG1*) had been cloned by Schatz, Oettinger and Baltimore in 1989 (18). This discovery was soon followed by the cloning of *RAG2* (19), thus allowing the identification of the two molecules that initiate V(D)J recombination and permit generation of T and B lymphocytes. In 1992, Mombaerts and coworkers demonstrated that *Rag1* gene-targeted mice lack T and B lymphocytes (20) and, in 1996, Klaus Schwarz identified *RAG* gene defects in patients with B$^-$ SCID (21). The search for the OS syndrome gene had started. My group joined forces with Anna Villa and Paolo Vezzoni in Milan. I had been collaborating with them on the molecular pathophysiology of the hyper-IgM syndrome, and we have remained good friends ever since. We hypothesized that OS resulted from hypomorphic mutations in the *RAG1/2* genes. Analysis of a series of samples from patients treated in Brescia proved that we were correct. To gain additional biochemical and molecular insight into the expression and function of the *RAG* mutants identified in patients with OS, Anna flew to New York, and worked at Mount Sinai Hospital in the lab of Eugenia Spanopoulou, an outstanding and generous investigator, whose life ended prematurely in a tragic airplane accident in 1998, the same year when we published our discovery (22). Similar results were obtained by Barbara Corneo and Jean-Pierre de Villartay in Alain Fischer's group in Paris (23). These data demonstrated that mutations that are potentially permissive for residual T (and less so, B) lymphocyte development may cause OS, but the molecular mechanism(s) underlying immune dysregulation in this disorder remained elusive.

A key discovery that shed light on autoimmunity in OS and other forms of combined immunodeficiency came with the identification of mutations in the autoimmune regulator (*AIRE*) gene in patients with autoimmune polyendocrinopathy candidiasis ectodermal dystrophy (APECED) syndrome (24, 25), a rare monogenic disorder dominated by autoimmune manifestations (26). The function of the AIRE transcription factor was elegantly unraveled by Mark Anderson in Diane Mathis's group, who demonstrated that AIRE is expressed predominantly by a subset of medullary thymic epithelial cells (mTECs), where it drives expression of tissue-restricted antigens (TRAs) that are assembled with major histocompatibility complex class II (MHC-II) molecules and presented to nascent T lymphocytes, enabling deletion of self-reactive T-cells (Fig. 4.1) (27). Studies in mice

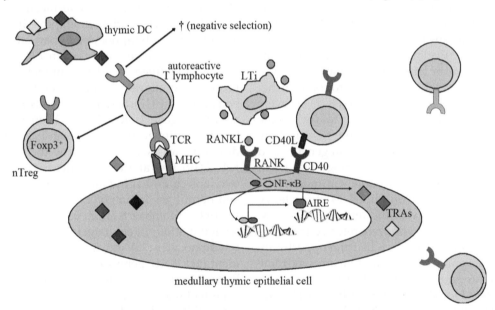

FIGURE 4.1 Mechanisms of T-cell tolerance in the thymus. Cross-talk between medullary thymic epithelial cells and lymphoid cells induces maturation of medullary thymic epithelial cells. In particular, secretion of RANK ligand by lymphoid tissue induced (LTi) cells secrete RANK ligand (RANKL) already at early stages during thymic development, and newly generated thymocytes express CD40 ligand (CD40L). These molecules bind to RANK and CD40, respectively, which are both expressed by medullary thymic epithelial cells, triggering activation of the NF-κB signaling pathway, ultimately inducing expression of the autoimmune regulator (*AIRE*) gene. The transcription factor AIRE drives expression of tissue-restricted antigens (TRAs), which are presented in association with major histocompatibility complex class II (MHC-II) molecules on the surface of medullary thymic epithelial cells or of thymic dendritic cells (DC). Recognition of TRA/MHC-II complex by self-reactive thymocytes that are generated in the thymus leads to clonal deletion or to conversion into natural regulatory T-cells (nTreg) expressing the FOXP3 transcription factor. This figure is reproduced in color in the color section.

have shown that signals delivered by thymocytes are crucial to induce maturation of mTECs and to maintain AIRE expression (28–30). Consistent with this, in collaboration with Fabio Facchetti, a prominent immunopathologist, we demonstrated that reduced ability to generate T-cells in OS due to hypomorphic *RAG* mutations compromises mTECs homeostasis, expression of AIRE and of TRAs (31). Thus, self-reactive T-cells that are generated because of the residual and stochastic V(D)J rearrangement process are not properly deleted in the thymus and may be exported to the periphery, contributing to the immunopathology of OS.

Identification of the molecular basis of another rare monogenic autoimmune disease also contributed to defining the pathophysiology of immune dysregulation of OS. Immune dysregulation polyendocrinopathy enteropathy X-linked (IPEX) syndrome is a severe disease characterized by early-onset enteropathy, type 1 diabetes, other autoimmune manifestations and lymphoid proliferation. IPEX is due to mutations of the Forkhead box protein 3 (*FOXP3*) gene (32–34), that encodes for a transcription factor required for the development of regulatory T (Treg)-cells (35). Generation of FOXP3$^+$ natural Treg (nTreg) cells in the thymus requires close interaction between developing T lymphocytes, mTECs and thymic dendritic cells (DCs) (36) (see Fig. 4.1). Patients with OS have an abnormal thymic architecture, with a lack of Hassall's corpuscles and fewer DCs, and fail to generate nTreg cells (37). Finally, while we and others were investigating mechanisms of immune dysregulation in patients with OS, it became clear that this condition may be due not just to *RAG* defects but to hypomorphic mutations in a variety of SCID-associated genes (38). Regardless of the nature of the genetic defect, impaired development of T-cells results in thymic architecture abnormalities and defective maturation of mTECs and thymic DCs, thus leading to a common mechanism that accounts for immune dysregulation associated with severe immune deficiency.

EXPANDING THE SPECTRUM OF AUTOIMMUNITY IN RAG DEFICIENCY

The study of OS proved that genetic defects that severely compromised T-cell development may also affect mechanisms of central tolerance. This was a groundbreaking discovery that led to a reconsideration of the relationship between immunodeficiency and autoimmunity; these were no longer considered manifestations at the opposite end of the spectrum of immune disorders. Indeed, that autoimmunity may represent a cardinal feature of combined immunodeficiency was confirmed by a series of more recent discoveries. Catharina Schuetz was the first to demonstrate hypomorphic *RAG* mutations in patients with delayed onset immunodeficiency associated with granulomatous lesions (39), and Suk See de Ravin extended these findings by describing a patient in whom RAG deficiency resulted in a disease phenotype resembling Wegener's granulomatosis and myasthenia gravis (40). Remarkably, this patient had received a thymectomy in the attempt to control myasthenia, indicating that the genetic defect was permissive for the development of a normal sized thymus. The occurrence of myasthenia gravis, a typical autoantibody-mediated disease, prompted a revisit of the notion that RAG deficiency causes virtual absence of B lymphocytes and profound hypogammaglobulinemia (other than elevated IgE). Indeed, several patients have been reported who have RAG-dependent autoimmune manifestations associated with detectable circulating B-cells and hypergammaglobulinemia (39–42). This, however, is a puzzling finding. Indeed, there are two main reasons why hypomorphic *RAG* mutations may also cause abnormalities of B-cell tolerance, and not just T-cell tolerance as reported above. First, a large fraction of immature B lymphocytes that are generated in the bone marrow recognize self-antigens (43); binding of self-antigens prolongs *RAG* expression and induces a secondary rearrangement of the immunoglobulin light chain genes. This process, known as receptor editing, represents a mechanism of central B-cell tolerance by reducing the proportion of naïve B-cells with self-reactive specificity (44). Consistent with the critical role of *RAG* in mediating V(D)J recombination, we demonstrated that hypomorphic *RAG* mutations also cause reduced receptor editing (42), and may thus contribute to persistence and peripheral export of stochastically-generated self-reactive specificities. These data were also confirmed in two knock-in mouse models of Omenn syndrome that had been generated by Anna Villa and Fred Alt (42, 45). Abnormalities of peripheral tolerance may also be involved in the immune dysregulation associated with hypomorphic *RAG* mutations. In particular, B-cell activating factor (BAFF) plays an important role in the survival of peripheral B-cells. Lower levels of BAFF receptor are expressed by anergic, self-reactive B-cells, which are therefore at a disadvantage to non-self-reactive B lymphocytes for a response to BAFF (46). Serum levels of BAFF are tightly regulated, and are higher in B-cell lymphopenic hosts, including patients with B^- or B^{low} immunodeficiency (42). Elevated BAFF levels had been postulated to play a role in several human conditions with immune dysregulation (47); moreover, it had been demonstrated that *in vivo* neutralization of BAFF ameliorates autoimmunity in a mouse model of diabetes (Fig. 4.2)

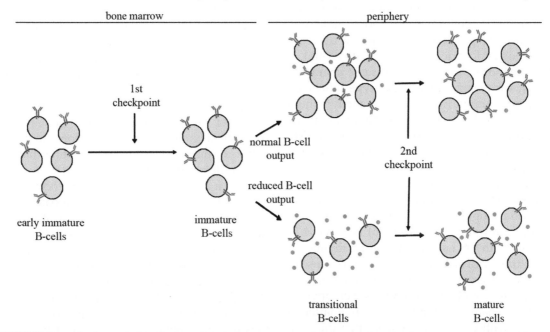

FIGURE 4.2   Checkpoints of B-cell tolerance in the bone marrow and in the periphery. Stochastic rearrangements of the V, D, and J elements of immunoglobulin genes in bone marrow B-cell progenitors allow generation of a diversified repertoire of early immature B lymphocytes, including a significant fraction of cells with self-reactive specificity (indicated by a blue trait on the immunoglobulin variable region). Prolonged expression of the RAG genes allows secondary V(D)J rearrangement that modifies the immunoglobulin specificity, thus leading to a decrease in the proportion of self-reactive cells among immature B-cells (first checkpoint). Upon egress to the periphery, B-cells respond to growth/survival factors, including BAFF (green solid circles). Anergic self-reactive B-cells express lower amounts of BAFF receptors. In the presence of normal B-cell output from the bone marrow, and of steady-state BAFF concentration in the serum, survival of non self-reactive mature B-cells is favored, resulting in a decrease in the proportion of mature self-reactive B-cells (secondary checkpoint). However, generation of a reduced number of transitional B-cells is associated with higher serum BAFF levels, thereby allowing rescue of self-reactive mature B-cells. This figure is reproduced in color in the color section.

(48). Not surprisingly, we found that patients with hypomorphic RAG mutations have high BAFF serum levels (42), and Anna Villa showed that administration of anti-BAFF monoclonal antibody to a mouse model of Omenn syndrome leads to the disappearance of anti-double-strand DNA antibodies and improves tissue infiltrates (45). Altogether, these data provided unanticipated evidence that abnormalities in B-cell tolerance are an important component of autoimmunity in leaky forms of SCID, and *RAG* deficiency in particular.

THE LESSON LEARNED: WHEN COMBINED IMMUNODEFICIENCY AND AUTOIMMUNITY GO HAND-IN-HAND

The journey through the immune dysregulation of Omenn syndrome and leaky SCID, two very rare conditions, had taken several years, but had led to novel and exciting findings. Importantly, it had helped to reject the hypothesis that autoimmunity and immunodeficiency are at the opposite end of conditions with abnormal immune function. Indeed, the very same mechanisms of impaired T-cell tolerance that had been identified in Omenn syndrome (defective recognition of tissue-restricted antigens in the context of a poorly organized thymic architecture, and impaired generation of Treg cells) have been demonstrated more recently in other forms of combined immunodeficiency with associated immune dysregulation. It is so for defects of calcium signaling, and STIM1 deficiency in particular (49). Moreover, immune dysregulation has also been described in other conditions of defective T-cell signaling, such as hypomorphic ZAP70 mutations (50), LCK deficiency (51) and ITK deficiency (52). Studies in a hypomorphic Zap70-mutated mouse model had shown that reduced strength of TCR-mediated signaling not only affects the robustness of T-cell generation, but also impairs the development of Treg cells,

hence favoring autoimmune manifestations (53). Moreover, mutations of the *macrophage stimulating 1 (MST1)* gene in humans cause a profound immunodeficiency with autoimmunity (54, 55). MST1 is involved in T-cell survival and regulation of migration, and defective migration of $Mst1^{-/-}$ thymocytes is associated with impaired recognition of self-antigens, deficient negative selection of self-reactive T-cells, and peripheral autoimmunity (56). Recently, dominant gain-of-function mutations in signal transducer and activator of transcription (STAT)1 were shown to cause an IPEX-like phenotype in patients with wild-type *FOXP3* who presented with a variety of mucosal and disseminated fungal infections in addition to polyendocrinopathy, enteropathy and dermatitis (57). With the growing number of novel genetic defects of the immune system identified through whole exome sequencing and whole genome sequencing, there is no doubt that this is just the beginning of a story yet to be written.

References

1. Bruton OC. Agammaglobulinemia. *Pediatrics* 1952;**9**: 722–8.
2. Glanzmann E, Riniker P. [Essential lymphocytophthisis; new clinical aspect of infant pathology]. *Ann paediatr Internat R Pediatrs* 1950;**175**:1–32.
3. Berendes H, Bridges RA, Good RA. A fatal granulomatosus of childhood: The clinical study of a new syndrome. *Minnesota Med* 1957;**40**:309–12.
4. Kostmann R. Infantile genetic agranulocytosis; agranulocytosis infantilis hereditaria. *Acta Paediatr Suppl* 1956; **45**:1–78.
5. Silverstein AM. Essential hypocomplementemia: Report of a case. *Blood* 1960;**16**:1338–41.
6. Mackay IR. Travels and travails of autoimmunity: A historical journey from discovery to rediscovery. *Autoimmun Revs* 2010;**9**:A251–258.
7. Boorman KE, Dodd BE, Loutit JF. Haemolytic icterus (acholuric jaundice) congenital and acquired. *Lancet* 1946;**1**:812–4.
8. Hargraves MM, Richmond H, Morton R. Presentation of two bone marrow elements; the tart cell and the l.E. Cell. *Procs Staff meetings Mayo Clin* 1948;**23**:25–8.
9. Casey TP. Aspects of autoimmunity. 3. Autoimmunity in strains of New Zealand mice. *NZ Med J* 1966;**65**:105–10.

10. Walport MJ, Davies KA, Morley BJ, Botto M. Complement deficiency and autoimmunity. *Ann NY Acad Sci* 1997;**815**:267–81.

11. Sneller MC, Strober W, Eisenstein E, Jaffe JS, Cunningham-Rundles C. NIH conference. New insights into common variable immunodeficiency. *Ann Iintern Med* 1993;**118**:720–30.

12. Sullivan KE, Mullen CA, Blaese RM, Winkelstein JA. A multiinstitutional survey of the Wiskott-Aldrich syndrome. *J Pediatrs* 1994;**125**:876–85.

13. Fischer A, Le Deist F, Hacein-Bey-Abina S, et al. Severe combined immunodeficiency. A model disease for molecular immunology and therapy. *Immunol Rev* 2005;**203**: 98–109.

14. Omenn GS. Familial reticuloendotheliosis with eosinophilia. *N Engl J Med* 1965;**273**:427–32.

15. Villa A, Notarangelo LD, Roifman CM. Omenn syndrome: Inflammation in leaky severe combined immunodeficiency. *J Allergy Clin Immunol* 2008;**122**:1082–6.

16. de Saint-Basile G, Le Deist F, de Villartay JP, et al. Restricted heterogeneity of T lymphocytes in combined immunodeficiency with hypereosinophilia (Omenn's Syndrome). *J Clin Invest* 1991;**87**:1352–9.

17. Tonegawa S. Somatic generation of antibody diversity. *Nature* 1983;**302**:575–81.

18. Schatz DG, Oettinger MA, Baltimore D. The V(D)J recombination activating gene, RAG-1. *Cell* 1989;**59**: 1035–48.

19. Oettinger MA, Schatz DG, Gorka C, Baltimore D. RAG-1 and RAG-2, adjacent genes that synergistically activate V(D)J recombination. *Science* 1990;**248**:1517–23.

20. Mombaerts P, Iacomini J, Johnson RS, Herrup K, Tonegawa S, Papaioannou VE. Rag-1-deficient mice have no mature B and T lymphocytes. *Cell* 1992;**68**:869–77.

21. Schwarz K, Gauss GH, Ludwig L, et al. RAG mutations in human B-cell-negative SCID. *Science* 1996;**274**:97–9.

22. Villa A, Santagata S, Bozzi F, et al. Partial V(D)J recombination activity leads to Omenn Syndrome. *Cell* 1998;**93**:885–96.

23. Corneo B, Moshous D, Gungor T, et al. Identical mutations in RAG1 or RAG2 genes leading to defective V(D)J recombinase activity can cause either T-B-severe combined immune deficiency or Omenn Syndrome. *Blood* 2001;**97**:2772–6.

24. Finnish-German APECED Consortium. An autoimmune disease, APECED, caused by mutations in a novel gene featuring two PHD-type zinc-finger domains. *Nat Genet* 1997;**17**:399–403.

25. Nagamine K, Peterson P, Scott HS, et al. Positional cloning of the apeced gene. *Nat Genet* 1997;**17**:393–8.

26. Moraes-Vasconcelos D, Costa-Carvalho BT, Torgerson TR, Ochs HD. Primary immune deficiency disorders presenting as autoimmune diseases: IPEX and APECED. *J Clin Immunol* 2008;**28**(Suppl 1):S11–19.

27. Anderson MS, Venanzi ES, Klein L, et al. Projection of an immunological self shadow within the thymus by the Aire protein. *Science* 2002;**298**:1395–401.

28. Akiyama T, Shimo Y, Yanai H, et al. The tumor necrosis factor family receptors RANK and CD40 cooperatively establish the thymic medullary microenvironment and self-tolerance. *Immunity* 2008;**29**:423–37.

29. Hikosaka Y, Nitta T, Ohigashi I, et al. The cytokine RANKL produced by positively selected thymocytes fosters medullary thymic epithelial cells that express autoimmune regulator. *Immunity* 2008;**29**:438–50.

30. Irla M, Hugues S, Gill J, et al. Autoantigen-specific interactions with CD4+ thymocytes control mature medullary thymic epithelial cell cellularity. *Immunity* 2008;**29**: 451–63.

31. Cavadini P, Vermi W, Facchetti F, et al. AIRE deficiency in thymus of 2 patients with Omenn Syndrome. *J Clin Invest* 2005;**115**:728–32.

32. Chatila TA, Blaeser F, Ho N, et al. JM2, encoding a fork head-related protein, is mutated in X-Linked Autoimmunity-Allergic Disregulation Syndrome. *J Clin Invest* 2000;**106**:R75–81.

33. Bennett CL, Christie J, Ramsdell F, et al. The Immune Dysregulation, Polyendocrinopathy, Enteropathy, X-Linked Syndrome (IPEX) is caused by mutations of FOXP3. *Nat Genet* 2001;**27**:20–1.

34. Wildin RS, Ramsdell F, Peake J, et al. X-linked neonatal diabetes mellitus, enteropathy and endocrinopathy syndrome is the human equivalent of mouse scurfy. *Nat Genet* 2001;**27**:18–20.

35. Zheng Y, Rudensky AY. Foxp3 in control of the regulatory T-cell lineage. *Nat Immunol* 2007;**8**:457–62.

36. Watanabe N, Wang YH, Lee HK, Ito T, Cao W, Liu YJ. Hassall's corpuscles instruct dendritic cells to induce CD4+CD25+ regulatory T-cells in human thymus. *Nature* 2005;**436**:1181–5.

37. Poliani PL, Facchetti F, Ravanini M, et al. Early defects in human T-cell development severely affect distribution and maturation of thymic stromal cells: Possible implications for the pathophysiology of Omenn Syndrome. *Blood* 2009;**114**:105–8.

38. Marrella V, Maina V, Villa A. Omenn syndrome does not live by V(D)J recombination alone. *Curr Opin Allergy Clin Immunol* 2011;**11**:525–31.

39. Schuetz C, Huck K, Gudowius S, et al. An immunodeficiency disease with RAG mutations and granulomas. *N Engl J Med* 2008;**358**:2030–8.

40. De Ravin SS, Cowen EW, Zarember KA, et al. Hypomorphic RAG mutations can cause destructive midline granulomatous disease. *Blood* 2010;**116**:1263–71.

41. Felgentreff K, Perez-Becker R, Speckmann C, et al. Clinical and immunological manifestations of patients with atypical severe combined immunodeficiency. *Clin Immunol* 2011;**141**:73–82.

42. Walter JE, Rucci F, Patrizi L, et al. Expansion of immuno-globulin-secreting cells and defects in B-cell tolerance in Rag-dependent immunodeficiency. *J Exp Med* 2010;**207**: 1541–54.

43. Wardemann H, Yurasov S, Schaefer A, Young JW, Meffre E, Nussenzweig MC. Predominant autoantibody production by early human B-cell precursors. *Science* 2003;**301**:1374–7.

44. Gay D, Saunders T, Camper S, Weigert M. Receptor editing: An approach by autoreactive B-cells to escape tolerance. *J Exp Med* 1993;**177**:999–1008.

45. Cassani B, Poliani PL, Marrella V, et al. Homeostatic expansion of autoreactive immunoglobulin-secreting cells in the Rag2 mouse model of Omenn Syndrome. *J Exp Med* 2010;**207**:1525–40.

46. Lesley R, Xu Y, Kalled SL, et al. Reduced competitiveness of autoantigen-engaged B-cells due to increased dependence on BAFF. *Immunity* 2004;**20**:441–53.

47. Moisini I, Davidson A. BAFF: A local and systemic target in autoimmune diseases. *Clin Exp Immunol* 2009;**158**: 155–63.

48. Zekavat G, Rostami SY, Badkerhanian A, et al. In vivoBLyS/BAFF neutralization ameliorates islet-directed autoimmunity in nonobese diabetic mice. *J Immunol* 2008; **181**:8133–44.

49. Picard C, McCarl CA, Papolos A, et al. STIM1 mutation associated with a syndrome of immunodeficiency and autoimmunity. *N Engl J Med* 2009;**360**:1971–80.

50. Turul T, Tezcan I, Artac H, et al. Clinical heterogeneity can hamper the diagnosis of patients with ZAP70 deficiency. *Eur J Pediatr* 2009;**168**:87–93.

51. Hauck F, Randriamampita C, Martin E, et al. Primary T-cell immunodeficiency with immunodysregulation caused by autosomal recessive LCK deficiency. *J Allergy Clin Immunol* 2012;**130**:1144–52 e11.

52. Huck K, Feyen O, Niehues T, et al. Girls homozygous for an IL-2-inducible T-cell kinase mutation that leads to protein deficiency develop fatal EBV-associated lymphoproliferation. *J Clin Invest* 2009;**119**:1350–8.

53. Hsu LY, Tan YX, Xiao Z, Malissen M, Weiss A. A hypomorphic allele of ZAP-70 reveals a distinct thymic threshold for autoimmune disease versus autoimmune reactivity. *J Exp Med* 2009;**206**:2527–41.

54. Nehme NT, Pachlopnik Schmid J, Debeurme F, et al. MST1 mutations in autosomal recessive primary immunodeficiency characterized by defective naive T-cell survival. *Blood* 2012;**119**:3458–68.

55. Abdollahpour H, Appaswamy G, Kotlarz D, et al. The phenotype of human STK4 deficiency. *Blood* 2012;**119**: 3450–7.

56. Ueda Y, Katagiri K, Tomiyama T, et al. Mst1 regulates integrin-dependent thymocyte trafficking and antigen recognition in the thymus. *Nat Commun* 2012;**3**:1098.

57. Uzel G, Sampaio EP, Lawrence MG, et al. Dominant gain-of-function STAT1 mutations in FOXP3 wild-type immune dysregulation-polyendocrinopathy-enteropathy-X-linked-like syndrome. *J Allergy Clin Immunol* 2013;**131**: 1611–23.

5

Immunological Tests – from the Microscope to Whole Genome Analysis

Thomas A. Fleisher

Chief, Department of Laboratory Medicine, NIH Clinical Center, National Institutes of Health, Bethesda, MD, USA

O U T L I N E

INTRODUCTION

Over the past six decades, the characterization of primary immunodeficiency disorders (PIDD) paralleled new discoveries in laboratory immunology and helped to develop this field into the formal laboratory discipline of diagnostic immunology. The underpinnings of developments in diagnostic testing depended on a series of discoveries that defined the specific components of the immune system both at the cellular and protein and, more recently, at the DNA level. As these advances were reported, astute investigators connected key clinical phenotypes characteristic of these "experiments of nature" with abnormalities detected in laboratory studies. Assembling

Primary Immunodeficiency Disorders: A Historic and Scientific Perspective
2014 Published by Elsevier Inc.

this clinical and laboratory information helped to define many of the critical immunological components necessary for the normal human host defense.

HISTORY OF THE EVALUATION OF HUMORAL IMMUNITY

The historical evolution of the "partnership" between laboratory testing and clinical discovery has as a prime example Bruton's description of what is now referred to as X-linked agammaglobulinemia (1) (see Chapter 11). The laboratory evaluation of this patient was driven by a series of findings that culminated with testing based on the development of serum protein electrophoresis (pioneered by Tiselius in the 1930s), together with related work that followed. Electrophoresis established the specific migration patterns of human serum proteins subjected to an electrical field and these include albumin, alpha globulins, beta globulin and gamma globulin. Additional work by Tiselius and Kabat proved that the gamma globulin component contained the majority of antibodies (2). This was followed by seminal work in the 1940s which focused on patients with streptococcal infections and demonstrated that the production of the specific antibodies in response to group A hemolytic streptococci was associated with increased gamma globulin levels (3). The focus of Bruton's publication was a child who had 19 episodes of sepsis over a four-and-one-half-year interval, ten of which yielded positive blood cultures for *S. pneumoniae* involving eight different serotypes. Extensive evaluation failed to identify an infectious focus and the presence of a positive Schick test demonstrated that the patient failed to produce neutralizing antibodies to diphtheria toxin, despite a history of repeated diphtheria toxoid vaccinations. These findings directed further studies using a newly available clinical laboratory test, serum protein electrophoresis. The anticipated result, according to Bruton, was increased levels of gamma globulin due to

the history of recurrent bacterial infections (4). However, the patient's serum did not demonstrate increased levels but rather total absence of gamma globulin while the other serum protein components were present in normal amounts. This finding clearly suggested that the patient's recurrent bacterial infections were associated with the lack of gamma globulin. Bruton then went on to administer exogenous immune serum globulin (Cohn fraction II preparation) subcutaneously to the patient and demonstrated that this led to the appearance of an identifiable gamma globulin tracing. Importantly, this therapy was associated with clinical improvement in the patient, while discontinuation of immune serum globulin therapy was associated with clinical deterioration as well as the loss of this protein component on repeat serum electrophoresis. In this case, an astute clinician armed with new laboratory testing identified a finding that helped to clarify the underlying basis of an immunological defect and effectively defined a new class of immunological disorders, the antibody deficiency syndrome.

Additional advancements in defining the basic biochemistry of gamma globulins (immunoglobulins) established sedimentation rate differences based on molecular size within this family of proteins (5). Applying these findings to the clinic led to the description of a patient with an increased level of 19S gamma globulins (IgM) but with decreased 7S gamma globulins (IgG, IgA). This important discovery by the group at the Children's Hospital Boston provided the first description of a patient with a hyper IgM syndrome (6) (see Chapter 16).

In 1964, the major isotypes (classes) of gamma globulin were identified as differing in their heavy chains and were classified as IgG, IgA and IgM (rather than using the less specific categories of 7S and 19S gamma globulins). This important discovery was followed shortly thereafter by the description of methods to quantitate specifically each of these immunoglobulin classes using an agar gel diffusion method referred to as radial immunodiffusion (RID) (7, 8) (Fig. 5.1). The

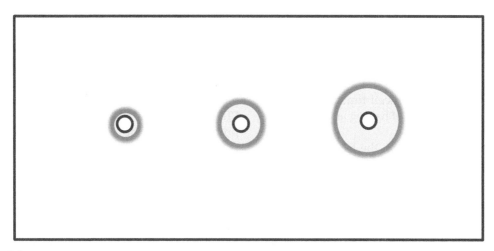

FIGURE 5.1 Radial immunodiffusion (RID) assay. In this example, anti-IgG is contained within the agar of this slide assay system and serum is placed in the pre-cut wells. IgG migrates radially from the well and precipitates when antigen–antibody equivalence has been reached. In this example of three serum samples, the precipitin ring diameter reflects the level of serum IgG. The actual concentration of IgG (or other immunoglobulin class, complement proteins, etc.) is determined by comparing the diameter of the precipitin ring from the patient sample to a series of standards. This figure is reproduced in color in the color section.

use of RID, whether the end point or fixed time method, became, in the late 1960s, the standard approach to quantitating the major classes of immunoglobulins in the evaluation of humoral immunity. The capacity to characterize the specific immunoglobulin classes led to the discovery of the most common primary immunodeficiency, selective IgA deficiency (9). It also allowed a more complete and quantitative assessment of patients with agammaglobulinemia and hypogammaglobulinemia when compared to using serum electrophoresis. The next major step in the laboratory evaluation of immunoglobulins was the development of automated nephelometry, a method that largely replaced RID owing to the speed and accuracy of this technique and which remains in routine use currently (10). The ability to quantitate the very low abundance immunoglobulin, IgE, awaited development of the more sensitive technique involving the immunoassay, and application of this method allowed the characterization of those primary immunodeficiencies associated with decreased or increased IgE levels (11, 12).

Another step in the evolution of laboratory assessment directed at antibody deficiencies involved the measurement of specific *in vivo* antibody responses. This had been available using various approaches, including the previously noted Schick test that evaluated for antibody to diphtheria toxoid, which when present neutralized cutaneously injected diphtheria toxin. Hence a negative Schick test was indicative of a normal antibody response to prior vaccination. Additional methods to evaluate specific antibody production were developed, but were generally cumbersome to perform and not generally offered. Following the development of the immunoassay, evaluation of specific antibody responses became readily available in the clinical laboratory (13). Initially, this testing was dependent on a radionuclide-based detection system, but this was by a methodological modification using an enzyme-substrate colorimetric detection method, the enzyme linked immunabsorbent assay or ELISA (14). The description of the ELISA method spawned the development of a large variety of antigen-specific antibody tests, some of which were designed to evaluate *in vivo* antibody responses to selected protein and carbohydrate vaccines. The assessment of specific post-vaccine antibody responses to both protein and carbohydrate is now a standard method to assess B-cell functional capacity in a patient suspected of having an antibody deficiency.

The next major step in the laboratory evaluation of antibody deficiencies was directed at characterizing circulating B-cells in affected patients. This began with the recognition that B-cells expressed surface immunoglobulin, which could be detected using fluorochrome conjugated polyclonal antibodies to a specific immunoglobulin heavy chain (e.g. gamma, mu) followed by B-cell quantitation using fluorescence microscopy. Application of this technique helped to characterize antibody deficiencies based on the absence or presence of circulating B-cells and later by examination of patient bone marrow for B-cells and their precursors (15).

The capacity for an expanded characterization of circulating lymphocytes, including B-cells, was driven by two major developments: the description of a method to generate monoclonal antibodies and the evolution of flow cytometry. In the 1970s, Kohler and Milstein published a landmark paper describing a method that allowed for the continuous production of an antibody with a defined specificity. They received the Nobel Prize for this work, which established the approach to generating monoclonal antibodies (16). This technique ultimately resulted in the availability of a wide range of monoclonal reagents, many of which specifically characterize lymphocytes and other cells of the immune system. It also led to the development of a classification process based on a cluster of differentiation (CD) system that identifies unique cell surface proteins and assigns a specific CD number (e.g. CD3) that is developed via a consensus process (17). Currently, there are more than 300 CD specificities, and these monoclonal reagents conjugated to specific fluorochromes aided in the routine clinical application of flow cytometry. This was further facilitated by technical developments in flow cytometry making this readily adaptable to clinical use (18). The combination of these two technologies has allowed the expanded immunophenotypic characterization of PIDDs, including those that impact antibody production (18). The advantage of this approach is that it facilitates evaluating large numbers of cells in solution and allows the generation of multiple characteristics (parameters) for each analyzed event (cell). In the case of antibody deficiencies, flow cytometry has enabled disease categorization based on the absence or presence of specific B-cell precursors in the bone marrow, the absence or presence of circulating B-cells as well as specific alterations in circulating B-cell subpopulations.

A more recent step in applying laboratory technology to better understand PIDDs was a product of the incredible progress in molecular genetics. This has allowed the routine characterization of genetic mutations using the method developed in the 1970s by Sanger, a discovery that resulted in his being awarded the Nobel Prize (19). Using this approach, the genetic defects for a number of PIDDs were identified in the early 1990s; since then progress has been absolutely stunning, and now more than 200 different genetic defects associated with PIDDs have been identified (20). Defining the molecular basis of antibody deficiencies followed a similar time-line, starting in the 1990s with the characterization of X-linked agammaglobulinemia as resulting from mutations in the gene encoding the protein, Bruton's tyrosine kinase (BTK) (21). Since this description, additional genetic defects which produce autosomal recessive forms of congenital agammaglobulinemia have been defined, as well as a variety of specific gene defects causing various types of hyper IgM syndromes (20). Defining the specific genetic defects in the most frequent antibody deficiency disorders, including common variable immune deficiency (CVID) and selective IgA deficiency, have remained elusive, but progress is being made and these will certainly be identified in the future, most likely as products of multiple different genetic defects.

HISTORY OF THE EVALUATION OF CELLULAR (T-CELL) IMMUNITY

The opening chapter of the clinical and laboratory evaluation of cellular immunodeficiencies was the product of linking the absence of lymphocytes with susceptibility to severe opportunistic infection early in life. This relationship was recognized in the 1950s by two different groups in Switzerland, who described a total of four infants who died of severe fungal and bacterial infections (22, 23). These investigators recognized that the severity of their disease, together with the marked lymphopenia, clearly distinguished these patients from the previous descriptions of agammaglobulinemia. The laboratory characterization of the patients beyond evaluating the absolute lymphocyte count awaited the description of a method to evaluate in vitro lymphocyte function. This was first reported in 1960, with the description of a serendipitously discovered method to evaluate lymphocyte proliferation in response to stimulation with the mitogen, phytohemagglutinin (PHA) (24). This important discovery was followed within a few years by the description of a technique to separate mononuclear cells rich in lymphocytes using density gradient centrifugation, a method still in use (25). Semi-automation of the evaluation of in vitro T-cell proliferation represented the next step which enabled these assays to be more readily adaptable for clinical use (26).

Testing in vitro T-cell function was next coupled with the capacity to specifically identify T-cells, based on the observation that thymic-derived lymphocytes spontaneously bind to sheep red blood cells, forming a "rosette" (Fig. 5.2) (27). The combination of evaluating in vitro T-cell proliferation and specifically

quantifying circulating T-cells enabled a more complete assessment of cellular immunity. These in vitro laboratory assays complemented the previously defined in vivo method for evaluating cellular immunity based on delayed type hypersensitivity reactions (DTHs) to specific recall antigens. The delayed cutaneous response was actually first recognized by Jenner in the 18th century, when a subject who had been successfully vaccinated with cowpox developed a cutaneous reaction upon cowpox re-exposure. Approximately a century later, Koch reported this same type of cutaneous immunological reaction when an extract of tubercle bacilli was injected intradermally into patients with tuberculosis. During the 20th century, DTH was proven to be mediated by lymphocytes through in vivo cell transfer experiments (28). This ultimately led to the clinical application of DTH as a method to assess prior exposure/infection with tuberculosis, and later as an approach to evaluating cellular immunity. The latter approach typically involved use of purified protein derivative (PPD) plus a combination of additional possible recall antigens (e.g. streptokinase/streptodornase (SK/SD), candidin, trichophytin) as a means to assess the functional capacity of the cellular immune system (29).

Testing the cutaneous immunological response to specific antigens also set the stage for development of in vitro testing using the same recall antigens as stimulants in a culture system similar to that used for mitogen-induced proliferation but extending the culture period to 6–7 days. This in vitro culture method was also developed to evaluate T-cell proliferation to alloantigens in an assay referred to as the mixed leukocyte culture (MLC) (30). This assay was initially developed as a means of testing major histocompatibility (MHC, human leukocyte antigen [HLA]) class II compatibility but it can also be used as a part of an in vitro assessment of T-cell function (30). Thus, by the mid-1970s, cellular immunity could be assessed both by quantitating total T-cells using the E-rosette assay,

A.

B.

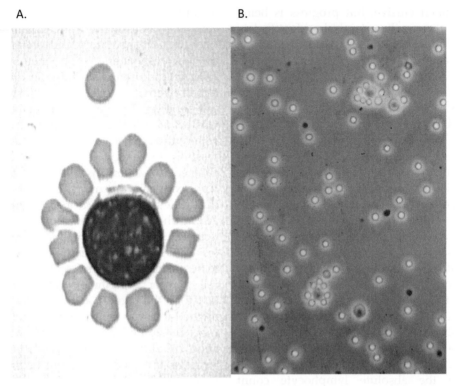

FIGURE 5.2   E-rosette. Panel A demonstrates a T-cell with sheep red cells forming a rosette highlighted by staining. Panel B demonstrates a direct preparation of peripheral blood mononuclear cells mixed with sheep red blood cells, demonstrating T-cell erythrocyte rosette formation; this represents the test system used to quantify T-cells. This figure is reproduced in color in the color section.

and by evaluating *in vitro* T-cell function based on the proliferative response to mitogens, antigens and alloantigens.

Another major step in the evaluation of cellular immunity occurred in the 1970s, as a product of work done by Zinkernagel and Doherty, which led to their receiving the Nobel Prize. Their studies established that MHC restriction was required for specific T-cell cytotoxicity directed against virally-infected target cells (31). This observation ultimately led to the definition of MHC (HLA) restriction for all T-cell-antigenic peptide (displayed by antigen presenting cells) interactions, and this together with the characterization of the T-cell antigen receptor (TCR) defined how T-cells specifically respond to antigen. The development of monoclonal antibodies directed to T-cell surface proteins facilitated the recognition that the TCR complex also included the various CD3 protein chains. This also helped to define the critical roles played by CD4 and CD8 as co-receptors for TCR-antigenic peptide interactions. Clarification of T-cell subsets was a direct product of the expanding menu of monoclonal antibodies, and this technique facilitated the definition of naïve and memory cells within the CD4 and CD8 subsets, a critical approach for evaluating certain T-cell immune deficiencies (32). There is now a large range of reagents which provides specific information regarding T-cell subsets that includes specific cell functional potential as well as state of differentiation and activation status.

Work also initiated in the 1970s identified a growth factor in the culture supernatant of stimulated T-cells that was initially classified as T-cell growth factor but which was ultimately identified as interleukin (IL)-2 (33). These studies, along with prior work characterizing viral and immune interferons, represented the earliest recognition that secreted molecules represent critical effectors in an immune response. This has evolved such that the soluble mediators of immunity are understood to include different classes of molecules such as interleukins, chemokines, interferons and hematopoietic growth factors. It is now common practice in the clinical evaluation of a patient suspected of having a cellular immune deficiency to quantitate *in vitro* cytokine production under various culture conditions, as well as evaluate the specific receptors linked to these mediators and, in certain settings, assess intracellular signaling induced by cytokine binding its receptor (34). This approach to laboratory testing has particular relevance in selected immunodeficiencies associated with defective cytokine receptors. In addition, there are also primary immunodeficiencies in which cytokine production is altered as a consequence of the underlying defect (20). The release of soluble mediators during an immune reaction is now recognized to constitute a complex interplay of multiple players, with more than 35 interleukins, 45 chemokines and three classes of interferons having been characterized.

The recognition that cytokine production is a critical part of the immunological response led to further definition of helper T-cell function during the 1980s. This initially involved identifying T-helper 1 (Th1) and Th2 cells based on unique patterns of cytokine production following lymphocyte activation (35). The complexity of this classification has expanded to include at least Th17, T-follicular helper (Tfh) cells and T-regulatory cells, plus additional categories that have been suggested (e.g. Th9, Th22). The division of labor between the different T-helper cell types can be identified following T-cell activation by evaluating the presence of characteristic intracellular cytokines using flow cytometry (36). One example of the utility of this approach relates to the susceptibility to *Candida* infection seen in the autosomal dominant form of hyper-IgE syndrome (Job's syndrome), which was ultimately linked to the demonstration of a deficiency in Th17-cell generation (37).

The capability to identify and quantify various T-cell subsets, assess T-cell function and evaluate cytokine and chemokine production has dramatically increased the capacity for evaluating the cellular arm of the adaptive immune system. Understanding the development and maturation of these critical cellular elements has been facilitated by the molecular characterization of a large number of genetic mutations associated with T-cell dysfunction (20). These efforts have defined multiple genetic defects that result in the development of severe combined immune deficiency (SCID). This work also has clarified that there can be a range of clinical phenotypes observed with mutations in the same gene. The new frontier in the genetic characterization of patients with immune deficiency will involve next-generation sequencing (38). In fact, a host of new genetic defects have already been identified based on the use of whole-exome sequencing, and the expectation is that whole-genome sequencing will be applied in the very near future to identifying even more genetic alterations which result in defective immune function and lead to altered host defense.

One of the vexing challenges in the diagnosis of serious congenital immune deficiencies is the lag that has so often been observed between the onset of symptoms and the final establishment of the diagnosis. This delay has proven to be particularly devastating in the setting of a severe T-cell deficiency, such as SCID (see Chapter 14), where the patient's outcome is compromised by a delay in providing curative therapy involving immune reconstitution (via hematopoietic stem cell transplantation). To address this, Chan and Puck successfully modified a previously described method that identifies recent thymic immigrants. This involved applying this test to genomic DNA obtained from the routine newborn screening (Guthrie) card (39). The actual method is called the T-cell receptor excision circle (TREC) assay, and it tests for the presence of circular DNA generated from the excised TCRδ gene during the productive rearrangement of the α/β TCR in the thymus. The generation of these specific TRECs occurs in approximately 70% of developing T-cells destined for release from the thymus into the circulation (39, 40). The TREC assay has now been adopted as a routine part of newborn screening in multiple states in the USA, with more than 2 000 000 newborns having been screened to date. The results have been extremely encouraging, with successful early detection and immunological reconstitution in those infants identified during the newborn period with SCID.

HISTORY OF THE EVALUATION OF THE IL12/23-INTERFERON-γ CIRCUIT

The critical role of the IL12/23-inteferon-γ circuit in host defense against non-tuberculous mycobacteria infection was clarified in the 1990s with the first report of a defect in this pathway that was associated with fatal Bacillus Calmette–Guérin (BCG) infection in an infant (41). This was followed by series of reports describing defects in additional receptor components, cytokines and intracellular signaling proteins involved in this circuit (42) (see Chapter 4). Evaluation of these patients depends on a combination of flow cytometry to assess the expression of specific surface proteins (e.g. interferon-γ receptor α chain, IL-12 receptor β1 chain), evaluation of cytokine secretion in response to specific stimuli, assessment of intracellular protein phosphorylation (e.g. STAT-1 phosphorylation in response to interferon-γ stimulation) as well as standard genetic evaluation for mutations in the genes linked to this clinical phenotype (18, 34).

HISTORY OF THE EVALUATION OF NK-CELL DEFECTS

The identification of the role of natural killer (NK) cells in host defense evolved more slowly than the characterization of B-cell and T-cell primary immunodeficiencies. Defects in NK-cell function were first documented in two publications in the 1980s. The first described three siblings with diminished NK-cell activity and increased susceptibility to serious Epstein–Barr virus (EBV) infection; the second reported an adolescent patient who had no detectable NK cells and developed recurrent herpes viral infections (43, 44). The critical role of NK cells in host defense was further clarified by the finding that patients with the X-linked lymphoproliferative syndrome associated with a profound defect in handling EBV infection were found to have progressive NK-cell deficiency and virtual NKT-cell absence (45). In addition, patients with familial hemophagocytic lymphohisticytosis (HLH) syndromes were found to have NK-cell (and cytotoxic T-cell) defects, suggesting a broader role for cytotoxic lymphocytes in immunoregulation (46). Most recently, a genetic defect in the transcription factor GATA-2 has been found to be associated with NK-cell dysfunction (47). The clinical phenotype of this disorder is complex, and goes beyond specific infectious susceptibility to include increased risk for malignancy as well as other somatic findings. Interestingly, evaluation of a tissue sample from the adolescent patient with recurrent herpes viral infections noted above identified a *GATA-2* mutation as the cause of her disease. The laboratory evaluation of NK cells generally includes phenotypic characterization by flow cytometry using a variety of specific monoclonal reagents, as well as functional testing for cytolytic activity using NK-specific targets such as the erythroleukemia cell line, K562, and IgG-coated target cells to assess antibody-dependent cellular cytotoxicity (ADCC) (18, 48). More recently, a flow-cytometry-based assay to evaluate NK-cell function has been described, in which the level of cell surface CD107a upregulation on NK cells can be assessed following *in vitro* exposure to specific target cells (49).

HISTORY OF THE EVALUATION OF TOLL-LIKE RECEPTOR DEFECTS

Seminal work during this millennium has defined two major categories of Toll-like receptor (TLR) defects that present either with disseminated pyogenic infections or herpes simplex encephalitis (50) (see Chapter 4). The evaluation of these patients typically involves *in vitro* stimulation of leukocytes (or other cell types including fibroblasts) with the various TLR ligands and measuring cytokine production in the culture supernatants. This is usually followed by standard mutation analysis of the specific genes associated with either clinical phenotype (18). A rapid screening method to assess TLR function using flow cytometry has also been described (51).

HISTORY OF THE EVALUATION OF NEUTROPHIL IMMUNITY

It has long been known that significant neutropenia, whether congenital or iatrogenic, increases the risk of bacterial and fungal infections (see Chapter 10). Therefore, with an appropriate clinical history, the initial neutrophil evaluation should always include one or more leukocyte counts with differential to assess the absolute neutrophil count (ANC). This should be accompanied by a review of neutrophil morphology, since certain defects associated with infectious susceptibility have abnormal morphology (e.g. Chediak–Higashi syndrome). If persistent or cyclical neutropenia is identified, the next step would involve mutation analysis for specific gene defects associated with congenital neutropenia.

One of the earliest examples of defining a congenital abnormality in neutrophil function based on an *in vitro* test was made by Paul Quie, Robert Good and colleagues in the late 1960s. This group documented defective bactericidal activity in males with recurrent infection and elevated gamma globulins who had been diagnosed with the recently defined disorder, chronic granulomatous disease (CGD) (52) (see Chapter 13). This observation was followed shortly thereafter by the description of a simple slide test to evaluate neutrophil oxidative burst using the nitroblue tetrazolium (NBT) test (53). In this assay, neutrophils are loaded with the dye and then, following cell activation, the colorless NBT is converted to blue formazan that can be easily detected by light microscopy (Fig. 5.3A). In contrast, cells from patients with CGD fail to reduce the dye and, as a consequence of their enzyme defect, no blue color is observed (Fig. 5.3B). Female carriers of the X-linked recessive disorder typically have two populations of neutrophils, one that is normal and one that is defective (Fig. 5.3C). This test has largely been supplanted by the more quantitative dihydrorhodamine (DHR) flow cytometry assay that also measures reactive oxygen products following neutrophil activation, and this assay has actually proven useful in predicting survival in patients with CGD (54, 55).

A second major congenital defect of neutrophil function was linked to abnormal cell migration (see Chapter 21) (56). This was ultimately identified as leukocyte adhesion deficiency (LAD), and the original patients identified had what is now known as LAD type I, linked to a defect in the cell surface expression of beta-2 integrins. The defect in this family of adhesion molecules prevents neutrophils from migrating from blood vessels to sites of infection (57). The definitive laboratory screening test to establish this diagnosis involves flow cytometric evaluation for neutrophil surface expression of CD18 (and CD11a, 11b, 11c) (18). LAD type I is typically accompanied by a marked neutrophilia that is

A.   B.   C.

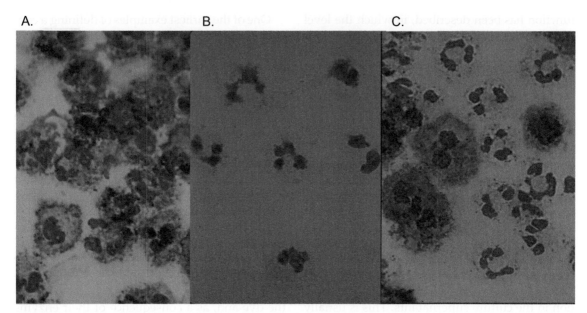

FIGURE 5.3   NBT test. Panel A demonstrates normal oxidase activity in neutrophils from a control with blue formazan dye clearly present. Panel B demonstrates absence of oxidase activity in the neutrophils of an X-linked CGD patient (no blue dye). Panel C demonstrates a combination of normal and abnormal neutrophils from an X-linked CGD carrier. This figure is reproduced in color in the color section.

the result of the inability of the cells to migrate into tissues. The carrier parents typically have lower level expression of CD18 on the surface of their neutrophils compared to controls, but they are generally clinically well. A second form of LAD (type II) has been identified associated with a defect in fucosylation, and this can be screened for by evaluating for the expression of CD15s (SLeX) on neutrophils (58). In both LAD type I and type II, the diagnosis should be confirmed by mutation analysis whenever possible.

An additional area of laboratory testing related to neutrophil function involves evaluation of neutrophil chemotaxis and bacterial killing. The former can be performed using a chemotactic (Boyden) chamber and a variety of different chemoattractants or, alternatively, in an agarose system again utilizing chemoattractants (59, 60). The bactericidal assay evaluates *in vitro* phagocytosis as well as the degree of neutrophil killing of the ingested microbes, however, this assay has significant biological variability and

has proven difficult to standardize. In general, both of these tests are performed only in very specialized laboratories and the results do not provide specific diagnostic information for characterizing the majority of congenital neutrophil defects.

HISTORY OF THE EVALUATION OF THE COMPLEMENT SYSTEM

The majority of defects in complement associated with increased susceptibility to infection fall into two categories: the early component defects that have a high incidence of systemic lupus erythematosus (SLE) or SLE-like disease as well as recurrent bacterial infections (similar to those seen with defects in antibody production). The other general category is associated with defects in the late complement components (terminal attack complex) as well as alternative pathway defects in Factor B and properdin, all of which are associated with increased susceptibility to neisserial infections (see Chapter 17). The simplest screening tests used to identify complement defects involve evaluating the classical complement pathway based on the total hemolytic complement (CH50) assay, together with another functional assay to evaluate the alternative complement pathway (AP50) (61, 62). In the setting of an appropriate clinical history, a low to absent CH50 with a normal AP50 is consistent with an early classical pathway component defect, while a low to absent CH50 and AP50 are characteristic of a late component defect. In any setting associated with an abnormal hemolytic assay of complement activity, the evaluation can be extended to test for specific complement components using component-specific immunoassays and, in some cases, using functional assays for individual complement components.

CONCLUSION

The application of laboratory testing has played and continues to play a critical role in the definition and diagnosis of PIDD. The past 60 years have witnessed an extraordinary increase in understanding the intricacies of host defense, which has been paralleled by an expansion in the number of defined primary immunodeficiencies. The link between basic discoveries in human immunology and the identification of disorders of human host defense has depended on the development of new methods to assess the immune system. This "partnership" between clinical immunology and diagnostic immunology appears to be reaching new heights with the recent, major advances in genetic testing, and these will likely herald an even more rapid pace of discovery in the field of primary immunodeficiencies.

Acknowledgment

This work was supported by the Intramural Research Program of the NIH Clinical Center, National Institutes of Health.

References

1. Bruton OC. Agammaglobulinemia. *Pediatrics* 1952;**9**: 722–8.
2. Tiselius A, Kabat E. Electrophoresis of immune serum. *Science* 1938;**87**:416.
3. Anderson HA, Kunkel HG, McCarty M. Quantitative anti-streptokinase studies in patients infected with group A hemolytic streptococci: a comparison with serum anti-streptolysin and gamma globulin levels with special reference to the occurrence of rheumatic fever. *J Clin Invest* 1948;**4**:425–34.
4. Smith CIE, Notarangelo LD. Molecular basis of X-linked immunodeficiencies. *Adv Genet* 1997;**35**:57–115.
5. Franklin EC, Kunkel HG. Immunologic differences between the 19 S and the 7 S components of normal human gamma globulin. *J Immunol* 1957;**78**:11–8.
6. Rosen FA, Kevy SV, Merler E, Janeway CA, Gitlin D. Recurrent bacterial infections and dysgammaglobulinemia: deficiency of 7S gamma-globulins in the presence of elevated 19S gamma-globulins. Report of two cases. *Pediatrics* 1961;**28**:182–95.
7. Fahey JL, McKelvey EM. Quantitative determination of serum immunoglobulins in antibody-agar plates. *J Immunol* 1965;**94**:84.
8. Mancini G, Carbonara AO, Heremans JF. Immunochemical quantitation of antigens by single radial immunodiffusion. *Immunochemistry* 1965;**2**:235.
9. Crabbe PA, Heremans JF. Selective IgA deficiency and steatorrhea. A new syndrome. *Am J Med* 1967;**42**:319–26.
10. Ritchie RF. A simple, direct and sensitive technique for measurement of specific proteins in dilute solutions. *J Lab Med* 1967;**70**:512–7.
11. Isazaka K, Ishizaka T. Immunoglobulin E: Current status and clinical laboratory applications. *Arch Pathol Lab Med* 1976;**100**:289–92.
12. Buckley RH, Fiscus SA. Serum IgD and IgE concentrations in immunodeficiency deficiencies. *J Clin Invest* 1975;**55**:157–65.
13. Yallow RS, Berson SA. Immunological specificity of human insulin: application to immunoassay of insulin. *J Clin Invest* 1961;**40**:2190–8.
14. Voller A, Bidwell D, Huldt G, Engvall E. A microplate method of enzyme-linked immunosorbent assay and its application in malaria. *Bull World Health Org* 1974;**51**: 209–11.
15. Cooper MD, Lawton AR. Circulating B-cells in patients with immunodeficiencies. *Am J Pathol* 1972;**69**:513–28.
16. Kohler G, Milstein C. Continuous cultures of fused cells secreting antibody of predefined specificity. *Nature* 1975; **256**:495–7.
17. Proceedings of the 9th international workshop on human leukocyte differentiation antigens. *Immunol Lett* 2011; **134**:103-118.

18. Oliveira JB, Fleisher TA. Molecular- and flow cytometry-based diagnosis of primary immunodeficiency disorders. *Curr Allergy Asthma Rep* 2010;**6**:460–7.

19. Sanger F, Nicklen S, Coulson AR. DNA sequencing with chain-terminating inhibitors. *Proc Natl Acad Sci USA* 1977;**74**:5463–7.

20. Al-Herz W, Bousfiha A, Casanova JL, et al. Primary immunodeficiency diseases: an update on the classification from the international union of immunological societies expert committee for primary immunodeficiencies. *Front Immunol* 2011;**2**:54.

21. Conley ME, Parolini O, Rohrer J, Campana D. X-linked agammaglobulinemia: new approaches to old questions based on the identification of the defective gene. *Immunol Rev* 1994;**138**:5–21.

22. Hitzig WH, Biro Z, Bosch H, Huser J. Agammaglobulinemia & alymphocytosis with atrophy of lymphatic tissue. *Helv Pediatr Acta* 1958;**13**:551–85.

23. Tobler R, Corrier H. Familial lymphopenia with agammaglobulinemia & severe moniliaisis: the essential lymphocytophthisis as a special form of early childhood agammaglobulinemia. *Helv Pediatr Acta* 1958;**13**:313–38.

24. Nowell PC. Phytohemagglutinin: an initiator of mitosis in cultures of normal human lymphocytes. *Cancer Res* 1960;**20**:462–6.

25. Boyum A. Ficoll Hypaque method for separating mononuclear cells and granulocytes from human blood. *Scan J Clin Lab Invest* 1966;**Suppl**:77.

26. Hartzman RJ, Bach ML, Bach FH, Thurman G, Sell KW. Precipitation of radioactivity labeled samples: a semiautomatic multiple sample processor. *Cell Immunol* 1972;**4**:182–7.

27. Wybran J, Fudenberg HH. Rosette formation, a test for cellular immunity. *Trans Assoc Am Physicians* 1971;**84**:239–47.

28. Lawrence HS. The cellular transfer of cutaneous hypersensitivity reactivity to hemolytic streptococci. *J Immunol* 1952;**68**:159–78.

29. Palmer DL, Reed DP. Delayed hypersensitivity skin testing. Clinical correlates and anergy. *J Infect Dis* 1974;**130**:138–43.

30. Bach FH, Bock H, Graupner K, Day E, Klostermann H. Cell kinetic studies in mixed leukocyte cultures: an in vitro model of homograft reactivity. *Proc Natl Acad Sci USA* 1969;**62**:377–84.

31. Zinkernagel RM, Doherty PC. Restriction of *in vitro* T-cell-mediated cytotoxicity in lymphocytic choriomeningitis within a syngeneic or semiallogeneic system. *Nature* 1974;**248**:701–2.

32. Sanders ME, Makgoba MW, Shaw S. Human naïve and memory T-cells: reinterpretation of helper-inducer and suppressor-inducer subsets. *Immunol Today* 1988;**9**:195–8.

33. Ruscetti FW, Gallo RC. Human T-lymphocyte growth factor: regulation of growth and function of T lymphocytes. *Blood* 1981;**57**:379–93.

34. Fleisher TA, Dorman SE, Anderson JA, Vail M, Brown MR, Holland SM. Detection of intracellular phosphorylated STAT-1 by flow cytometry. *Clin Immunol* 1999;**90**:425–30.

35. Mosmann TR, Cherwinski H, Bond MW, Giedin MA, Coffman RL. Two types of murine helper T-cell clone. I. Definition according to profiles of lymphokine activities and secreted proteins. *J Immunol* 1986;**136**:2348–57.

36. Foster B, Prussin C, Liu F, Whitmire JK, Whitton JL. Detection of intracellular cytokines by flow cytometry. *Curr Protoc Immunol* 2007;**6**:6.24.

37. Milner JD, Brenchley JM, Laurence A, et al. Impaired T(H)17 cell differentiation in subjects with autosomal dominant hyper-IgE syndrome. *Nature* 2008;**452**:773–6.

38. Chou J, Ohsumi TK, Geha RS. Use of whole exome and genome sequencing in the identification of genetic causes of primary immunodeficiencies. *Curr Opin Allergy Clin Immunol* 2012;**12**:623–8.

39. Chan K, Puck JM. Development of a population-based screening for severe combined immunodeficiency. *J Allergy Clin Immunol* 2005;**115**:391–8.

40. Puck JM. Laboratory technology. For population-based screening for severe combined immunodeficiency in neonates: the winner is T-cell receptor excision circles. *J Allergy Clin Immunol* 2012;**129**:607–16.

41. Joungay E, Altare F, Lamhamedi S, et al. Interferon-gamma receptor deficiency in an infant with fatal bacilli Calmette–Guerin infection. *N Engl J Med* 1996;**335**:1956–61.

42. Holland SM. Interferon gamma, IL-12, IL-12R, STAT-1 immunodeficiency diseases: disorders of the interface of innate and adaptive immunity. *Immunol Res* 2007;**38**:342–6.

43. Fleisher G, Starr S, Koven N, Kamiya H, Douglas SD, Henle WA. A non X-linked syndrome with susceptibility to severe Epstein–Barr virus infections. *J Pediatr* 1982;**100**:727–30.

44. Biron CA, Byron KS, Sullivan JL. Severe herpesvirus infections in an adolescent without natural killer cells. *N Engl J Med* 1989;**320**:1731–5.

45. Parolini R, D'Andrea E, Poletti A, et al. X-linked lymphoproliferative disease: 2B4 molecules displaying inhibitory rather than activating function are responsible for the inability of natural killer cells to kill Epstein–Barr virus infected cells. *J Exp Med* 2000;**192**:337–46.

46. Russell JH, Ley TJ. Lymphocyte mediated cytotoxicity. *Annu Rev Immunol* 2002;**20**:323–70.

47. Mace EM, Hsu AP, Monaco-Shawver l, et al. Mutations in GATA2 cause human NK cell deficiency with specific loss of the CD56 (bright) subset. *Blood* 2013;**121**:2669–77.

48. Orange JS. Formation and function of the lytic NK-cell immunologic synapse. *Nat Rev Immunol* 2008;**8**:713–25.

49. Alter G, Malenfant JM, Altfeld M. CD107a as a functional marker for the identification of natural killer cell activity. *J Immunol Methods* 2004;**294**:15–22.

50. Casanova JL, Abel L, Quintana-Murci L. Human TLRs and IL-1Rs in host defense: natural insights from evolutionary, epidemiological and clinical genetics. *Annu Rev Immunol* 2001;**29**:447–91.

51. Von Bernuth H, Ku CL, Rodriguez-Gallego C, et al. A fast procedure for the detection of defects in Toll-like receptor signaling. *Pediatrics* 2006;**118**:2498–503.

52. Quie PG, White JG, Holmes B, Good RA. In vitro bactericidal capacity of human polymorphonuclear leukocytes: diminished activity in chronic granulomatous disease of childhood. *J Clin Invest* 1967;**46**:668–79.

53. Baehner RL, Nathan DG. Quantitative nitroblue tetrazolium test in chronic granulomatous disease. *N Engl J Med* 1968;**278**:971–6.

54. Vowells SJ, Fleisher TA, Sekhsaria S, Alling DW, Maguire TE, Malech HL. Genotype-dependent variability in flow cytometric evaluation of reduced nicotinamide adenine dinucleotide phosphate oxidase function in patients with chronic granulomatous disease. *J Pediatr* 1996;**128**:104–7.

55. Kuhns DB, Albord WG, Heller T, et al. Residual NADPH oxidase survival in chronic granulomatous disease. *N Engl J Med* 2010;**363**:2600–10.

56. Hayward AR, Harvey BA, Leonard J, Greenwood MC, Wood CB, Soothill JF. Delayed separation of the umbilical cord widespread infections, and defective neutrophil mobility. *Lancet* 1979;**1**:1099–101.

57. Anderson DC, Schmalsteig FD, Finegold MJ, et al. The severe and moderate phenotypes of heritable Mac-1, LFA-1 deficiency; their quantitative definition and relation to leukocyte dysfunction and clinical features. *J Infect Dis* 1985;**152**:668–89.

58. Etzioni A, Frydman M, Pollack S, et al. Brief report: recurrent severe infections caused by a novel leukocyte adhesion deficiency. *N Engl J Med* 1992;**327**:1789–92.

59. Madeerazo EG, Woronick CL, Ward PA. Micropore filter assay of human granulocyte locomotion: problems and solutions. *Clin Immunol Immunopathol* 1978;**11**:196–211.

60. Nelson RD, Quie PG, Simmons RL. Chemotaxis under agarose: a new and simple method for measuring chemotaxis and spontaneous migration of human polymorphonuclear leukocytes and monocytes. *J Immunol* 1975;**115**:1650–5.

61. Norman ME, Gall EP, Taulor A, Laster L, Nilsson UR. Serum complement profiles in infants and children. *J Pediatr* 1975;**87**:912–6.

62. Takada A, Imamura Y, Takada Y. Relationships between the haemolytic activities of the human complement system and complement components. *Clin Exp Immunol* 1979;**35**:324–8.

Primary Immunodeficiency in the Developing Countries

Aziz A. Bousfiha[1], Leila Jeddane[1], Antonio Condino-Neto[2]

[1]Clinical Immunology Unit, IbnRoshd University Hospital.
King Hassan II University, Casablanca, Morocco
[2]Department of Immunology, Institute of Biomedical Sciences,
University of São Paulo, Brazil

INTRODUCTION

The history of primary immunodeficiency (PID) in emerging countries illustrates the challenges faced by experts in the field who were charged with developing expensive and specialized care in countries with low incomes and a lack of medical and scientific infrastructure. Here, we present some examples of these difficulties, and provide an account of the history and development of the medical field of PID in Africa, the Middle East and Latin America.

Out of Africa, we can report that forming an association to support PID patients not only promoted patient care in the organization's home country, but also triggered an unprecedented collaboration of African nations by facilitating the first congress of the African Society for Immune Deficiencies (ASID) in 2008 in Morocco.

In Latin America, the interest in PID started during the 1980s, when the AIDS epidemic directed the attention of physicians to immunological disorders. Physicians and researchers from several Latin American countries started to identify PID patients, began to develop awareness campaigns and, in 1993, formed the Latin American Group for Immune Deficiencies (LAGID), which became a Society (LASID) in 2009. The field advanced continuously in the continent through educational activities provided by medical societies and patient associations.

In the Middle East, the differences between rich and poor countries led to considerable variability in the chronology and pace of progress in developing awareness and generating centers devoted to the diagnosis and treatment of PID patients. While Saudi Arabia and Iran progressed rapidly and competitively at all levels, other countries are struggling to catch up.

HISTORY OF PRIMARY IMMUNODEFICIENCY IN AFRICA

Reflecting on the history of PID in Africa, we can distinguish three periods of development. During the first period, individuals in scattered locations became interested in PID, often after returning from training abroad. During the second phase, the development of specialized centers occurred at a national level. Finally, in 2008, a small pan-African group succeeded in creating their own society, ASID.

Early Initiatives by Individual Visionaries

In the beginning, several initiatives were started in a few African countries to develop diagnostic centers for PID.

In Egypt, one of the first people to develop an interest in PID was Dr Aysha El Marsafy. The first laboratory explorations began there in 2003 with his measurements of C3, C4 and serum Ig levels.

Professor David Beatty, an immunologist at the University of Cape Town, South Africa, published the first case series in 1976 (1), focusing on chronic granulomatous disease. His laboratory offered comprehensive diagnostic testing as early as 1983, including *in vitro* lymphocyte proliferation to mitogens and antigens, immunoglobulin levels, complement components and the

nitroblue tetrazolium (NBT) test. Subsequently, other physicians, e.g. Monika Esser, who organized the third ASID meeting in 2013 in South Africa, and Brian Elley, joined the Cape Town group.

In Tunisia, Mohamed Bejaoui, a hemato-oncologist pediatrician, received specialty training at Hôpital Necker-Enfants Malades, Paris, under the supervision of Professor Claude Griscelli. After returning to Tunis in April 1988, he collaborated with Professor Koussay Dellagi from the Pasteur Institute of Tunis to identify and treat PID patients in his clinic. To complete the team, two geneticists, one trained in the USA and the other in France, and a biologist (Mohamed Ridha Barbouch), were asked to join.

In Morocco in the 1980s, Abdellah Benslimane, head of the Laboratory of Immunology at the King Hassan II University, established a diagnostic platform to identify PID patients, which included measurement of serum immunoglobulin levels by Mancini's radial immunodiffusion, lymphocyte subsets by immunofluorescence, CH50 and complement component quantitation, as well as the NBT test. However, this major effort failed to identify PID patients because clinicians with expertise in PID did not exist at that time in Morocco, and the whole project gradually disappeared. Twelve years later, in 1997, Ahmed Aziz Bousfiha, a young pediatrician from the Department of Pediatric Infectious Diseases, IbnRushd University Hospital, Casablanca, underwent training in the Pediatric Immuno-Hematology Unit, Hôpital Necker-Enfants Malades, Paris, under the supervision of Alain Fischer and Jean-Laurent Casanova. Back in Casablanca, he began to identify patients suspected of having PID and sent specimens for laboratory analysis abroad. However, because of financial problems, many families could not afford to pay for these tests. This prompted the creation, in 1998, of a foundation to help children living with PID. This foundation was named Hajar (www.hajar-maroc.org) which is the name of a Moroccan girl who, needing bone marrow transplantation for HLA II deficiency, died due to lack of resources.

Algeria is just starting to set up laboratory facilities to diagnose PID with pediatrician Professor Rachida Boukari spearheading the group.

Establishment of Centers Specializing in PID

Proud of the accomplishments achieved in the late 1990s, the pioneers developed national teams of individuals deeply committed to PID. Awareness campaigns for physicians and the general public were initiated, and training centers for young clinicians and biologists were established in a number of countries. In parallel, specific facilities to treat PID patients with immunoglobulin substitution were gradually established in several African countries.

In Tunisia, Professor Bejaoui developed a national network that included all university centers and the major hospitals of Tunisia to increase PID awareness and recruitment of patients. This initiative was successful and, after eight years of data collection, resulted in the publication of a series of 152 patients with PID (2). By 2011, more than 700 Tunisian patients had been enrolled in the network (Fethi Mellouli, personal communication). In Egypt, although no official network exists, friendly personal communications have developed. Two centers are presently caring for PID patients, both located in Cairo: the Immunology Clinic at the Pediatric Department of Cairo University, in existence since 2003, and a unit at Ain-Shams University.

The foundation of the Moroccan Society for Primary Immunodeficiency (MSPID www.pid-moroccasociety.org) and the publication of several awareness articles for pediatricians (3–7) led to an exponential increase in the number of patients, forcing the hospital administration in Casablanca to provide funds for setting up a laboratory platform to perform on-site immunological testing for PID. Importantly, the MSPID decided to invite to each of the congress members of scholarly societies of adult medical disciplines that share similar interests: gastro-enterologists, hemato-oncologists, pulmonologists and dermatologists. This strategy led to a marked increase in the number of adult PID patients in Morocco.

Efforts to Organize: The Foundation of ASID

The idea of forming the ASID was conceived in 2007 in Tunisia during a brainstorming session which included Aziz Bousfiha, Mohamed Bejaoui, Ridha Barbouche, Fethi Mellouli and Jean-Laurent Casanova, then President of the European Society for Immundeficiencies (ESID). The first ASID congress was held in October 2008 in Casablanca, with the strong support of ESID and the Association HAJAR.

The ASID Congress in Casablanca provided an opportunity to bring together more than 30 delegates interested in PID from Morocco, Tunisia, South Africa, Egypt, Algeria, Mauritania and Guinea (Fig. 6.1). The event was chaired by Professor Claude Griscelli, currently ASID honorary president, Professor Jean-Laurent Casanova, then ESID president and Professor Amos Etzioni, then future ESID president and speakers came from Africa, America, Asia and Europe. This first Congress of ASID was supported by the International Nursing Group for Immunodeficiencies (INGID), the International Patient Organization for Primary Immunodeficiencies (IPOPI) and the Jeffrey Modell Foundation (JMF).

ASID (www.asid.ma) became the spokesperson of Africa for issues related to PID, has joined the International Union of Immunological Societies (IUIS) PID expert committee and supports INGID and IPOPI. ASID organizes regional activities, and maintains a biennial continental PID congress. Training schools which focus on PID were organized in Dakar, in Egypt, in South Africa and in Tunisia. The second ASID Congress took place in Hammamet,

FIGURE 6.1 ASID Inaugural meeting in Casablanca, October 2008. Aziz Bousfiha (arrow) and, on his right Claude Griscelli, Luigi Notarangelo (Boston) and Bouchra Benhayoun (Hajar Association). Jean-Laurent Casanova (ESID president, star) in the middle of several African delegates but also several other experts from the five continents. This figure is reproduced in color in the color section.

Tunisia in March 2012 and the third ASID Congress was held in South Africa in 2013. ASID has facilitated publications by its members at national level, including studies of the genetic characterization and the incidence of PID (8–12). Importantly, ASID members from different regions are beginning to collaborate. Admou et al. published an approach for PID diagnosis in resource-limited countries (13) and Barbouche et al. reported a multicenter series on PID in North Africa (14). A collaborative project is in progress between Pasteur Institutes in Morocco, Algeria and Tunisia focusing on genetic characterization of PID. Recently, ASID initiated a project with the aim of collecting data on all PID patients in Africa. This registry is based on the computerized registry already existing in Morocco and modeled on the ESID registry.

PID Management in Africa

Medical care for PID patients is one of the great challenges in Africa. Immunoglobulin (Ig) substitution therapy began in 1988 in Tunisia, with the successful use of IM preparations given subcutaneously. An intravenous immunoglobulin (IVIG) formulation, introduced in 1990, was produced from placenta-derived Ig, which was replaced in 1992 by serum-derived Ig. In Morocco, IVIG was not affordable for most patients. This prompted health authorities to sign an agreement between the Blood Transfusion National Center and the LFB industry to subcontract IVIG production in France using Moroccan plasma. This arrangement reduced the cost by 66%. Hematopoietic stem cell transplantation (HSCT) poses another challenge in PID management. The first HLA-identical HSCT was carried out in South Africa in 1996 on a child with X-linked severe combined immunodeficiency (X-SCID) at the Groote Schuur Hospital BMT unit. In Tunisia, the first HSCT was performed in 1998 on a baby with Omenn syndrome (15). Efforts are now underway to improve the outcomes of HSCT, particularly for patients with MHC II deficiency, which is endemic in North Africa. In Morocco, ten patients have undergone HSCT since the creation of the Hajar association, but only two of these procedures were performed in Morocco, both in 2010. In response to the great demand, young physicians are being sent for training to centers with experience in non-malignant HSCT.

PID in Africa: Concluding Remarks

The story of PID in Africa has taught us that the development of tools to diagnose these diseases and to effectively manage their medical problems requires strong collaboration between clinicians, laboratory teams and government agencies. Based on the Moroccan and Tunisian experience, certain logistical rules to guarantee success can be identified. The first step is to build medical teams who are capable of recruiting patients and requesting meaningful laboratory evaluations. The second step is to set up special laboratory teams that are capable of exploring the immune defects of individual patients. Good communication between these two teams is required for accurate diagnosis and adequate treatment. Unfortunately, the necessary facilities for the diagnosis and treatment of PID patients are available in only about a dozen of the 54 African countries that form this vast continent of over one billion citizens. Local action and international support are needed to assist

ASID in encouraging the authorities and the African scientific community to promote public awareness of PID, which we believe significantly contributes to death from infectious diseases, especially among children in countries with a high rate of consanguinity.

HISTORY OF PRIMARY IMMUNODEFICIENCY IN THE MIDDLE EAST

Review of the first publications in this field written by Middle Eastern scientists and listed on PubMed allowed us to distinguish three steps in Middle Eastern PID history. First, a few physicians, in collaboration with colleagues from abroad, published selected case reports and mini-case series, mainly on diseases with a characteristic clinical phenotype. In the late 1990s, a limited number of Middle Eastern physicians were sent abroad to train in specialized centers. Back in their own countries, they published case reports and formed specialized teams. More recently, we witnessed the creation of PID-focused centers and the emergence of registries. As a result, PID management has greatly improved and research has flourished – with interesting results.

First Step: Case Reports

It was not until the late 1980s–early 1990s that Saudi physicians began to publish case reports, in collaboration with foreign physicians working in their centers (16–19). These early publications were limited to easily diagnosed PID disorders (20–23). The same pattern can be recognized in other Middle Eastern countries, though the timeline may differ (24–28).

Second Step: Physician Training

In the late 1990s, Saudi physicians in particular, but also medical doctors and biologists from countries in the region other than Saudi Arabia were sent abroad by their governments to be trained in the field of PID. These highly specialized centers in Paris, Boston, Toronto, Montreal and other cities welcomed and trained these physicians and scientists. Back home, for instance at the King Faiçal Hospital, these newly trained individuals formed specialized teams, and began to diagnose and treat PID patients. Series of PID patients were soon published, examples being the first report on primary antibody deficiency in Arabs (29) and the results from sophisticated projects, such as preimplantation diagnosis of ataxia telangiectasia (30). In Iran and Kuwait, newly established PID teams also began to publish, and national registries were implemented (31–33).

Third Step: Development of PID Centers and Registries

In recent years, the PID field greatly advanced throughout the Middle East. Keeping patient care in mind, highly specialized centers were established, aiming for accurate diagnosis and adequate management of PID patients. Series on the outcome of HSCT, or other therapeutic interventions, were published (34–37). Given the high rate of consanguinity in these countries, it was not surprising to discover unique phenotypes of autosomal recessive disorders in the region. Indeed, several unique PIDs were first described in the Middle East, examples being DOCK8 deficiency, CD40 deficiency and IRAK-4 deficiency (38–40). Interestingly, small countries suddenly began to publish their series of PID patients. However, only Iran and Kuwait have established national registries, which they have used to publish several reports (41–45). Recently, two single-center series were published providing epidemiological PID profiles for Qatar and Oman (46–47).

The history of PID in Iran is of special interest and can serve as an example for the Middle East region. Again, three periods can be distinguished (48). The first period began in 1978 with the establishment of the Division of Clinical Immunology and Allergy at the Children's

Medical Center in Teheran by Professor Abolhassan Farhoudi, who trained in the UK. The second period lasted from 1988 to 1997, with the establishment of a training program in pediatric allergy and immunology, allowing extension of clinics for PID patients and the establishment of a unit dedicated to IVIG infusions. In 1997, a group of students and junior and senior faculty members established the Iranian Primary Immunodeficiency Registry (IPIDR) (33, 41). National and international collaborations led to a large number of publications (48). Finally, the Iranian Primary Immunodeficiency Association (IPIA) and the Immunology, Asthma and Allergy Research Institute (IAARI) were established, resulting in improved care of patients with PIDs. The first International Congress of Immunodeficiencies was organized in Tehran in early 2005.

PID in the Middle East: Concluding Remarks

Due to the high degree of consanguinity in the Middle East, autosomal recessive PID conditions are much more common than in other parts of the world. While, until the late 1980s, PID was almost unrecognized in most countries, the field has increased tremendously in recent years. There are now qualified physicians in most countries, and laboratory techniques available that help diagnose and treat immunodeficiency conditions. Centers for HSCT now exist in many places. Still more work is needed to increase the awareness of PID in the Middle East and to help provide early and adequate therapy.

HISTORY OF PRIMARY IMMUNODEFICIENCY IN LATIN AMERICA

The recognition of PID in Latin America began in the 1980s, when the AIDS epidemic escalated into a public health issue, and physicians became aware of immunological disorders.

Two of the early pioneers in this field were Charles Naspitz from São Paulo and Renato Berron from Mexico City, both physicians interested in allergy, rheumatology and immunology, who, as early as the 1980s, trained the first generation of clinical immunologists in their countries. During the third Asociación Latino Americana de Inmunología (ALAI) meeting in Santiago, Chile, a group with interest in PID, including Marta Zelaszko from Argentina, Magda Carneiro-Sampaio from Brazil, Monika Cornejo from Chile, Diana Garcia-Olarte from Columbia and Ricardo Sorensen from the University of Louisiana, USA discussed the formation of an organization with a focus on PID.

Efforts to Organize: The Formation of LAGID and LASID

The small inaugural group that formed in Chile elected a "coordinator", Ricardo Sorensen (of Chilean origin), then at the University of Louisiana, and a secretary, Lilly Leiva of El Salvador origin and also working at the University of Louisiana. The mission of this Latin American Group of Immune Deficiency (LAGID) was to promote awareness of PID, attract doctors and other health professionals to the field, promote physician and patient education, improve PID management, stimulate clinical and basic research, start a registry and support the creation of patient associations. LAGID grew rapidly and eventually included 14 countries, helped to establish patient associations, and created an online registry. A number of scientific publications resulted from these efforts. In 2009, LAGID became a society (LASID) with its own website (www.lasid.org), rapidly expanding throughout Latin America. As a result, a second generation of Latin American doctors was created, cutting edge research became possible and the management of PID patients improved.

The formation of LASID allowed the countries of South and Central America to focus on several important aspects.

Educational Activities and Awareness Campaign

LAGID organized yearly meetings, attendance at which increased from its inception until 2009, when it became a professional society, LASID (Latin American Society for Immunodeficiencies). A major aim of LASID is to promote biannual meetings with a focus on education and diagnoses and treatment of PID. To ensure state of the art programming and to initiate scientific collaboration, faculty from Latin America, the USA (Clinical Immunology Society [CIS]) and Europe (ESID) are regularly invited to these meetings.

These initiatives led to an increase in the quality and quantity of scientific publications by Latin American scientists who, together with their North American and European colleagues, profoundly influenced research and the diagnostic capabilities of Latin American PID centers. These interactions provided new opportunities for the second generation of Latin American physicians and investigators to join advanced PID centers in the USA and Europe. Most importantly, the centers with expertise in PID that had developed in Brazil, Argentina, Colombia, Chile, Mexico and Costa Rica now strongly interact among themselves and have begun to attract doctors and investigators from other Latin American countries for training in the field of PID.

The attendance at the LAGID/LASID meetings grew from an average of 150 to 600 people at recent meetings. Starting in 2009, junior members of LASID were encouraged to present their work. Simultaneously, the number of younger North American and European investigators attending the LASID meetings has increased, resulting in scientific interactions with Latin American physicians and scientists. Interestingly, it was noticed that young European investigators participating in these meetings elected to become PhD students or post-doctorate researchers in Latin American laboratories under the supervision of senior Latin American investigators.

To broaden awareness of PID, LASID members have organized local educational courses in their own countries, and collaborated with sister societies representing pediatrics, infectious disease, or allergy/immunology. A Summer School program structured like the courses offered by ESID and CIS started in 2006 and has been offered biannually. Because of widespread interest in the program, LASID, in addition to students and fellows, opened the sessions to senior physicians expressing interest in PID. These educational activities have been supported by regional research agencies such as Fundação de Amparo a Pesquisa do Estado de Sao Paulo (www.fapesp.br), Conselho Nacional de Desenvolvimento Científico e Tecnológico (www.cnpq.br), Consejo Nacional de Investigaciones Científicas y Técnicas, Administrativo de Ciencia Tecnologia e Innovacíon (http://www.colciencias.gov.co), Fundación Mexicana para Niños y Niñas con Inmunodeficiencia Primaria (http://fumeni.org.mx) and by the Jeffrey Modell Foundation and the pharmaceutical industry. The LASID Fellowship program, started in 2011, is supported by Baxter Bioscience and provides funding for five fellows biannually.

Thus, LAGID/LASID developed a successful educational and scientific strategy throughout the continent, attracted many physicians and allied professionals and promoted collaboration between Latin American physician scientists and their colleagues in Europe and North America. The results of this strategy have translated into better patient care, an increase in the diagnoses of PID patients and the generation of high quality scientific publications.

Progress in Diagnosis and Treatment

In recent years, diagnostic and treatment centers have been established at public and private medical schools and hospitals in Brazil, Argentina, Chile, Colombia, Mexico and Costa Rica offering immunological testing, including biochemical and flow cytometric assays, as well as molecular/genetic diagnoses. The costs of these tests are covered by government, private health insurance companies or agencies providing research funding, depending on each country's regulations.

Intravenous or subcutaneous immunoglobulin therapy is available in several countries (Brazil, Argentina, Chile, Colombia, Mexico, Costa Rica, Peru and Ecuador), paid for by government or private insurance companies. However, little information as to the availability of immunoglobulin therapy is available for other Latin American countries. Bone marrow transplantation is performed in centers located in Brazil, Chile, Argentina and Mexico. However, the number of hospital beds necessary to support the growing number of PID patients in need of hematopoietic stem cell transplantation is insufficient and the capacity and quality of bone marrow and cord blood banks vary considerably. LASID supports the introduction of gene therapy for certain immune defects in the hope of solving the problem of insufficient bone marrow and cord blood banks.

PID Registry

LAGID members started to register patients in 1993. An online LASID Registry using the ESID software was activated in 2009 in São Paulo, designed to support clinical-epidemiological research. The first LAGID-supported report was published in 1998 and included 1428 patients from Argentina, Brazil, Chile, Colombia, Costa Rica, Mexico, Paraguay and Uruguay (49), followed by a second publication in 2007 documenting 3321 cases from 12 Latin American countries (50). As reported by other registries, the most frequent conditions observed in Latin America are antibody deficiencies. By April 2013, the Registry had collected 4239 cases from 11 countries and 78 participating centers. A number of Registry-inspired manuscripts are currently in progress, including the collection of complications caused by immunizing PID patients with BCG. Thus, the LASID Registry has become a central program, stimulating scientific exchange among various Latin American centers, providing data regarding the epidemiology and molecular defects of PID that are unique to the continent (50–54).

Newborn Screening

As demonstrated in the USA, newborn screening (NBS) for SCID patients is highly effective, and is supported by LASID. So far, NBS has been initiated in Mexico and Brazil. In Mexico, NBS was implemented in the State of Vera Cruz with government support, and coordinated by the Fundación Mexicana para Niños y Niñas con Inmunodeficiencia Primaria. In Brazil, a pilot program was started with the support of Fundação de Amparo a Pesquisa do Estado de São Paulo and the Ministry of Health, involving the Institute of Biomedical Sciences at the University of São Paulo, Escola Paulista de Medicina at the Federal University of São Paulo, and the Associação dos Pais e Amigos dos Excepcionais (APAE-SP), an institution that performs newborn screening in São Paulo, Brazil on over 200 000 babies per year.

Interaction of LASID with International Organizations

Solid relations with ESID and CIS have been fundamental for the development of LAGID/LASID. Members from these sister societies have supported Latin American physicians/scientists in organizing educational events, developing collaborative research and training the up and coming LASID generation. These efforts have succeeded in improving diagnostic capabilities, boosted the number and quality of scientific publications that have uncovered novel PID phenotypes/genotypes, and elucidated disease mechanisms. Interaction with pediatric, infectious disease and allergy/clinical immunology societies from individual Latin American countries, and with

ALAI have helped enormously to disseminate awareness of PID and to start scientific interactions. The IUIS also includes Latin American Representatives on its Expert Committee for PID.

The Jeffrey Modell Foundation (JMF) has been a major partner of LASID, establishing the first JMF center in Latin America at Escola Paulista de Medicina, Federal University of São Paulo in 2007, followed by additional JMF centers in Mexico, Columbia, Argentina and Chile. IPOPI includes Latin American representatives and has helped develop local patient associations.

Concluding Remarks

While most Latin American countries have banded together in LASID and have made their own contributions to the field of PID, they are facing challenges, some being unique to the continent, some identical to those troubling the rest of the world. The level of education, the extent of funding for health care and for biological research and the prevalence of infectious disease varies from one Latin American country to another and within regions of a given state. The high mortality rate from infections affecting the general population in some regions of Latin America makes it difficult to suspect a diagnosis of primary immunodeficiency. The universal use of BCG to provide protection from tuberculosis has created major problems for patients with a T-cell deficiency or a neutrophil-killing defect. Although polio vaccine using mitigated live virus has been associated with paralytic polio in patients with antibody deficiency, live polio vaccine is still used in some parts of Latin America. Awareness of PID, although improved in recent years, is still marginal in many regions. Throughout Latin America, the number of diagnostic centers is often insufficient and the availability of optimal therapies limited by financial considerations. There is, however, remarkable progress in the awareness, diagnosis and treatment of PID. Pilot programs for screening newborns for SCID are currently in progress in Brazil and Mexico. Centers providing medical care for PID patients have been developed in Brazil, Argentina, Mexico, Columbia, Chile and Costa Rica, and competitive translational and basic research in this field is being performed at an increasing number of universities and medical schools. On the other hand, there are countries in South and Central America which lack the financial means and political will to establish the educational, diagnostic and therapeutic standards set by the developed world. Only when these goals are accomplished can we expect local governments to comprehend the benefits of sophisticated diagnostic strategies, newborn screening, hematopoietic stem cell transplantation and gene therapy.

References

1. Haddad HL, Beatty DW, Dowdle EB. Chronic granulomatous disease of childhood. *S Afr Med J* 1976;**50**: 2068–72.
2. Bejaoui M, Barbouche MR, Sassi A, et al. Primary immunodeficiency in Tunisia: study of 152 cases. *Arch Pediatr* 1997;**4**:827–31.
3. Bousfiha AA, Darif M, Abid A. Les déficits immunitaires primitifs : une pathologie à évoquer. *Espérance Méd* 1997;**4**:421–4.
4. Bousfiha AA, Debre M, Abid A. Utilisation des immunoglobulines humaines chez l'enfant. *Mar Méd* 2000;**22**:669.
5. Bousfiha AA, Abid A, Benslimane A. Primary immunodeficiency diseases in Casablanca. *Acta Pediatr Espan* 2002;**60**:429.
6. Bousfiha AA, Ailal F, Abid F, et al. Les déficits immunitaires primitifs au Maroc : A propos de 73 cas. *Rev Mar Mal Enf* 2003;**1**:19–24.
7. Bousfiha AA, et al. Evoquer, explorer et diagnostiquer un déficit immunitaire primitif au Maroc. *Rev Mar Mal Enf* 2005;**5**:64–70.
8. Naamane H, El Maataoui O, Ailal F, et al. The 752 delG26 mutation in the RFXANK gene associated with major histocompatibility complex class II deficiency: evidence for a founder effect in the Moroccan population. *Eur J Pediatr* 2010;**169**:1069–74.
9. Jeddane L, Ailal F, Dubois-d'Enghien C, et al. Molecular defects in Moroccan patients with ataxia-telangiectasia. *Neuromol Med* 2013;**15**(2):288–94.
10. Driss N, Ben-Mustapha I, Mellouli F, et al. High susceptibility for enterovirus infection and virus excretion features in Tunisian patients with primary immunodeficiencies. *Clin Vaccine Immunol* 2012;**19**:1684–9.
11. Ben-Mustapha I, Ben-Farhat K, Guirat-Dhouib N, et al. Clinical, immunological and genetic findings of a large Tunisian series of major histocompatibility complex class II deficiency patients. *J Clin Immunol* 2013;**33**(4):865–70.
12. Naidoo R, Ungerer L, Cooper M, et al. Primary immunodeficiencies: a 27-year review at a tertiary paediatric hospital in Cape Town, South Africa. *J Clin Immunol* 2011;**31**:99–105.
13. Admou B, Haouach K, Ailal F, et al. Primary immunodeficiencies: Diagnosis approach in emergent countries. *Immunol Biol Spec* 2010;**25**:257–65.
14. Barbouche MR, Galal N, Ben-Mustapha I, et al. Primary immunodeficiencies in highly consanguineous North African populations. *Ann NY Acad Sci* 2011;**1238**:42–52.
15. Mellouli F, Torjmen L, Ksouri H, et al. Bone marrow transplantation without conditioning regimen in Omenn syndrome: a case report. *Pediatr Transplant* 2007;**11**:922–6.
16. Raziuddin S, Bilal N, Benjamin B. Transient T-cell abnormality in a selective IgM-immunodeficient patient with Brucella infection. *Clin Immunol Immunopathol* 1988;**46**:360–7.
17. Raziuddin S, Danial BH. OKT4+ T cell deficiency and an association of immunoglobulin deficiency in systemic lupus erythematosus. *Clin Exp Immunol* 1988;**72**:446–9.
18. Raziuddin S, Elawad ME, Benjamin B. T-cell abnormalities in antibody deficiency syndromes. *Scand J Immunol* 1989;**30**:419–24.
19. Hugosson C, Harfi H. Disseminated BCG-osteomyelitis in congenital immunodeficiency. *Pediatr Radiol* 1991;**21**: 384–5.
20. Brismar J, Harfi HA. Partial albinism with immunodeficiency: a rare syndrome with prominent posterior fossa white matter changes. *Am J Neuroradiol* 1992;**13**:387–93.
21. Harfi HA, Malik SA. Chediak-Higashi syndrome: clinical, hematologic, and immunologic improvement after splenectomy. *Ann Allergy* 1992;**69**:147–50.
22. Joshi RK, al Asiri RH, Haleem A, et al. Cutaneous granuloma with ataxia telangiectasia – a case report and review of literature. *Clin Exp Dermatol* 1993;**18**:458–61.
23. Knox-Macaulay HH, Bashawri L, Davies KE. X linked recessive thrombocytopenia. *J Med Genet* 1993;**30**:968–9.
24. Osundwa VM, Dawod ST. The occurrence of ataxia-telangiectasia and common variable immunodeficiency in siblings: case report. *Ann Trop Paediatr* 1994;**14**:71–3.
25. Mohammed SH, Vyas H. Chronic granulomatous disease with renal stones. *Pediatr Radiol* 1992;**22**:596–7.
26. el Noor IB, Venugopalan P, Johnston WJ, Froude JR. Ventricular aneurysm and myocarditis in a child with the hyperimmunoglobulin E syndrome. *Eur Heart J* 1995;**16**:714–5.
27. Shome DK, Al-Mukharraq H, Mahdi N, et al. Clinicopathological aspects of Chediak-Higashi syndrome in the accelerated phase. *Saudi Med J* 2002;**23**:464–6.

28. Abu Jawdeh L, Haidar R, Bitar F, et al. Aspergillus vertebral osteomyelitis in a child with a primary monocyte killing defect: response to GM-CSF therapy. *J Infect* 2000;**41**:97–100.

29. al-Attas RA, Rahi AH. Primary antibody deficiency in Arabs: first report from eastern Saudi Arabia. *J Clin Immunol* 1998;**18**:368–71.

30. Hellani A, Laugé A, Ozand P, et al. Pregnancy after pre-implantation genetic diagnosis for ataxia telangiectasia. *Mol Hum Reprod* 2002;**8**:785–8.

31. White AG, Raju KT, Abouna GM. A six-year experience with recurrent infection and immunodeficiency in children in Kuwait. *J Clin Lab Immunol* 1988;**26**:97–101.

32. Al-Herz W. Primary immunodeficiency disorders in Kuwait: first report from Kuwait National Primary Immunodeficiency Registry (2004-2006). *J Clin Immunol* 2008;**28**:186–93.

33. Aghamohammadi A, Moein M, Farhoudi A, et al. Primary immunodeficiency in Iran: first report of the National Registry of PID in Children and Adults. *J Clin Immunol* 2002;**22**:375–80.

34. Geha RS, Malakian A, LeFranc G, et al. Immunologic reconstitution in severe combined immunodeficiency following transplantation with parental bone marrow. *Pediatrics* 1976;**58**:451–5.

35. Abu Jawdeh L, Haidar R, Bitar F, et al. Aspergillus vertebral osteomyelitis in a child with a primary monocyte killing defect: response to GM-CSF therapy. *J Infect* 2000;**41**:97–100.

36. Salehi T, Fazlollahi MR, Maddah M, et al. Prevention and control of infections in patients with severe congenital neutropenia; a follow up study. *Iran J Allergy Asthma Immunol* 2012;**11**:51–6.

37. Al-Wahadneh AM, Khriesat IA, Kuda EH. Adverse reactions of intravenous immunoglobulin. *Saudi Med J* 2000;**21**:953–6.

38. Engelhardt KR, McGhee S, Winkler S, et al. Large deletions and point mutations involving DOCK8 in the autosomal recessive form of the hyper-IgE syndrome. *J Allergy Clin Immunol* 2009;**124**:1289.

39. Ferrari S, Giliani S, Insalaco A, et al. Mutations of CD40 gene cause an autosomal recessive form of immunodeficiency with hyper IgM. *Proc Natl Acad Sci USA* 2001;**98**:12614–9.

40. Picard C, Puel A, Bonnet M, et al. Pyogenic bacterial infections in humans with IRAK-4 deficiency. *Science* 2003;**299**:2076.

41. Rezaei N, Pourpak Z, Aghamohammadi A, et al. Consanguinity in primary immunodeficiency disorders; the report from Iranian Primary Immunodeficiency Registry. *Am J Reprod Immunol* 2006;**56**:145–51.

42. Moin M, Aghamohammadi A, Kouhi A, et al. Ataxia-telangiectasia in Iran: clinical and laboratory features of 104 patients. *Pediatr Neurol* 2007;**37**:21–8.

43. Rezaei N, Mohammadinejad P, Aghamohammadi A. The demographics of primary immunodeficiency diseases across the unique ethnic groups in Iran, and approaches to diagnosis and treatment. *Ann NY Acad Sci* 2011;**1238**:24–32.

44. Al-Herz W, Moussa MA. Survival and predictors of death among primary immunodeficient patients: a registry-based study. *J Clin Immunol* 2012;**32**:467–73.

45. Al-Herz W, Naguib KK, Notarangelo LD, et al. Parental consanguinity and the risk of primary immunodeficiency disorders: Report from the Kuwait National Primary Immunodeficiency Disorders Registry. *Int Arch Allergy Immunol* 2011;**154**:76–80.

46. Ehlayel MS, Bener A, Laban MA. Primary immunodeficiency diseases in children: 15-year experience in a tertiary care medical center in Qatar. *J Clin Immunol* 2012;**33**(2):317–24.

47. Al-Tamemi S, Elnour I, Dennison D. Primary immunodeficiency diseases in Oman: five year's experience at Sultan Qaboos University Hospital. *World Allergy Organ J* 2012;**5**:52–6.

48. Aghamohammadi A, Moin M, Rezaei N. History of primary immunodeficiency diseases in Iran. *Iran J Pediatr* 2010;**20**:16–34.

49. Zelazko M, Carneiro-Sampaio M, Cornejo de Luigi M, et al. Primary immunodeficiency diseases in Latin America: first report from eight countries participating in the LAGID. Latin American Group for Primary Immunodeficiency Diseases. *J Clin Immunol* 1998;**18**:161–6.

50. Leiva LE, Zelazco M, Oleastro M, et al. Primary immunodeficiency diseases in Latin America: the second report of the LAGID registry. *J Clin Immunol* 2007;**27**:101–8.

51. Condino-Neto A, Franco JL, Espinosa-Rosales FJ, et al. Advancing the management of primary immunodeficiency diseases in Latin America: Latin American Society for Immunodeficiencies (LASID) initiatives. *Allergol Immunopathol (Madr)* 2012;**40**:187–93.

52. Errante PR, Franco JL, Espinosa-Rosales FJ, Sorensen R, Condino-Neto A. Advances in primary immunodeficiency diseases in Latin America: epidemiology, research, and perspectives. *Ann NY Acad Sci* 2012;**1250**:62–72.

53. Leiva LE, Bezrodnik L, Oleastro M, et al. Primary immunodeficiency diseases in Latin America: proceedings of the Second Latin American Society for Immunodeficiencies (LASID) Advisory Board. *Allergol Immunopathol (Madr)* 2011;**39**:106–10.

54. Condino-Neto A, Franco JL, Trujillo-Vargas C, et al. Critical issues and needs in management of primary immunodeficiency diseases in Latin America. *Allergol Immunopathol (Madr)* 2011;**39**:45–51.

7

Jeffrey Asked Us to "Do Something"! Our Journey

Vicki Modell and Fred Modell

Jeffrey Modell Foundation

Vicki and Fred Modell with the picture of their son Jeffery. This figure is reproduced in color in the color section.

On September 25th, 1970 our only son, Jeffrey, was born. He came into our lives very wanted and very cherished. How could we know that he would only be with us for 15 years? Or how he would change our lives forever, along with the lives of thousands of people he would never meet?

Jeffrey's first ten months were peaceful and happy. He plunged into life with joy, and quickly

Primary Immunodeficiency Disorders: A Historic and Scientific Perspective

developed the cheerful, outgoing personality that everyone remembers about him.

But then things started to go very, very wrong. Just before his first birthday, Jeffrey developed a hepatitis-like condition and was hospitalized with high fevers, jaundice and an enlarged spleen. The doctors ordered extensive tests. There was an unbearable wait, and then the diagnosis, hypogammaglobulinemia.

In an instant our world was shattered. On the outside, Jeffrey was so beautiful and looked so normal. But inside, his immune system, his life line, was seriously flawed. The doctors explained the medical aspects of his illness, and told us the prognosis for his condition was uncertain.

This alone would have thrown our lives into turmoil. But there were the added anxieties caused by limited medical knowledge at that time. After all, who had ever heard of primary immunodeficiency? What were the causes? What effects would it have on Jeffrey, and our family? How could we explain to him, a child, why he kept getting ill? We could barely understand it ourselves. We sought the best medical opinions and he had the best medical care, but science had not caught up to Jeffrey.

And the lack of knowledge was only part of the problem. In the whole time Jeffrey was ill, we never met another family and he never met another child living with primary immunodeficiency. Not one! Virtually every other illness had support and information networks that anyone could access. Primary immunodeficiency had none.

And so life went on. Jeffrey had the recurring infections typical of primary immunodeficiency, and our lives became a whirlwind of hospitals and doctors' offices. We saw hepatologists, hematologists, immunologists, rheumatologists, oncologists and just about every other "ologist" that exists. Every succeeding illness disrupted his young life. There were so many lonely nights in the hospital and so many lonely days when he was sick at home.

There were some treatments available, but they brought problems of their own. At three or four years of age, Jeffrey had to endure painful, intramuscular injections of gamma globulin given with huge needles that left him sore for days. By the time he was ten or 11, he was started on intravenous immunoglobulin (IVIG) and his first treatment was given to him by Dr Max Cooper in Birmingham, Alabama. He was then enrolled into the first clinical trial for IVIG at Memorial Sloan Kettering, in the care of Dr Robert Good. These treatments sometimes improved his life for weeks at a time. But time will never erase the memories of the infusions with their many side effects, including chills, fever, headaches and nausea.

Through it all, Jeffrey managed to thrive. He reached normal height and weight and kept up in school, camp and sports. But throughout the years, Jeffrey's problems never went away.

Courage and humor were Jeffrey's great assets. His determination was inspiring. But the time bomb kept ticking inside him. One story sums up this whole middle part of his short life.

One day, Jeffrey came home from school with a permission slip, bubbling with excitement. His eighth grade class was planning a three-day "whaling expedition" on Cape Cod in April. Jeffrey told us to "just sign the slip", that he desperately wanted to go, that it meant everything in the world to him. He was old enough and smart enough to know what the consequences might be, but he said "it was worth it".

I could only imagine how cold and damp it is on Cape Cod in April – the rustic, unheated cabins they would sleep in, the wet clothing from the whaling boat. I knew I would not have control of what he wore, or if he would be careful. After all, he was a teenager and wanted to be just like the rest of his friends.

Fred and I carefully weighed the pros and cons. We knew he loved the sea, boating and being with his friends. With great trepidation, we signed the permission slip and crossed our fingers. I thought about asking the pediatrician his opinion, but I knew what he'd say, so I didn't. Off Jeffrey went.

Jeffrey called home several times to say he was having fun and feeling great. He was obviously the greatest actor of all time – an Oscar winning role. Three days later, the bus rolled in and he came off laughing and excited, dirty and grimy, with bright blue eyes and shiny red cheeks. I knew those cheeks were a tell-tale sign of fever. Sure enough – 104°F (40°C).

That trip, those three or four days, were the most memorable, enjoyable, fulfilling days of his life; something he could share with his friends. Several of the kids landed up with a cold, but Jeffrey landed up in the hospital. He told the nurses and doctors, who had come to be his best friends, about his trip – the whale sightings, the girls. He laughed off the fact that he was now paying for it.

Jeffrey survived and bounced back. But his doctors, who compared him to a cat with nine lives, told us he had already used up seven or eight. One day, without warning, he developed hepatitis, and a few months later, pneumocystis pneumonia. He had two more life-threatening bouts with pneumocystis, and the third one, sadly, took his precious life in 1986.

Throughout his life, Jeffrey would say to us, "Do Something! You went to college. You're smart!" And to his doctors, he would say, "Do Something. You're smart. You went to medical school!" How heartbreaking it was to hear your child say that. Little did he know, his doctors and his parents did everything they could, but sadly, science had not caught up to Jeffrey.

When he died, he took with him our fear, our dread, our loneliness, our despair. But we could not let his life be in vain. We had to "Do Something" to fulfill his unanswered prayers.

Jeffrey never knew how he would change our lives forever, and the lives of so many thousands of children he would never meet. But in the spirit of his optimism and courage, we created the Jeffrey Modell Foundation – not in memory of his death, but in celebration of his life, and to give life.

Throughout his lifetime, we were privileged to have consulted with four of the most important immunologists of the 20th century: Max Cooper, Bob Good, Fred Rosen and Walter Hitzig. Each of them spent countless hours, days and weeks trying to unravel the mysteries of Jeffrey's condition. We and Jeffrey were so fortunate to have met the most pre-eminent immunologists of the time – each of whom made a profound impact upon the diagnosis, treatment and understanding of primary immunodeficiency – each of whom contributed to keeping Jeffrey alive for 15 years. And when we thought about establishing the Jeffrey Modell Foundation, those four giants, together with their colleagues, set an agenda for this new entity that has carried us to this day.

The Jeffrey Modell Foundation is a public charity devoted to early and precise diagnosis, meaningful treatments, and ultimately, cures – through clinical and basic research, physician education, patient support, advocacy, public awareness and newborn screening.

Our first project, in the early 1990s, was to develop the "Ten Warning Signs of Primary Immunodeficiency" with the American Red Cross and our newly established Medical Advisory Board. Since then, the "Ten Warning Signs" have been translated into more than 40 languages and have been distributed throughout the world. In 2010, the Warning Signs were slightly revised, as the medical community developed greater knowledge of primary immunodeficiencies.

To date, we have named and funded 125 Jeffrey Modell Diagnostic and Research Centers, and today, the Jeffrey Modell Centers Network includes 518 expert physicians, at 196 academic teaching hospitals, serving 191 cities, 68 countries and spanning six continents. Remarkably, more than 44 new genes have been discovered at Jeffrey Modell Centers in the past five years.

The Jeffrey Modell Foundation (JMF) has organized more than 40 scientific symposia worldwide. In celebration of its 25th year, the Foundation hosted a World Immunology Conference "A

Global Get-Together", attended by 300 researchers and clinicians from 62 countries.

The Foundation has partnered with government agencies, including a multimillion dollar research collaboration with the National Institutes of Health. For the past seven years, JMF has collaborated with the US Centers for Disease Control and Prevention to conduct a Public Awareness and Physician Education Program, which has resulted in a ten-fold increase in the number of patients identified, referred, diagnosed and treated. Public Service Advertising appears on television, radio, magazines, billboards, airports, malls and bus shelters throughout the USA. Donated media have reached over $250 million.

Patients worldwide are supported by JMF with Kids Days and through our World Immunodeficiency Network (WIN); physicians and nurses receive travel grant support to attend important conferences and meetings through winMD and winRN.

Over the past decade, we have appeared at the United States Congress to advocate for newborn screening for severe combined immune deficiency (SCID). In 2010, the US Secretary of Health and Human Services recommended SCID to the National Core Panel, the first condition to be added in the last 12 years. JMF has implemented and funded programs to screen for SCID in many states. Hundreds of newborn babies with life-threatening conditions have been screened, diagnosed, transplanted and cured!

To date, the Foundation has funded 25 postdoctoral fellowships and four Jeffrey Modell Endowed Chairs in Pediatric Immunology Research. In addition, we established the Jeffrey Modell Immunology Center, a newly constructed building, at the Harvard Medical School.

Recently, the Bill and Melinda Gates Foundation announced a collaboration with JMF, in which the two foundations agreed to work together to eradicate polio worldwide.

Twenty-six years after its creation, the Jeffrey Modell Foundation continues its mission of hope, advocacy and action by vigorously supporting physicians, researchers and the ever expanding global patient community – directing its efforts toward early diagnosis, meaningful treatments, equal access to care and, ultimately, cures for primary immunodeficiency. During our journey, we have had the privilege of working alongside many courageous and inspiring individuals, who have also created patient advocacy organizations in honor of or in memory of a child...specifically, the Immune Deficiency Foundation (IDF), SCID Angels for Life, Wiskott–Aldrich syndrome (WAS) Foundation, Chronic Granulomatous Disease (CGD) Foundation, the International Patient Organization for Primary Immunodeficiency (IPOPI), as well as all of the other dedicated patient organizations throughout the world. Together and united, we have shaped one strong message and one strong voice that will resonate for generations to come.

As we look back, it was Jeffrey who created the passion for our work. In a small way, we hope we have been facilitators and perhaps the "glue" for advancing science and health policy in all regions of the world.

By supporting scientific forums, by funding research, by supporting professors and researchers, by providing funds that enable the best and brightest to choose immunology as a life-long career, we hope to have been a critical catalyst for the increased pace of scientific discovery, and we hope that we have influenced and caused positive change for health policy around the world. Whether it's funding for newborn screening or support for pediatric immunology, we have advocated before health departments and health ministries across the continents.

We have seen great success throughout the course of our journey, but we still have miles to go. None of our success could have happened without the brilliance and compassion of the doctors, the tender, loving expert care of the nurses, the resolve and dedication of the researchers, the bravery and courage of the families, the vision and constant support of our donors, the perseverance of the quiet, unsung heroes working days and nights in their labs, and most of all, the patients who inspire us, propel us, and compel us to continue our incredible journey.

We believe that all of us together – *we can, we will, and we are* – making a difference. We can meet the challenges, capture the opportunities, and give all children for generations to come a brighter, healthier future.

Yes, Jeffrey, you set us on our journey alone, but indeed we are "Doing Something".

You taught us that to succeed in life you need three things: a wish bone, a back bone and a funny bone. The wish bone is for the ideals, the goals, and the dreams; the back bone is for the fortitude and the courage to pursue them; and the funny bone is for the laughter and tears that are necessary for the little bumps along the road.

We wish everyone a lifetime filled with more laughter than tears, as we all continue to pave new paths, brave new worlds, and walk new roads along this extraordinary journey…

Jeffrey Modell
1970-1986

If I can stop one heart from breaking
I shall not live in vain,
If I can ease one life the aching,
Or cool one pain,
Or help one fainting robin unto his nest again,
I shall not live in vain.

-Emily Dickinson

Finally Found: The Ataxia-Telangiectasia Gene and its Function

Amos Etzioni[1], Hans D. Ochs[2], Deborah McCurdy[3], Richard A. Gatti[4]

[1]The Rappaport Medical School, Technion, Haifa, Israel
[2]University of Washington School of Medicine, Seattle, Washington, USA
[3]Department of Pediatrics, David Geffen School of Medicine at UCLA, Los Angeles, California, USA
[4]Departments of Pathology and Laboratory Medicine, and Human Genetics, David Geffen School of Medicine at UCLA, Los Angeles, CA, USA

OUTLINE

DEFINING THE CLINICAL CHARACTERISTICS OF A-T, A SLOW EVOLUTION DRIVEN BY SERENDIPITY AND METHODOLOGY

Based on its characteristic clinical phenotype when fully expressed, ataxia telangiectasia (A-T) can be diagnosed without hesitation by an experienced clinician. The syndrome was reported in 1926 by two Czech neurologists, Ladislav Syllaba and Kamil Henner (Fig. 8.1), who described three siblings with progressive choreo-athetosis ("athetose double") and ocular telangiectasia (Fig. 8.2) (1). In 1968, Henner provided an update (2), indicating that his patients' symptoms were identical (except for pulmonary

Professor Ladislav Syllaba (1920) : 1868 - 1930

Professor Kamil Henner (1965): 1895 - 1967

FIGURE 8.1   Ladislav Syllaba and Kamil Henner, neurologists at Charles University of Prague who first described the syndrome in 1926 (1) as athétose double idiopathique et congénital.

1926

FIGURE 8.2   Ocular telangiectasia described by Syllaba and Henner as "triangular network of hypertrophic/hyperplastic blood vessels of the bulbar conjunctiva." Ink drawing, patient Anezka Ch., Figure 2 in (1).

infections) to those reported in 1957–1958 by Boder and Sedgwick (3, 4). Sadly, as Henner recalls in his 1968 follow-up report "the economic situation of this Czech family had deteriorated and the locomotion of the patients had become almost impossible; after some time, each of the three affected siblings resorted to suicide".

The second clinical report of the syndrome was published in 1941, by Madame Denise Louis-Bar, at that time a resident at the Institute Bunge at Antwerp, Belgium, who described in great detail a 9-year-old boy with a cerebellar syndrome characterized by ataxia and telangiectasia who was cared for by Professor Ludo van Bogaert at the neurological clinic of the same Institute (5). A

follow-up report of the same case was provided in 1966 by Lilly Martin who indicated that, when 15 years of age, the patient's telangiectases had extended, the speech had become guttural and the patient walked with difficulty (6). By 22 years of age, the telangiectases had spread further to the thighs and calves, his speech was almost incomprehensible, and he was unable to feed himself. Three years later, he was bedridden, showed diffuse muscle wasting and extreme hypotonia, but had no pulmonary symptoms and no evidence of lung disease. He died at age 27 of "cardiac syncope". This retro-analysis of the three Czech and one Belgian patient, in the 1960s, did indeed confirm that the diagnoses were A-T.

At first sight, this slow paced and often halting progress in recognizing such a distinct disease entity is surprising. But it is not unique to A-T. The Wiskott–Aldrich syndrome was first described as a hematological disease in 1937, rediscovered in 1954, and recognized as primary immunodeficiency (PIDD) in 1968 (7–9) (see Chapter 9). There is a common thread: no autopsies were performed in these early reports; most affected children must have died prematurely of infections (which was not an uncommon event in those days) and, on the political front, the world was in upheaval in the first half of the 20th century. In contrast, when Elena Boder and Robert Sedgwick (Fig. 8.3) studied A-T in the early 1950s, medical research was flourishing in the USA, large medical centers had been established, referrals were the norm, and scientific meetings supported the frequent exchange of observations and opinions. The extensive reports published by Boder and Sedgwick (3, 4) and, independently by two other groups, Biemond (10) and Wells and Shy (11), included a total of 14 cases with three autopsies. Boder, perhaps in a philosophical mood when presenting her material at a WHO-sponsored PIDD meeting organized by Robert Good, suggested that the explanation for this halting progress and slow start may have been "partly in the nature of the syndrome itself" (12). She speculated that before the advent of antibiotics, with most A-T children being susceptible to infections, few were likely to live long enough to manifest the full A-T phenotype. This interpretation is supported by the striking clinical distinction between the A-T patients reported in 1957–58, who all suffered from recurrent pulmonary infections (3, 4, 10, 11), and those described earlier (1, 5).

FIGURE 8.3 Elena Boder and Robert P. Sedgwick, neurologists from Children's Hospital, Los Angeles, who described the syndrome as ataxia-telangiectasia in 1957 (3). This figure is reproduced in color in the color section.

UNDERSTANDING THE IMMUNE DEFICIENCY IN A-T

There was one additional explanation for the expanded interest in A-T after the initial recognition of the phenotype by neurologists. With the new methodology of serum electrophoresis, many patients were found to have abnormal immunoglobulin levels (reviewed in 12, 13) and, most importantly, the thymus was noted to be absent in one of the original A-T patients reported by Boder and Sedgwick (3). This caught the interest of the Minnesota group that had formed around Robert Good and which had recently described a two-compartment immune system by studying thymectomized and bursectomized chickens (see Chapter 2). By 1964, Good and his team ("the Good guys") had successfully categorized A-T as a PIDD with a combined cellular and humoral defect, based on clinical and autopsy findings (13, 14). They reported that delayed-type hypersensitivity testing and skin homograft rejection were impaired in most A-T patients studied and that at autopsy, the thymus was either absent or dysplastic/epithelial-stromal, lacking Hassall's corpuscles and thymic lymphocytes (13). The humoral defect was observed by a number of investigators and subsequently characterized as hypogammaglobulinemia or dysgammaglobulinemia (15) with frequent deficiencies of serum IgA and IgG/IgG subclasses (13, 16–19), occasionally with elevated IgM (14, 17–20) and absent antibody responses to recall and neoantigens (13, 14, 21–26).

When summarizing their clinical observations based on the original A-T patients, Boder and Sedgwick reported that, in all their patients, ataxia was the first recognizable symptom, with onset in infancy and usually apparent to the parents when the infant began to walk (27). The ataxia was cerebellar in nature, was initially noticed to affect the gait (or sometimes at rest) and later became associated with intention tremor of movement and slurred speech. In most cases, the telangiectatic lesions appeared during late infancy, around the age of four years, first noticed in the bulbar conjunctivae and often progressing to the face, arms and hands. Sinopulmonary infections and bronchiectasis were progressive and sometimes resulted in clubbing. The reasons for the susceptibility to infection remained puzzling until immunologists became interested in the immunobiology of the pleiotropic syndrome.

After the rediscovery of the syndrome in 1957, a succession of A-T-related publications appeared and by 1963 more than 100 cases had been identified, autosomal recessive inheritance had been established, and lymphoreticular malignancy was recognized as the second most frequent cause of death (14). In 1964, the Good group suggested that the triad of a defective thymus, abnormal humoral immunity and increased incidence of malignancies characteristic of A-T patients may be a critical key to the understanding of the complex host factors that link immune deficiency and malignancy (13). By 1978, a higher incidence of cancer in relatives of A-T patients was established, suggesting that asymptomatic heterozygous individuals are also prone to developing cancer (28–32). Today, with diagnostic confirmation possible by DNA sequencing of the *ATM* gene, other clinical variants of A-T are being described, sometimes in individuals with neither ataxia nor telangiectasia who develop bilateral breast cancer after 50 years of age that may sometimes be complicated by adverse reactions to radiation therapy and chemotherapy and perhaps neurological symptoms appearing after radiation or chemotherapy (32–34). Other A-T variants, such as in Mennonites, present primarily with early-onset dystonia and brain cancer with late-onset ataxia (35). On the other hand, it is sometimes difficult to discern whether some of the dystonias described may not be confounded with the choreoathetosis that is seen in most A-T patients (4).

During the early 1980s, A-T had generated enough interest among immunologists, neurologists, biologists and oncologists to initiate a small international meeting held in Solvang,

California in January 1984. Referees were invited, each experts in their own areas (e.g. David Botstein, Philip Hanawalt, Robert Painter, Pasko Rakic, Winston Salser), and were asked to unbiasedly weigh the data presented and help to frame the future of A-T research. During the four days of discussions, a roadmap was designed for studies aimed at resolving key molecular and genetic interactions. Boder led off the workshop by characterizing A-T as a multisystem disorder that included developmental delay, skin and hair changes and progeric signs in addition to cerebellar symptoms, immune deficiency and malignancies (27). Ten years later, the *ATM* gene had been mapped and cloned, largely as a result of those discussions. With the exception of cerebellar circuitry, each of the milestones set at that workshop was significantly clarified during the next 20 years; the neuroscience has proven the most difficult; hence its omission from this historical review.

While the diagnosis of A-T was initially clinical and based on the manifestation of signs of cerebellar ataxia, telangiectasia, recurrent respiratory infections and increased susceptibility to malignancies, laboratory tests eventually became available to support the diagnosis; these included elevated α-feto-protein levels (36), increased radiation sensitivity (37–40) and chromosomal instability (41). A high rate of chromosomal breaks occurs and these are specific to lymphocytes, not to fibroblasts (41), with translocation breakpoints that involve the sites of B-cell receptor genes (at 2p12; 14q32; 22p11) and T-cell receptor genes (at 7p14; 7q35; 14q11) (41–44). Following identification of the *ATM* gene in 1995, a definitive diagnosis of A-T could be made by: (1) immunoblotting to measure the ATM protein (45); and (2) sequencing the *ATM* gene for disease-causing mutations, but many suspected A-T diagnoses were made by alert cytogeneticists (42).

An explanation for the very frequent dysgammaglobulinemia in A-T was provided after the discovery of *ATM* as the gene responsible for A-T. The Nussenzweig group showed that *ATM* is required for recombination between immunoglobulin switch regions (46) and, more recently, a Dutch group demonstrated that the antibody deficiency observed in A-T patients correlates with abnormal B- and T-cell homeostasis, resulting in reduced immune repertoire diversity (47). This idea had already been proposed ten years earlier by Pan-Hammarström and colleagues (48).

While development of the thymus seemed consistently defective in A-T, only a third of patients displayed lymphopenia (49, 50). Biggar and Good, in a 1975 review of the immunodeficiency in A-T, reported that early in the course of the disease, cellular immunity is only minimally defective (51) and, as the disease progresses, cellular immune defects may appear to be more striking and include general lymphoid hypoplasia and structural abnormalities; however, this was not as described by Nowak-Wegrzyn et al. (49). McFarlin and colleagues noticed variability in the proliferative response to various mitogens even in the same individual, and suggested that a plasma inhibitor could be dampening the response (52). Later studies concurred that the peak response to lymphocyte stimulation by mitogens is delayed from 48 hours to 72 hours in most patients, causing some confusion. Early on, this property of A-T cells often prevented cytogeneticists from harvesting sufficient numbers of lymphoblasts for studying chromosomes in metaphase. Slow colony growth of A-T cells is also manifested in colony survival assays (whether using fibroblasts, lymphoblasts or induced pluripotent stem cells derived from A-T fibroblasts) (38–40, 53).

Treatment is limited to symptomatic measures. Those A-T patients with recurrent pulmonary infections may need prophylactic antibiotics. If antibody deficiency is confirmed, intravenous immunoglobulin (IVIG) replacement therapy should be considered. Physiotherapy may temporarily delay the dependence on a wheelchair. Patients should be advised to avoid X-radiation and chemotherapeutic agents.

Hematopoietic stem cell transplantation is not indicated. Gene therapy for A-T will have to wait until new technologies for gene correction are developed.

DISCOVERY OF THE GENETIC DEFECT IN A-T

Although the family reported by Syllaba and Henner included three siblings, two boys and one girl, and subsequent reports pointed out "a strong familial incidence", it took 40 years to demonstrate convincingly that A-T is indeed an autosomal recessive disease caused by a single gene defect. The problem was that after years of descriptive observations, the A-T phenotype based on clinical and laboratory features had become quite variable and the possibility that A-T could be genetically heterogeneous and consist of more than one genetic disease was considered (54, 55). Without a gene in hand, it was difficult to develop exclusion criteria and, thereby, establish the true incidence of A-T.

In the early 1980s, it was realized that, in the USA, the rate of consanguinity in parents of an individual with A-T was not greater than that of the normal population, and thus the mutant gene may be more widely distributed in the general population than had been previously recognized. It was estimated that at gene equilibrium, heterozygotes may comprise from 1 to 7% of the general population (29). The apparent increase in the incidence of A-T in the post-World War II era was attributed to the use of antibiotics, as previously mentioned, which increased the life expectancy of those A-T patients with chronic lung infections. In 1975, the incidence of A-T in California was estimated to be 2–3 per 100 000 (12), similar to today's estimate of 1/40 000–100 000 worldwide. However, the true incidence figures for A-T are about to be challenged by the newly mandated "Newborn Screening for SCID" programs that are being instituted worldwide, based on T-cell receptor excision

circles (TRECs) in lymphocytes, which have been shown to identify infants with A-T (56). At first glance, these programs appear to be identifying more A-T cases than expected in newly born children from well-defined populations (e.g. per individual state of the USA) and using very stringent criteria for confirming a diagnosis of A-T, such as: (1) absence of ATM protein by immunoblotting (45) and (2) confirmation of *ATM* mutations by genomic sequencing. In rare situations where two disease-causing mutations are not found, cellular radiosensitivity testing is added as a functional assay to confirm a diagnosis in an otherwise asymptomatic infant (39, 40).

In the 1980s, complementation studies were performed by fusing fibroblasts from pairs of A-T patients (called heterodykaryons) and measuring their response to irradiation (57, 58). This approach was based on the hypothesis that a genetic defect in the repair of damaged DNA might be corrected by DNA from the other A-T patient's DNA (59, 60). Using these techniques, at least four complementation groups, designated A, C, D and E, were assigned by several laboratories collaborating with coded fibroblast cultures (57–61).

These complementation groups were not distributed equally among A-T patients, and no correlation was found between clinical phenotypes and complementation groups. The possibilities that the complementation groups represented different AT genes (*ATA, ATC, ATD* and *ATE*) or different mutations within a single gene capable of intragenic complementation were discussed extensively and they vastly complicated the design of genetic linkage studies (62, 63).

Beginning in 1981, a different approach was pursued by Richard Gatti and Ken Lange at UCLA, who were the first to apply a new array of biallelic genetic markers, developed by David Botstein at MIT and Ray White and his team at the University of Utah (64), to search for the genomic location of the A-T gene. After testing 171 markers that covered approximately 30% of the physical map of the human genome – work that extended to 1988 and included a 61-member Amish A-T pedigree and 30 additional A-T families of American and Turkish ancestries – two markers were identified at chromosome 11q22-23, THY1 and pYNB3.12; their alleles co-segregated with the inheritance of A-T (65–67). All eight of the families that had been typed as Complementation Group A and two families that had been typed as Complementation Group C supported a linkage of an A-T gene to the chromosome 11q22-23 region. (No doubt, some of the other 20 families supporting this linkage would also have typed as Group A but those studies were still in progress.) It later became clear that the Nijmegen Breakage Syndrome families did *not* link to this region (68). However, this mapping strategy did not put to rest the possibility that other distinct A-T genes might be identified at other sites in the genome, or be clustered within the 11q22-23 region, a distance of about 8 centiMorgans. Further support for 11q22-23 being the likely location of an A-T gene was provided by the gene transfer experiments of Ejima and colleagues, which showed that introduction of a wild-type human chromosome 11 into SV40-transformed fibroblasts, derived from an A-T patient, restored normal radiation resistance (69).

The next step was to clone at least the one A-T gene within the 11q22-23 region, and this could only be done by increasing the resolution of the mapping markers ("fine mapping"), in preparation for physical cloning. This strategy required many more families. Forty healthy three-generation families were obtained through a collaboration with CEPH (Centre d'Etude du Polymorphisme Humain) (70) and 176 A-T families were contributed over the next seven years through an international consortium organized by Richard Gatti with the laboratories of Malcolm Taylor (UK), Yosef Shiloh (Israel), Oscar Porras (Costa Rica), Ozden Sanal (Turkey), Luciana Chessa (Italy) and others. The data generated by the consortium were able to narrow the localization of the A-T gene to an ≈500 Kb interval of chromosome 11q23.1 (71, 72). This location was gene

TABLE 8.1   International A-T workshops

Sussex, England	November 5–7, 1980
Solvang, California, USA	January 16–20,1984
Newport Beach, California, USA	February 22–24, 1987
Newport Beach, California, USA	May 21–24,1989
Newport Beach, California, USA	May 17–20,1992
Birmingham, England	March 22–25,1994
Clermont-Ferrand, France	November 22–24,1997
Las Vegas, Nevada, USA	February 14–17, 1999
Fraser Island, Queensland, Australia	September 10–14, 2003
Belgirate, Lago Maggiore, Italy	June 8–11, 2005
Banff, Alberta, Canada	September 8–12, 2006
Kyoto, Japan	April 22–26, 2008
Redondo Beach, California, USA	April 11–14, 2010
New Delhi, India	February 7–11, 2012
Birmingham, England	July 28–31, 2013

rich, making the hunting for a specific gene challenging. The intra-consortium competition only ended when, in 1995, the Israeli group identified a gene in which they could identify disease-causing mutations in all families tested, regardless of complementation group (73–75). The gene was designated as *ATM* (*A-T mutated*). Thus, one of the major milestones of the Solvang 1984 Workshop had been realized! Many other workshops would follow (Table 8.1).

ATM is a large gene that codes for a serine/threonine protein kinase (66 exons encoding 3056 amino acids), similar to other protein PI3 kinases that are involved in mitogenic signal transduction, meiotic recombination and cell cycle control (Fig. 8.4). *ATM*'s primary role appears to be that of a chief mobilizer of the nuclear response to double-strand breaks (DSB) of DNA which, if not repaired within minutes, result in severely disrupted DNA replication, leading to the death of replicating cells

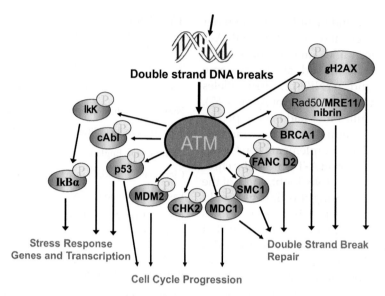

FIGURE 8.4 The multiple functions of ATM in response to double-stranded DNA breaks, including the rapid repair of DNA breaks, and its impact on cell cycle progression, stress response, genes and transcription. This figure is reproduced in color in the color section.

or chromosomal translocations that set the stage for leukemia or cancer (21, 76–82). In response to DSBs, a highly ordered cascade of events occurs, characterized by the recruitment to the site of a set of proteins collectively called "sensors" or "mediators", which form protein complexes (including the BRCA1 "surveillance" complex and the nuclear magnetic resonance (NMR) complex proteins) around DSBs. These complexes attract and activate (phosphorylate) ATM which, as a kinase with an extensive range of downstream substrates to phosphorylate, has a wide range of effects in the recognition, repair and rejoining of broken ends of DNA (83).

While DNA damage repair accounts for most of the characteristic abnormalities seen in A-T patients, it is important to realize that ATM plays many important cytoplasmic roles as well, including the phosphorylation of multiple protein substrates in response to oxidative stress and mitochondrial insufficiency (80). Neurons are highly differentiated post-mitotic cells that, according to conventional wisdom, cannot be

replenished. Being highly active metabolically, neurons generate large amounts of reactive oxygen species, leading to damage of genomic DNA. Lack of ATM may directly result in neuronal cell death, including that of cerebellar Purkinje neurons (84). There is also some evidence to suggest that neurons may re-enter the cell cycle when under extreme stress (85).

ATM MUTATIONS WORLDWIDE

To date, >800 unique mutations in the *ATM* gene have been identified. Most are unique to single families, and the majority of patients in Europe and North America are compound heterozygotes. Founder effects and haplotypes have been described for A-T patients of many ethnic groups, including Costa Rican, Amish, Mennonites, Moroccan Jews, Japanese, British, French, Norwegian, German, Polish, Russian, Brazilian, Spanish, Italian, Turkish, Iranian and American (86–107). Although the mutation frequencies of different groups vary considerably because of founder mutation effects, the worldwide distribution of *ATM* mutations can be approximated as follows: 40% frameshifts (including mainly small nucleotide insertions and deletions); 26% splicing; 21% primary nonsense; 9% missense, 2% in-frame deletions; and 4% other mutations. Genotype–phenotype correlations were examined in a recent study that analyzed a large cohort of 240 patients from France. The Kaplan–Meier 20-year survival revealed a life expectancy that was lower (54.7%) among patients with *ATM* null mutations than that (64.1%) observed in A-T patients with hypomorphic (i.e. mild) mutations, reportedly due to earlier onset of cancer (102).

The A-T-patient mutation spectrum should be contrasted with another very different spectrum of heterozygous *ATM* mutations that has been observed among patients with breast cancer (31, 105, 106). These patients do not manifest A-T; thus, the breast cancer *ATM* mutation spectrum primarily reflects mutations that express a dominant form of disease, in a specific tissue, and most likely by a pathogenesis that differs significantly from that of A-T – a mechanism that almost certainly centers on cerebellar circuitry and the welfare of Purkinje neurons. The *ATM*/breast cancer mutation spectrum differs most notably from the A-T mutation spectrum in that the frequency of *ATM* missense "mutations" in breast cancer patients is very high whereas it is only ≈9% in A-T patients. The frequency of missense *ATM* mutations may be much higher in the general population than the 1% that is commonly cited for A-T carriers – Swift et al. suggested a figure as high as 7% (29).

CONCLUDING REMARKS

After a delayed discovery and little interest from the medical establishment for four decades, A-T was rediscovered in 1957 and quickly caught the interest of neurologists, immunologists, oncologists and geneticists alike. As is often the case, the first phase of investigation was descriptive and explored the clinical phenotypes, the pathologic abnormalities and the immunological defects. The next phase of investigation focused on finding a biological explanation for the clinical triad: immune deficiency, susceptibility to malignancies and cerebellar dysfunction. This effort led to the discovery that A-T fibroblasts were radiation sensitive and prone to developing double-strand DNA breaks. The new concept resulted in the formation of a consortium that organized bi-annual international workshops dedicated to A-T research, identification of A-T pedigrees, shared results of hypothesis-driven experiments and, after a period of trial and error, resulted in the localization and then cloning of the *ATM* gene that was consistently mutated in all A-T families studied. The ATM protein is a very large protein kinase that plays a critical role in the recognition and repair of double-strand DNA breaks, but also participates in the activation of multiple signaling pathways and in mitochondrial function. At present, the treatment of A-T is solely symptomatic and includes immunoglobulin replacement for patients with low IgG levels. Ionizing irradiation should be avoided whenever possible.

Because hematopoietic stem cell transplantation is not an option for A-T patients, future treatment may depend on developing chemical compounds that will correct or abrogate the primary pathogenesis of A-T by manipulating the *ATM* mutations, starting with splicing and nonsense mutations (108, 109). This challenging project, already under way in the Gatti lab, will require specially designed drugs for each unique splicing mutation. For nonsense mutations, a more generic readthrough drug may be sufficient; however, to be fully therapeutic, it will have to cross the blood–brain barrier to reach the cerebellum.

Acknowledgments

We thank the investigators who went before us, our mentors who took such pains to describe clearly the A-T syndrome(s), the A-T families for their unswerving faith in our efforts and patience with the slow pace of discovery, and those among them that formed parent-guided organizations and raised funds for further research. This work was partially supported by APRAT Foundation (Clermont-Ferrand, France). We also thank Dawn Marie Pares for editing and finalizing the manuscript.

References

1. Syllaba L, Henner K. Contribution in l'indépendence de l'athéltose double idiopathique et congénital. *Rev Neurol* 1926;**1**:541–62.
2. Henner K. A propos de la description par MMe Louis-Bar de l'Ataxia teleangiectasia. Priorité de la description, par Lad. Syllaba et K. Henner en 1926, de réseau vasculaire conjonctival. *Soc Franc Neurol* 1968;**118**:60–3.
3. Boder E, Sedgwick RP. Ataxia-telangectasia. A familial syndrome of progressive cerebella ataxia, oculocutaneous telangectasia and frequent pulmonary infection. A preliminary report on 7 children, an autopsy and a case history. *Univ South Calif Med Bull* 1957;**9**:15.
4. Boder E, Sedgwick R. Ataxia-telangectasia. A familial syndrome of progressive cerebella ataxia, oculocutaneous telangectasia and frequent pulmonary infection. *Pediatrics* 1958;**21**:526–54.
5. Louis-Bar D. Sur un syndrome progressif comprenant des télangecasies capillaries cutanées et conjoctivales symétriques, à disposition naevoïde et de trouble cérébelleux. *Confir Neurol* 1961;**4**:32–42.
6. Martin L. The nosological position of ataxiatelangiectasia retrospective study of the first observation. *J Neurol Sci* 1966;**3**:2–9.
7. Wiskott A. Familiärer, angeborener Morbus Werlhofii? *Monatsschr Kinderheilkd* 1937;**68**:212–6.
8. Aldrich RA, Steinberg AG, Campbell DC. Pedigree demonstrating a sex-linked recessive condition characterized by draining ears, eczematoid dermatitis and bloody diarrhea. *Pediatrics* 1954;**13**:133–9.
9. Cooper MD, Chae HP, Lowman JT, Krivit W, Good RA. Wiskott-Aldrich syndrome. An immunologic deficiency disease involving the afferent limb of immunity. *Am J Med* 1968;**44**:499–513.
10. Biemond A. Palaeocerebellar atrophy with bronchiectasis and telangiectasis of the conjunctiva bulbi as a familial syndrome. In: Bogaert Lv, Radermecker J, editors. *1st Intern Congr of Neurological Sciences.* London: Pergamon Press; 1957. p. 206.
11. Wells CE, Shy GM. Progressive familial choreoathetosis with cutaneous telangiectasis. *J Neurol Neurosurg Psychiatr* 1957;**20**:98–104.

12. Boder E. Ataxia-telangiectasia: some historic, clinical and pathologic observations. *Birth Defects Orig Artic Ser* 1975;**11**:255–70.
13. Peterson RD, Kelly WD, Good RA. Ataxia-telangiectasia. its association with a defective thymus, immunologicaldeficiency disease, and malignancy. *Lancet* 1964;**1**: 1189–93.
14. Peterson RD, Cooper MD, Good RA. Lymphoid tissue abnormalities associated with ataxia-telangiectasia. *Am J Med* 1966;**41**:342–59.
15. Gutmann L, Lemli L. Ataxia-telangiectasia associated with hypogammaglobulinemia. *Arch Neurol* 1963;**8**: 318–27.
16. Thieffry S, Arthuis M, Aicardi J, Lyon G. [Ataxiatelangiectasis. (7 personal cases)]. *Rev Neurol (Paris)* 1961;**105**:390–405.
17. Rivat-Peran L, Buriot D, Salier JP, Rivat C, Dumitresco SM, Griscelli C. Immunoglobulins in ataxia-telangiectasia: evidence for IgG4 and IgA2 subclass deficiencies. *Clin Immunol Immunopathol* 1981;**20**:99–110.
18. Oxelius VA, Berkel AI, Hanson LA. IgG2 deficiency in ataxia-telangiectasia. *N Engl J Med* 1982;**306**:515–7.
19. Gatti RA, Bick M, Tam CF, et al. Ataxia-telangiectasia: a multiparameter analysis of eight families. *Clin Immunol Immunopatho* 1982;**23**:501–16.
20. Etzioni A, Ben-Barak A, Peron S, Durandy A. Ataxiatelangiectasia in twins presenting as autosomal recessive hyper-immunoglobulin M syndrome. *Isr Med Assoc J* 2007;**9**:406–7.
21. Gatti RA. Ataxia-telangiectasia. In: Scriver CR, Beaudet AL, Sly WS, Valli D, editors. *The Metabolic and Molecular Bases of Inherited Disease.* 8th edn New York: McGraw-Hill; 2001. p. 705–32.
22. Ochs HD, Davis SD, Wedgwood RJ. Immunologic responses to bacteriophage phi-X 174 in immunodeficiency diseases. *J Clin Invest* 1971;**50**:2559–68.
23. Peterson RD, Blaw M, Good RA. Ataxia-telangectasia; a possible clinical counterpart of the animals rendered immunologically incompetent by thymectomy. *J Pediatr* 1963;**63**:701–3.
24. Sanal O, Ersoy F, Yel L, et al. Impaired IgG antibody production to pneumococcal polysaccharides in patients with ataxia-telangiectasia. *J Clin Immunol* 1999;**19**: 326–34.
25. Sanal O, Ozbas-Gerceker F, Yel L, et al. Defective antipolysaccharide antibody response in patients with ataxia-telangiectasia. *Turk J Pediatr* 2004;**46**:208–13.
26. Guerra-Maranhao MC, Costa-Carvalho BT, Nudelman V, et al. Response to polysaccharide antigens in patients with ataxia-telangiectasia. *J Pediatr (Rio J)* 2006;**82**: 132–6.
27. Boder E. Ataxia tenenagiectasia: an overview. In: Gatti RA, Swift M, editors. *Ataxia-Telangiectasia: Genetics, Neuropathology, and Immunology of a Degenerative Disease of*

Childhood – Proceedings of a Conference held in Solvang, California, January 16-20, 1984. New York: Alan R. Liss, Inc; 1985. p. 1–63.

28. Chen PC, Lavin MF, Kidson C, Moss D. Identification of ataxia telangiectasia heterozygotes, a cancer prone population. *Nature* 1978;**274**:484–6.

29. Swift M, Sholman L, Perry M, Chase C. Malignant neoplasms in the families of patients with ataxia-telangiectasia. *Cancer Res* 1976;**36**:209–15.

30. Swift M, Morrell D, Massey RB, Chase CL. Incidence of cancer in 161 families affected by ataxia-telangiectasia. *N Engl J Med* 1991;**325**:1831–6.

31. Gatti RA, Good RA. Occurrence of malignancy in immunodeficiency diseases. A literature review. *Cancer* 1971;**28**:89–98.

32. Concannon P. ATM heterozygosity and cancer risk. *Nat Genet* 2002;**32**:89–90.

33. Saunders-Pullman RJ, Gatti R. Ataxia-telangiectasia: without ataxia or telangiectasia? *Neurology* 2009;**73**:414–5.

34. Byrd PJ, Srinivasan V, Last JI, et al. Severe reaction to radiotherapy for breast cancer as the presenting feature of ataxia telangiectasia. *Br J Cancer* 2012;**106**:262–8.

35. Saunders-Pullman R, Raymond D, Stoessl AJ, et al. Variant ataxia-telangiectasia presenting as primary-appearing dystonia in Canadian Mennonites. *Neurology* 2012;**78**:649–57.

36. Waldmann TA, McIntire KR. Serum-alpha-fetoprotein levels in patients with ataxia-telangiectasia. *Lancet* 1972;**2**:1112–5.

37. Taylor AM, Harnden DG, Arlett CF, et al. Ataxia telangiectasia: a human mutation with abnormal radiation sensitivity. *Nature* 1975;**258**:427–9.

38. Painter RB, Young BR. Radiosensitivity in ataxia-telangiectasia: a new explanation. *Proc Natl Acad Sci USA* 1980;**77**:7315–7.

39. Huo YK, Wang Z, Hong JH, et al. Radiosensitivity of ataxia-telangiectasia, X-linked agammaglobulinemia, and related syndromes using a modified colony survival assay. *Cancer Res* 1994;**54**:2544–7.

40. Sun X, Becker-Catania SG, Chun HH, et al. Early diagnosis of ataxia-telangiectasia using radiosensitivity testing. *J Pediatr* 2002;**140**:724–31.

41. Kojis TL, Schreck RR, Gatti RA, Sparkes RS. Tissue specificity of chromosomal rearrangements in ataxia-telangiectasia. *Hum Genet* 1989;**83**:347–52.

42. Kojis TL, Gatti RA, Sparkes RS. The cytogenetics of ataxia telangiectasia. *Cancer Genet Cytogenet* 1991;**56**:143–56.

43. Davis MM, Gatti RA, Sparkes RS. Neoplasia and chromosomal breakage in ataxia-telangiectasia: a 2:14 translocation. *Kroc Found Ser* 1985;**19**:197–203.

44. Russo G, Isobe M, Gatti R, et al. Molecular analysis of a t(14;14) translocation in leukemic T-cells of an ataxia telangiectasia patient. *Proc Natl Acad Sci USA* 1989;**86**:602–6.

45. Chun HH, Sun X, Nahas SA, et al. Improved diagnostic testing for ataxia-telangiectasia by immunoblotting of nuclear lysates for ATM protein expression. *Mol Genet Metab* 2003;**80**:437–43.

46. Reina-San-Martin B, Chen HT, Nussenzweig A, Nussenzweig MC. ATM is required for efficient recombination between immunoglobulin switch regions. *J Exp Med* 2004;**200**:1103–10.

47. Driessen GJ, Ijspeert H, Weemaes, et al. Antibody deficiency in patients with ataxia telangiectasia is caused by disturbed B- and T-cell homeostasis and reduced immune repertoire diversity. *J Allergy Clin Immunol* 2013;**131**:1367–75.

48. Pan-Hammarstrom Q, Dai S, Zhao Y, et al. ATM is not required in somatic hypermutation of VH, but is involved in the introduction of mutations in the switch mu region. *J Immunol* 2003;**170**:3707–16.

49. Nowak-Wegrzyn A, Crawford TO, Winkelstein JA, Carson KA, Lederman HM. Immunodeficiency and infections in ataxia-telangiectasia. *J Pediatr* 2004;**144**:505–11.

50. Eisen AH, Karpati G, Laszlo T, Andermann F, Robb JP, Bacal HL. Immunologic deficiency in ataxia telangiectasia. *N Engl J Med* 1965;**272**:18–22.

51. Biggar WD, Good RA. Immunodeficiency in ataxia-telangiectasia. *Birth Defects Orig Artic Ser* 1975;**11**:271–6.

52. McFarlin DE, Strober W, Waldmann TA. Ataxia-telangiectasia. *Medicine (Baltimore)* 1972;**51**:281–314.

53. Lee P, Martin NT, Nakamura K, et al. SMRT compounds abrogate cellular phenotypes of ataxia telangiectasia in neural derivatives of patient-specific hiPSCs. *Nat Commun* 2013;**4**:1824.

54. Huang PC, Sheridan 3rd RB. Genetic and biochemical studies with ataxia telangiectasia. *A review. Hum Genet* 1981;**59**:1–9.

55. Woods CG, Bundey SE, Taylor AM. Unusual features in the inheritance of ataxia telangiectasia. *Hum Genet* 1990;**84**:555–62.

56. Mallott J, Kwan A, Church J, et al. Newborn screening for SCID identifies patients with ataxia telangiectasia. *J Clin Immunol* 2013;**33**:540–9.

57. Jaspers NG, Painter RB, Paterson MC, Kidson C, Inoue T. Complementation analysis of ataxia-telangiectasia. *Kroc Found Ser* 1985;**19**:147–62.

58. Gatti RA. Molecular genetics of ataxia-telangiectasia. In: de Jong JMBV, editor. *Handbook of Clinical Neurology*. Amsterdam: Elsevier Science Publ; 1991. p. 425–31.

59. Murnane JP, Painter RB. Complementation of the defects of DNA synthesis in irradiated and unirradiated ataxia-telangiectasia cells. *Proc Natl Acad Sci USA* 1982;**79**:1960–3.

60. Jaspers NG, Bootsma D. Genetic heterogeneity in ataxia-telangiectasia studied by cell fusion. *Proc Natl Acad Sci USA* 1982;**79**:2641–4.

61. Jaspers NG, Gatti RA, Baan C, Linssen PC, Bootsma D. Genetic complementation analysis of ataxia telangiectasia and Nijmegen breakage syndrome: a survey of 50 patients. *Cytogenet Cell Genet* 1988;**49**:259–63.

62. Botstein D. Considerations affecting the feasibility of mapping a single-gene disorder using restriction fragment polymorphisms, with special references to ataxia-telangiectasia. In: Gatti RA, Swift M, editors. *Ataxia-telangiectasia: Genetics, Neuropathology, and Immunology of a Degenerative Disease of Childhood*. New York: Alan R. Liss Inc; 1985. p. 121–31.

63. Botstein D. General Discussion. In: Gatti RA, Swift M, editors. *Ataxia-telangiectasia: Genetics, Neuropathology, and Immunology of a Degenerative Disease of Childhood*. New York: Alan R. Liss Inc; 1985. p. 215–21.

64. Botstein D, White RL, Skolnick M, Davis RW. Construction of a genetic linkage map in man using restriction fragment length polymorphisms. *Am J Hum Genet* 1980;**32**:314–31.

65. Nakamura Y, Leppert M, O'Connell P, et al. Variable number of tandem repeat (VNTR) markers for human gene mapping. *Science* 1987;**235**:1616–22.

66. Gatti RA, Shaked R, Wei S, Koyama M, Salser W, Silver J. DNA polymorphism in the human Thy-1 gene. *Hum Immunol* 1988;**22**:145–50.

67. Gatti RA, Berkel I, Boder E, et al. Localization of an ataxia-telangiectasia gene to chromosome 11q22-23. *Nature* 1988;**336**:577–80.

68. Stumm M, Gatti RA, Reis A, et al. The ataxia-telangiectasia-variant genes 1 and 2 are distinct from the ataxia-telangiectasia gene on chromosome 11q23.1. *Am J Hum Genet* 1995;**57**:960–2.

69. Ejima Y, Oshimura M, Sasaki MS. Establishment of a novel immortalized cell line from ataxia telangiectasia fibroblasts and its use for the chromosomal assignment of radiosensitivity gene. *Int J Radiat Biol* 1990;**58**:989–97.

70. Dausset J, Cann H, Cohen D, Lathrop M, Lalouel JM, White R. Centre d'etude du polymorphisme humain (CEPH): collaborative genetic mapping of the human genome. *Genomics* 1990;**6**:575–7.

71. Foroud T, Wei S, Ziv Y, et al. Localization of an ataxia-telangiectasia locus to a 3-cM interval on chromosome 11q23: linkage analysis of 111 families by an international consortium. *Am J Hum Genet* 1991;**49**:1263–79.

72. Lange E, Borresen AL, Chen X, et al. Localization of an ataxia-telangiectasia gene to an approximately 500 Kb interval on chromosome 11q23.1: linkage analysis of 176 families by an international consortium. *Am J Hum Genet* 1995;**57**:112–9.

73. Savitsky K, Bar-Shira A, Gilad S, et al. A single ataxia telangiectasia gene with a product similar to PI-3 kinase. *Science* 1995;**268**:1749–53.

74. Uziel T, Savitsky K, Platzer M, et al. Genomic Organization of the ATM gene. *Genomics* 1996;**33**:317–20.

75. Platzer M, Rotman G, Bauer D, et al. Ataxia-telangiectasia locus: sequence analysis of 184 Kb of human genomic DNA containing the entire ATM gene. *Genome Res* 1997;**7**:592–605.

76. Gatti RA, Becker-Catania S, Chun HH, et al. The pathogenesis of ataxia-telangiectasia. Learning from a Rosetta Stone. *Clin Rev Allergy Immunol* 2001;**20**:87–108.

77. Chiarle R, Zhang Y, Frock RL, et al. Genome-wide translocation sequencing reveals mechanisms of chromosome breaks and rearrangements in B cells. *Cell* 2011;**147**:107–19.

78. McKinnon PJ. ATM and the molecular pathogenesis of ataxia telangiectasia. *Annu Rev Pathol* 2012;**7**:303–21.

79. Shiloh Y, Ziv Y. The ATM protein kinase: regulating the cellular response to genotoxic stress, and more. *Nat Rev Mol Cell Biol* 2013;**14**:197–210.

80. Ambrose M, Gatti RA. Pathogenesis of ataxia-telangiectasia: the next generation of ATM functions. *Blood* 2013;**121**:4036–45.

81. Bradshaw PS, Stavropoulos DJ, Meyn MS. Human telomeric protein TRF2 associates with genomic double-strand breaks as an early response to DNA damage. *Nat Gene* 2005;**37**:193–7.

82. Demuth I, Bradshaw PS, Lindner A, et al. Endogenous hSNM1B/Apollo interacts with TRF2 and stimulates ATM in response to ionizing radiation. *DNA Repair (Amst)* 2008;**7**:1192–201.

83. Matsuoka S, Ballif BA, Smogorzewska A, et al. ATM and ATR substrate analysis reveals extensive protein networks responsive to DNA damage. *Science* 2007;**316**:1160–6.

84. Barzilai A, Biton S, Shiloh Y. The role of the DNA damage response in neuronal development, organization and maintenance. *DNA Repair (Amst)* 2008;**7**:1010–27.

85. Barzilai A, McKinnon PJ. Genome maintenance in the nervous system; insight into the role of the DNA damage response in brain development and disease. *DNA Repair (Amst)* 2013;**12**:541–2.

86. Mitui M, Nahas SA, Chun HH, Gatti RA. Diagnosis of ataxia-telangiectasia: ATM mutations associated with cancer. In: Nakamura R, Grody W, Wu S, Nagle D, editors. *Cancer Diagnostics: Current and Future Trends*. Humana Press; 2004. p. 473–87.

87. Telatar M, Teraoka S, Wang Z, et al. Ataxia-telangiectasia: identification and detection of founder-effect mutations in the ATM gene in ethnic populations. *Am J Hum Genet* 1998;**62**:86–97.

88. Telatar M, Wang S, Castellvi-Bel S, et al. A model for ATM heterozygote identification in a large population: four founder-effect ATM mutations identify most of Costa Rican patients with ataxia telangiectasia. *Mol Genet Metab* 1998;**64**:36–43.

89. Uhrhammer N, Lange E, Porras O, et al. Sublocalization of an ataxia-telangiectasia gene distal to D11S384 by ancestral haplotyping in Costa Rican families. *Am J Hum Genet* 1995;**57**:103–11.

90. Gilad S, Bar-Shira A, Harnik R, et al. Ataxia-telangiectasia: founder effect among North African Jews. *Hum Mol Genet* 1996;**5**:2033–7.

91. Ejima Y, Sasaki MS. Mutations of the ATM gene detected in Japanese ataxia-telangiectasia patients: possible preponderance of the two founder mutations 4612del165 and 7883del5. *Hum Genet* 1998;**102**:403–8.

92. Stankovic T, Kidd AM, Sutcliffe A, et al. ATM mutations and phenotypes in ataxia-telangiectasia families in the British Isles: expression of mutant ATM and the risk of leukemia, lymphoma, and breast cancer. *Am J Hum Genet* 1998;**62**:334–45.

93. Laake K, Telatar M, Geitvik GA, et al. Identical mutation in 55% of the ATM alleles in 11 Norwegian AT families: evidence for a founder effect. *Eur J Hum Genet* 1998;**6**:235–44.

94. Sandoval N, Platzer M, Rosenthal A, et al. Characterization of ATM gene mutations in 66 ataxia telangiectasia families. *Hum Mol Genet* 1999;**8**:69–79.

95. Mitui M, Bernatowska E, Pietrucha B, et al. ATM gene founder haplotypes and associated mutations in Polish families with ataxia-telangiectasia. *Ann Hum Genet* 2005;**69**:657–64.

96. Birrell GW, Kneebone K, Nefedov M, et al. ATM mutations, haplotype analysis, and immunological status of Russian patients with ataxia telangiectasia. *Hum Mutat* 2005;**25**:593.

97. Coutinho G, Mitui M, Campbell C, et al. Five haplotypes account for fifty-five percent of ATM mutations in Brazilian patients with ataxia telangiectasia: seven new mutations. *Am J Med Genet A* 2004;**126A**:33–40.

98. Mitui M, Campbell C, Coutinho G, et al. Independent mutational events are rare in the ATM gene: haplotype prescreening enhances mutation detection rate. *Hum Mutat* 2003;**22**:43–50.

99. Chessa L, Piane M, Magliozzi M, et al. Founder effects for ATM gene mutations in Italian Ataxia Telangiectasia families. *Ann Hum Genet* 2009;**73**:532–9.

100. Cavalieri S, Funaro A, Pappi P, Migone N, Gatti RA, Brusco A. Large genomic mutations within the ATM gene detected by MLPA, including a duplication of 41 Kb from exon 4 to 20. *Ann Hum Genet* 2008;**72**:10–8.

101. Eng L, Coutinho G, Nahas S, et al. Nonclassical splicing mutations in the coding and noncoding regions of the ATM Gene: maximum entropy estimates of splice junction strengths. *Hum Mutat* 2004;**23**:67–76.

102. Micol R, Ben Slama L, Suarez F, et al. Morbidity and mortality from ataxia-telangiectasia are associated with ATM genotype. *J Allergy Clin Immunol* 2011;**128**:382–9.

103. Nakamura K, Du L, Tunuguntla R, et al. Functional characterization and targeted correction of ATM mutations identified in Japanese patients with ataxia-telangiectasia. *Hum Mutat* 2012;**33**:198–208.

104. Nakamura K, Fike F, Haghayegh S, et al. A-T$_{winnipeg}$: Pathogenesis of ATM missense mutation c.6200C>A with decreased protein expression and downstream signaling, early-onset dystonia, cancer, and life-threatening radiotoxicity. *Mol Genet Genomic Med* In Press.

105. Gatti RA, Tward A, Concannon P. Cancer risk in ATM heterozygotes: a model of phenotypic and mechanistic differences between missense and truncating mutations. *Mol Genet Metab* 1999;**68**:419–23.

106. Tavtigian SV, Oefner PJ, Babikyan D, et al. Rare, evolutionarily unlikely missense substitutions in ATM confer increased risk of breast cancer. *Am J Hum Genet* 2009;**85**:427–46.

107. Mitui M, Nahas SA, Du LT, et al. Functional and computational assessment of missense variants in the ataxia-telangiectasia mutated (ATM) gene: mutations with increased cancer risk. *Hum Mutat* 2009;**30**:12–21.

108. Du L, Pollard JM, Gatti RA. Correction of prototypic ATM splicing mutations and aberrant ATM function with antisense morpholino oligonucleotides. *Proc Natl Acad Sci USA* 2007;**104**:6007–12.

109. Gatti RA. SMRT compounds correct nonsense mutations in primary immunodeficiency and other genetic models. *Ann NY Acad Sci* 2012;**1250**:33–40.

Wiskott–Aldrich Syndrome: from a Fatal Hematologic Disorder to a Curable Immunodeficiency

Hans D. Ochs[1], Bernd H. Belohradsky[2]

[1]Professor of Pediatrics, University of Washington School of Medicine,
Seattle Children's Research Institute, Seattle, WA, USA
[2]Professor of Pediatrics, Haunerschen Kinderspital, Ludwig Maximilians University,
München, Germany

INTRODUCTION

The clinical symptoms of classic Wiskott–Aldrich syndrome (WAS) – early onset of eczema, recurrent ear infections, bloody diarrhea and thrombocytopenia with small platelets – are difficult to overlook. Why, then, was this entity not recognized until 1937, when Alfred Wiskott (Fig. 9.1), a pediatrician with an interest in pulmonary diseases, observed three brothers who presented in infancy with this symptomatology, while their sisters had none of these pathological findings (1)? It took 17 years for another pediatrician, Robert Aldrich (Fig. 9.1), to rediscover this entity while studying a large family with multiple affected males who had all died in infancy (2). At a time of social unrest, without a safe food supply and before the era

FIGURE 9.1   Left: Alfred W. Wiskott, at the time of retirement as chairman of Pediatrics at the von Haunersche Kinderspital, Ludwig Maximilians University of Munich. Right: Robert A. Aldrich, pictured while Professor of Preventive Medicine/Public Health at the University of Washington, Seattle.

of antibiotics, child mortality was high and rare cases of immune deficiencies did not reach the attention of the medical profession. Although both Wiskott and Aldrich noticed that the affected boys had problems with infections, including otitis media and sepsis, the idea of an immune defect was not considered until 1968, when Mike Blaese (3) and Max Cooper (4) announced that WAS qualified as a primary immune deficiency disease (PIDD). Finally, in 1994, the causative gene, designated *WAS*, was identified by positional cloning (5), allowing a systematic dissection of the molecular basis of this puzzling syndrome.

THE DISCOVERY OF WAS AS A NOVEL CLINICAL ENTITY

The life and work of Alfred Wilhelm Wiskott, who differentiated WAS from idiopathic thrombocytopenia (ITP), was influenced by the turbulent times in early 20th century Germany. He was born in 1898 in Nordrhein-Westfalen into a coal mining family. Before he could finish high

school, at the age of 17, he volunteered to join the German Army at the height of World War I. During the battle of Verdun, he was wounded and lost his leg beneath the knee to amputation. He was released from the trenches of war and started medical school at the Ludwig Maximilians University of Munich. After graduation, he joined the Department of Pediatrics at the Dr von Haunersche Kinderspitalin Munich as a resident, and in 1932 received his habilitation with a study entitled "Contribution to the pathogenesis and clinical presentation of pneumonia during early infancy". In 1939, he was elected chairman of the Department of Pediatrics at the Ludwig Maximilians University in Munich, a position he kept until his retirement in 1968. The years before, during and after World War II were difficult for the Haunersche Children's hospital, as well as for its chairman. A large number of the medical staff consisted of Jewish pediatricians who went into exile (6) shortly after Hitler's "Machtergreifung" (power grab), and most of the male physicians were drafted at the beginning of the war. As an employee of the government, Wiskott and most other employees of the

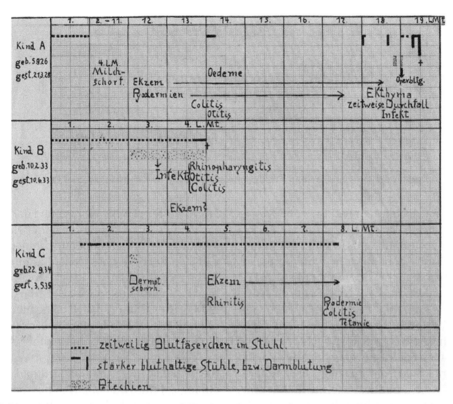

FIGURE 9.2 Wiskott's handwritten summary of the clinical course of the three brothers (A, B, C) presented in 1936 at the Pediatric meeting in Wuerzburg and published one year later in *Monatsschrift für Kinderheilkunde* (1). LM = Lebensmonat (month of life), ZeitweiseDurchfall = intermittent diarrhea, Zeitweilige Blutfäserchenim Stuhl = intermittent bloody strings in stool. This figure is reproduced in color in the color section.

hospital had to join the NSDAP (the Nazi party) and, when the American military took over in 1945, Wiskott was temporarily dismissed until his political rehabilitation. As recounted years later by David Smith, a geneticist working during the 1960s and 1970s at the University of Washington, Wiskott welcomed the young American medical officer to join him at rounds at the Haunersche Kinderspital and enlisted Dr Smith's help in getting the hospital back on its feet after World War II.

Wiskott was a young pediatrician at the beginning of his academic career when he met a family with a medical problem he had never seen before: each of the three boys was found to have early onset thrombocytopenia that did not fit the ITP phenotype, known in Germany as Morbus Werlhofii or Morbus maculosus haemorrhagicus. Each of the three brothers presented during the first few weeks of life with petechiae, eczema, bloody diarrhea and otitis media with mucoid discharge (Fig. 9.2).

In spite of diet adjustments, each of the boys died in infancy while the four sisters remained healthy. Wiskott presented his observations in 1936 during the annual meeting of the German Pediatric Society in Würzburg, pointing out that each affected boy had: (1) a peculiar tendency for bleeding, specifically in the gastrointestinal tract; (2) a surprisingly persistent anergy with lack of resistance to infections; (3) a tendency to develop edema, weight loss and eczema.

Vierte Sitzung
am Donnerstag, den 23. Juli 1936, 15,15 Uhr.

_____

Familiärer, angeborener Morbus Werlhofii?
Herr A. Wiskott (München).

Mit 1 Textabbildung.

Wir hatten Gelegenheit, ein eigenartiges, offenbar übereinstim-
mendes Krankheitsgeschehen bei allen drei männlichen Sprossen
einer Familie (die 4 Schwestern sind krankheitsfrei) zu verfolgen.

FIGURE 9.3   Wiskott's oral presentation in 1936 at the annual meeting of the German Society of Pediatrics in Wuerzburg.

He noticed that each infant had thrombocytopenia, small platelets by smear, lymphopenia and a transient eosinophilia; however, the megakaryocytes in the bone marrow were normal in number. To indicate his doubt that this was ITP, he added a question mark to his oral presentation (Fig. 9.3) entitled, "Familiärer, angeborener Morbus Werlhofii?"(Familial congenital ITP?) and published a full article with the same title in 1937 (1). One could speculate that he avoided the term "inherited" or "genetic", since inherited diseases were a political liability at that time, and Wiskott may have intended to protect the family. The medical records, still available at Children's Hospital in Munich, clearly indicate that Wiskott recognized this entity as a genetic disease (Fig. 9.4A).

One of the authors (BHB) resumed contact with the family when a descendent delivered a boy with symptoms resembling those of his distant cousins. The discovery of the WAS mutation, a two base pair deletion resulting in frameshift and premature termination of transcription (c.73_74 del AC) confirmed the diagnosis and brought this family's heartbreaking story to a conclusion (7) (see Fig. 9.4B).

The syndrome remained unnoticed until 1954, when Aldrich recognized and reported a large family with a sex-linked recessive condition characterized by draining ears, eczematoid dermatitis and bloody diarrhea. Robert Aldrich, born in 1917 in Illinois as the son of a nationally known pediatrician, studied medicine at

Northwestern University, trained in Pediatrics at the University of Minnesota and, before starting his career as an administrator, joined the Mayo Clinic in Rochester as a consultant in pediatrics. As he recounted later, Aldrich took turns attending the inpatient service at St Mary's Hospital in Rochester, Minnesota when he met a 6-month-old boy who was admitted in serious condition with a clinical picture that neither Aldrich nor his colleagues had ever seen or read about in the pediatric literature. Puzzled and looking for etiological clues that might come from the child's environment or the history of the family, Aldrich invited the mother to sit down with him once more to go into more detail. Aldrich recalled that "when the mother arrived, accompanied by her own mother, I spent at least an hour of questioning about the environment but failed to discover any leads". Then Aldrich began asking about relatives who might have had a similar illness, when the child's grandmother exclaimed sadly, "Just like all the rest of them". This led to a search for each male death in the family and to establishing the X-linked nature of this new syndrome (Fig. 9.5). With the help of Dutch pediatricians, Aldrich was able to trace the female carrier, who had come from the Netherlands six generations earlier to live in Iowa. Later, Aldrich pointed out that he and his colleagues did not publish their findings for more than a year after the research study was completed, since they could not believe that this entity had not been previously described. "Finally, together with my

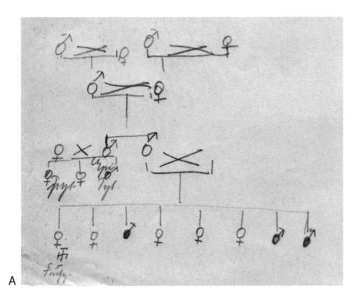

A

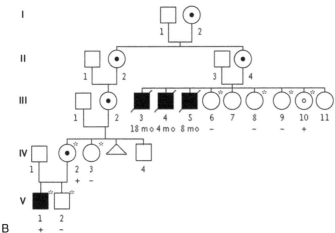

B

FIGURE 9.4 Pedigrees of the Bavarian family with three affected boys: (A) Wiskott's hand drawn family tree; he seems to emphasize the father's side of the family. The eldest daughter is indicated to have died of prematurity. (B) The updated version of the Bavarian pedigree (6), showing an additional affected male, a distant cousin to the three brothers originally reported by Wiskott (1). Sequence analysis of DNA from this new patient allowed the identification of the *WAS* mutation (c.73_74delAC), resulting in premature termination of transcription nine codons downstream of the deletion. *(From Binder V. et al.,* New England Journal of Medicine, *2006, with permission; 7.)* This figure is reproduced in color in the color section.

colleagues, Arthur Steinberg and Donald Campbell, an account was published in *Pediatrics* (2); it was several months later that a prominent German pediatrician wrote, directing my attention to a short paper in a journal by Wiskott (1937) describing precisely the clinical picture in three brothers." It is only fair to point out that Aldrich and his coworkers made a truly novel observation, the recognition of the X-linked inheritance of the syndrome by exploring in detail the family history and exploiting cutting edge genetics.

The published pedigree (see Fig. 9.5) illustrates the effort taken by these authors to generate a story that goes back generations and across the Atlantic. Aldrich recognized the fact that the affected boys had in common draining ears, eczema and bloody diarrhea, while the females transmitting the disease had no symptoms. In the resulting paper (2), he and his co-workers described in detail the pattern of inheritance, discussing the newly recognized fact that "each human being has '24' pairs of chromosomes,

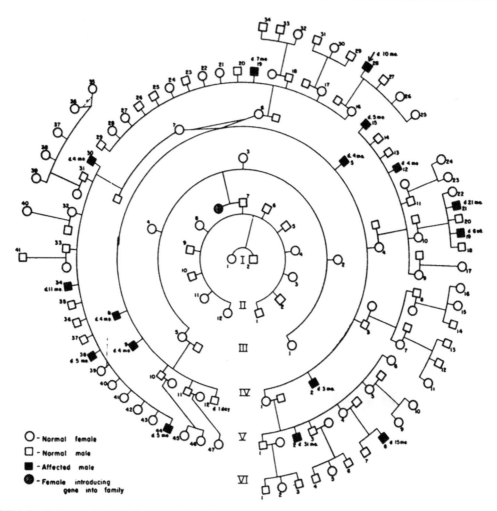

FIGURE 9.5   Pedigree of the family reported by Aldrich in 1954 (2), showing six generations with affected males, clearly demonstrating X-linked recessive inheritance. *(From* Pediatrics *with permission.)*

that in females, the members of each of the '24' pairs are morphologically alike, whereas in the male the members of only 23 of the pairs are alike". They further explain that in males, "the 24th pair consists of one chromosome, called the X chromosome, which is much larger than the other, called the Y chromosome and that the females have two X chromosomes and no Y chromosome, whereas males have one of each". When reading this story, one of the authors (HDO) was reminded of an exciting hour in

biology class while in high school back in 1955, when the biology teacher, an elderly gentleman who had a reputation for painfully dull classes, discussed with excitement the facts of sex determination by explaining the roles of X and Y chromosomes, which form one of the "24" pairs of human chromosomes. He ended his lecture with the statement, "How interesting that humans have 48 chromosomes, exactly the same as the tomato". That was the dictum in 1954/1955. It was two more years later when Joe-Hin Tjio, while working in Lundt, Sweden and after developing new photographic techniques, discovered that the number of human chromosomes is 46 and not 48 (8), as had been believed for some 50 years, based on the work of Theophilus Painter, who in 1923 studied human testicular tissues and came up with a human chromosome number of 48 (9).

Interestingly, neither Alfred Wiskott nor Robert Aldrich had a chance to follow up on their finding/discovery of WAS. Wiskott, who had been invited in 1964 to give a lecture during a hematology meeting in Freiburg, Germany, started his presentation with the disclaimer:

> I agreed to the invitation by the chairman to talk about the Wiskott–Aldrich syndrome with some hesitation. For once, I do not want to cause the impression that I put myself into the limelight as the discoverer of this disease. In addition, I have to acknowledge that I personally have not had new experiences related to the syndrome. The cases I reported in 1937 have been the only WAS patients observed in Germany. However, I concluded that at this point I could, at least, renew interest in this syndrome with its interesting problematic. The disease has been recognized in many countries since the publication by Aldrich in 1954, some 42 patients in the USA, eight in Switzerland, four in Italy, three in Holland, two in England and a few single cases in Denmark, Greece, Sweden and Spain.

Wiskott's appeal for "renewed interest in this syndrome" was not in vain: 14 years later, an international group of immunologists from France, Germany and the USA published a 100 page monograph that included the clinical course, laboratory data and long-term prognosis of 138 WAS patients (10). Aldrich continued his career as an administrator, becoming Chairman of Pediatrics at the University of Washington in Seattle for a few years, then moving to the NIH to establish the National Institute of Child Health and Human Development at the invitation of John F. Kennedy, and later taking up positions as Professor of Preventive Medicine/Public Health at the University of Colorado and at the University of Washington. Aldrich died in Seattle at the age of 80.

WAS IS RECOGNIZED AS A COMBINED IMMUNE DEFICIENCY

It took 14 more years after Aldrich's report to discover the immunological abnormalities in patients with classic WAS. In 1968, Mike Blaese and colleagues at NIH suggested a defect in antigen processing or recognition (3), and Max Cooper and the Minnesota group described a defect involving the afferent limb of immunity (4). These reports kindled the interest of clinical and basic immunologists, resulting in new biologic concepts and curative treatments. Both T and B lymphocyte functions were found to be affected. While the number of circulating lymphocytes is normal or only moderately decreased during infancy, older children and adults often present with lymphopenia, and reduced numbers of B-cells in lymph node follicles and T-cells in the interfollicular area (11). This may be in part due to accelerated cell death observed in peripheral blood lymphocytes from patients with classic WAS (12). Abnormal T-cell function is suggested by diminished but not absent lymphocyte responses to mitogens (4), depressed proliferative responses to allogenic cells (13) and immobilized anti-CD3 monoclonal antibody (14). In a retrospective study of 154 patients, Sullivan and coworkers (15) reported depressed proliferation to mitogens and to allogenic cells in approximately 50% of the patients, while skin tests for delayed-type hypersensitivity were abnormal in 90%. The increased incidence of pneumocystis jirovecii pneumonia also points to a significant T-cell defect in classic WAS. Isohemagglutinin titers are frequently low, and antibody responses to recall and neo-antigens are variable (13, 15). A consistent finding is a markedly depressed response to polysaccharide antigen (4, 13). While antibody responses to a variety of protein antigens, including diphtheria and tetanus toxoid, have been reported to be abnormal in more than 50% of WAS patients, those to live virus vaccines were normal (15). Responses to intravenous immunization with bacteriophage φX174 are severely depressed in patients with classic WAS, whose neutralizing titers fail to amplify during re-exposure to the antigen and to switch from IgM to IgG, suggesting abnormal class-switch recombination and lack of somatic hypermutation, either due to defective T–B-cell interaction or an intrinsic B-cell defect (13). There is indeed evidence that B-cell function is also affected, possibly due to abnormal cell motility and decreased ability to adhere and form cytoplasmic protrusions after stimulation with anti-CD40 and interleukin 4 (IL-4) (16). The susceptibility of WAS patients to viral infections and to malignancies may be a direct consequence of cytolytic or cytotoxic dysfunction. Natural killer (NK) cells from patients with classic WAS who lack WAS protein have a markedly reduced accumulation of F-actin in the immunologic synapse which may impair NK cell–target cell interaction. As a result, all patients with classic WAS and most with mutations associated with the XLT phenotype have defective cytolytic NK cell function (17, 18), while the percentage of NK cells is normal or increased. This inhibition of NK cell-mediated cytotoxicity is associated with a reduced ability of WAS/XLT-NK cells to form conjugates with susceptible target cells and to accumulate actin at the synapse. Furthermore, regulatory T-cells (Tregs) of WAS protein (WASp)-deficient individuals and wasp$^{-/-}$ mice fail to suppress effector cells *in vitro* and are incapable of controlling autoimmunity in mouse models (19–21). Abnormal chemotaxis of neutrophils and monocytes (13, 22), associated with failure to assemble podosomes (23), results in a severe defect of cell adhesion and motility.

Thrombocytopenia and small platelet volume are consistent findings in patients with mutations in the *WAS* gene, except those with mutations that interfere with self-inhibition. Platelet counts vary from patient to patient as well as in affected individuals, from 5000/mm^3 to 50 000/mm^3 or higher. Intermittent thrombocytopenia has been observed in males of two otherwise asymptomatic families with unique amino acid

substitutions in WASp (24). In most patients, the mean platelet volume (MPV) is half that of normal control subjects (3.5 to 5.0 fl versus 7.1 to 10.5 fl in normal controls), resulting in a thrombocrit of approximately 0.01% compared with a normal range of 0.14% to 0.31% (13). After splenectomy, platelet counts and platelet volume increase but are still less than that of normal controls (25). Increased expression of phosphatidylserine on circulating platelets from WAS patients (26) and the presence of opsonizing antiplatelet antibodies in approximately half of wasp-deficient mice (27) have been suggested as a possible cause of increased phagocytosis of platelets. In a recently published retrospective analysis of data collected by the French registry of patients with WAS/XLT, a specific subset of infants less than two years of age was identified. These infants had an extremely poor prognosis due to a high incidence of autoimmunity, inflammation and malignancies; they often presented with severe, refractory thrombocytopenia, and were suspected, but not proven, to have antiplatelet autoantibodies (28), demonstrating that autoimmune thrombocytopenia can be a complication of WAS, especially in the very young. Additionally, it has been suggested that ineffective thrombocytopoiesis is at least in part responsible for the low platelet count (13).

WAS, AUTOIMMUNITY AND MALIGNANCY

As early recognition and better conservative therapy improved survival from early death during infancy (4) to teenage years and older (15), additional complications were recognized.

Autoimmune diseases are frequent complications, having been reported in 40% of a large cohort of WAS patients (15). The most common autoimmune manifestation is hemolytic anemia, followed by vasculitis involving both small and large vessels (29, 30), renal disease, Henoch–Schönlein-like purpura and inflammatory bowel disease. In a French study, 40 out of 55 WAS patients had at least one autoimmune or inflammatory complication (31). Autoimmune hemolytic anemia developed in 20 patients with onset before the age of five years. Arthritis was present in 29%, neutropenia in 25%, vasculitis including cerebral vasculitis in 29%, inflammatory bowel disease in 9% and renal disease in 3%. A high serum IgM concentration was a significant risk factor for the development of autoimmune disease or early death (31). The incidence of autoimmune diseases in XLT patients is substantially less frequent than in classic WAS (32).

Malignancies in WAS can occur during childhood but are more frequent in adolescents and young adults. In a large North American cohort, malignancies were present in 13% (15), the average age at onset of malignancies being 9.5 years. Considering the increased life expectancy, it is reasonable to assume that the incidence of malignancies will further increase as untransplanted WAS patients get older. The most frequent malignancy reported is lymphoma, usually an Epstein–Barr (EBV)-positive B-cell lymphoma, which suggests a direct relationship with a defective immune system. WAS-associated malignancies have a poor prognosis, as illustrated by the fact that only one of 21 patients who developed a malignancy was alive more than two years after establishing the diagnosis (15). The incidence of malignancies in patients with the XLT phenotype is unknown but is less than in classic WAS. In a retrospective study, nine of 173 XLT patients (5.2%) developed malignancies (32). The prognosis is grim: six of the nine patients had died of malignancies at the time of analysis.

THE DISCOVERY OF THE MOLECULAR BASIS OF WAS/XLT

In the early 1990s, the WAS gene was mapped to the region Xp11.22–Xp11.3 (33) and sequenced three years later by positional cloning (5). The WAS gene consists of 12 exons, and codes for a protein with a molecular weight of 54 kDa. The WAS protein was found to be constitutively expressed in all hematopoietic stem cell-derived lineages and located predominantly in the cytoplasmic compartment, with the highest protein density being along the cell membrane (34). The WASp homology 1 domain, located at the N-terminus of the WAS gene, allows interaction of WASp with the WASp-interacting protein (WIP) (35). The GTPase-binding domain (GBD) (coded by exons 7 and 8) provides autoinhibitory contact between the GBD and the carboxy terminal region of WASp, which can be released by the activated (GTP) form of Cdc42, a member of the Rho family of GTPases (36). It is believed that the GBD has a direct effect on actin polymerization and an indirect effect on the interaction of the C-terminus of WASp with the Arp2/3 actin nucleating complex (37).

The verproline/cofilin-acidic region domain, located in the C-terminus of WASp, plays a key role in the regulation of actin polymerization (38). If WASp is activated by GTP-Cdc42, the C-terminal region is released from its autoinhibitory contact and binds to the actin-related proteins (Arp) 2 and Arp3, forming the Arp2/3 complex and leading to actin nucleation (39). Thus, WASp is a key member of a family of proteins that link signaling pathways to actin cytoskeleton reorganization by activating Arp2/3-mediated actin polymerization (39). The proline-rich domain of WASp interacts with SH3 domains of selected cytoplasmic proteins, suggesting that WASp participates in the intracellular signaling of hematopoietic cells. This mechanism depends on auto- and allophosphorylation, as WASp by itself undergoes tyrosine (Tyr291) phosphorylation, e.g. when platelets adhere to collagen (40).

Since the identification of the WAS gene, over 400 unique WAS mutations have been identified. The most common are missense mutations, followed by splice-site mutations, short deletions, and nonsense mutations. In addition, there are insertions, complex mutations, and large deletions which comprise about 12% of the mutations identified. Most deletions and insertions result in frameshift and early termination of transcription. In addition, unique missense mutations within the GBD/Cdc42 binding domain have been reported to prevent the self-inhibition of WASp and to cause X-linked neutropenia, with a clinical phenotype very different from classic WAS/XLT (41). These activating (gain-of-function) mutations interfere with the autoinhibitory contact of the C-terminus of WASp with the GBD, resulting in a permanently active configuration of WASp (42). Six hotspot mutations were identified, defined as occurring in over 2.5% of all mutations. Three were point mutations within the coding region and the other three involved splice sites (43). A strong genotype–phenotype correlation has been reported by a number of investigators (34, 43–45). Mutations affecting the WAS gene were found to cause three distinct phenotypes: (1) the classic WAS phenotype, consisting of the triad of thrombocytopenia with microplatelets, recurrent infections as a result of immunodeficiency, eczema and often complicated by autoimmune diseases and malignancies; (2) the milder XLT variant, characterized predominantly by thrombocytopenia and small platelets (46), which can be intermittent (24); (3) congenital neutropenia without the characteristic clinical findings of WAS or XLT (41, 42). A simple scoring system has been designed to define these clinical phenotypes (11, 34). A most consistent phenotype–genotype correlation was observed when the patients were divided into two categories: WASp-positive if the mutated protein was present and of normal size, and WASp-negative if the protein was absent or truncated. Patients with mutations that allow the expression of normal-sized mutated protein, often in reduced quantity, predominantly developed the XLT phenotype; whereas those patients whose lymphocytes could not express WASp or expressed only truncated WASp were more likely to have the classic

WAS phenotype (43, 44). Progression to a score of five, due to either autoimmune disease or malignancy was observed in both groups but was far more frequent in WASp-negative patients with an initial score of three or four (32, 43).

The quantity of WASp can be estimated by flow cytometry or by Western blotting, and the protein size can be determined by Western blot. Somatic mosaicism due to spontaneous reversions of the causative mutations or second site mutations that restore WASp expression seem to be more frequent in WAS patients than other PIDDs (47–49). However, it is unclear whether the restored or altered WAS protein leads to improved immune function (50, 51).

Several symptomatic female patients have been identified as being heterozygous for mutations in the WAS gene. They presented with either a classic WAS phenotype (52, 53) or an XLT phenotype (54, 55). In most cases, the X chromosome with the normal WAS gene remains inactive.

Splenectomy, sometimes recommended for XLT patients, effectively stops the bleeding tendency by increasing the number of circulating platelets. However, splenectomy increases the risk of potentially fatal septicemia and, if performed, requires lifelong antibiotic prophylaxis (32).

Hematopoietic stem cell transplantation, using as source bone marrow, peripheral blood or cord blood, is considered a curative therapy for WAS. To prevent rejection, a conditioning regimen of busulfane in combination with cyclophosphamide, fludarabine or antithymocyte globulin with or without irradiation is recommended (56).

Based on cellular and murine models, a clinical gene therapy protocol using the transfer of autologous hematopoietic stem cells transduced with a WASp-expressing retroviral vector was developed and initially, in 2006, used in two WAS patients (57). Prior to infusion of the genetically modified hematopoietic progenitor cells, the patients underwent partial ablation with busulfan at a dose of 4 mg/kg of body weight on days −3 and −2. Expression of WASp in lymphoid and myeloid cells was observed within three to six months after gene therapy, and sustained increases in platelet numbers as well as in WASp⁺ thrombocytes were achieved while the clinical condition of the patients improved (57). In 2009, the trial was expanded to include eight additional WAS patients. Comprehensive insertion-site analysis showed vector integration that targeted multiple genes controlling growth and immunological responses and a persistently polyclonal hematopoiesis. Disappointingly, as of late 2013, eight of ten gene-treated patients had developed leukemia, demonstrating that the retroviral vector is not safe for WAS patients.

A second gene therapy trial using a lentiviral vector containing the *WAS* gene has been carried out in Italy. Expression of WASp in lymphocytes, myeloid cells and platelets was excellent and so far, none have reported any adverse events (58). Additional clinical gene therapy trials are underway in several centers and are expected eventually to make gene therapy an accepted alternative curative form of therapy for WAS/XLT patients.

WIP DEFICIENCY – AN AUTOSOMAL RECESSIVE DISORDER WITH A WAS PHENOTYPE

A female infant with symptoms and laboratory findings characteristic of WAS was found to have a mutation in *WIPF1* that encodes WASp-interacting protein (WIP) (59). The patient presented at 11 days of age with eczema, vascular and ulcerative lesions of the skin and mucosa, poor weight gain and thrombocytopenia (but with normal platelet volume). Family history was notable for a female sibling who had vesicular and ulcerative skin lesions and died of sepsis at four months of age. The WIP-deficient patient's immune abnormality included defective chemotaxis, lack of T-cell proliferation in response to co-culture with anti-CD3, abnormal NK cell

function and mildly elevated IgE. Because of failure to thrive and chronic infections, she underwent hematopoietic stem cell transplantation at 4.5 months of age. Analysis of the *WIPF1* gene revealed a homozygous nonsense mutation in exon six, located upstream of the region coding the WASp-binding domain of WIP. Both WIP and WASp were absent in the patient's leukocytes. This observation illustrates that WAS cannot be diagnosed solely on the basis of lack of WASp expression, but may require sequence analysis of the *WAS* gene. WIP deficiency should be suspected in male and female patients with features of WAS in whom WASp is absent but *WAS* sequence and mRNA levels are normal.

CONCLUDING REMARKS

The reports of the first two families with classic WAS, published 17 years apart by general pediatricians, were clinically accurate, correctly recognized the syndrome as a genetic disease, but were largely ignored by contemporary hematologists, geneticists and immunologists. Only after the concept of primary immunodeficiency disorders had been established and the interest of immunologists ignited did this rare disease receive attention and undergo systematic clinical, immunological and hemato/oncological evaluation and molecular analysis. Because of the severity of symptoms that, in the 1960s, caused inevitable early death, novel therapeutic options were explored including successful attempts to cure the disease with hematopoietic stem cell transplantation (56, 60), and later with gene therapy (57, 58). The discovery of the molecular defect provided investigators with a novel protein responsible for multiple functions, explaining in part the different clinical phenotypes associated with *WAS* gene mutations. The discovery that *WIP* mutations can cause a phenotype resembling WAS illustrates the need for diagnostic strategies that are based on molecular facts.

References

1. Wiskott A. Familiärer, angeborener Morbus Werlhofii? *Monatsschr Kinderheilkd* 1937;**68**:212–6.

2. Aldrich RA, Steinberg AG, Campbell DC. Pedigree demonstrating a sex-linked recessive condition characterized by draining ears, eczematoid dermatitis and bloody diarrhea. *Pediatrics* 1954;**13**:133–9.

3. Blaese RM, Strober W, Brown RS, Waldmann TA. The Wiskott-Aldrich syndrome. A disorder with a possible defect in antigen processing or recognition. *Lancet* 1968;**1**:1056–61.

4. Cooper MD, Chae HP, Lowman JT, Krivit W, Good RA. Wiskott-Aldrich syndrome. An immunologic deficiency disease involving the afferent limb of immunity. *Am J Med* 1968;**44**:499–513.

5. Derry JM, Ochs HD, Francke U. Isolation of a novel gene mutated in Wiskott-Aldrich syndrome. *Cell* 1994;**78**:635–44.

6. Autenrieth A, Rosenecker J. Ärzte am Dr. von Haunerschen Kinderspital, die Opfer der nationalsozialistischen Verfolgung wurden. *Hauner J* 2010; NA.

7. Binder V, Albert MH, Kabus M, Bertone M, Meindl A, Belohradsky BH. The genotype of the original Wiskott phenotype. *N Engl J Med* 2006;**355**:1790–3.

8. Tjio JH, Levan A. The chromosome number in man. *Hereditas* 1956;**42**:1.

9. Painter TS. Studies in mammalian spermatogenesis. II. The spermatogenesis of man. *J Exp Zool* 1923;**37**:291.

10. Belohradsky BH, Griscelli C, Fundenberg HH, Marget W. [The Wiskott-Aldrich syndrome]. *Ergeb Inn Med Kinderheilkd* 1978;**41**:85–184.

11. Ochs HD, Notarangelo L. Primary immunodeficiency diseases. In: Ochs HD, Smith CIE, Puck JM, editors. *Primary Immunodeficiency Diseases, A Molecular and Genetic Approach*. 3rd edn New York: Oxford University Press; 2014. p. 531–56.

12. Rengan R, Ochs HD, Sweet LI, et al. Actin cytoskeletal function is spared, but apoptosis is increased, in WAS patient hematopoietic cells. *Blood* 2000;**95**:1283–92.

13. Ochs HD, Slichter SJ, Harker LA, Von Behrens WE, Clark RA, Wedgwood RJ. The Wiskott-Aldrich syndrome: studies of lymphocytes, granulocytes, and platelets. *Blood* 1980;**55**:243–52.

14. Molina IJ, Sancho J, Terhorst C, Rosen FS, Remold-O'Donnell E. T-cells of patients with the Wiskott-Aldrich syndrome have a restricted defect in proliferative responses. *J Immunol* 1993;**151**:4383–90.

15. Sullivan KE, Mullen CA, Blaese RM, Winkelstein JA. A multiinstitutional survey of the Wiskott-Aldrich syndrome. *J Pediatr* 1994;**125**:876–85.

16. Westerberg L, Larsson M, Hardy SJ, Fernandez C, Thrasher AJ, Severinson E. Wiskott-Aldrich syndrome protein deficiency leads to reduced B-cell adhesion, migration, and homing, and a delayed humoral immune response. *Blood* 2005;**105**:1144–52.

17. Orange JS, Ramesh N, Remold-O'Donnell E, et al. Wiskott-Aldrich syndrome protein is required for NK cell cytotoxicity and colocalizes with actin to NK cell-activating immunologic synapses. *Proc Natl Acad Sci USA* 2002;**99**:11351–6.

18. Gismondi A, Cifaldi L, Mazza C, et al. Impaired natural and CD16-mediated NK cell cytotoxicity in patients with WAS and XLT: ability of IL-2 to correct NK cell functional defect. *Blood* 2004;**104**:436–43.

19. Maillard MH, Cotta-de-Almeida V, Takeshima F, et al. The Wiskott-Aldrich syndrome protein is required for the function of CD4(+)CD25(+)Foxp3(+) regulatory T-cells. *J Exp Med* 2007;**204**:381–91.

20. Humblet-Baron S, Sather B, Anover S, et al. Wiskott-Aldrich syndrome protein is required for regulatory T-cell homeostasis. *J Clin Invest* 2007;**117**:407–18.

21. Marangoni F, Trifari S, Scaramuzza S, et al. WASP regulates suppressor activity of human and murine CD4(+)CD25(+)FOXP3(+) natural regulatory T-cells. *J Exp Med* 2007;**204**:369–80.

22. Altman LC, Snyderman R, Blaese RM. Abnormalities of chemotactic lymphokine synthesis and mononuclear leukocyte chemotaxis in Wiskott-Aldrich syndrome. *J Clin Invest* 1974;**54**:486–93.

23. Linder S, Nelson D, Weiss M, Aepfelbacher M. Wiskott-Aldrich syndrome protein regulates podosomes in primary human macrophages. *Proc Natl Acad Sci USA* 1999;**96**:9648–53.

24. Notarangelo LD, Mazza C, Giliani S, et al. Missense mutations of the WASP gene cause intermittent X-linked thrombocytopenia. *Blood* 2002;**99**:2268–9.

25. Litzman J, Jones A, Hann I, Chapel H, Strobel S, Morgan G. Intravenous immunoglobulin, splenectomy, and antibiotic prophylaxis in Wiskott-Aldrich syndrome. *Arch Dis Child* 1996;**75**:436–9.

26. Shcherbina A, Rosen FS, Remold-O'Donnell E. Pathological events in platelets of Wiskott-Aldrich syndrome patients. *Br J Haematol* 1999;**106**:875–83.

27. Marathe BM, Prislovsky A, Astrakhan A, Rawlings DJ, Wan JY, Strom TS. Antiplatelet antibodies in WASP(−) mice correlate with evidence of increased in vivo platelet consumption. *Exp Hematol* 2009;**37**:1353–63.

28. Mahlaoui N, Pellier I, Mignot C, et al. Characteristics and outcome of early-onset, severe forms of Wiskott-Aldrich syndrome. *Blood* 2013;**121**:1510–6.

29. Ilowite NT, Fligner CL, Ochs HD, et al. Pulmonary angiitis with atypical lymphoreticular infiltrates in Wiskott-Aldrich syndrome: possible relationship of lymphomatoid granulomatosis and EBV infection. *Clin Immunol Immunopathol* 1986;**41**:479–84.

30. McCluggage WG, Armstrong DJ, Maxwell RJ, Ellis PK, McCluskey DR. Systemic vasculitis and aneurysm formation in the Wiskott-Aldrich syndrome. *J Clin Pathol* 1999;**52**:390–2.

31. Dupuis-Girod S, Medioni J, Haddad E, et al. Autoimmunity in Wiskott-Aldrich syndrome: risk factors, clinical features, and outcome in a single-center cohort of 55 patients. *Pediatrics* 2003;**111**:e622–7.

32. Albert MH, Bittner TC, Nonoyama S, et al. X-linked thrombocytopenia (XLT) due to WAS mutations: clinical characteristics, long-term outcome, and treatment options. *Blood* 2010;**115**:3231–8.

33. Kwan SP, Lehner T, Hagemann T, et al. Localization of the gene for the Wiskott-Aldrich syndrome between two flanking markers, TIMP and DXS255, on Xp11.22-Xp11. 3. *Genomics* 1991;**10**:29–33.

34. Zhu Q, Watanabe C, Liu T, et al. Wiskott-Aldrich syndrome/X-linked thrombocytopenia: WASP gene mutations, protein expression, and phenotype. *Blood* 1997;**90**:2680–9.

35. Ramesh N, Geha R. Recent advances in the biology of WASP and WIP. *Immunol Res* 2009;**44**:99–111.

36. Kim AS, Kakalis LT, Abdul-Manan N, Liu GA, Rosen MK. Autoinhibition and activation mechanisms of the Wiskott-Aldrich syndrome protein. *Nature* 2000;**404**:151–8.

37. Hall A. Rho GTPases and the actin cytoskeleton. *Science* 1998;**279**:509–14.

38. Miki H, Miura K, Takenawa T. N-WASP, a novel actin-depolymerizing protein, regulates the cortical cytoskeletal rearrangement in a PIP2-dependent manner downstream of tyrosine kinases. *EMBO J* 1996;**15**:5326–35.

39. Takenawa T, Suetsugu S. The WASP-WAVE protein network: connecting the membrane to the cytoskeleton. *Nat Rev Mol Cell Biol* 2007;**8**:37–48.

40. Oda A, Ochs HD, Druker BJ, et al. Collagen induces tyrosine phosphorylation of Wiskott-Aldrich syndrome protein in human platelets. *Blood* 1998;**92**:1852–8.

41. Devriendt K, Kim AS, Mathijs G, et al. Constitutively activating mutation in WASP causes X-linked severe congenital neutropenia. *Nat Genet* 2001;**27**:313–7.

42. Ancliff PJ, Blundell MP, Cory GO, et al. Two novel activating mutations in the Wiskott-Aldrich syndrome protein result in congenital neutropenia. *Blood* 2006;**108**:2182–9.

43. Jin Y, Mazza C, Christie JR, et al. Mutations of the Wiskott-Aldrich Syndrome Protein (WASP): hotspots, effect on transcription, and translation and phenotype/genotype correlation. *Blood* 2004;**104**:4010–9.

44. Imai K, Morio T, Zhu Y, et al. Clinical course of patients with WASP gene mutations. *Blood* 2004;**103**:456–64.

45. Lemahieu V, Gastier JM, Francke U. Novel mutations in the Wiskott-Aldrich syndrome protein gene and their effects on transcriptional, translational, and clinical phenotypes. *Hum Mutat* 1999;**14**:54–66.

46. Villa A, Notarangelo L, Macchi P, et al. X-linked thrombocytopenia and Wiskott-Aldrich syndrome are allelic diseases with mutations in the WASP gene. *Nat Genet* 1995;**9**:414–7.

47. Ariga T, Yamada M, Sakiyama Y, Tatsuzawa O. A case of Wiskott-Aldrich syndrome with dual mutations in exon 10 of the WASP gene: an additional de novo one-base insertion, which restores frame shift due to an inherent one-base deletion, detected in the major population of the patient's peripheral blood lymphocytes. *Blood* 1998;**92**:699–701.

48. Wada T, Konno A, Schurman SH, et al. Second-site mutation in the Wiskott-Aldrich syndrome (WAS) protein gene causes somatic mosaicism in two WAS siblings. *J Clin Invest* 2003;**111**:1389–97.

49. Davis BR, Candotti F. Revertant somatic mosaicism in the Wiskott-Aldrich syndrome. *Immunol Res* 2009;**44**:127–31.

50. Boztug K, Germeshausen M, Avedillo Diez I, et al. Multiple independent second-site mutations in two siblings with somatic mosaicism for Wiskott-Aldrich syndrome. *Clin Genet* 2008;**74**:68–74.

51. Trifari S, Scaramuzza S, Catucci M, et al. Revertant T lymphocytes in a patient with Wiskott-Aldrich syndrome: analysis of function and distribution in lymphoid organs. *J Allergy Clin Immunol* 2010;**125**:439–48.

52. Parolini O, Ressmann G, Haas OA, et al. X-linked Wiskott-Aldrich syndrome in a girl. *N Engl J Med* 1998;**338**:291–5.

53. Boonyawat B, Dhanraj S, Al Abbas F, et al. Combined de-novo mutation and non-random X-chromosome inactivation causing Wiskott-Aldrich syndrome in a female with thrombocytopenia. *J Clin Immunol* 2013;**33**:1150–5.

54. Inoue H, Kurosawa H, Nonoyama S, et al. X-linked thrombocytopenia in a girl. *Br J Haematol* 2002;**118**:1163–5.

55. Lutskiy MI, Sasahara Y, Kenney DM, Rosen FS, Remold-O'Donnell E. Wiskott-Aldrich syndrome in a female. *Blood* 2002;**100**:2763–8.

56. Moratto D, Giliani S, Bonfim C, et al. Long-term outcome and lineage-specific chimerism in 194 patients with Wiskott-Aldrich syndrome treated by hematopoietic cell transplantation in the period 1980-2009: an international collaborative study. *Blood* 2011;**118**:1675–84.

57. Boztug K, Schmidt M, Schwarzer A, et al. Stem-cell gene therapy for the Wiskott-Aldrich syndrome. *N Engl J Med* 2010;**363**:1918–27.

58. Aiuti A, Biasco L, Scaramuzza S, et al. Lentiviral hematopoietic stem cell gene therapy in patients with Wiskott-Aldrich syndrome. *Science* 2013;**341**:1233151.

59. Lanzi G, Moratto D, Vairo D, et al. A novel primary human immunodeficiency due to deficiency in the WASP-interacting protein WIP. *J Exp Med* 2012 6 ;**209**: 29-34.

60. Bach FH, Albertini RJ, Joo P, Anderson JL, Bortin MM. Bone-marrow transplantation in a patient with the Wiskott-Aldrich syndrome. *Lancet* 1968;**2**:1364–6.

Neutropenia – More Genetic Defects Than Ever Expected

Karl Welte

Department of Molecular Hematopoiesis, Medical School Hannover,
Hannover, Germany

HISTORICAL MILESTONES (FIG. 10.1)

In 1922, the German hematologist Werner Schultz reported five cases under the heading "Gangrenous processes and injury of the granulocytic system" in the *Deutsche medizinische Wochenschrift* (Fig. 10.1) (1). These patients suffered from stomatitis, gingivitis, tonsillitis, pharyngitis and laryngitis. There was absence of granulocytes in the circulating blood. Death occurred within two weeks. At autopsy, the only abnormal finding was the great diminution of granulocytes and myelocytes in the bone marrow. Schultz applied to this syndrome for the first time the term "agranulocytosis" (1) and,

in some subsequent publications, the term "Schultz syndrome" was used. Others, like Friedemann in 1923 (2) and Prendergast in 1927 (3) reported similar cases which they called "agranulocytic angina". All these publications included clinical syndromes which today we would name "servere chronic neutropenia". The mortality was as high as 90% (4) and did not change until antibiotics became available in the 1950s. Even then with antibiotic therapies, more than 50% of the patients died from severe bacterial infections, and in the textbooks of Hematology from the 1960s and 1970s, congenital neutropenia or agranulocytosis were reported as life-threatening disorders of hematopoiesis for which only symptomatic treatment was

Primary Immunodeficiency Disorders: A Historic and Scientific Perspective

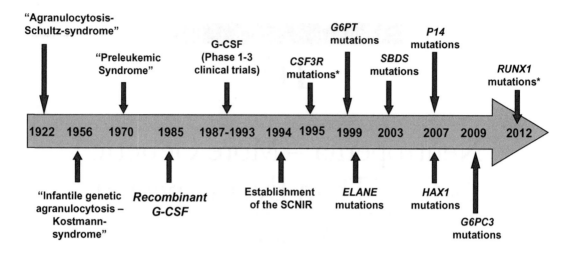

FIGURE 10.1   Milestones in the history of congenital neutropenias. This figure is reproduced in color in the color section.

available. Gilman et al. (5) mentioned that of the 36 cases of genetic agranulocytosis reported in the 1950s and 1960s, only seven (19%) survived past three years of age. In 1950 and 1956, in a Swedish journal (*Svenska Läkartidningen*) (6) and in *Acta Paediatrica Scandinavica* (7), Rolf Kostmann reported the autosomal recessive inheritance of familial agranulocytosis in patients living in a small parish in Northern Sweden and introduced the term "infantile genetic agranulocytosis". Subsequently, the term "Kostmann syndrome" was used for patients with congenital neutropenia. Twelve of the 14 patients reported by Kostmann died of infection before the age of six months. In 1959, in *Helvetica Medica Acta* (8), Walter Hitzig reported a family in whom the father and two children suffered from "familial neutropenia with dominant inheritance". The difference between the patients reported by Kostmann and Hitzig was the recessive versus dominant inheritance. Gilman et al. (5) and Miller (9) were the first to recognize cases of terminal leukemia

in patients with congenital agranulocytosis, suggesting that congenital agranulocytosis is a preleukemic syndrome. The first cloning experiments of patients' bone marrow cells in soft agar were performed in the 1970s (10, 11), demonstrating that human spleen-conditioned media or colony-stimulating activities can induce the formation of granulocyte colonies with normal maturation, suggesting that granulocyte precursors from congenitally neutropenic patients are potentially capable of normal proliferation and differentiation provided that they are supplied with a proper inducer. However, Robert Good's group observed abnormal *in vitro* differentiation of cells from patients with congenital neutropenia, suggesting an intrinsic cell defect in their cohort (12). Based on these contradictory results, some investigators proposed that the complex nomenclature used to differentiate chronic neutropenic states be discarded until a better basis for classification was to become available (13).

A breakthrough in understanding the pathological mechanisms and in the treatment of congenital neutropenias occurred with the purification, biochemical characterization (14) and cloning (15) of granulocyte colony-stimulating factor (G-CSF) in 1985 and 1986 at the Memorial Sloan-Kettering Cancer Center (MSKCC), New York, and at Amgen, Thousand Oaks, CA, USA (see Fig. 10.1). At the time we succeeded in generating pure G-CSF, initially called Pluripoietin (14), a patient with Kostmann syndrome was admitted to the transplant unit of the MSKCC, headed by Richard O'Reilly, to receive a bone marrow transplantation. However, HLA typing did not reveal a suitable donor, allowing us to investigate whether his bone marrow cells might respond to this growth factor of granulopoiesis. They did respond, at least partially, encouraging us to plan a Phase I study with G-CSF in congenital neutropenia patients. Before this study was initiated, we could demonstrate *in vivo* that non-human primates did increase neutrophilic granulocytes in a dose-dependent manner without adverse events (16). The first attempt to induce neutrophil differentiation in patients with congenital neutropenia was initiated in 1987, resulting in an increase in the absolute neutrophil count (ANC) to above 1000 cells/μL or more in all five patients studied (see Fig. 10.1) (17). However, the dose requirements varied. There were no adverse effects on other cell lineages. Pre-existing chronic infections resolved clinically, and the number of new infectious episodes and the requirement for intravenous antibiotics decreased (17).

At the same time, our group in Hannover, Germany, initiated the first clinical trial in Europe, comparing granulocyte-macrophage colony-stimulating factor (GM-CSF) with G-CSF in patients with congenital neutropenia (18). We first treated five patients using recombinant human granulocyte-macrophage colony-stimulating factor (rhGM-CSF) for 42 days. In all patients, we observed a specific, dose-dependent increase in the absolute white blood cell counts.

However, in four patients this was due to an increase in eosinophils, and only one patient experienced an increase in the ANC. Subsequently, all patients received recombinant human G-CSF at a dose of 3–15 μg/kg/d, subcutaneously. In contrast to GM-CSF treatment, all five patients responded to G-CSF during the first six weeks of treatment with an increase in the ANC to above 1000/μL (18). The level of ANC could be kept up during maintenance treatment. In one patient, the increase in ANC was associated with improvement of a severe pneumonitis caused by *Peptostreptococcus* and resistant to antibiotic treatment. No severe bacterial infections occurred in any of the patients during CSF treatment. All patients tolerated GM-CSF and G-CSF treatment without severe side effects. These results demonstrated the beneficial effect of rhG-CSF in severe congenital neutropenia (SCN) patients in contrast to GM-CSF. These findings confirmed that rhG-CSF is the most promising of all available treatments for congenital neutropenia (18). The correction of neutropenia resulting in improvement of clinical status dramatically changed the high morbidity, and therefore the quality of life in these patients.

David Dale and collegues (19) conducted a randomized controlled phase III trial with G-CSF for chronic neutropenia. One hundred and twenty-three patients with recurrent infections and severe chronic neutropenia (absolute neutrophil counts below 500/μL) were enrolled in this multicenter trial. They were randomized to either immediately beginning recombinant human G-CSF (filgrastim) (3.45–11.50 μg/kg/d, subcutaneously) or to enter a four-month observation period followed by G-CSF administration. Blood neutrophil counts, bone marrow cell histology and incidence and duration of infection-related events were monitored. Of the 123 patients enrolled, 120 received G-CSF. On therapy, 108 patients reached a median absolute neutrophil count of >1500/μL. Infection-related events were significantly decreased

($P < 0.05$) with approximately a 50% reduction in the incidence and duration of infection-related events and an almost 70% reduction in the duration of antibiotic use. Asymptomatic splenic enlargement occurred frequently; other adverse events reported were bone pain, headache and rash, which were generally mild and easily manageable. These data indicate that G-CSF treatment of patients with severe chronic neutropenia induces the production and maturation of neutrophils, increases circulating neutrophils in the bone marrow and reduces infection-related events.

In 1994, we established the Severe Chronic Neutropenia International Registry (SCNIR) to monitor the clinical course, treatment and disease outcome in patients with severe chronic neutropenia (see Fig. 10.1) (20, 21). The Registry has now the largest collection of long-term data collected worldwide from patients with this condition. As of 2014, we have collected data from more than 1500 patients from the USA, Canada, Australia and Europe. Participation in the Registry benefits patients, their families and the physicians involved in the care of these patients by providing the most up-to-date information on the natural history, disease mechanisms and treatment of congenital neutropenia. The Registry was initially sponsored by a grant from Amgen. In 2000, the SCNIR became an independent foundation with headquarters at the University of Washington, Seattle, WA, USA and the Medical School Hannover, Germany with David Dale, Seattle, and Karl Welte, Hannover as co-directors.

PATHOLOGICAL MECHANISMS

Our current knowledge of the pathological mechanisms of severe congenital neutropenia suggests a heterogeneous group of disorders with a common hematological and clinical phenotype characterized by a maturation arrest of myelopoiesis at the level of the

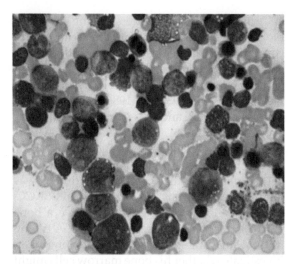

FIGURE 10.2 Typical bone marrow of a patient with congenital neutropenia. This figure is reproduced in color in the color section.

promyelocyte/myelocyte stage (Fig. 10.2), resulting in a peripheral blood ANC $<500/\mu L$ and early onset of bacterial infections. Congenital neutropenia patients produce their own physiological amounts of G-CSF, but it does not lead to normal differentiation of myeloid precursor cells to mature neutrophils and, therefore, treatment with pharmalogical dosages of recombinant G-CSF is required (see below). Also the number and distribution of G-CSF receptors on myeloid precursor cells are normal, suggesting that the G-CSF receptor-downstream signaling pathways are defective (20, 21). Congenital neutropenia (CN) follows an autosomal dominant or autosomal recessive pattern of inheritance. Over the last ten years, remarkable progress has been achieved with respect to the identification of genetic defects causing CN. The underlying genetic defect of Kostmann syndrome has recently been identified as mutations in the *HAX1* gene (22, 23). Genetic analyses in autosomal dominant and sporadic cases of CN indicate that the majority of these cases are attributable to mutations in the elastase 2 (*ELANE*) gene encoding neutrophil elastase (24, 25). Interestingly, patients with cyclic neutropenia also harbor mutations within the *ELANE* gene. In addition, mutations in a number of rarely affected genes have been identified in autosomal recessive CN (see below). CN needs to be differentiated from secondary causes of neutropenia such as autoimmune neutropenia, infection or drug-induced neutropenia.

CN Patients With *ELANE* Mutations (ELA-CN)

ELANE encodes neutrophil elastase (NE), a protease synthesized and packaged in the primary granules of neutrophil precursors at the promyelocyte stage of development. The gene is not expressed in other tissues. Heterozygous mutations in the *ELANE* gene are the most common genetic abnormality, found in approximately 40–50% of CN patients and in more than 80% of cyclic neutropenia patients, suggesting a dominant mechanism of action (24, 25). However, some cases of CN with *ELANE* mutations arise sporadically, consistent with its transmission as an autosomal dominant disorder (25). More than 100 mutations, located in all five exons of the *ELANE* gene, have been described in CN patients resulting in proteins with a wide range of enzyme activities. Two case reports of paternal mosaicism for an *ELANE* mutation also provide evidence for autosomal dominant inheritance and indicating that mutant NE protein has no effect on wild-type neutrophils (26, 27). Evidence for the causative role of *ELANE* mutations has been derived from a recently published study of five unrelated children from healthy mothers, who have been impregnated with the semen from the same sperm donor (28). Furthermore, mutations in the *ELANE* gene are responsible for the clinical phenotype of cyclic neutropenia, another autosomal dominant inherited disorder with regular neutropenic phases due to the cycling of blood cells, but less severe infections. Genetic analysis has shown that patients with congenital and cyclic neutropenia can comprise

ELANE mutations at the same site of the gene. However, the mechanism(s) responsible for the development of the different phenotypes, congenital or cyclic neutropenia, are not yet understood. The diversity of mutations within the *ELANE* gene in CN and the lack of any consistent effect of the mutated NE protein on its enzymatic properties led us and others to hypothesize that structural rather than functional enzymatic characteristics of the mutated NE protein may be responsible for the neutropenia. Three recent studies propose that mutations in the *ELANE* gene result in the production of misfolded NE protein in the endoplasmic reticulum (ER) which induces an unfolded protein response (UPR) leading to subsequent UPR-dependent apoptosis (29–31). Indeed, the chaperone family member HsP70 protein BiP, which is associated with misfolded-protein-induced ER stress, is upregulated in CN (29). An as yet unexplained finding is the observation that the expression of NE protein is severely downregulated in CN, independent of whether CN patients have *ELANE* mutations or not (32, 33). As to the role *ELANE* mutations play in causing CN, additional studies are required to understand the pathological mechanism(s) leading to maturation arrest at the promyelocyte stage in these patients.

CN Patients With *HAX1* Mutations (HAX1-CN)

Biallelic *HAX1* mutations were detected in patients belonging to the original pedigree described by Kostmann (6, 7), and were also found in CN patients of consanguineous parents (22, 23), demonstrating that HAX1-CN is an autosomal recessive form of CN. The HAX1 protein is ubiquitously expressed and localizes mainly to the mitochondrial membrane and, to a lesser extent, to the endoplasmatic reticulum and nuclear envelope. To date, the generally held hypothesis of the mechanism of neutropenia in CN patients carrying *HAX1* mutations has been that mutated HAX1 proteins cause apoptosis of myeloid cells via defective mitochondrial membrane potential (22). Since HAX1 is a ubiquitously expressed protein, it was unclear why in CN patients with *HAX1* mutations only granulocytic precursors undergo apoptosis leading to isolated neutropenia. An explanation is the finding that in patients with *HAX1* mutations, G-CSF failed to induce the expression and function of the hematopoietic-specific HAX1 binding partner HCLS1 *in vitro* and *in vivo* (34), suggesting that both HAX1 and HCLS1 are required for G-CSF-receptor signaling, and that downregulation of HCLS1 by defective HAX1 expression caused isolated myeloid defects in CN patients. G-CSF leads to phosphorylation of HCLS1 protein via the interaction of HCLS1 with G-CSF receptor-associated tyrosine kinases Lyn and Syk. HCLS1 binds to the LEF-1 transcription factor, allowing the transport of LEF-1 into the nucleus upon G-CSF stimulation (34). Moreover, we could show that LEF-1 is the master regulator of myelopoiesis by controlling expression of myelopoiesis-specific transcription factors such as C/EBP-alpha (35). Patients with SCN due to mutations in the *HAX1* gene have a profound defect in the G-CSF-triggered expression and phosphorylation of the HCLS1 protein. As a consequence, autoregulation and expression of LEF-1 are abrogated, leading to defective granulopoiesis (34). *In vitro* inhibition of HCLS1 or HAX1 by short hairpin RNA (shRNA) severely disrupted granulocytic differentiation of CD34+ cells, due to defective translocation of LEF-1 to the nucleus and subsequent lack of LEF-1 and C/EBPa expression, demonstrating a significant role of HCLS1 and HAX1 in G-CSF-triggered granulopoiesis (34). Thus, we have demonstrated that the HCLS1 binding protein HAX1 is essential for HCLS1 and LEF-1 interaction. Intriguingly, central nervous system involvement appears to be restricted to those patients in whom the *HAX1* mutation affects both alternatively spliced isoforms of HAX1 (36, 37).

CN Patients With *G6PC3* Mutations (G6PC3-CN)

We have recently identified a previously unrecognized syndrome associated with CN and variable extra-hematopoietic features, caused by biallelic mutations in the gene encoding the glucose-6-phosphate catalytic subunit 3 (*G6PC3*) (38). Eleven of 12 patients also had additional organ involvement, including structural heart defects, urogenital defects and increased visibility of superficial veins. The role of G6PC3 in granulopoiesis is supported by the finding that *g6pc3−/−* knockout mice also reveal neutropenia and increased myeloid cell apoptosis (39). In contrast to glycogen storage disease type Ib, which is also associated with severe neutropenia, G6PC3-CN patients do not reveal glycogen storage symptoms. Since in glycogen storage disease type Ib, the glucose-6-phosphate transporter (G6PT) is defective, and since G6PT forms a complex with G6PC3, it is very likely that the G6PT/G6PC3 complex is needed to maintain myeloid cell viability including neutrophil homeostasis. The underlying pathway leading to neutropenia in the absence of G6PC3 involves increased ER-stress and increased apoptosis (38).

Other Rare Causes of CN

Neutropenia caused by mutations in other genes such as *GFI-1* (40), *WASp* (41, 42), p14 (43), *TAZ* (44, 45), etc. are very rare and have been reported in few patients or families only.

Common Pathological Mechanisms

Patients with *ELA2-*, *HAX1-*, or *G6PC3-*mutations reveal similar morphological (see Fig. 10.2) and clinical phenotypes in terms of myelopoiesis and response to G-CSF, suggesting that there are common downstream molecular events caused by these mutations. Both misfolded elastase protein and G6PC3 protein have in common that they cause ER stress and UPR (31, 38). The UPR induces a general decrease of protein synthesis through inhibition of translation initiation, increase of chaperone protein expression and activation of ER-associated protein degradation. If these responses are inadequate to compensate for the quantity of misfolded proteins, they induce apoptosis through increased caspase expression. *HAX1* mutation leading to lack of a functional HAX1 protein compromises vital function of mitochondria (22), also leading to increased apoptosis. Downstream signals secondary to these events lead to the lack of expression of the transcription factor LEF-1 and subsequent C/EBP-alpha downregulation, thus reducing the survival and differentiation of granulocytic progenitors and leading to the "maturation arrest" of granulopoiesis in bone marrow and resulting in lower numbers of blood neutrophils (35). C/EBP-alpha, however, is the transcription factor required for steady state neutrophilic granulopoiesis (46). Therefore, the question arises, which transcription factor(s) are responsible for the *in vivo* response of CN patients to pharmacological doses of G-CSF? We identified nicotinamide-phosphoribosyltransferase (Nampt) as a novel and essential enzyme mediating this process by determining the signaling pathways activated during G-CSF-triggered granulopoiesis in healthy individuals and in patients with CN (47). Intracellular Nampt and NAD+ levels in myeloid cells, as well as in plasma, were significantly higher in long-term G-CSF-treated CN patients (47). The molecular events triggered by elevated Nampt levels include NAD+/SIRT1-dependent activation, subsequent induction of C/EBP-alpha and C/EBP-beta and, ultimately, upregulation of G-CSF synthesis and G-CSF receptor expression. G-CSF, in turn, further increases Nampt levels. Therefore, G-CSF acts in CN patients through Nampt to increase intracellular NAD+ which, in combination with SIRT1, binds and activates C/EBP-beta leading to the so-called

"emergency" neutrophil granulopoiesis (46, 47) answering the question: how does G-CSF lead to an increase of neutrophils in CN patients?

Acquired G-CSF Receptor Mutations

Mutations in a region of the G-CSF-receptor (*CSF3R*) gene coding for the intracytoplasmatic domain of the G-CSFR were discovered in 1994 by Ivo Touw's group (48), and were initially suggested to be the cause of CN. Subsequently, it became clear that these mutations were somatic mutations acquired during the life of CN patients (49, 50). All mutations introduce a stop codon predicted to lead to the truncation of the intracellular part of the G-CSFR protein with the loss of the carboxy-terminal-negative regulatory domain that includes loss of one or more tyrosines (49). However, in the majority of patients, only one allele of *CSF3R* is affected. Expression of these mutations in myeloid cell lines leads to enhanced proliferation, resistance to apoptosis and increased cell survival (51–53). Knock-in mice bearing receptor mutants equivalent to those found in CN patients revealed a hyperproliferative response to exogenous G-CSF, but despite prolonged G-CSF administration, none of the mice developed leukemia (53). In a recent study (54), *CSF3R* mutation analysis was performed in 148 CN patients of whom 61 (41%) harbored *CSF3R* mutations. Interestingly, five out of 30 patients tested prior to G-CSF treatment revealed *CSF3R* mutations, suggesting that G-CSF administration is not responsible for these mutations. Of note, no mutations were detected in patients with cyclic neutropenia or other types of chronic neutropenia. The incidence of *CSF3R* mutations in CN patients who have not developed leukemia so far was 34% (43 out of 125) and 78% in patients who have transformed into acute myeloid leukemia (AML) or acute lymphoblastic leukemia (ALL) (18 out of 23 patients). *CSF3R* mutations were detected at approximately 20 different nucleotide positions (54). Some patients displayed a *CS-F3R* mutation at one nucleotide position only, whereas other patients were found to harbor mutations at two or more nucleotide positions. In many cases, increasing numbers of different mutations developed during the course of observation, whereas low percentages of clones were detected at initial analysis with increasing mutation frequency (number of different nucleotides and number of positive clones) over several years (54, 55) suggesting a genetic instability of at least the *CSF3R* gene.

Based on the clonal succession of various *CSF3R* mutations, we conclude that the acquisition of a *CSF3R* mutation is an event that does occur prior to malignant transformation (54). The highly elevated risk of leukemic progression in patients who have acquired a *CSF3R* mutation argues for a significant contribution of these mutations in leukemogenesis.

The question of how *CSF3R* mutations are acquired during the course of life and how the affected cells contribute to the progression to leukemia is not yet understood. One possible explanation is that the truncated G-CSFR protein affects G-CSFR endocytosis and G-CSFR signaling (56–59). Indeed, cells harboring mutated G-CSFR protein reveal activated (phosphorylated) STAT5 protein, most likely by defective endocytosis of G-CSFR, thereby prolonging the G-CSFR signaling. Ivo Touw and colleagues suggested that SOCS3 efficiently suppresses STAT3 and STAT5 activation by G-CSFR in wild-type cells, whereas in cells harboring truncated G-CSFR, SOCS3 still inhibits STAT3, but the inhibition of STAT5 activation is completely lost (57–59). SOCS3-induced inhibition of STAT5 requires mechanisms controlled by the G-CSFR C-terminus, which is missing in the G-CSFR (CSF3R) mutated protein. The increased STAT5/STAT3 activation ratio results in signaling abnormalities of truncated G-CSFR (CSF3R) potentially being leukemogenic (59). We and others have shown that constitutive activation of STAT5 alters myelopoiesis by

downregulation of myeloid-associated differentiation factors such as C/EBP-alpha (60, 61). We found that G-CSF-treated myeloid progenitor cells from CN patients express high levels of activated phospho-STAT5 protein (61). These data were further confirmed by findings in a recently published mouse model (62). Studies of these mice have shown that G-CSFR (*CSF3R*) mutations confer a strong clonal advantage of hematopoietic stem cells via activation of STAT5 that is dependent on exogenous G-CSF (62). Activated STAT5 multimerizes and then translocates to the nucleus to function as a transcription factor. Defining promoter binding sites unique to stem cells bearing the *CSF3R* mutation may help to identify gene products that are directly involved in clonal expansion of hematopoietic stem cells and/or leukemic clones (63). A possible mechanism of leukemogenesis could be the prolonged activation of STAT5 in G-CSFR (*CSF3R*) mutation-bearing cells, rendering them susceptible to additional molecular aberrations.

In a recent study, we were able to demonstrate that *CSF3R* mutations are not restricted to the myeloid compartment but are also detectable in lymphoid cells, although in lower percentages (64). In this study, we also could show that the mutated G-CSFR-receptor-bearing cells have an *in vivo* growth advantage over cells with wild-type G-CSFR. The altered signaling in CSF3R mutated cells which confer a G-CSF dependent competitive advantage of hematopoietic progenitors may allow the acquisition of additional mutations or lead to chromosomal aberrations which ultimately cause overt leukemia.

Intriguingly, the majority of patients with *CSF3R* mutations (75%) had also acquired *RUNX1* mutations. The high frequency of *RUNX1* mutations in CN patients who developed leukemia and the high incidence of *CSF3R* mutations preceding these mutations suggests a unique molecular pathway of leukemia development (65).

CLINICAL PRESENTATION

At diagnosis, ANC, site of infections and bone marrow morphology (see Fig. 10.2) do not distinguish between patients harboring *ELANE* or *HAX1* mutations. However, family history of consanguineous parenthood is indicative for the autosomal recessive subtypes with *HAX1* or *G6PC3* mutations. In the course of the disease, all patient groups independent of the underlying genetic defect respond well to treatment with the hematopoietic growth factor G-CSF. The ANCs of patients on G-CSF treatment (filgrastim, lenograstim) were recorded over a time period of up to 20 years and are shown in Figs. 10.3 and 10.4. In HAX1-deficient patients, the onset of neurological symptoms during childhood (cognitive defects, mental retardation, epilepsy) may be associated with congenital neutropenia (36, 37). Recent findings suggest a genotype–phenotype correlation, since central nervous system involvement appears to be restricted to those patients in whom the *HAX1* mutation affects both alternatively spliced isoforms of HAX1 (36, 37, 66, 67).

The distribution of the genetic defects determined in more than 300 CN patients from the SCNIR (European branch) tested for mutations in *ELA2*, *HAX1*, *G6PC3* and other genes is shown in Figure 10.5. Of the 226 patients in whom we detected mutations, 91 patients (40%) harbor *ELANE* mutations, 32 (14%) *HAX1* mutations, and 10 patients *G6PC3* mutations. Currently, a comprehensive genetic analysis of all registered patients within the SCNIR is ongoing, in order to determine the prevalence of *HAX1/ELA2* and other gene mutations. We anticipate that the frequency of mutations in various neutropenia-related genes will be dependent on ethnic background and consanguinity. Both HAX1- and ELA2- patient groups have an increased risk of leukemic transformation, whereas in none of the other patient groups with known mutations (e.g. *G6PC3*, *WASp*, *p14*, *GFI1*) have leukemias been reported so far. Under cytokine treatment,

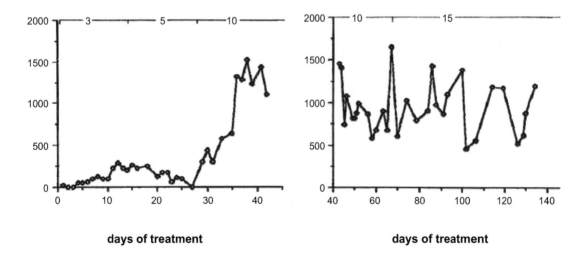

days of treatment   **days of treatment**

Increasing dosages of G-CSF (3, 5,10,15 µg/kg/d s.c.) were used until the ANC of at least 1000/µl was maintained

FIGURE 10.3   Neutrophil counts/µL (ANC) in Patient #1 with congenital neutropenia.

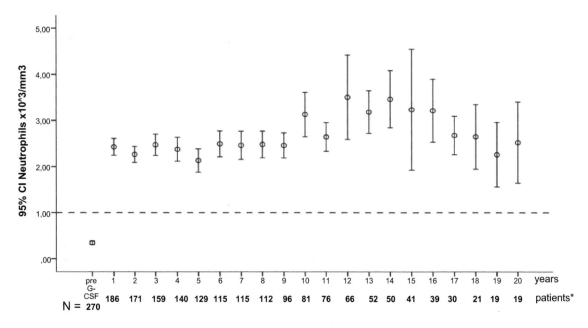

*total number of patients evaluated = 289, the number in each year represent the number of patients evaluated in
the individual years of treatment (19 patients are treated for more than 20 years)

FIGURE 10.4 Absolute neutrophil counts in patients with severe congenital neutropenias.

the most important parameter for the risk of bacterial infections is the neutrophil count and not the genetic subtype. Therefore, maintenance of a sufficient neutrophil count (more than $1000/\mu L$) is essential to minimize the risk of severe infections. However, even with these blood counts, patients may still suffer from gingivitis and develop early parodontitis even during childhood. A current study is evaluating the correlation between genetic subtype, ANC response and dental care. Preliminary results suggest no difference by genetic subtype, but lower risk with sufficiently high mean ANCs and continuous good dental hygiene (68, 69). The development of early onset osteopenia and osteoporosis has been reported in approximately 40% of patients (70). However, the underlying mechanisms of bone abnormalities are still unclear. The development of osteopenia also seems to be

independent of the underlying genetic defect, but further genotype–phenotype correlation analysis is required.

In addition to the hematopoietic manifestations, some genetic subtypes may present with additional clinical features, such as organ anomalies involving heart and urogenital tract in G6PC3-CN and neurological abnormalities, like seizures and delayed mental development, in a proportion of HAX1-CN patients. It is known from other bone marrow failure syndromes associated with neutropenia, such as Shwachman–Diamond syndrome or Barth syndrome, that the clinical presentation may be extremely heterogeneous. However, in the CN subtypes described above, the clinical presentation is not yet fully defined and with new patients identified, further clinical features may be described.

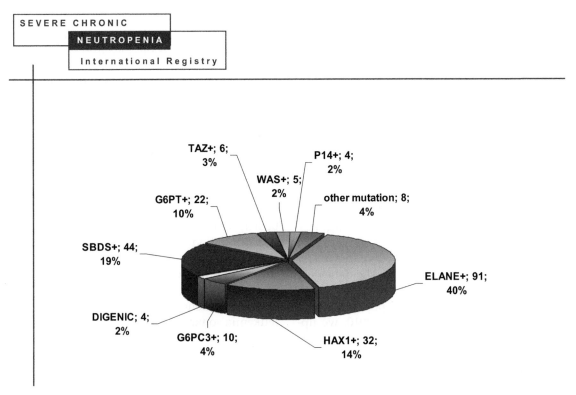

FIGURE 10.5 Distribution of gene mutations in European congenital neutropenia patients. This figure is reproduced in color in the color section.

TREATMENT

As mentioned above, recombinant human granulocyte-colony stimulating factor (G-CSF) (filgrastim or lenograstim) has been available since 1987 for treatment of CN (15–18). Usually the initial dose is 3 μg/kg/d for 14 days; if the ANC does not increase to above 1000/μL blood, the dose is increased and adjusted accordingly to maintain a complete response, which is defined as peripheral blood neutrophil counts of above 1000/μL (see Figs. 10.3 and 10.4). An extensive study on the first 374 patients demonstrated that the median pretreatment ANC was 129/mL and that the ANC increased upon G-CSF treatment to a median of 2125/mL (71). The G-CSF dose to achieve and maintain an ANC above 1000/mL varied between 1 and 120 μg/kg/d, with the majority of patients responding to G-CSF dosages below 20 μg/kg/d (71). Discrimination of G-CSF dosages by genotype shows that HAX1 patients require a similar median dose (5.5 μg/kg/d) as compared to ELA2 patients with 5 μg/kg/day, suggestive of an equal severity of the underlying defect of granulocyte differentiation. Patients with G6PC3 mutations require 4 μg/kg/d. However, statistical significance is not reached due to the limited patient numbers in the subgroups.

For patients who do not respond to G-CSF treatment at dosages of more than 80 μg/kg/d, hematopoietic stem cell transplantation (HSCT) is the only currently available treatment (72). Following successful transplantation, patients normalize their peripheral blood counts and do not require further G-CSF treatment.

LEUKEMIA SECONDARY TO CN

It is now well accepted that CN is a pre-leukemic syndrome, and a significant proportion of CN patients develop leukemia (5, 9, 20, 21, 54, 56, 65, 71–77).

The SCNIR recently reported on the first 374 CN patients (enrolled 1987–2000) on long-term G-CSF therapy in more detail to identify the risk of leukemic transformation (73). The cumulative incidence of leukemia was 21% after ten years (73). Intriguingly, the risk of leukemia increased with the dose of G-CSF. Less responsive patients, who require more than 8 μg/kg/day of G-CSF, had a cumulative incidence of leukemia of 40% after ten years, compared to 11% in more responsive patients (73). The data were interpreted as indicating that a poor response to G-CSF defines an "at risk" CN population and predicts an adverse outcome. In 2010, we updated this study on long-term G-CSF patients enrolled in the SCNIR. The annual risk of myelodysplastic syndrome (MDS)/AML reached a plateau with 2.3%/year after ten years. This risk now appears similar to, rather than higher than, the risk of AML in Fanconi anemia and dyskeratosis congenita (74).

Independent of the genetic subtype, conversion to leukemia in CN patients was associated with one or more cellular genetic abnormalities, for example monosomy 7, RAS mutations, trisomy 21, or G-CSF receptor (CSF3R) mutations, which may be diagnostically useful in identifying subgroups of patients at high risk of developing leukemia. Interestingly, marrow cells from nearly 80% of the CN patients who transformed to leukemia showed somatic point mutations in the CSF3R gene, suggesting that these mutations may play an important role in leukemogenesis. In patients who have not yet developed leukemia, the frequency of CSF3R mutations is much lower (approximately 30%) (54). In both ELA2 and HAX1 patients, crucial mutations occurred during the course of the disease. The time course between acquisition of a CSF3R mutation and development of leukemia varies considerably. In some patients, CSF3R gene mutations are only present in the leukemic cells. Other patients show single or multiple mutations of the CSF3R gene for several years prior to leukemic transformation (54, 55). Interestingly, CSF3R mutations can be detected in patients transforming not only into AMLs but also ALL and chronic myelomonocytic leukemia (54, 76, 77). G-CSFR (CSF3R) analyses cannot be used for diagnostic purposes for the underlying disease, but may be helpful in screening for the risk of malignant transformation – for growth of preleukemic cell clones and for overt leukemia. Interestingly, in the small group of patients with G6PC3 mutations, none of the patients acquired CSF3R mutations and none of the patients has developed leukemia so far.

Twenty of 31 patients (64.5%) with AML secondary to CN were found to have mutations within RUNX1 (65). Intriguingly, the majority of patients with RUNX1 mutations (75%) had acquired CSF3R mutations. The high frequency of RUNX1 mutations in CN patients who developed leukemia and the high incidence of CSF3R mutations preceding these mutations suggest a unique molecular pathway of leukemia development (65).

A recent report hypothesized a link between G-CSF administration and the development of monosomy 7 in CN patients (78), documenting that monosomy 7 cells are abnormally sensitive to high concentrations of G-CSF and that G-CSF leads to an expansion of a pre-existing monosomy 7 clone (79).

The development of leukemia is a multistep process characterized by a series of cellular genetic changes suggesting a genetic predisposition to malignant transformation. If and how G-CSF affects this predisposition remains unclear; furthermore, there are no historical controls for comparison to resolve the issue. Long-term therapy, however, with pharmacological doses of G-CSF may cause genomic instability due to increased pressure on cell division and DNA replication. Moreover, G-CSF may lead to a preferential secondary outgrowth of a pre-existing cell clone with mutations for example in *CSF3R* or/ and *RUNX1*.

For patients who have developed leukemia, hematopoietic stem cell transplantation is currently the only available treatment (72). HSCT should be performed as early as possible and intensive AML-type chemotherapy should in general not be given prior to conditioning for HSCT to avoid toxicities. The survival post-HSCT is approximately 80%.

CONCLUSION

Gene mutations involved in the pathophysiology of CN include *ELANE* mutations in autosomal dominant CN and *HAX1-* and *G3PC3* mutations in autosomal recessive CN. Clinical presentation and bone marrow morphology do not discriminate between genetic subtypes in early childhood, but additional symptoms, like epilepsy, which are typical of a subgroup of patients with *HAX1* mutations, may occur later in childhood. Patients with *G6PC3* or *TAZ* mutations present with additional heart and urogenital anomalies. The use of G-CSF remains first-line treatment for all CN patients. Maintenance of ANC above 1000/μL is important for the prevention of severe infections. Regardless of genetic subtype, CN patients are at risk of leukemic transformation. The acquisition of *CSF3R* mutations occurs irrespective of the underlying genetic defect and identifies patients at risk for leukemic transformation. Patients who develop *RUNX1* mutations, monosomy 7, other significant chromosomal abnormalities or MDS/ leukemia should urgently proceed to HSCT. Further genotype–phenotype correlation analysis is required to identify subgroup specific risks.

Acknowledgments

We are indebted to all the patients and their families who agreed to participate in these studies. We are also grateful to the many physicians worldwide who faithfully and generously submitted data on their patients. We thank all colleagues of the Severe Chronic Neutropenia International Registry at the University of Washington, Seattle, Washington, USA (Director: Dr David Dale), and the Severe Chronic Neutropenia European Registry (SCNER) at the Medizinische Hochschule Hannover, Germany (Dr Cornelia Zeidler).

References

1. Schultz W. Gangräneszierende Prozesse und Defekt des Granulozytensystems. *Dtsche Med Wochensch* 1922;**48**:1495–6.
2. Friedemann V. Agranulocytic angina. *Med Klin* 1923;**19**:1957.
3. Prendergast D. A case of agranulocytic angina. *Can Med Assoc J* 1927;**17**:446–7.
4. Chown G, Gelfand AS. Agranulocytosis. *Can Med Assoc J* 1933;**29**:128–34.
5. Gilman PA, Jackson DP, Guild HC. Congenital agranulocytosis: prolonged survival and terminal acute leukemia. *Blood* 1970;**36**:576–85.
6. Kostmann R. Hereditär reticulos – en ny systemsjukdom. *Svenska Läkartidningen* 1950;**47**:2861–8.
7. Kostmann R. Infantile genetic agranulocytosis. *Acta Paediatr Scand* 1956;**45**(Suppl. 105):1–78.
8. Hitzig WH. Familiäre Neutropenie mit dominantem Erbgang und Hypergammaglobulinämie. *Helvet Med Acta* 1959;**26**:779–84.
9. Miller RW. Childhood cancer and congenital defects. A study of US death certificates during the period 1960–66. *Pediatr Res* 1969;**3**:389.
10. Barak Y, Paran M, Levin S, Sachs L. In vitro induction of myeloid proliferation and maturation in infantile genetic agranulocytosis. *Blood* 1971;**38**:74–80.
11. Amato D, Freedman MH, Saunders EF. Granulopoiesis in severe congenital neutropenia. *Blood* 1976;**47**:531–8.
12. Zucker-Franklin D, L'Esperance PL, Good R. Congenital neutropenia: an intrinsic cell defect demonstrated by electron microscopy of soft agar colonies. *Blood* 1977;**49**:425–36.
13. Pincus SH, Boxer LA. Stossel TP Chronic neutropenia in childhood. Analysis of 16 cases and review of the literature. *Am J Med* 1976;**61**:849–61.
14. Welte K, Platzer E, Gabrilove JL, Levi E, Mertelsmann R, Moore MA. Purification and biochemical characterization of human pluripotent hematopoietic colony-stimulating factor. *Proc Natl Acad Sci USA* 1985;**82**:1526–30.

15. Souza LM, Boone TC, Gabrilove J, et al. Recombinant human granulocyte colony-stimulating factor: effects on normal and leukemic myeloid cells. *Science* 1986;**232**:61–5.

16. Welte K, Bonilla MA, Gillio AP, et al. Recombinant human granulocyte colony-stimulating factor. Effects on hematopoiesis in normal and cyclophosphamide-treated primates. *J Exp Med* 1987;**165**:941–8.

17. Bonilla MA, Gillio AP, Ruggeiro, M, et al. Effects of recombinant human granulocyte colony-stimulating factor on neutropenia in patients with congenital agranulocytosis. *N Engl J Med* 1989;**320**:1574–80.

18. Welte K, Zeidler C, Reiter A, et al. Differential effects of granulocyte-macrophage colony-stimulating factor and granulocyte colony-stimulating factor in children with severe congenital neutropenia. *Blood* 1990;**75**:1056–63.

19. Dale DC, Bonilla MA, Davis MW, et al. A randomized controlled phase III trial of recombinant human granulocyte colony-stimulating factor (Filgrastim) for treatment of severe chronic neutropenia. *Blood* 1993;**81**:2496–502.

20. Welte K, Boxer L. Severe chronic neutropenia: Pathophysiology and therapy. *Semin Hematol* 1997;**34**:267–78.

21. Welte K, Zeidler C, Dale DC. Severe congenital neutropenia. *Semin Hematol* 2006;**43**:189–95.

22. Klein C, Grudzien M, Appaswamy G, et al. HAX1 deficiency causes autosomal recessive severe CN (Kostmann disease). *Nat Genet* 2007;**39**:86–92.

23. Germeshausen M, Grudzien M, Zeidler C, et al. Novel HAX1 mutations in patients with severe congenital neutropenia reveal isoform-dependent genotype-phenotype associations. *Blood* 2008;**111**:4954–7.

24. Dale DC, Person DE, Bolyard AA, et al. Mutations in the gene encoding neutrophil elastase in congenital and cyclic neutropenia. *Blood* 2000;**96**:2317–22.

25. Horwitz MS, Duan Z, Korkmaz B, et al. Neutrophil elastase in cyclic and severe congenital neutropenia. *Blood* 2007;**109**:1817–24.

26. Ancliff PJ, Gale RE, Watts MJ, et al. Paternal mosaicism proves the pathogenic nature of mutations in neutrophil elastase in severe congenital neutropenia. *Blood* 2002;**100**:707–9.

27. Germeshausen M, Schulze H, Ballmaier M, et al. Mutations in the gene encoding neutrophil elastase (ELA2) are not sufficient to cause the phenotype of congenital neutropenia. *Br J Haematol* 2001;**115**:222–4.

28. Boxer LA, Stein S, Buckley D, et al. Strong evidence for autosomal dominant inheritance of severe congenital neutropenia associated with ELA2 mutations. *J Pediatr* 2006;**148**:633–6.

29. Köllner I, Sodeik B, Schreek S, et al. Mutations in neutrophil elastase causing congenital neutropenia lead to cytoplasmic protein accumulation and induction of the unfolded protein response. *Blood* 2006;**108**:493–500.

30. Grenda DS, Murakami M, Ghatak J, et al. Mutations of the ELA2 gene found in patients with severe congenital neutropenia induce the unfolded protein response and cellular apoptosis. *Blood* 2007;**110**:4179–87.

31. Xia J, Link DC. Severe congenital neutropenia and the unfolded protein response. *Curr Opin Hematol* 2008;**15**:1–7.

32. Kawaguchi H, Kobayashi M, Nakamura K, et al. Dysregulation of transcriptions in primary granule constituents during myeloid proliferation and differentiation in patients with severe congenital neutropenia. *J Leukoc Biol* 2003;**73**:225–34.

33. Skokowa J, Fobiwe JP, Dan L, Thakur BK, Welte K. Neutrophil elastase is severely down-regulated in severe congenital neutropenia independent of ELA2 or HAX1 mutations but dependent on LEF-1. *Blood* 2009;**114**:3044–51.

34. Skokowa J, Klimiankou M, Klimenkova O, et al. Interactions among HCLS1, HAX1 and LEF-1 proteins are essential for G-CSF-triggered granulopoiesis. *Nat Med* 2012;**18**:1550–9.

35. Skokowa J, Cario G, Uenalan M, et al. LEF-1 is crucial for neutrophil granulocytopoiesis and its expression is severely reduced in congenital neutropenia. *Nat Med* 2006;**12**:1191–7.

36. Carlsson G, van't Hooft I, Melin M, et al. Central nervous system involvement in severe congenital neutropenia: neurological and neuropsychological abnormalities associated with specific HAX1 mutations. *J Intern Med* 2008;**264**:388–400.

37. Germeshausen M, Grudzien M, Zeidler C, et al. Novel HAX1 mutations in patients with severe congenital neutropenia reveal isoform-dependent genotype-phenotype associations. *Blood* 2008;**111**:4954–7.

38. Boztug K, Appaswamy G, Ashikov A, et al. A novel syndrome with severe congenital neutropenia is caused by mutations in G6PC3. *N Engl J Med* 2009;**360**:32–43.

39. Cheung YY, Kim SY, Yiu WH, et al. Impaired neutrophil activity and increased susceptibility to bacterial infections in mice lacking glucose-6-phosphate-beta. *J Clin Invest* 2007;**117**:784–93.

40. Person RE, Li FQ, Duan Z, et al. Mutations in protooncogen GFI1 causes human neutropenia and target ELA2. *Nat Genet* 2003;**34**:308–12.

41. Devriendt K, Kim AS, Mathijs G, et al. Constitutive activating mutations in WASP causes X-linked severe congenital neutropenia. *Nat Genet* 2001;**27**:313–7.

42. Ancliff PJ, Blundell MP, Cory G, et al. Two novel activating mutations in the Wiskott–Aldrich syndrome protein result in congenital neutropenia. *Blood* 2006;**108**:2182–9.

43. Bohn G, Allroth A, Brandes G, et al. A novel human primary immunodeficiency syndrome caused by deficiency of the endosomal adaptor protein p14. *Nat Med* 2007;**13**:38–45.

44. Barth PG, Scholte HR, Berden JA, et al. An X-linked mitochondrial disease affecting cardiac muscle, skeletal muscle and neutrophil leucocytes. *J Neurol Sci* 1983;**62**:327–55.

45. van Raam BJ, Kuijpers TW. Mitochondrial defects lie at the basis of neutropenia in Barth syndrome. *Curr Opin Hematol* 2009;**16**:14–9.

46. Hirai H, Zhang P, Dayaram T, et al. C/EBPbeta is required for 'emergency' granulopoiesis. *Nat Immunol* 2006;**7**:732–9.

47. Skokowa J, Lan D, Thakur BK, et al. Nampt is essential for the G-CSF induced myeloid differentiation via a novel NAD+/SIRT1-dependent pathway. *Nat Med* 2009;**15**:151–8.

48. Dong F, Hoefsloot LH, Schelen AM, et al. Identification of a nonsense mutation in the granulocyte-colony-stimulating factor receptor in severe congenital neutropenia. *Proc Natl Acad Sci USA* 1994;**91**:4480–4.

49. Dong F, Brynes RK, Tidow N, et al. Mutations in the gene for the granulocyte-colony stimulating factor receptor in patients with acute myeloid leukemia preceded by severe CN. *N Engl J Med* 1995;**333**:487–93.

50. Tidow N, Pilz C, Teichmann B, et al. Clinical relevance of point mutations in the cytoplasmatic domain of the granulocyte-colony stimulating factor gene in patients with severe CN. *Blood* 1997;**88**:2369–75.

51. McLemore ML, Poursine-Laurent J, Link DC. Increased granulocyte colony-stimulating factor responsiveness but normal resting granulopoiesis in mice carrying a targeted granulocyte colony-stimulating factor receptor mutation derived from a patient with severe congenital neutropenia. *J Clin Invest* 1998;**102**:483–92.

52. Hunter MG, Avalos BR. Granulocyte colony-stimulating factor receptor mutations in severe congenital neutropenia transforming to acute myelogenous leukemia confer resistance to apoptosis and enhance cell survival. *Blood* 2000;**95**:2132–7.

53. Hermans MH, Antonissen C, Ward AC, et al. Sustained receptor activation and hyperproliferation in response to granulocyte colony-stimulating factor (G-CSF) in mice with a severe congenital neutropenia/acute myeloid leukemia-derived mutation in the G-CSF receptor gene. *J Exp Med* 1999;**189**:683–92.

54. Germeshausen M, Ballmaier M, Welte K. Incidence of CSF3R mutations in severe CN and relevance for leukemogenesis: Results of a long-term survey. *Blood* 2007;**109**:93–9.

55. Tschan CA, Pilz C, Zeidler C, et al. Time course of increasing numbers of mutations in the granulocyte colony-stimulating factor receptor gene in a patient with CN who developed leukemia. *Blood* 2001;**97**:1882–4.

56. Ancliff PJ, Gale RE, Liesner R, et al. Long-term follow-up of granulocyte colony-stimulating factor receptor mutations in patients with severe congenital neutropenia: implications for leukaemogenesis and therapy. *Br J Haematol* 2003;**120**:685–90.

57. Hermans MH, van de Geijn GJ, Antonissen C, et al. Signaling mechanisms coupled to tyrosines in the granulocyte colony-stimulating factor receptor orchestrate G-CSF-induced expansion of myeloid progenitor cells. *Blood* 2003;**101**:2584–90.

58. van de Geijn GJ, Gits J, Aarts LH, et al. G-CSF receptor truncations found in SCN/AML relieve SOCS3-controlled inhibition of STAT5 but leave suppression of STAT3 intact. *Blood* 2004;**104**:667–74.

59. Irandoust MI, Aarts LH, Roovers O, et al. Suppressor of cytokine signaling 3 controls lysosomal routing of G-CSF receptor. *EMBO J* 2007;**26**:1782–93.

60. Wierenga AT, Schepers H, Moore MA, et al. STAT5-induced self-renewal and impaired myelopoiesis of human hematopoietic stem/progenitor cells involves down-modulation of C/EBPalpha. *Blood* 2006;**107**:4326–33.

61. Gupta K, Kuznetsova I, Klimenkova O, et al. Bortezomib induces granulocytic differentiation of CD34+ cells from congenital neutropenia patients by reversing hyperactivate-STAT5a-dependent downregulation of LEF-1. *Blood* 2014;**123**:2550–61.

62. Liu F, Kunter G, Krem MM, et al. Csf3r mutations in mice confers a strong clonal HSC advantage via activation of Stat5. *J Clin Invest* 2008;**118**:946–55.

63. Bagby GC. Discovering early molecular determinants of leukemogenesis. *J Clin Invest* 2008;**118**:847–50.

64. Germeshausen M, Welte K, Ballmaier M. In vivo expansion of cells expressing acquired CSF3R mutations in patients with severe congenital neutropenia. *Blood* 2009;**113**:668–70.

65. Skokowa J, Steinemann D, Katsman-Kuipers JE, et al. Co-operativity of RUNX1 and CSF3R mutations in the development of leukemia in severe congenital neutropenia: a unique pathway in myeloid leukemogenesis. *Blood* 2014;**123**:2229–37.

66. Matsubara K, Imai K, Okada S, et al. Severe developmental delay and epilepsy in a Japanese patient with severe congenital neutropenia due to HAX1 deficiency. *Haematologica* 2007;**92**:123–5.

67. Rezaei N, Chavoshzadeh Z, Alaei OR, et al. Association of HAX1-deficiency with neurological disorder. *Neuropediatrics* 2007;**38**:261–3.

68. Carlsson G, van't Hooft I, Melin M, et al. Central nervous system involvement in severe congenital neutropenia: neurological and neuropsychological abnormalities associated with specific HAX1 mutations. *J Intern Med* 2008;**264**:388–400.

69. Schilke R, Stanulla M, Finke CH, et al. Influence of periodontitis co-factors in patients with severe chronic neutropenia. *J Dent Res* 2006;**85**:1138.

70. Yakisan E, Schirg E, Zeidler C, et al. High incidence of significant bone loss in patients with severe congenital neutropenia (Kostmann's syndrome). *J Pediatr* 1997;**131**:592–7.

71. Zeidler C, Germeshausen M, Klein C, Welte K. Clinical implications of ELA2-, HAX1-, and G-CSF-receptor (CSF3R) mutations in severe congenital neutropenia. *Br J Haematol* 2009;**144**:459–67.

72. Zeidler C, Welte K, Barak Y, et al. Stem cell transplantation in patients with severe CN without evidence of leukemic transformation. *Blood* 2000;**95**:1195–8.
73. Rosenberg PS, Alter BP, Bolyard AA, et al. Severe Chronic Neutropenia International Registry. The incidence of leukemia and mortality from sepsis in patients with severe CN receiving long-term G-CSF therapy. *Blood* 2006;**107**:4628–35.
74. Rosenberg PS, Zeidler C, Bolyard AA, et al. Stable long-term risk of leukaemia in patients with severe congenital neutropenia maintained on G-CSF therapy. *Br J Haematol* 2010;**150**:196–9.
75. Donadieu J, Leblanc T, Bader Meunier B, et al. Analysis of risk factors for myelodysplasia/leukemia and infectious death among patients with CN: experience of the French Severe Chronic Neutropenia Study Group. *Haematologica* 2005;**90**:45–53.
76. Germeshausen M, Ballmaier M, Schulze H, et al. Granulocyte colony-stimulating factor receptor mutations in a patient with acute lymphoblastic leukemia secondary to severe CN. *Blood* 2001;**97**:829–30.
77. Germeshausen M, Schulze H, Kratz C, et al. An acquired G-CSF receptor mutation results in increased proliferation of CMML cells from a patient with severe CN. *Leukemia* 2005;**19**:611–7.
78. Göhring G, Karow A, Steinemann D, et al. Chromosomal aberrations in congenital bone marrow failure disorders – an early indicator for leukemogenesis? *Ann Hematol* 2007;**86**:733–9.
79. Sloand EM, Yong AS, Ramkissoon S, et al. Granulocyte colony-stimulating factor preferentially stimulates proliferation of monosomy 7 cells bearing the isoform IV receptor. *Proc Natl Acad Sci USA* 2007;**103**:14483–8.

B-Cell Defects: From X-linked Recessive to Autosomal Recessive Agammaglobulinemia

Jerry A. Winkelstein[1], CI Edvard Smith[2]

[1]Emeritus Professor of Pediatrics, Medicine and Pathology, Johns Hopkins University School of Medicine, Baltimore, Maryland, USA

[2]Professor of Molecular Genetics, Clinical Research Center, Department of Laboratory Medicine, Karolinska Institutet, Stockholm/Huddinge, Sweden

INTRODUCTION

Agammaglobulinemia holds a special place in the history of the primary immunodeficiency diseases (PIDD). It was Bruton's original case report of a young boy with agammaglobulinemia in 1952 that stands as the first description of a primary immunodeficiency disease (1). Although the clinical characteristics of two other disorders now classified as PIDD, Wiskott–Aldrich syndrome and ataxia telangiectasia, had been described before Bruton's original description of agammaglobulinemia, including the patient's increased susceptibility to infection, in neither disease was evidence of an immunodeficiency documented at the time of the original description (2, 3).

BRUTON'S ORIGINAL CASE REPORT

Bruton's paper described an 8-year-old boy who had at least ten different episodes of pneumococcal bacteremia/sepsis, some with the same serotype, beginning at age four years (1) (Fig. 11.1). He had been well for his first four years and there was no family history of an increased susceptibility to infection. To quote Bruton's own words:

When he repeated the same type organism...it was suggested that he failed to build antibodies for that particular organism and an autogenous vaccine was prepared of pneumococci. This was given over a period of five months, but no antibody titer could be demonstrated in the blood serum at the end of this period. ...

His blood likewise failed to show typhoid antibodies following administration of typhoid vaccine in the usual manner... With this constant inability to produce antibodies it was suggested that one might expect some derangement in the gammaglobulin fraction. By electophorectic analysis of his blood serum, his blood repeatedly gave completely negative results for gamma globulin (Fig. 11.2)... The absence of gamma globulin gave the most hopeful clue to a possible prophylaxis. Accordingly, he was given subcutaneously human immune serum globulin...For the past 14 months the patient has had monthly injections of gamma globulin, without benefit of other prophylactic measures and has suffered no attack of sepsis.

The ability of Bruton to identify and treat the first patient with agammaglobulinemia was made possible by a number of important advances in biology and medicine in the late 1930s

AGAMMAGLOBULINEMIA

By COL. OGDEN C. BRUTON, M.C., U.S.A.
Washington, D.C.

THE complete absence of gamma globulin in human serum with a normal total protein as determined by electrophoretic analysis does not appear to have as yet been reported in the literature. Stern[1] mentions two cases of hypoproteinemia in children who had "almost complete absence of gamma globulin and were singularly free from infection." Schick[2] reported a similar congenital case without nephrosis with a review of the literature in which the total protein was low, the gamma globulin fraction low, and edema present. The latter findings in nephrosis are well known. Krebs[3] reported a case in which there was a "depression or gamma globulin in hypoproteinemia due to malnutrition." The present author had the opportunity of following a patient without nephrotic syndrome, with normal nutrition, with complete absence of the gamma globulin fraction and normal total serum protein through several years of many infections, including 19 episodes of clinical sepsis in which some type pneumococcus was recovered by blood culture 10 times. This entity, which, it was found, could be controlled by supplying gamma globulin as contained in concentrated immune human serum globulin, appears to be unique.

FIGURE 11.1   Bruton's original article.

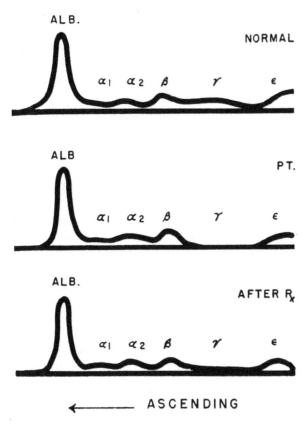

FIGURE 11.2 The protein electrophoresis of Bruton's original case demonstrating absence of the gammaglobulin fraction of serum (middle panel labeled PT).

for the disorder (*BTK*) (9, 10) and the defective/deficient protein (Bruton's tyrosine kinase) (9, 10) are now all named in honor of Bruton. However, it cannot be determined with certainty whether Bruton's first case of agammaglobulinemia was in fact the X-linked recessive form of the disease. Bruton himself raised the question in the discussion of his article. "The fact that the child survived 4-1/2 years without severe infections makes the first (hypothesis that he had a congenital deficiency) unlikely". However, we now know that a significant, but small, number of patients with XLA present first with clinical symptoms after four years of age (11) and that a significant percentage of patients are the result of new mutations (12, 13) so neither his age at presentation nor the lack of family history preclude the diagnosis of XLA.

Charles A. Janeway from Boston Children's Hospital made the following remark in 1954 (14), further demonstrating the unique contribution of Bruton:

> To begin with, we should have been, but we were not, bright enough to discover the disease, because the first patient with symptom complex, the basic cause of which was the inability to produce gamma globulins was seen by us as long ago as 1942.. . Then I received a letter in 1950 by Colonel Bruton, Chief of Pediatrics at the Walter Reed Army Hospital. He wrote me a long letter, describing a case which since has been published in *Pediatrics*.

and early 1940s. The development and availability of penicillin (4) enabled the affected child to survive long enough to be diagnosed. The recognition that the gammaglobulin fraction of serum contained antibody activity (5) and the clinical application of protein electrophoresis (6) together made it possible to recognize agammaglobulinemia as his underlying immune deficiency. And, finally, the availability of gammaglobulin injections (7) made it possible for him to be successfully treated.

The X-linked recessive form of agammaglobulinemia (XLA), also known as Bruton's agammaglobulinemia (8), the gene responsible

Although the report on the first case of agammaglobulinemia is his best known paper, Bruton was an accomplished clinician who published a total of 32 clinical articles involving a number of important areas of clinical pediatrics, including hernia repairs, aortic stenosis, cystic fibrosis, salicylate ingestion and hemophilia. At the time of the case report, Bruton was a Colonel in the United States Army stationed at Walter Reed Army Hospital in Washington, DC (Fig. 11.3). He remained a quiet, unassuming and very likable pediatrician even after his landmark discovery (15).

FIGURE 11.3   Colonel Bruton while at Walter Reed Army Hospital.

ELUCIDATION OF THE PATHOPHYSIOLOGY

At the time of Bruton's description of the first case of XLA, there was not enough knowledge about the immune system in general or the cellular basis for antibody formation specifically, to allow elucidation of the pathophysiological basis of the disease. Astrid Fagraeus had demonstrated in the late 1940s that plasma cells could make antibodies (16). Although there was a concept of humoral and cell-mediated immunity, the identification of B- and T-cells did not exist until the end of the 1960s (17). With the subsequent identification of cell surface immunoglobulins on B-cells, and the elucidation of the different stages of B-cell development, it became possible to assign the defect to early stages of B-cell development (18–20). Over time, the cellular basis underlying the differentiation defect was further delineated, including the fact that the defect was intrinsic to B-cells (21, 22) and identification of the precise stages of B-cell development affected by the molecular

genetic defect (23). From a clinical point of view, since B-cells make up the major cell type in germinal follicles, in XLA, patients' lymph nodes, tonsils and adenoids are very small, and germinal follicles are absent on histological analysis (24, 25). Moreover, in the lamina propria of the gut and bone marrow, plasma cells are typically absent (24–26).

IDENTIFICATION OF THE MOLECULAR GENETIC DEFECT

In the early 1960s, it was reported that the XLA gene was not linked to the Xg blood group determinants (27), but it was not until the mid-1980s that its location was revealed with high resolution. Thus, the XLA gene was first mapped to the Xq21.3-Xq22 region of the long arm (28, 29) and, later, with the aid of new markers (30, 31), its position was determined with greater precision. However, while the final aim of the mapping studies was the identification of the mutated gene, as it often turned out, it was not the XLA gene mappers who succeeded in cloning of disease gene, and the corresponding protein remained elusive. At least two papers suggested the possibility of an enzyme defect prior to the cloning, but these proposals were simply hypotheses (32, 33).

Instead, in 1991–1992, two groups independently initiated work which subsequently would result in the identification of the disease gene. In Europe, a collaborative effort between researchers in the UK and Sweden first resulted in the identification of an unrecombined yeast artificial chromosome (YAC) y178-3 from the implicated region on the X chromosome. Subsequently, the y178-3 clone was used to enrich cDNAs from B-lineage cells. The enriched cDNAs thereafter served as probes in Southern blots made from restriction enzyme-digested genomic DNA obtained from a number of XLA patients in order to identify alterations suggestive of a mutation. The implicated mutations in a novel gene were verified by sequencing, and "Patient A", the first patient with an identified mutation, was a child originating from Southern Sweden (9). Simultaneously at UCLA in the USA, in an effort to identify new tyrosine kinases in B-cell progenitors, an unknown transcript was identified and the corresponding gene found to locate on the X chromosome. Because of the many forms of X-linked recessive human disorders, which almost exclusively affect males, this immediately made the corresponding gene a candidate for a human disease. While the exact mutations were not pinpointed in this report, reduction in, or the absence of mRNA, protein expression, and kinase activity were observed in XLA pre-B- and B-cell lines (10). The two papers complemented each other and received major attention far outside the immunodeficiency arena, owing to the fact that this was one of the first examples of a tyrosine kinase implicated in human inherited disease. The cloning of the *BTK* gene took place during an exciting era when it was still possible to identify new genes by linkage analysis and positional cloning. According to Francis S. Collins, director of the Human Genome Project, *BTK* became the 17th gene identified by positional cloning (23). In the immunodeficiency field, it was preceded only by the *CYBB* gene (causing chronic granulomatous disease), which was detected through a combination of linkage analysis and large gene deletions on the short arm of the X chromosome (34).

When the corresponding papers were in press, one of us, Edvard Smith from the European team, learned that the same gene might also have been cloned in the USA. The first investigator he contacted was not involved, but mentioned Owen N. Witte as a likely candidate. Owen N. Witte was reached by phone, and it was realized that both teams presumably had identified the very same gene. After the phone call, in this pre-Internet era, the manuscripts were simultaneously faxed between UCLA and Karolinska Institutet, and the identity of the gene was conclusively verified. The European group had named the gene *ATK* (agammaglobulinemia tyrosine kinase), whereas the UCLA team called it *BPK* (B-cell progenitor kinase). When the phone call was made it was already too late to adopt a common name to be used in the cloning articles, but a novel name was agreed on only a month later, when the defect in Xid mice was found to originate from a missense mutation in the corresponding mouse gene (35, 36). The new name *BTK*, Bruton tyrosine kinase, makes use of two letters from both the original abbreviations and, in this way, the unnecessary use of two parallel names was avoided. *BTK* also became the official gene name. The gene product is involved in transducing signals from the cell membrane, with the B-cell receptor (BCR) and the pre-BCR as major origins of such signaling.

Immediately prior to the first public oral report of the gene discovery, which would take place at the B-cell Keystone Meeting in Taos, New Mexico in the beginning of 1993, Smith visited Boston and met with Fred Rosen, one of the pioneers in the immunodeficiency field. When asked about Dr Bruton's current whereabouts, Fred Rosen mentioned that he might still be living close to Bethesda. Coincidentally, the trip to the Keystone conference had one more scheduled stop, namely NIH, Bethesda, for a meeting with Jeffrey D. Thomas and William E. Paul on the collaborative Xid gene analysis (35). Upon arrival in Bethesda, Smith, with great curiosity, checked the hotel phonebooks and eventually found "Ogden C. Bruton, MD", with an address in Silver Spring, Maryland, located just east of Bethesda. Two days later, more than 40 years after the original patient description appeared in print, together with Jeffrey Thomas, he met with Dr Bruton and his wife Kathryn, who were already a couple at the time of the seminal 1952 paper. Interestingly, Kathryn Bruton mentioned the importance of the fact that her husband went ahead and published his early observations as a single author. She was of the opinion that if Bruton had waited, she was certain that his contribution to the discovery would

FIGURE 11.4   One of the discoverers of the genetic defect meets the discoverer of the disease in 1993. Dr Bruton is seated, his wife Kathryn D Bruton is standing to his left, Dr Jeffrey D Thomas is standing to his right and Dr CI Edvard Smith is standing behind him. *(Reproduced with permission from (37).)* This figure is reproduced in color in the color section.

have gone essentially unnoticed, and that scholars in Boston with a stronger academic standing would have received the credit. A photograph of the participants documented this meeting (Fig. 11.4) (37). It was during the NIH visit in the early 1993 that the name change to BTK was agreed upon.

THE EVOLVING CLINICAL PICTURE OF AGAMMAGLOBULINEMIA

Soon after Bruton's original description of what is now referred to as XLA, a number of additional cases were described (38–42, reviewed in 23). As expected, the clinical expression initially reported mirrored the first case suffering from blood-borne infections. However, it was but a few years later that respiratory infections and diarrhea were noted as well. While Bruton,

together with Apt, Gitlin and Janeway published a second paper on agammaglobulinemia in 1952 suggesting a congenital defect, they did not mention the possibility of X-linked inheritance (42). The first suggestion that the disorder was inherited as an X-linked recessive trait was instead provided by Janeway the next year, when he described three additional males with agammaglobulinemia and noted that two of them had maternal male relatives who had died from infection in childhood (38).

One of the limitations in developing a comprehensive clinical picture of XLA had been the fact that the patients are uncommon and most physicians and/or centers had a limited number of cases, at least during the first several years after the initial description of the disease. Often, patients with the most dramatic and/or serious clinical manifestations of XLA, those with unusually mild clinical phenotypes and patients with atypical presentations or unusual complications were reported as single cases or in limited series. They had something novel that was of special interest! This is in contrast to certain other genetically determined diseases like cystic fibrosis or sickle cell disease, which are much more common and thus afford clinicians a better opportunity to generate a comprehensive clinical picture of the disease from their own clinical experience.

In order to overcome this limitation, in the 1980s and 1990s, a number of authors, cooperative groups and/or registries began to assemble larger series of patients with XLA, usually from a number of different institutions (11, 43–46). These series differed in many ways, including whether they focused on a single primary immunodeficiency disease, such as XLA, whether they required definitive criteria to assign a specific diagnosis or whether they assigned all patients with "agammaglobulinemia" to a single diagnostic category. Nevertheless, in spite of these inherent limitations, many of these series have contributed significantly to advancing the clinical picture of XLA. In addition, as an increasing number of patients were reported with unusual clinical manifestations, such as late onset of symptoms, uncommon infections or complications, they also provided a more comprehensive clinical picture of XLA.

As mentioned above, the initial reports focused on children with blood-borne infections such as bacteremia/sepsis, meningitis, septic arthritis and osteomeyelitis. However, in the six decades since Bruton's original report, the presenting infections have changed. Although blood-borne infections were very common clinical presentations in the first decade after the initial description of XLA, respiratory infections such as sinusitis, otitis media and pneumonia are much more common initial presentations now (11, 43–46). In fact, recurrent pneumonia often leads to chronic lung disease (11, 43–46). The bacteria usually responsible were encapsulated organisms, such as the pneumococci, streptococci, *Haemophilus influenzae* and *Pseudomonas* species, reflecting the critical role of opsonizing

antibodies in resistance to these organisms. In addition, some uncommon bacterial pathogens, such as *Helicobacter* species, have emerged more recently (47).

One of the more interesting features of infections in patients with XLA, and patients with other primary immunodeficiency diseases, is the fact that infections with "common" organisms in these patients can present with atypical or "uncommon" clinical manifestations. For example, a mere four years after XLA was first described, a number of patients were reported who had a variety of disorders which did not appear to have an infectious etiology. Prominent among these was arthritis of the large joints of the extremities, such as knees, resembling juvenile rheumatoid arthritis (48). Fortunately, the arthritis usually responded to increasing doses of gammaglobulin (49, 50). Some were later shown to be caused by infection with *Mycoplasma* and *Ureaplasma* species (51).

A more puzzling clinical manifestation of XLA was a disorder that resembled dermatomyositis. It was first described a few years after Bruton's original case (48, 52, 53). Its clinical presentation resembled classic dermatomyositis with muscle edema and inflammation and rash. Elevated levels of creatinine phosphokinase were frequently absent, in contrast to immunocompetent individuals with dermatomyositis. The disease was usually fatal, even when treated with high doses of gammaglobulin. Although initial reports of infections in XLA emphasized that the patients were not unduly susceptible to the usual "childhood" viruses, the dermatomyositis-like illness was eventually shown to be an unusual manifestation of a chronic systemic infection with enteroviruses (e.g. ECHO virus and coxsackie virus) (54–56). In most patients, the chronic enteroviral infections persist for months or years and usually are associated with chronic hepatitis and/or meningoencephalitis (55, 56). In more recent years, with earlier diagnosis and improved therapy with the advent of intravenous gammaglobulin, systemic, chronic enteroviral infections have become much less frequent (56). The fact that common enteroviral infections in XLA patients could cause clinical manifestations so different from the limited acute illness seen in immunocompetent hosts exemplifies the way that the clinical expression of infections depends in part on the status of the host's immune system.

DIAGNOSIS

In the decade after Bruton's description of the first case, any male with very low immunoglobulins was considered a case of "Bruton's agammaglobulinemia", especially if there was a maternal male relative with a similar condition. As technology advanced and the pathophysiology of the disorder was elucidated, the absence of plasma cells in the bone marrow, lymph nodes and gut and absent B lymphocytes in the blood and peripheral lymphoid tissue became necessary for diagnosis (19, 25, 26). Finally, once the molecular genetic basis for the disorder was identified, mutations in the *BTK* gene and/or absent *BTK* mRNA and/or protein became the "gold standard" for diagnosis (9, 10, 13).

It is likely that in the first years after its initial description, many, if not most, patients with XLA were not identified as immunodeficient. The disorder was considered a medical rarity, quantitative serum immunoglobulin analysis was not widely available and patients with less dramatic infections were not considered to fit the clinical picture of XLA. However, by the end of the 1960s, serum immunoglobulins could be measured in the diagnostic laboratories of most large hospitals. In addition, as more and more PIDD were identified and brought to physicians' and public attention, and as the AIDS epidemic became a public health problem, the awareness of immune deficiencies increased. The development of routine diagnostic tests, coupled with increasing physician awareness of the PIDD, is likely to have led to more frequent and earlier diagnosis of XLA. In fact, two different series

of patients with documented XLA, spanning 40 years, revealed that patients born more recently (e.g. the 1990s) were diagnosed at an earlier age than those born earlier (the 1960s) (11, 45). This clearly indicates that the medical community has improved with respect to awareness and diagnosis of immune deficiency diseases.

The presence of a positive family history for this X-linked disease should be an important clue for diagnosis. Unfortunately, that is not always the case. A study performed in 1985, reflecting patients diagnosed in the 1960s and 1970s, examined how often patients were diagnosed solely because of a positive family history, before the child develops any clinical symptoms relating to XLA. Unfortunately, only 12% of such patients in that study were diagnosed based solely on their positive family history. This indicates that the presence of a positive family history did not contribute as often as it could have to an early diagnosis in pre-symptomatic children (43). Fortunately, in a more recent series published some 20 years later, 35% of patients with a positive family history were diagnosed before clinical symptoms developed (11). Thus, it appears that physicians are paying more attention to family history now than they did previously as an aid to diagnosing XLA.

TREATMENT

From the first patient described with agammaglobulinemia in 1952, immunoglobulin replacement therapy has always been the mainstay of treatment of XLA (1). During the Second World War, techniques were developed to fractionate plasma into some of its major constituents for use as plasma substitutes. As a result, Edsall and Cohn developed alcohol fractionation of plasma (reviewed in 7 and 57 and in Chapter 23). Cohn fraction II contains IgG and was used immediately after the war as a prophylaxis against some viral infections such as poliomyelitis. It was this preparation that Bruton used to treat his first patient (1). This form of passive immunity, although life saving, had some inherent limitations: it did not contain biologically significant amounts of IgA and IgM and it did not provide any opportunity for rapidly increasing titers of antibody against an infecting microorganism during infection, as would occur in active immunity.

This preparation could only be given intramuscularly or subcutaneously, since it contained aggregated IgG, which could lead to intravascular complement activation and the generation of other harmful inflammatory mediators. In fact, Bruton's original patient received his gammaglobulin subcutaneously (1), but this route was subsequently abandoned for many years. Unfortunately, intramuscular administration had significant limitations. The first was that the intramuscular route only allowed for limited amounts to be given to the patient at one time, limiting the dose of replacement gammaglobulin to approximately 100 mg/kg monthly. This is significantly below the amount given today. Equally important to the patient was, that even to approach a monthly dose of 100 mg/kg, the volume of the intramuscular preparation was very uncomfortable, if not quite painful, for many days after the injection. Most patients, especially children and their parents, dreaded seeing the doctor, since these trips meant considerable discomfort and pain. In fact, most children required more than one person to "hold them down" while the nurse injected them in multiple sites!

In the late 1970s and early 1980s, preparations of IgG suitable for intravenous infusion (IVIG) were prepared using a variety of techniques to eliminate aggregates, such as proteolytic enzymes (e.g. pepsin), reduction and alkylation and incubation at low pH (7). These preparations retained their protective functions as antibodies, had reasonable half-lives after infusion and allowed patients to receive doses that restored their serum levels of IgG to near normal or normal levels. Soon after the introduction of IVIG, the subcutaneous route for infusion was further developed, allowing patients to receive immunoglobulin in full therapeutic doses, but without the need for intravenous access (58).

The development of preparations of IgG suitable for either intravenous or subcutaneous infusion provided the patient with another benefit, the opportunity to receive replacement therapy at home! Not having to travel to have infusions/injections done in the doctor's office provided one more opportunity to "normalize" their already complicated lives (58–60).

PROGNOSIS

The chapter on XLA in the leading textbook dealing with primary immunodeficiency diseases in 1973 declared: "The outlook for these patients is grim" (61). This pessimistic statement reflected a number of factors in the early years after the disease was first identified. There undoubtedly was a selection bias in the reported cases up to that time, with the more severe and unusual patients being overrepresented. But the statement likely reflected the reality of the times. Diagnosis was often delayed until after the recurrent infections had caused irreparable damage to lungs and other organs. The available preparations of gammaglobulin could only be given intramuscularly, severely limiting the replacement dose, with normal levels of serum IgG not being reached. In addition, diagnosis of specific etiologies for infections of the lower respiratory tract was difficult and the number of antibiotics available for their treatment was limited.

Since those early years, the prognosis has improved significantly (11, 45, 62, 63). The brighter outlook for these patients has resulted from earlier diagnosis, better identification of infecting microorganisms, a wider menu of antibiotics and the ability to deliver enough gammaglobulin. Indeed, it is now possible to diagnose and treat patients before infections have resulted in permanent damage to their lungs and other vital organs. These patients are likely to survive well into adulthood and lead productive and fulfilling lives. In fact, in two studies examining the quality of life in adults with XLA, it was found that they married, had children and had higher levels of education and income than the general public (64, 65). In one registry of XLA patients in the USA, 25% of patients were over 21 years of age (11). In the same registry, when followed for five years, only three of 80 patients had died; one following a bone marrow transplant, one from hepatitis C contracted from a contaminated lot of gammaglobulin and one from sepsis as a complication of renal failure secondary to membrano-proliferative glomerulonephritis. Thus, two of the three deaths in that five year period were from iatrogenic causes.

AUTOSOMAL RECESSIVE AGAMMAGLOBULINEMIA

The fact that there are other forms of agammaglobulinemia than XLA was apparent when females lacking antibodies, and patients with adult onset of disease were described in the 1950s (reviewed in 23). Unfortunately, at the time of their original description, the tools needed to elucidate the pathogenesis were not available, making it difficult retrospectively to draw conclusions regarding the etiology of the agammaglobulinemia from many of the original patient descriptions.

The genetic abnormalities responsible for the autosomal recessive agammaglobulinemias (ARA) were all identified in the search for mutations in genes known to be involved in pre-BCR/BCR signaling from both biochemical studies and from the inactivation of the corresponding genes in mice. The first molecularly characterized form of ARA was described by Conley and co-workers in 1996, and is the result of mutations in the immunoglobulin heavy constant μ-chain gene, with the gene symbol IGHM (66). Interestingly, when larger numbers of such patients were studied (67), it became evident that Jan Waldenström, renowned for his work on Waldenström's macroglobulinemia, had described Swedish male siblings with agamma-

globulinemia in 1955 (68) that later were identified as ARA (67). As a curiosity, Waldenström taught Smith Internal Medicine at Serafimerlasarettet in Stockholm, when his regular teacher, the king's physician, accompanied Gustaf Adolf VI during archeological excavations in Italy. Forty-seven years after Waldenström's original report in 1955, the siblings were found to carry a splice site mutation in the IGHM gene (67). Waldenström died in 1996, the same year that mutations in the IGHM gene were first described, but prior to the identification of mutations in the patients he described in the 1950s. Subsequently, many additional causes of ARA have been identified as the result of mutations in many other genes, namely IGLL1, CD79A, CD79B, BLNK and PIK3R1 as described and reviewed in several recent articles (69–72).

The different ARAs are less frequent than XLA. However, patients' symptoms, signs and susceptibility to infections do not significantly differ from XLA, as could be expected from the fact that all have blocked signaling via the pre-BCR/BCR. While there is a certain degree of leakiness with regard to the B-cell differentiation block in XLA, this phenomenon is frequently less obvious in ARA (69). Based on the phenotype of various gene knockout mice, it can be expected that additional forms of genetic defects causing ARA will be identified in humans (72). However, even if the immunological phenotype is restricted to B-cell development, for some of these ARAs, it is likely that developmental defects in non-immune cells will also be found.

References

1. Bruton OC. Agammaglobulinemia. *Pediatrics* 1952;**9**:722–7.
2. Wiskott A. Familiärer, angeborener Morbus Werlhofii? *Mschr Kiderheilk* 1937;**68**:212–6.
3. Syllaba L, Henner K. Contribution a l'independence de l'athetose double idiopathique et congenitale: Atteinte familiale, syndrome dystrophique, signe du reseau vasculaire conjonctival, integrite psychique. *Rev Neurol (Paris)* 1926;**1**:541–62.

4. Fleming A. Penicillin: The Robert Campbell Oration. *Ulster Med J* 1944;**13**:95–122.

5. Tiselius A, Kabat EA. Antibodies as globulins. *J Exp Med* 1939;**69**:119–24.

6. Tiselius A. Electrophoresis of serum globulins. Electrophoretic analysis of normal and immune sera. *Biochem J* 1937;**31**:1464–77.

7. Stiehm ER, Orange JS, Ballow M, Lehman H. Therapeutic use of immunoglobulins. *Adv Pediatr* 2010;**57**:185–218.

8. World Health Organization Scientific Group. Primary immunodeficiency diseases. *Immunodef Rev* 1992;**3**: 195–236.

9. Vetrie D, Vorechovsky I, Sideras P, et al. The gene involved in X-linked agammaglobulinemia is a member of the src family of protein-tyrosine kinases. *Nature* 1993;**361**:226–33.

10. Tsukada S, Saffran DC, Rawlings DJ, et al. Deficient expression of a B-cell cytoplasmic tyrosine kinase in X-linked agammaglobulinemia. *Cell* 1993;**72**:279–90.

11. Winkelstein JA, Marino MK, Lederman HM, et al. X-linked agammaglobulinemia: Report on a registry of 201 patients. *Medicine* 2006;**85**:193–207.

12. Conley ME, Mathias D, Treadaway J, Minegishi Y, Rohrer J. Mutations in btk in patients with presumed X-linked agammaglobulinemia. *Am J Hum Genet* 1998;**62**:1034–43.

13. Väliaho J, Smith CIE, Vihinen M. BTKbase: mutation database for X-linked agammaglobulinemia. *Hum Mut* 2006;**27**:1209–17.

14. Janeway CA. Cases from the medical grand rounds. Massachusetts general hospital. *Am Pract Digest Treat* 1954;**5**:487–92.

15. Lawrence Pakula, MD. Personal communication.

16. Fagraeus A. Antibody production in relation to the development of plasma cells; in vivo and in vitro experiments. *Acta Med Scand* 1948;**130**(Suppl):204.

17. Mitchell GF, Miller JFAP. Cell to cell interaction in the immune response. II. The source of hemolysin-forming cells in irradiated mice given bone marrow and thymus or thoracic duct lymphocytes. *J Exp Med* 1968;**128**:821–37.

18. Naor D, Bentwich Z, Cividalli G. Inability of peripheral lymphoid cells of agammaglobulinaemic patients to bind radioiodinated albumins. *Aust J Exp Biol Med Sci* 1969;**47**:759–61.

19. Pearl ER, Vogler LB, Okos AJ, Crist WM, Lawton III AR, Cooper MD. B lymphocytes precursors in human bone marrow; an analysis of normal individuals and patients with antibody deficiency states. *J Immunol* 1978;**120**:1169–75.

20. Cooper MD, Lawton AR. Circulating B-cells in patients with immunodeficiency. *Am J Pathol* 1972;**69**:513–28.

21. Conley ME, Brown P, Pickard AR, et al. Expression of the gene defect in X-linked agammaglobulinemia. *N Engl J Med* 1986;**315**:564–7.

22. Fearon ER, Winkelstein JA, Civin CI, Pardoll DM, Vogelstein B. Carrier detection in X-linked agammaglobulinemia by analysis of X-chromosome inactivation analysis. *N Engl J Med* 1987;**316**:427–31.

23. Sideras P, Smith CIE. Molecular and cellular aspects of X-linked agammaglobulinemia. *Adv Immunol* 1995;**59**: 135–223.

24. Good RA. Clinical investigations in patients with agammaglobulinemia. *J Lab Clin Med* 1954;**44**:803.

25. Good RA. Studies on agammaglobulinemia. II. Failure of plasma cell formation in bone marrow and lymph nodes of patients with agammaglobulinemia. *J Lab Clin Med* 1955;**46**:167–81.

26. Ochs HD, Ament ME, Davis SD. Giardiasis with malabsorption in X-linked agammaglobulinemia. *N Engl J Med* 1972;**287**:341–2.

27. Sanger R, Race RR. The Xg blood groups and familial hypogammaglobulinemia. *Lancet* 1963;**7286**:859–60.

28. Kwan S-P, Kunkel L, Bruns G, Wedgwood RJ, Latt S, Rosen FS. Mapping of the X-linked agammaglobulinemia locus by use of restriction fragment-length polymorphism. *J Clin Invest* 1986;**77**:649–52.

29. Mensink EJBM, Thompson A, Schot JDL, van de Greef WMM, Sandkuyl LA, Schuurman RKB. Mapping of a gene for X-linked agammaglobulinemia and evidence for genetic heterogeneity. *Hum Genet* 1986;**73**:327–32.

30. Guioli S, Arveiler B, Bardoni B, et al. Close linkage of probe p212 (DXS178) to X-linked agammaglobulinemia. *Hum Genet* 1989;**84**:19–21.

31. Kwan S-P, Terwilliger J, Parmley R, et al. Identification of a closely linked DNA marker, DXS178, to further refine the X-linked agammaglobulinemia locus. *Genomics* 1990;**6**:238–42.

32. Henley WL. Hypogammaglobulinemia and hypergammaglobulinemia. *J Mt Sinai Hosp NY* 1959;**26**:138–59.

33. Smith CIE, Hammarström L. Cellular basis of immunodeficiency. *Ann Clin Res* 1987;**19**:220–9.

34. Royer-Pokora B, Kunkel LM, Monaco AP, et al. Cloning the gene for an inherited human disorder –chronic granulomatous disease – on the basis of its chromosomal location. *Nature* 1986;**322**:32–8.

35. Thomas JD, Sideras P, Smith CI, Vorechovský I, Chapman V, Paul WE. Colocalization of X-linked agammaglobulinemia and X-linked immunodeficiency genes. *Science* 1993;**261**:355–8.

36. Rawlings DJ, Saffran DC, Tsukada S, et al. Mutation of unique region of Bruton's tyrosine kinase in immunodeficient XID mice. *Science* 1993;**261**:358–61.

37. Smith CIE, Notarangelo LD. Molecular basis for X-linked immunodeficiencies. *Adv Genet* 1997;**35**:57–115.

38. Janeway CA, Apt I, Gitlin D. Agammaglobulinemia. *Trans Assoc Am Phys* 1953;**66**:100–2.

39. Jean R. Hypo- ou agamma-globulinémie isolée chez l'enfant. *Presse Med* 1953;**61**:828–9.

40. Hayles AB, Stickler GB, McKensie BF. Decrease in serum gamma globulin (agammaglobulinemia): Report of 3 cases. *Pediatrics* 1954;**14**:449–54.

41. Marin NH. Agammaglobulinemia: A congenital defect. *Lancet* 1954;**ii**:1094–5.

42. Bruton OC, Apt L, Gitlin D, Janeway CA. Absence of serum gamma globulins. *Am J Dis Child* 1952;**84**:632–6.

43. Lederman HM, Winkelstein JA. X-linked agammaglobulinemia: An analysis of 96 patients. *Medicine* 1985;**64**:145–56.

44. Hermaszewski RA, Webster AD. Primary hypogammaglobulinemia: A survey of clinical manifestations and complications. *Q J Med* 1993;**86**:31–42.

45. Plebani A, Soresina A, Rondelli R, et al. Clinical, immunological and molecular analysis in a large cohort of patients with X-linked agammaglobulinemia: An Italian multicenter study. *Clin Immunol* 2002;**104**:221–30.

46. Conley ME, Howard V. Clinical findings leading to the diagnosis of X-linked agammaglobulinemia. *J Pediatr* 2002;**141**:566–71.

47. Turvey SE, Leo SH, Boos A, et al. Successful approach to treatment of Helicobacter bilis infection in X-linked agammaglobulinemia. *J Clin Immunol* 2012;**32**:1404–8.

48. Janeway CA, Gitlin D, Craig JM, Grice DS. Collagen disease in patients with congenital agammaglobulinemia. *Trans Assoc Am Phys* 1956;**69**:93–7.

49. Gitlin D, Janeway CA, Apt L, Craig JM. Agammaglobulinemia. In: Lawrence HS, editor. *Cellular and Humoral Aspects of the Hypersensitivity States. A Symposium*. New York: Hoeber-Harper; 1959. p. 375–441.

50. Hansel TT, Haeney MR, Thompson RA. Primary hypogammaglobulinemia and arthritis. *Br Med J* 1987;**295**:174–5.

51. Franz A, Webster AD, Furr PM, Taylor-Robinson D. Mycoplasmal arthritis in patients with primary immunoglobulin deficiency: Clinical features and outcome in 18 patients. *Br J Rheumatol* 1997;**36**:661–8.

52. Page AR, Hansen AE, Good RA. Occurrence of leukemia and lymphoma in patients with agammaglobulinemia. *Blood* 1963;**21**:97–100.

53. Gotoff SP, Smith RD, Sugar O. Dermatomyositis and cerebral vasculitis in a patient with agammaglobulinemia. *Am J Dis Child* 1972;**123**:53–7.

54. Bardelas JA, Winkelstein JA, Seto DSY, Tsai T, Rogal A. Fatal ECHO 24 infection in a patient with hypogammaglobulinemia: Relationship to dermatomyositis-like syndrome. *J Pediatr* 1977;**90**:396–9.

55. Wilfert CM, Buckley RH, Mohanakumar T, et al. Persistent and fatal central nervous system ECHOvirus infections in patients with agammaglobulinemia. *N Engl J Med* 1977;**296**:1485–9.

56. Halliday E, Winkelstein JA, Webster AD. Enteroviral infection in primary immunodeficiency disease: a survey of morbidity and mortality. *J Infect* 2003;**46**:1–8.

57. Janeway CA. Plasma fractionation. *Adv Intern Med* 1949;**3**:295–307.

58. Gardulf A, Hammarstrom L, Smith CIE. Home treatment of hyopogammaglobulinemia with subcutaneous gammaglobulin by rapid infusion. *Lancet* 1991;**338**:162–6.

59. Kobayashi RH, Kobayashi AD, Lee N, Fischer S, Ochs HD. Home self administration of intravenous immunoglobulin therapy to children. *Pediatrics* 1990;**85**:705–9.

60. Daly PB, Evans JH, Kobayashi RH, et al. Home-based immunoglobulin infusion therapy: quality of life and patient health perceptions. *Ann Allergy* 1991;**67**:504–10.

61. Davis SD. Antibody deficiency diseases. In: Stiehm ER, Fulginiti V, editors. *Immunologic Disorders in Infants and Children*. Philadelphia: WB Saunders Company; 1973. p. 184.

62. Ochs HD, Smith CI. X-linked agammaglobulinemia. A clinical and molecular analysis. *Medicine (Baltimore)* 1996;**75**:287–99.

63. Quartier P, Debra M, De Blic J, et al. Early and prolonged intravenous immunolgobulin therapy in childhood agammaglobulinemia: A retrospective survey of 31 patients. *J Pediatr* 1999;**134**:589–96.

64. Howard V, Greene JM, Pahwa S, et al. The health status and quality of life of adults with X-linked agammaglobulinemia. *Clin Immunol* 2006;**118**:201.

65. Winkelstein JA, Conley ME, James C, Howard V, Boyle J. Adults with X-linked agammaglobulinemia: Impact of disease on daily lives, quality of life, educational and socioeconomic status, knowledge of inheritance and reproductive attitudes. *Medicine* 2008;**87**:253–8.

66. Yel L, Minegishi Y, Coustan-Smith E, et al. Mutations in the mu heavy-chain gene in patients with agammaglobulinemia. *N Engl J Med* 1996;**335**:1486–93.

67. Lopez Granados E, Porpiglia AS, Hogan MB, et al. Clinical and molecular analysis of patients with defects in mu heavy chain gene. *J Clin Invest* 2002;**110**:1029–35.

68. Kulneff N, Pedersen KO, Waldenström J. Drei Falle von Agammaglobulinamie; Ein klinischer, genetischer und physikalischchemischer Beitrag zur Kenntnis des Proteinstoffwechsels. *Schweiz Med Wsch* 1955;**85**:363–8.

69. Conley ME, Dobbs AK, Farmer DM, et al. Primary B-cell immunodeficiencies: comparisons and contrasts. *Annu Rev Immunol* 2009;**27**:199–227.

70. Lougaris V, Ferrari S, Plebani A. Ig beta deficiency in humans. *Curr Opin Allergy Clin Immunol* 2008;**8**:515–9.

71. Conley ME, Dobbs AK, Quintana AM, et al. Agammaglobulinemia and absent B lineage cells in a patient lacking the p85α subunit of PI3K. *J Exp Med* 2012;**209**:463–70.

72. Berglöf A, Turunen JJ, Bestas B, Gissberg O, Blomberg KEM, Smith CIE. Agammaglobulinemia: causative mutations and their implications for therapy. *Expert Rev Clin Immunol* 2013; in press.

The Discovery of the Familial Hemophagocytosis Syndromes

Geneviève de Saint Basile[1,2,3], Alain Fischer[1,3,4,5]

[1]Inserm U1163, Paris, France
[2]CEDI, Assistance Publique-Hôpitaux de Paris, Paris, France
[3]Univ. Paris Descartes, Sorbonne Paris Cité, Imagine Institute, Hôpitaux de Paris, Paris, France
[4]Unité d'Immunologie et Hématologie Pédiatrique, Hôpital Necker Enfants Malades, Assistance Publique-Hôpitaux de Paris, Paris France
[5]Collège de France, Paris, France

INTRODUCTION

Hemophagocytic lymphohistiocytosis (HLH) can be sporadic or inherited as familial hemophagocytic lymphohistiocytosis (FHL), and is a unique medical entity that was not considered as a primary immunodeficiency until about 20 years ago, when a defect in the granule-dependent cytotoxic function of T and natural killer (NK) cells was identified in familial cases. Defects in T and NK cytotoxicity are now used as a diagnostic test, and their recognition has changed the treatment of HLH.

DESCRIPTION: THE EARLY YEARS

The first defining description of HLH was provided by Farquhar and colleagues in 1952 (1), who coined the name familial hemophagocytic reticulosis and considered this a distinct entity, although closely related to Letterer–Siwe syndrome. Most of the presently known clinical manifestations of HLH were recognized by these authors including fever, hepatosplenomegaly, bleeding, pancytopenia and a familial occurrence. They reported that the lesions were apparently not infectious and without granulomata while there was a "cellular infiltration diffuse in character" that showed a remarkable and almost pathognomonic hemophagocytosis (Fig. 12.1). The fatal character of the syndrome early in life was also described, with an addi-

tional third case in the same family reported in 1958 (2). A few years later, Nelson and colleagues provided a very precise description of the syndrome under the name of generalized lymphohistiocytic infiltration, by reporting three additional cases (3) who presented with fever, anemia, leukopenia, thrombocytopenia, hepatosplenomegaly and the new findings of meningitis or encephalitis, as characteristic clinical features. This is the first report to accurately describe lymphocytic as well as histiocytic infiltration of many organs with a "remarkable involvement of the central nervous system". Interestingly, they ignored the publication by Farquhar et al., but quoted a report published in 1951 by Reese and Levy (4) describing atypical cases of Letterer–Siwe syndrome that very likely represented HLH cases. Nelson et al. also noticed similarities between the manifestations they observed and those characteristic of Chediak–Higashi syndrome (CHS) (5, 6), stating that the only differences were the albinism and anomalous leukocytic inclusions

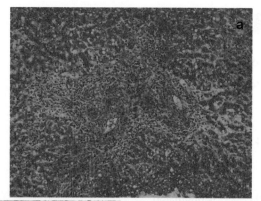
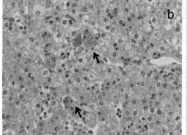
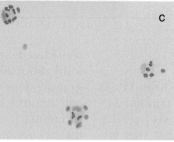

FIGURE 12.1 Cellular infiltration and hemophagocytosis: (A) H&E staining shows massive infiltration of portal tract and lobules of liver biopsy by mononuclear cells; (B) active phagocytosis of erythrocytes (arrow) by macrophages; (C) hemophagocytosis of blood elements by macrophages in the cerebrospinal fluid. *(Courtesy of Pr Nicole Brousse, Hôpital Necker, Paris, France.)* This figure is reproduced in color in the color section.

that characterize the CHS syndrome (3). Thus, by 1961, it was established that a condition with Mendelian inheritance causes early onset and fatal disease characterized by a non-infectious, diffuse tissue infiltration by lymphocytes and macrophages. The description was virtually complete. Subsequent reports introduced the term lymphohistiocytosis in 1963 and familial lymphohistiocytosis (FLH) in 1968 (7, 8), adding biochemical abnormalities such as hypertriglyceridemia and low fibrinogen levels (9). One hundred cases were listed in the first review of the FHL entity published in 1983 (10). The first hint regarding the pathogenesis came with the report of defective NK-cell cytotoxicity in patients with HLH (see below).

FHL AND OTHER INHERITED HLH CONDITIONS: THE SEVENTIES

Besides FHL, several other inherited conditions were increasingly recognized as causing a similar HLH entity, including Chediak–Higashi syndrome (6), Griscelli disease (11) and the X-linked proliferative syndromes where HLH is often, although not always, related to Epstein–Barr virus (EBV) infection (12, 13). In the 1970s, acquired HLH was described as "virus-associated hemophagocytic syndrome" in the absence of obvious family history (14). This was followed by many reports describing HLH in a range of conditions such as leukemia, lymphoma, inflammation and autoimmunity. A possible genetic susceptibility to "acquired HLH" still remains an open question.

ASSESSMENT OF IMMUNOLOGIC FUNCTIONS: TURN OF THE CENTURY

A defective immune system had long been suspected in the various inherited forms of HLH (15). As additional cases were analyzed,

the humoral immune response appeared intact in most affected patients, and the defective T-cell proliferation and interferon production of early reports were inconsistent. With time, these features were considered to be more likely secondary to the disease than a primary immune disorder. Two main features which underlie the immunopathology of HLH were progressively recognized:

1. In 1991, Henter and coworkers first reported the hypercytokinemia which characterizes the course of FHL (16) and is more generally seen in the expanded group of HLH. This study points out the high serum levels of inflammatory cytokines, such as tumor necrosis factor (TNF), interleukin 6 (IL-6) and interferon γ (IFN-γ) during the active phase of FHL, and the striking resemblance between the biological consequences induced by these inflammatory cytokines and the clinical manifestations of FHL. A direct effect of TNF, also known as cachectin, on the development of cachexia and muscular weakness often observed in FHL patients was postulated, while hypofibrinogenemia and hypertriglyceridemia were connected to the procoagulant and the lipoprotein lipase inhibitory activity of TNF (16–19). Subsequent studies reported the massive infiltration of the liver by IFN-γ-producing cytotoxic lymphocytes and by activated macrophages producing proinflammatory cytokines (20). A role for activated T lymphocytes in the pathogenesis of HLH was also implied by the detection of a high serum level of the IL-2-receptor (16, 21). Since activated lymphocytes were shown not to be the only source of soluble IL-2-receptor (22), surface expression of HLA-DR on T lymphocytes was used by subsequent investigators as an alternative marker of *in vivo* T-cell activation (23).

2. Impaired cytotoxic activity of NK cells from CHS patients was reported in 1980 (24)

following the discovery of this defect in the beige murine counterpart (25). Using a single cell cytotoxic assay, it was shown that the defect was not in the NK cell-target binding but rather in the effector cytotoxic ability of NK cells (26), an assumption corroborated by other studies, hence the nickname of "lazy" NK cells (27). A partial restoration of patients' NK-cell cytotoxicity was shown by IFN-γ treatment *in vitro* (27). A first report in 1984 described NK cytotoxic impairment during the acute phase of FHL, although normalization of the defect was reported during the remission phase (28). This was not confirmed in subsequent studies in which the defect persisted through all phases of the disease (29, 30).The low number of circulating NK cells frequently observed during HLH, and the difficulty in discriminating between primary and secondary HLH in the absence of a family history, may account for the discrepancy of these early findings. Cytotoxic dysfunction of NK cells appeared progressively as a relevant diagnostic marker of FHL, and a way to distinguish primary from secondary HLH (31, 32). A recently reported degranulation test that focuses only on the cytotoxic lymphocyte population allows the lytic granule secretion ability of these cells to be determined independently of the total peripheral blood mononuclear cell (PBMC) count (33). A defect in cytotoxic T-cell activity was subsequently reported (34). However, until the identification of the first FHL gene, it was unclear whether the global cytotoxic dysfunction observed was primary or secondary to the pathophysiological process (034).

In CHS, a defect in phagocytosis, bactericidal activity and chemotactic response was described by a number of authors. Early studies reported that the granulocytes of CHS are abnormal in their capacity to migrate in a Boyden chamber (35), to fuse their enlarged granules with phagosomes (36), or to kill organisms such as *Escherichia coli* with normal kinetics (37). These defective neutrophil functions have been subsequently confirmed and likely account for the higher susceptibility to bacterial infections including chronic periodontal disease observed in CHS patients.

The recognition of the diverse clinical and biological manifestations of HLH has led to generally accepted diagnostic criteria and to the establishment of mechanisms to share treatment and outcome data (38).

THE GENETICS OF HLH: AN ONGOING PROCESS

Autosomal recessive inheritance of FHL was established by early family studies (1, 2). Genetic markers of FHL have been unsuccessfully investigated for many years, and occasional reports of chromosomal aberrations were not confirmed in larger studies (39, 40). With the development of hundreds of microsatellites covering the whole genome, linkage analysis for homozygosity mapping became possible in cohorts of families with FHL. In 1999, a locus designated FHL was identified on chromosome 9q21.3–22 in four inbred FHL families of Pakistani descent (41). However, to date, no causative gene has been associated with this locus, and the robustness of these data has been questioned. Simultaneously, a second locus was found on chromosome 10q21–22 (FHL2), and evidence of genetic heterogeneity of this condition was provided (42). Shortly thereafter, gene candidate screening within the FHL2 locus allowed Stepp and colleagues to identify the first gene for FHL that encodes the cytolytic effector, perforin (Fig. 12.2) (43). This rather unexpected finding was decisive because it directly linked the function of the granule-dependent cytotoxic pathway to the pathophysiology of HLH and suggested additional etiologies for FHL. Similar genetic approaches identified three additional causes of FHL, all of

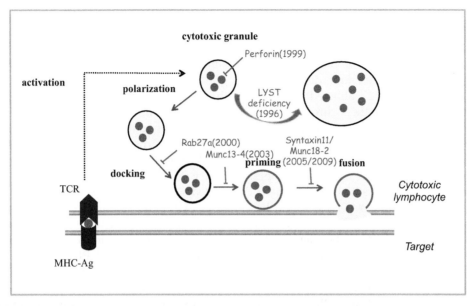

FIGURE 12.2 Inborn errors in the cytotoxic activity of lymphocytes causing HLH: the genetic defects causing hemophago-cytic lymphohistiocytic syndrome (HLH) affect discrete steps of the cytotoxic machinery, i.e. granule biogenesis/morphology, granule content, docking, priming or fusion. In brackets, year of gene discovery. This figure is reproduced in color in the color section.

them affecting the exocytosis of lymphocyte cytotoxic granules (Fig. 12.2). In 2003, FHL3 was associated with mutations in *UNC13-D* located on chromosome 17q25 (44). This gene encodes for Munc13-4, which primes the fusion of cytotoxic granules at the plasma membrane (Fig. 12.2). Priming function requires that Munc13 proteins interact with syntaxin, a member of the SNARE machinery. Predictably, it was a deficiency in a syntaxin family member, syntaxin-11, that was identified two years later as an additional cause of FHL (FHL4) (45) (Fig. 12.2). Finally, a defect in the syntaxin binding protein STXBP2/Munc18-2, that binds to syntaxin 11, was found to cause FHL5 (46, 47) (Fig. 12.2). With the identification of this fourth FHL gene in 2009, approximately 10% of the familial cases of FHL still remain molecularly undefined as of 2014 in spite of advances in genetic screening technology. There may be other genes regulating the cytotoxic function of lymphocytes that remain to be associated with these cases.

Some of the earliest descriptions of FLH commented on phenotypic similarities with the so-called accelerated phase observed in the syndromes described by Chediak and Higashi (3), and by Griscelli and coworkers (11, 48, 49). This relevant observation now has a molecular basis since, in all of these conditions, the defects functionally impair the same cytotoxic pathway. The small GTPase Rab27a, shown to be defective in Griscelli syndrome, regulates the docking step of cytotoxic granules at the plasma membrane (50). Surprisingly, although CHS was the first HLH to be genetically characterized, it remains the least understood. Seventeen years ago, an elegant approach, combining reverse genetics with the correction of the lysosome morphology of deficient fibroblasts by yeast artificial chromosomes (YAC), led to the gene identification in the beige mouse (*lyst*) and subsequently in its human counterpart, CHS (*LYST*) (51–53). The precise function of the CHS protein remains enigmatic. However, in light of the other identified HLH defects, impaired cytotoxic activity, described many years ago in this condition (24), can now be viewed as the cause of the accelerated phase of the syndrome.

The inheritance pattern of the lymphoproliferative (XLP) syndrome, first described by David Purtilo in 1975 (12), allowed an instant assignment of the locus to the X chromosome. In 1987, using restriction fragment length polymorphisms, Skare and colleagues localized the causative gene to Xq24-26 (54). Progressively, the critical region was narrowed by the use of overlapping cytogenetically visible interstitial deletions in XLP patients, and overlapping YAC clones and bacterial contigs (55). This multistep genetic mapping approach led in 1998 to the identification by Coffey et al. of the *SH2D1A* gene (56), while the same gene was simultaneously identified by Sayos and coworkers (57), while exploring the signaling pathway downstream of the cell-surface lymphocyte co-receptor SLAM. Indeed, SLAM-associated protein (SAP), the gene product of *SH2D1A*, a small adaptor protein, couples SLAM family receptors to a downstream signaling pathway. These receptors mediate an array of activating and/or regulatory signals including cytotoxic responses toward B-cells expressing ligands consisting of SLAM family molecules: NTBA and CD48 (58). Additional studies revealed that in some XLP families the disease segregation was compatible with the XLP1 region in Xq25, but *SH2D1A* was found to be wild type. In 2006, Rigaud et al. identified a second gene (*XIAP*), for XLP (XLP2), encoding an inhibitor of apoptosis, XIAP, in the same genetic region (13). Even now, some eight years after this discovery, it is not understood how the anti-apoptotic function of XIAP fits the paradigm of HLH and the defect in the cytotoxic pathway.

PATHOGENESIS: FROM MACROPHAGES TO T- AND NK CELLS

HLH was initially viewed as a "histiocytic proliferation reactive in nature" or a highly aggressive neoplastic disorder with some form of underlying immunodeficiency. The concept of an underlying immunodeficiency was first introduced by Nezelof's group, who noted a similarity between the histology of HLH and graft-versus-host disease (15). They postulated that, as in severe combined immunodeficiencies, foreign HLA mismatched T-cells could trigger a new immunopathological reaction. Although this theory turned out to be wrong, it suggested a link between HLH and primary immunodeficiency (PID). In 1978, Ladisch and coworkers reported several functional defects in lymphocytes from FHL patients that eventually appeared to be secondary (59) but, together with the report of DeVictor et al. (49), led to the view that FHL/HLH disorders were indeed PIDs. Recognition that hypercytokinemia was part of these syndromes illustrates the importance of macrophage activation (16). The role of T-cells was ascertained by the demonstration of activation markers such as a high level of soluble CD25 in plasma (16) and the identification of interferon-γ producing T-cells as the major infiltrating cells in the liver (20). These observations placed T-cells at the center of the pathogenesis, a concept that led to novel therapeutic options (see below). Finally, the last etiological mechanism was put in place by the progressive identification of defects in NK/T-cell cytotoxicity and recognized as a feature common to all inherited conditions leading to HLH (see Fig. 12.2).

THERAPY: FROM IMMUNOSUPRESSION AND CHEMOTHERAPY TO STEM CELL TRANSPLANTATION

FHL/HLH was initially reported as rapidly fatal, although spontaneous remissions were occasionally noticed. Because of the non-infectious, rather inflammatory aspect of the lesions, the treatments tried initially included adrenocorticotrophic hormone (ACTH) (1) and corticosteroids – leading at best to a mild and transient effect (10). Splenectomy was proposed because of the enlarged spleen but efficacy was again at best transient. Ladisch et al. reported occasional benefit from exchange transfusions (60). The apparent similarity of HLH with histiocytosis X/Letterer–Siwe syndrome led to treatment with vinblastine combined with corticosteroids, but only partial remission was achieved, and only in a minority of cases. Vinblastine-loaded platelets were tried with the aim of intoxicating hemophagocytic macrophages (61), again with only transient and partial effect.

A breakthrough in treating FHL/HLH was the introduction in 1980 of etoposide (VP-16) by Ambruso and colleagues (62): this led to complete remission in two FHL patients. To treat the accompanying CNS disease, Lilleyman used, in analogy to the treatment of leukemia, intrathecal injection of methotrexate in combination with vinblastine, corticosteroids and splenectomy, achieving remission for up to eight months (63).

It is interesting that, in the 1970s, pediatric onco-hematologists in charge of patients with FHL/HLH attempted to get rid of the lympho-histiocytic infiltrates by selecting chemotherapy and that they published the results in journals dealing with cancer (62, 63). Chemotherapy became popular because VP-16 injections were reproducibly shown to induce, in a majority of patients, complete remission of FHL/HLH (31, 64). This observation led to an international protocol (HLH-94, followed by HLH-2004) in which a primary eight-week course of VP-16 is given in combination with corticosteroids (dexamethasone) and cyclosporine (see below). This regimen led to sustained remission in approximately 80% of patients, a remarkable achievement considering the poor prognosis of untreated FHL/HLH (65–67). However, early toxicity leading to aplasia resulted in significant infectious morbidity and mortality (67).

Based on the concept that cytokines released by defective T-cells are central drivers of HLH, cyclosporine A was introduced as maintenance therapy (65). Our group hypothesized that it should be possible to abandon chemotherapy to reduce toxicity, and this could be replaced by

reagents targeting T-cells. Using anti-thymocyte globulins, we were able to induce full remission in 73%, and partial remission in 24% of 38 patients, first reported in 1993 and updated in 2007 (68, 69).

A recently reported alternative therapy includes the anti-T-cell monoclonal antibody campath-1 (alemtuzumab), which has the advantage of not activating T-cells before killing them, and has been proposed as salvage (70) or as first-line therapy. Whatever therapy had been used to control familial HLH, relapses occurred in every case, since chemotherapy or immunosuppression has only a temporary, not curative, effect. Treatment of EBV-driven HLH, such as in XLP1 patients, has benefited from the introduction of the monoclonal anti-CD20 antibody (rituximab) that is used to eliminate the EBV-infected B-cells that cause unwanted T-cell activation (71).

Because FHL/HLH is a disease of the hematopoietic system, allogeneic hematopoietic stem cell transplantation (HSCT) has been considered as a curative treatment. The first success of HSCT was reported in 1986 (72) in a patient who, 27 years later, is still alive and well. The effectiveness of HSCT has been confirmed by many groups. Up to 66% of the 124 HLH patients treated with HSCT as part of the HLH-94 protocol went into long-term remission and were considered cured. Haploidentical HSCT is also effective as long as the patients are in remission at the time of HSCT (73,74). Nevertheless, HSCT carries a high risk of toxicity directly related to the myeloablative regimen, especially in young children with inflammatory complications. The London group has successfully introduced HSCT with reduced intensity conditioning for the treatment of HLH, with up to 84% of these patients still living (75, 76) and a median follow-up of three years. These favorable results were confirmed by the Cincinnati group (77). HSCT has also been reported to cure XLP1- (and XLP2) associated HLH, with an 81% survival rate, versus 18% for XLP patients with HLH who were not transplanted (78). Analysis of the data suggests that a donor chimerism of T-cells above 10–20% is sufficient to achieve sustained remission (74), demonstrating the trans-regulatory effect of cytotoxic T-cells in controlling HLH.

The issue of long-term cognitive/neurological impairment caused by CNS-HLH has been pointed out in several series (66, 79, 80), paving the way for early treatment, while the role of intrathecal treatment with methotrexate remains under debate.

Further advances in the treatment of HLH may arise from better targeted, less toxic therapy such as neutralization of interferon-γ based on the experimental work in a murine model of HLH, demonstrating prevention and treatment of HLH by anti-interferon-γ antibody (81, 82). Gene therapy targeting either peripheral T-cells or hematopoietic stem cells may become an option in the years to come (83).

CONCLUSION

The path followed over the last 60 years that has led to the description of multiple HLH conditions, unraveling the biology, and molecular genetics and the understanding of the pathophysiology of FHL/HLH is remarkable and has led to significant advances in therapy. FHL/HLH can now be cured, although there is still room for improvement in the diagnosis and management of these disorders.

Acknowledgments

The authors thank Ms Tiouri for secretarial assistance and Dr Hayward for English editing.

References

1. Farquhar JW, Claireaux AE. Familial haemophagocytic reticulosis. *Arch Dis Child* 1952;**27**:519–25.
2. Farquhar JW, Macgregor AR, Richmond J. Familial haemophagocytic reticulosis. *Br Med J* 1958;**2**:1561–4.
3. Nelson P, Santamaria A, Olson RL, Nayak NC. Generalized lymphohistiocytic infiltration. A familial disease not previously described and different from Letterer-Siwe disease and Chediak-Higashi syndrome. *Pediatrics* 1961;**27**:931–50.
4. Reese AJ, Levy E. Familial incidence of non-lipoid reticuloendotheliosis (Letterer-Siwe disease). *Arch Dis Child* 1951;**26**:578–81.
5. Donohue WL, Bain HW. Chediak-Higashi syndrome; a lethal familial disease with anomalous inclusions in the leukocytes and constitutional stigmata: report of a case with necropsy. *Pediatrics* 1957;**20**:416–30.
6. Blume RS, Wolff SM. The Chediak-Higashi syndrome: studies in four patients and a review of the literature. *Medicine* 1972;**51**:247–80.
7. Macmahon HE, Bedizel M, Ellis CA. Familial erythrophagocytic lymphohistiocytosis. *Pediatrics* 1963;**32**:868–79.
8. Petersen RA, Kuwabara T. Ocular manifestations of familial lymphohistiocytosis. *Arch Ophthalmol* 1968;**79**:413–6.
9. McClure PD, Strachan P, Saunders EF. Hypofibrinogenemia and thrombocytopenia in familial hemophagocytic reticulosis. *J Pediatr* 1974;**85**:67–70.
10. Janka GE. Familial hemophagocytic lymphohistiocytosis. *Eur J Pediatr* 1983;**140**:221–30.
11. Griscelli C, Durandy A, Guy-Grand D, Daguillard F, Herzog C, Prunieras M. A syndrome associating partial albinism and immunodeficiency. *Am J Med* 1978;**65**:691–702.
12. Purtilo DT, Cassel CK, Yang JP, Harper R. X-linked recessive progressive combined variable immunodeficiency (Duncan's disease). *Lancet* 1975;**1**:935–40.
13. Rigaud S, Fondaneche MC, Lambert N, et al. XIAP deficiency in humans causes an X-linked lymphoproliferative syndrome. *Nature* 2006;**444**:110–4.
14. Risdall RJ, McKenna RW, Nesbit ME, et al. Virus-associated hemophagocytic syndrome: a benign histiocytic proliferation distinct from malignant histiocytosis. *Cancer* 1979;**44**:993–1002.
15. Nezelof C, Eliachar E. [Familial lymphohistiocytosis: a report of three cases with general review. Possible links with secondary syndromes (author's transl)]. *Nouve Rev Franc Hematol* 1973;**13**:319–37.
16. Henter JI, Elinder G, Soder O, Hansson M, Andersson B, Andersson U. Hypercytokinemia in familial hemophagocytic lymphohistiocytosis. *Blood* 1991;**78**:2918–22.
17. Spriggs DR, Sherman ML, Frei 3rd E, Kufe DW. Clinical studies with tumour necrosis factor. *Ciba Foundation Symp* 1987;**131**:206–27.
18. Henter JI, Carlson LA, Soder O, Nilsson-Ehle P, Elinder G. Lipoprotein alterations and plasma lipoprotein lipase reduction in familial hemophagocytic lymphohistiocytosis. *Acta Paediatr Scand* 1991;**80**:675–81.
19. Nawroth PP, Stern DM. Modulation of endothelial cell hemostatic properties by tumor necrosis factor. *J Exp Med* 1986;**163**:740–5.
20. Billiau AD, Roskams T, Van Damme-Lombaerts R, Matthys P, Wouters C. Macrophage activation syndrome:

characteristic findings on liver biopsy illustrating the key role of activated, IFN-gamma-producing lymphocytes and IL-6- and TNF-alpha-producing macrophages. *Blood* 2005;**105**:1648–51.

21. Komp DM, McNamara J, Buckley P. Elevated soluble interleukin-2 receptor in childhood hemophagocytic histiocytic syndromes. *Blood* 1989;**73**:2128–32.

22. Ina Y, Takada K, Sato T, Yamamoto M, Noda M, Morishita M. Soluble interleukin 2 receptors in patients with sarcoidosis. Possible origin. *Chest* 1992;**102**:1128–33.

23. Feldmann J, Le Deist F, Ouachee-Chardin M, et al. Functional consequences of perforin gene mutations in 22 patients with familial haemophagocytic lymphohistiocytosis. *Br J Haematol* 2002;**117**:965–72.

24. Haliotis T, Roder J, Klein M, Ortaldo J, Fauci AS, Herberman RB. Chediak-Higashi gene in humans I. Impairment of natural-killer function. *J Exp Med* 1980;**151**:1039–48.

25. Roder J, Duwe A. The beige mutation in the mouse selectively impairs natural killer cell function. *Nature* 1979;**278**:451–3.

26. Katz P, Zaytoun AM, Fauci AS. Deficiency of active natural killer cells in the Chediak-Higashi syndrome. Localization of the defect using a single cell cytotoxicity assay. *J Clin Invest* 1982;**69**:1231–8.

27. Targan SR, Oseas R. The "lazy" NK cells of Chediak-Higashi syndrome. *J Immunol* 1983;**130**:2671–4.

28. Perez N, Virelizier JL, Arenzana-Seisdedos F, Fischer A, Griscelli C. Impaired natural killer activity in lymphohistiocytosis syndrome. *J Pediatr* 1984;**104**:569–73.

29. Arico M, Nespoli L, Maccario R, et al. Natural cytotoxicity impairment in familial haemophagocytic lymphohistiocytosis. *Arch Dis Child* 1988;**63**:292–6.

30. Eife R, Janka GE, Belohradsky BH, Holtmann H. Natural killer cell function and interferon production in familial hemophagocytic lymphohistiocytosis. *Pediatr Hematol Oncol* 1989;**6**:265–72.

31. Arico M, Janka G, Fischer A, et al. Hemophagocytic lymphohistiocytosis. Report of 122 children from the International Registry. FHL Study Group of the Histiocyte Society. *Leukemia* 1996;**10**:197–203.

32. Henter JI, Arico M, Elinder G, Imashuku S, Janka G. Familial hemophagocytic lymphohistiocytosis: Primary hemophagocytic lymphohistiocytosis. *Hematol Oncol Clin N Am* 1998;**12**:417–33.

33. Bryceson YT, Pende D, Maul-Pavicic A, et al. A prospective evaluation of degranulation assays in the rapid diagnosis of familial hemophagocytic syndromes. *Blood* 2012;**119**:2754–3263.

34. Egeler RM, Shapiro R, Loechelt B, Filipovich A. Characteristic immune abnormalities in hemophagocytic lymphohistiocytosis. *J Pediatr Hematol Oncol* 1996;**18**:340–5.

35. Clark RA, Kimball HR. Defective granulocyte chemotaxis in the Chediak-Higashi syndrome. *J Clin Invest* 1971;**50**:2645–52.

36. Padgett GA. Neutrophilic function in animals with the Chediak-Higashi syndrome. *Blood* 1967;**29**:906–15.

37. Root RK, Rosenthal AS, Balestra DJ. Abnormal bactericidal, metabolic, and lysosomal functions of Chediak-Higashi Syndrome leukocytes. *J Clin Invest* 1972;**51**:649–65.

38. Henter JI, Horne A, Arico M, et al. HLH-2004: Diagnostic and therapeutic guidelines for hemophagocytic lymphohistiocytosis. *Pediatr Blood Cancer* 2007;**48**:124–31.

39. Gilgenkrantz S, Gregoire MJ, Chery M, Bordigoni P, Olive D. [Fragile site on chromosome 2 (q11) in a case of familial lymphohistiocytosis]. *J Genet Hum* 1984;**32**:209–19.

40. Hasle H, Brandt C, Kerndrup G, Kjeldsen E, Sorensen AG. Haemophagocytic lymphohistiocytosis associated with constitutional inversion of chromosome 9. *Br J Haematol* 1996;**93**:808–9.

41. Ohadi M, Lalloz MR, Sham P, et al. Localization of a gene for familial hemophagocytic lymphohistiocytosis at chromosome 9q21.3-22 by homozygosity mapping. *Am J Hum Genet* 1999;**64**:165–71.

42. Dufourcq-Lagelouse R, Jabado N, Le Deist F, et al. Linkage of familial hemophagocytic lymphohistiocytosis to 10q21-22 and evidence for heterogeneity. *Am J Hum Genet* 1999;**64**:172–9.

43. Stepp S, Dufourcq-Lagelouse R, Le Deist F, et al. Perforin gene defects in familial hemophagocytic lymphohistiocytosis. *Science* 1999;**286**:1957–9.

44. Feldmann J, Callebaut I, Raposo G, et al. Munc13-4 is essential for cytolytic granules fusion and is mutated in a form of familial hemophagocytic lymphohistiocytosis (FHL3). *Cell* 2003;**115**:461–73.

45. zur Stadt U, Schmidt S, Kasper B, et al. Linkage of familial hemophagocytic lymphohistiocytosis (FHL) type-4 to chromosome 6q24 and identification of mutations in syntaxin 11. *Hum Mol Genet* 2005;**14**:827–34.

46. zur Stadt U, Rohr J, Seifert W, et al. Familial hemophagocytic lymphohistiocytosis type 5 (FHL-5) is caused by mutations in Munc18-2 and impaired binding to syntaxin 11. *Am J Hum Genet* 2009;**85**:482–92.

47. Cote M, Menager MM, Burgess A, et al. Munc18-2 deficiency causes familial hemophagocytic lymphohistiocytosis type 5 and impairs cytotoxic granule exocytosis in patient NK cells. *J Clin Invest* 2009;**119**:3765–73.

48. Rubin CM, Burke BA, McKenna RW, et al. The accelerated phase of Chediak-Higashi syndrome. An expression of the virus-associated hemophagocytic syndrome? *Cancer* 1985;**56**:524–30.

49. Devictor D, Fischer A, Mamas S, et al. [Immunologic study of familial lymphohistiocytosis. Eight new case reports (author's transl)]. *Arc Franc Pediatr* 1982;**39**:135–40.

50. Ménasché G, Pastural E, Feldmann J, et al. Mutations in RAB27A cause Griscelli syndrome associated with hemophagocytic syndrome. *Nat Genet* 2000;**25**:173–6.

51. Perou CM, Moore KJ, Nagle DL, et al. Identification of the murine beige gene by YAC complementation and positional cloning. *Nat Genet* 1996;**13**:303–8.

52. Nagle DL, Karim AM, Woolf EA, et al. Identification and mutation analysis of the complete gene for Chediak-Higashi syndrome. *Nat Genet* 1996;**14**:307–11.

53. Barbosa MDFS, Nguyen QA, Tchernev VT, et al. Identification of the homologous beige and Chediak-Higashi syndrome genes. *Nature* 1996;**382**:262–5.

54. Skare JC, Milunsky A, Byron KS, Sullivan JL. Mapping the X-linked lymphoproliferative syndrome. *Proc Natl Acad Sci USA* 1987;**84**:2015–8.

55. Skare J, Wu BL, Madan S, et al. Characterization of three overlapping deletions causing X-linked lymphoproliferative disease. *Genomics* 1993;**16**:254–5.

56. Coffey AJ, Brooksbank RA, Brandau O, et al. Host response to EBV infection in X-linked lymphoproliferative disease results from mutations in an SH2-domain encoding gene [see comments]. *Nat Genet* 1998;**20**:129–35.

57. Sayos J, Wu C, Morra M, et al. The X-linked lymphoproliferative-disease gene product SAP regulates signals induced through the co-receptor SLAM. *Nature* 1998;**395**:462–9.

58. Palendira U, Low C, Chan A, et al. Molecular pathogenesis of EBV susceptibility in XLP as revealed by analysis of female carriers with heterozygous expression of SAP. *PLoS Biol* 2011;**9**:e1001187.

59. Ladisch S, Poplack DG, Holiman B, Blaese RM. Immunodeficiency in familial erythrophagocytic lymphohistiocytosis. *Lancet* 1978;**1**:581–3.

60. Ladisch S, Ho W, Matheson D, Pilkington R, Hartman G. Immunologic and clinical effects of repeated blood exchange in familial erythrophagocytic lymphohistiocytosis. *Blood* 1982;**60**:814–21.

61. Woo SY, Klappenbach RS, McCullars GM, Kerwin DM, Rowden G, Sinks LF. Familial erythrophagocytic lymphohistiocytosis: treatment with vinblastine-loaded platelets. *Cancer* 1980;**46**:2566–70.

62. Ambruso DR, Hays T, Zwartjes WJ, Tubergen DG, Favara BE. Successful treatment of lymphohistiocytic reticulosis with phagocytosis with epipodophyllotoxin VP 16-213. *Cancer* 1980;**45**:2516–20.

63. Lilleyman JS. The treatment of familial erythrophagocytic lymphohistiocytosis. *Cancer* 1980;**46**:468–70.

64. Fischer A, Virelizier JL, Arenzana-Seisdedos F, Perez N, Nezelof C, Griscelli C. Treatment of four patients with erythrophagocytic lymphohistiocytosis by a combination of epipodophyllotoxin, steroids, intrathecal methotrexate, and cranial irradiation. *Pediatrics* 1985;**76**:263–8.

65. Henter JI, Samuelsson-Horne A, Arico M, et al. Treatment of hemophagocytic lymphohistiocytosis with HLH-94 immunochemotherapy and bone marrow transplantation. *Blood* 2002;**100**:2367–73.

66. Horne A, Trottestam H, Arico M, et al. Frequency and spectrum of central nervous system involvement in 193 children with haemophagocytic lymphohistiocytosis. *Br J Haematol* 2008;**140**:327–35.

67. Trottestam H, Horne A, Arico M, et al. Chemoimmunotherapy for hemophagocytic lymphohistiocytosis: long-term results of the HLH-94 treatment protocol. *Blood* 2011;**118**:4577–84.

68. Stephan JL, Donadieu J, Ledeist F, Blanche S, Griscelli C, Fischer A. Treatment of familial hemophagocytic lymphohistiocytosis with antithymocyte globulins, steroids, and cyclosporin A. *Blood* 1993;**82**:2319–23.

69. Mahlaoui N, Ouachee-Chardin M, de Saint Basile G, et al. Immunotherapy of familial hemophagocytic lymphohistiocytosis with antithymocyte globulins: a single-center retrospective report of 38 patients. *Pediatrics* 2007;**120**:e622–628.

70. Marsh RA, Allen CE, McClain KL, et al. Salvage therapy of refractory hemophagocytic lymphohistiocytosis with alemtuzumab. *Pediatr Blood Cancer* 2013;**60**:101–9.

71. Milone MC, Tsai DE, Hodinka RL, et al. Treatment of primary Epstein-Barr virus infection in patients with X-linked lymphoproliferative disease using B-cell-directed therapy. *Blood* 2005;**105**:994–6.

72. Fischer A, Cerf-Bensussan N, Blanche S, et al. Allogeneic bone marrow transplantation for erythrophagocytic lymphohistiocytosis. *J Pediatr* 1986;**108**:267–70.

73. Jabado N, de Graeff-Meeder ER, Cavazzana-Calvo M, et al. Treatment of familial hemophagocytic lymphohistiocytosis with bone marrow transplantation from HLA genetically nonidentical donors. *Blood* 1997;**90**: 4743–8.

74. Ouachee-Chardin M, Elie C, de Saint Basile G, et al. Hematopoietic stem cell transplantation in hemophagocytic lymphohistiocytosis (HLH): a single report of 48 patients. *Pediatrics* 2006;**117**:e743–50.

75. Cooper N, Rao K, Gilmour K, et al. Stem cell transplantation with reduced-intensity conditioning for hemophagocytic lymphohistiocytosis. *Blood* 2006;**107**:1233–6.

76. Cooper N, Rao K, Goulden N, Webb D, Amrolia P, Veys P. The use of reduced-intensity stem cell transplantation in haemophagocytic lymphohistiocytosis and Langerhans cell histiocytosis. *Bone Marrow Transplant* 2008;**42**(Suppl 2):S47–50.

77. Marsh RA, Vaughn G, Kim MO, et al. Reduced-intensity conditioning significantly improves survival of patients with hemophagocytic lymphohistiocytosis undergoing allogeneic hematopoietic cell transplantation. *Blood* 2010;**116**:5824–31.

78. Booth C, Gilmour KC, Veys P, et al. X-linked lymphoproliferative disease due to SAP/SH2D1A deficiency: a multicenter study on the manifestations, management and outcome of the disease. *Blood* 2011;**117**:53–62.

79. Haddad E, Sulis ML, Jabado N, Blanche S, Fischer A, Tardieu M. Frequency and severity of central nervous system lesions in hemophagocytic lymphohistiocytosis. *Blood* 1997;**89**:794–800.

80. Jackson J, Titman P, Butler S, et al. Cognitive and psychosocial function post hematopoietic stem cell transplantation in children with hemophagocytic lymphohistiocytosis. *J Allergy Clin Immunol* 2013;**132**:889–95.

81. Jordan MB, Hildeman D, Kappler J, Marrack P. An animal model of hemophagocytic lymphohistiocytosis (HLH): CD8+ T-cells and interferon gamma are essential for the disorder. *Blood* 2004;**104**:735–43.

82. Pachlopnik Schmid J, Ho CH, et al. Neutralization of IFNgamma defeats haemophagocytosis in LCMV-infected perforin- and Rab27a-deficient mice. *EMBO Mol Med* 2009;**1**:112–24.

83. Rivat C, Booth C, Alonso-Ferrero M, et al. SAP gene transfer restores cellular and humoral immune function in a murine model of X-linked lymphoproliferative disease. *Blood* 2013;**121**:1073–6.

84. de Saint Basile G, Menasche G, Fischer A. Molecular mechanisms of biogenesis and exocytosis of cytotoxic granules. *Nat Reviews Immunol* 2010;**10**:568–79.

Chronic Granulomatous Disease – from a Fatal Disease to a Curable One

Steven M. Holland

Laboratory of Clinical Infectious Diseases, National Institute of Allergy and Infectious Diseases, National Institutes of Health, Bethesda, Maryland, USA

INTRODUCTION

While we all are students of history in some respect, it is of value to wonder why *this* particular history is worthy of study. Chronic granulomatous disease (CGD) was one of the earliest immunodeficiencies recognized. It was one of the first (if not the first) recognized due to a functional defect despite the normal size, shape and numbers of cells. Its fundamental mechanism

Primary Immunodeficiency Disorders: A Historic and Scientific Perspective
2014 Published by Elsevier Inc.

was adduced long before its genes were cloned. Its genetics were inferred early on, allowing it to serve as the first target for positional genetic cloning. Subsequently, it has been at the forefront of what we have learned about antibacterial prophylaxis, antifungal prophylaxis, cytokine immunotherapy, bone marrow transplantation and gene therapy. In addition to its major contributions to immunology, CGD has helped define certain aspects of infectious disease, such as invasive fungal susceptibility and very selected bacterial invasive infections. In common with so many immunodeficiencies, CGD embodies the Janus-like character of increased susceptibility to infection along with impaired regulation of inflammation. For all these reasons, the study of the history of CGD allows us to walk in the footsteps of giants, highlighting the importance of astute clinical observation coupled to careful laboratory investigation. Finally, the history of the discovery of CGD is a potent reminder of the fact that while not all clinical investigation uncovers that which it initially sought, the prepared mind of the serious clinician will find that which is there to be found. In this chapter, I will try to outline the critical steps in the now-detailed elucidation of CGD. I will track the first ≈35 years of its history chronologically, from clinical recognition to molecular cloning, and then will depart to a more interwoven approach. Sadly, not all critical contributions can be mentioned, for which I apologize at the beginning.

ORIGINAL DESCRIPTIONS

CGD was first described by Dr Charles Janeway and colleagues from Boston Children's Hospital at the 64th Annual Meeting of the American Pediatric Society in 1954 (1). The description of the disease we now know as X-linked agammaglobulinemia in 1952 (2) sparked a surge of interest in immunoglobulins as essential for fighting infection, and triggered a broad national search for more cases of agammaglobulinemia

manifesting as recurrent infection. While searching for cases of *hypo*gammaglobulinemia, Janeway and colleagues encountered five cases of *hyper*gammaglobulinemia associated with hepatomegaly, splenomegaly, diffuse lymphadenopathy and recurrent infection. Their report of these children elicited a chorus of agreement from the audience, each citing their own similar cases of hypergammaglobulinemia and recurrent infections. Reading the vigorous discussion that accompanies that first paper you can almost hear the straining at cause, mechanism and therapy. Even then, in its first formal mention, the disease we now know as CGD was compared to cystic fibrosis and had already spawned discussions of the advisability of chronic antibiotic use. It is noteworthy that many groups had already recognized the same disease entity, but none had crystallized it into a single syndrome. However, despite the recognition of the clinical entity, the majority of the discussants still believed that this was likely to represent a primary immunoglobulin dysfunction.

In 1957, Berendes et al. (3) and in 1959, Bridges et al. (4) reported four boys from the University of Minnesota with a similar disease, characterized by early onset of infections, diffuse granulomatous inflammation, hypergammaglobulinemia and inexorably progressive suppuration, culminating in death in childhood. However, they also noted that the episodes of suppurative infection were separated by periods of apparent good health. The spectrum of infections was quite limited in these children, mostly due to staphylococci and a few Gram-negative rods. White cell studies showed normal chemotaxis and normal pneumococcal killing, leaving them with the impression of a reticuloendothelial disease with features reminiscent of Hodgkin disease and Wegener granulomatosis. They coined the term "fatal chronic granulomatous disease of childhood". Landing and Shirkey (5) reported two boys with a similar presentation who had numerous lipid-laden macrophages in tissues, but they doubted that this was the same disease as described by Berendes and Bridges and coworkers, as the immunoglobulins were not dysfunctional enough. Critically, the group at Minnesota led by Robert A. Good decided to perform experiments with some of the typical organisms from CGD infections (staphylococci), rather than organisms that were common causes of disease in hypogammaglobulinemia (pneumococci). Staphylococci proved to be a prescient choice. In 1966, Holmes et al. (6) found that staphylococcal killing by CGD neutrophils was markedly impaired compared to normal, but it was not ablated. That is, CGD neutrophils prevented the increase in staphylococcal numbers but could not reduce them by anything like the same extent as normal neutrophils (Fig. 13.1).

In 1967, the same investigators also demonstrated that while the phagocytosis of staphylococci was normal, it was the intracellular killing of organisms that was defective (7). Therefore, the events of ingestion and digestion were separate and distinct, and patients with CGD had defects only in the latter. These studies identified the CGD metabolic defect as being downstream of phagocytosis, and critical for staphylococcal killing but dispensable for pneumococcal killing. But where and what was this critical component that was independent of antibody and discriminated between staphylococci and streptococci?

THE DEFECT

Neutrophil metabolism was known to be quite dynamic; the addition of bacteria to leukocytes had been shown in 1933 to induce a dramatic consumption of oxygen, the "respiratory burst" (8). In 1967, Baehner and Nathan (9) used nitroblue tetrazolium, a dye that changed color on reduction, to show that there was a defect in neutrophil oxidative activity in response

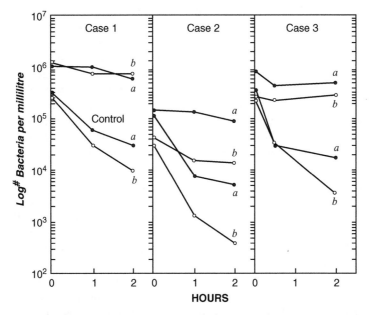

FIGURE 13.1 Impaired staphylococcal killing by CGD neutrophils. Note that CGD neutrophils were able to prevent staphylococcal outgrowth (stable staphylococcal numbers) but were unable significantly to reduce bacterial numbers over two hours. This is a reflection of the impaired, but not absent, bactericidal activity of CGD neutrophils, a critical differentiation of CGD from other conditions, such as neutropenia. (Figure by permission of *The Lancet* (6).)

to ingestion of bacteria. This observation simultaneously indicated a potential target for the underlying defect in CGD and also provided a diagnostic test that allowed clinical screening. They suggested that this might be due to an impaired NADH oxidase pathway. Around the same time, Holmes and colleagues also demonstrated the same defect in oxidative metabolism but were skeptical about the identification of NADH as the target of the oxidase, favoring NADPH, for which this inferred oxidase complex seemed to have greater affinity (10).

In 1965, Carson et al. (11) described 13 boys from several kindreds with what they called "progressive septic granulomatosis". These cases were notable for the existence of lipid-laden macrophages in many tissues as well as the first broad description of the disease as including lymphadenitis, pneumonia, hepatosplenomegaly, perirectal abscesses and fistulae and inflammatory bowel disease. This report linked together the cases of Landing and Shirkey (5) and the cases from the Good group at the University of Minnesota.

Until 1967, all the identified cases had been in boys, and some of the pedigrees were clearly consistent with an X-linked pattern. The development of the NBT test allowed demonstration of carrier mothers and sisters and normal fathers, further confirming the X-linked nature of the disease (12). However, the discovery of full-blown disease in girls in 1968 led to the recognition that there were autosomal recessive forms as well, suggesting that NADPH oxidase was likely a multicomponent enzyme (13–15).

The search for the underlying defect had to focus on patient cells, since there were no spontaneous animal models and this long predated the development of homologous recombination technology. It seemed relatively clear that this entailed a defect in oxidative metabolism, but where and when? The abnormalities were clearest after phagocytosis, but were they due to it? Clinical observation and *in vitro* work so far had suggested that some organisms were handled

normally by CGD neutrophils (e.g. streptococci, lactobacilli), while others were not (e.g. staphylococci, *Serratia*). In 1968, Baehner and Nathan (16) showed that both NADH oxidase and NADPH oxidase activities were depressed in CGD cells, and suggested that NADH activity was the likely culprit in CGD. The following year, Seymour Klebanoff and Lon White (17) addressed this issue by examining the ability of CGD cells to iodinate ingested bacteria, a process that depends on the generation of hydrogen peroxide and its modification by myeloperoxidase (MPO). They found that CGD cells were unable to iodinate *Serratia marsescens*, a typical CGD pathogen, but they were able to iodinate *Lactobacillus acidophilus*, an organism that is not a pathogen in CGD. Normal neutrophils could iodinate both organisms easily. Because of the necessity of hydrogen peroxide for the activity of MPO, they reasoned that CGD cells were not making endogenous hydrogen peroxide. However, in the presence of a bacterium that was making hydrogen peroxide, the cell could use the bacterial hydrogen peroxide to support iodination and facilitate bacterial killing. In contrast, if a bacterium was not making hydrogen peroxide (or destroyed what it made through the agency of catalase), then it would not provide the substrate for MPO activity and the bacterium would survive. This observation was supported by the fact that almost all pathogens in CGD are catalase producing, meaning that they degrade their own hydrogen peroxide and thereby do not support iodination. However, this logical explanation of a clinical observation turned out to be misleading. This could only be determined after the development of molecular tools that enabled directed knockouts in microorganisms and of mouse models of CGD (18, 19).

Despite the disagreements about whether the critical enzyme was an NADH or an NADPH oxidase, there was broad agreement that the critical defect was in the generation of hydrogen peroxide, and that this activity was cyanide resistant and linked to the hexose monophosphate shunt (HMS). Elegant proofs of this fact were provided in 1970 with the repair of the defect in hydrogen peroxide production by the ingestion of latex beads coated with glucose oxidase. When these beads were ingested by CGD neutrophils in the presence of glucose, their ability to kill bacteria (20) and reduce NBT (21) was restored. These experiments clearly proved that the generation of hydrogen peroxide was the central defect in CGD.

The next few years saw expansion of the clinical phenotype of CGD, including the recognition of Kell negative McLeod syndrome with transfusion sensitization in CGD in 1971 (22, 23), retinal involvement in 1972 (24), gastrointestinal involvement reminiscent of Crohn's disease (25) and obstructive uropathic lesions such as granulomatous cystitis (26). This period also saw the development of therapeutic approaches, including granulocyte transfusions (27).

Curnutte and colleagues provided a critical element in 1974 by showing that the generation of superoxide (O_2^-) could be measured in whole cells as the reduction of ferricytochrome c (28). This provided a tool to show that CGD cells were markedly impaired in the production of superoxide, and that this could be both a bactericidal agent on its own as well as the precursor for the generation of hydrogen peroxide. Therefore, the process had moved one step more proximal in the pathway. But which pathway was it? The final resolution of the NADH/NADPH question came in 1975. Using the cell free assays and ferricytochrome c reduction, Hohn and Lehrer (29) and Curnutte and Babior (30, 31) were able to show clearly that NADPH was the most important substrate for the oxidase enzyme, confirming the enzyme as NADPH oxidase.

THE PARTS

So now we have an NADPH oxidizing enzyme complex that creates superoxide, generates hydrogen peroxide and is linked to bacterial and fungal killing. It has at least two genetic forms, one recessive form on the X chromosome and another recessive one on at least one autosome. With a reasonably strong clinical phenotype that allows case finding, the tools are prepared for the next breakthroughs: cell-free systems, sensitive detectors, live cell systems and families with mutations in different genes.

In a series of elegant papers beginning in 1976, Tony Segal and coworkers identified a specific cytochrome spectrum from neutrophil membranes that was missing in CGD granulocytes, indicating that the protein complex was involved, now called cytochrome b, was essential to at least the X-linked forms of disease and was retained in some of the recessive cases of CGD (32–34). In 1980, Bass et al. recognized a role for arachidonate in the activation of NADPH oxidase using cell-free systems (35). Now began the race to identify the multiple proteins, and therefore the genes, that must be creating this enzyme complex.

EMERGING CLINICAL REPORTS

At the same time that the basic scientific landscape was being explored and charted, clinical oysters were open and the pearls were many for the taking. The first prenatal diagnosis using fetal cord blood for NBT testing was done in 1979 (36). Chang recognized the critical role of hydrogen peroxide in leishmanicidal activity, suggesting that the superoxide pathway was important for a wide range of intracellular pathogens (37). The association of the X-linked carrier state with high rates of discoid lupus was appreciated (38, 39). The first cases of *Chromobacterium violaceum* bacteremia in CGD were reported, suggesting that CGD might be a critical underlying factor in severe *C. violaceum* infections (40). In 1975, a syndrome of diffuse pulmonary infiltrates with fever and hypoxia was noted after smoking moldy marijuana (41). A similar syndrome was identified, over 30 years later, after exposure to mulch (42). These clinical entities are caused by the reaction of CGD phagocytes to fungi, both alive and dead, leading to exuberant inflammation, fever and hypoxia.

Trimethoprim-sulfamethoxazole was shown to be an effective long-term antibacterial in a small clinical report (43). Jacobs and Wilson studied the intra- and extracellular survival of different antibiotics in the setting of staphylococcal infection *in vitro* (44). They found excellent intracellular penetration of trimethoprim-sulfamethoxazole with restoration of bacterial killing by CGD cells to normal levels, giving further support to the long-term use of this compound. In 1983, Murray and colleagues (45) showed that interferon gamma was the critical cytokine that mediated leishmania killing in human macrophages, and its activity was correlated with the induction of hydrogen peroxide formation, starting the pursuit of interferon gamma as a restorative in CGD patients.

The inflammatory aspects of CGD had been appreciated from the earliest recognition of the disease, and the mechanisms had long been speculated to involve persistent infection or persistent antigen presentation. Henderson and Klebanoff added a new facet to the inflammatory predilection in CGD by documenting impaired leukotriene metabolism by CGD cells, allowing prolonged persistence of inflammatory mediators (46).

CELL FUSIONS AND CELL-FREE ASSAYS

In 1984, Hamers et al. performed a critical experiment by fusing monocytes from patients with cytochrome b positive and cytochrome b negative forms of CGD (47). They found that while the cells were individually incapable of NBT reduction, after fusion, normal NADPH oxidase capacity was restored, clearly showing that at least two of the factors that made up X-linked and autosomal recessive CGD could

complement each other, and that they had to be in the same or overlapping pathways. In 1985, using the same technique, they demonstrated a recessive form of cytochrome b negative CGD, bringing the recognized forms of disease to three (48). The demonstration that Epstein–Barr (EBV)-transformed B-cell lines were faithful surrogates for superoxide production opened a new opportunity for molecular and cellular manipulation of the different types of CGD in order to find the genes and understand their function (49). Finally, the development of a reliable cell-free system that allowed activation of NADPH oxidase activity *in vitro* created the opportunity to add or subtract specific components and begin to home in on the specific proteins and non-protein mediators controlling this pathway (50).

THE GENES

I have tried thus far to set the stage for the burst of discovery that propelled CGD from a rare and often fatal immunodeficiency into a major scientific and medical entity that rapidly became and still serves as a paradigm for bench to bedside integration. Using the growing knowledge of chromosomal arrangement and molecular genetics, Orkin and his group cloned the X-linked CGD gene using only a positional approach (51). They had previously identified the region Xp21 as the site of the X-linked CGD gene by fine mapping (23, 52) based on children who had X-CGD and Duchenne muscular dystrophy (DMD). One of the astute clinicians who recognized the possible physical link between two rare X-linked diseases was Boris Kousseff at University of South Florida (53). Orkin's group took advantage of immortalized EBV B cells from the patient reported by Kousseff (53) and the patient described by Francke et al. (23), recognizing that these patients were necessarily missing the DNA (and therefore the RNA) for the X-CGD gene. They performed subtraction of patient RNA from normal, and were left with a novel RNA that they surmised was the X-CGD gene. Controversy remained, but eventually both Orkin's and Segal's groups concurred that the gene cloned from Xp21 was indeed the one which encodes the beta subunit of the cytochrome b of the neutrophil (54, 55). Tantalizingly, Segal had also shown that the absence of the beta subunit of cytochrome b led to loss of the alpha subunit as well, suggesting that they were co-dependent for expression in some way (56). The X-linked protein was known to be lodged in membranes and to move via vesicles, while the majority of the recessive forms were thought to be cytoplasmic. But what were they?

In 1985, Segal's group had noted that during neutrophil activation for production of the oxidative burst a cytosolic protein of ≈44 kD was phosphorylated in normals and in the neutrophils of those with X-linked cytochrome b deficient CGD. However, they recognized that this same protein was not phosphorylated in those with recessive forms of CGD (57). Heyworth and Segal and the group of Babior confirmed that there were differences in pathways of activation for the phosphorylation of this protein (58, 59). In 1987, Curnutte and colleagues (60, 61) used the cell-free system to prove that there were purifiable cytosolic factors that were absent in CGD patients and of only half abundance in obligate carriers, consistent with an autosomal recessive structural protein. Close on the heels of this proof, the groups of Bob Clark (62) and Harry Malech (63) identified the proteins as the p47 [neutrophil cytosolic factor (NCF) 1] and p67 (NCF2) components in 1988. These observations allowed a broad national survey of autosomal recessive CGD by Clark et al. (64), showing that about one-third of CGD patients had defects in p47/NCF1, while only about 5% had defects in p67/NCF2. The cloning and characterization of p47/NCF1 (65, 66) and p67/NCF2 (67) followed in 1989 and 1990.

At this point, the vast majority of cases of CGD were clearly identified at a protein and molecular level. However, there were a few cases of CGD in patients who clearly had a recessive pattern of disease but were missing the cytochrome. Since those with the cytosolic factor defects were cytochrome positive, cytoplasmic defects seemed unlikely as a cause. On the other hand, both the Segal (56) and Orkin (68) groups had shown that loss of either component of the cytochrome led to loss of expression of the other one. Therefore, it was in a search for the remaining genetic form of CGD that Mary Dinauer in the Orkin group found recessive mutations in the alpha subunit of the cytochrome, now known as CYBA or p22 (69).

THE FUNCTION

The races to identify and clone the genes were over by 1989, leaving the work of understanding how the genes functioned and how to manipulate them still to be done. The groups of Curnutte and coworkers (70) and Rotrosen and Leto (71) clearly showed that p67 and p47 are phosphorylated during neutrophil activation and that they complex together to move from the cytosol to their rendezvous with cytochrome b at the membrane. Thus, the following scenario has been built by this point (72): cytochrome b is composed of a beta subunit (X-CGD) and an alpha subunit and embedded in the wall of the secondary granule. On cellular activation, the secondary granule fuses with the vacuolar membrane, orienting the cytochrome inward. In the cytosol, p47 and 67 join together and are phosphorylated; p47 binds to the cytochrome complex. Arachidonate is part of the complex as well. The assembly of this NADPH oxidase allows the harvest of the electron from NADPH that creates superoxide in the vacuole (or in the extracellular space if the membrane is fused with the plasma membrane). This superoxide is then turned into hydrogen peroxide by superoxide dismutase, and hydrogen peroxide is turned into hypohalous acid by MPO (MPO can use any of several halides, but in the neutrophil chloride is the standard one), yielding bleach.

It would be 19 years until the next, and as far as we can tell the last, gene involved in CGD, p40/NCF4, would be identified by Juan Matute in Mary Dinauer's group (73). However, this form of CGD is very rare (two cases identified so far) and appears to cause only gastrointestinal disease. These patients have a much higher respiratory burst by DHR than do those with other forms of CGD, perhaps accounting for its distinct phenotype.

Over the 60 years since its description, several main themes have overlapped in our understanding of CGD. I have tried to outline some of the very early discoveries, ranging from clinical recognition to molecular characterization to development of a conceptual model. Along the way, CGD has evolved from a disease with early and common fatality to one of effective management with high survival but persisting morbidities (74, 75).

While impairments of the NADPH oxidase typically present as phagocyte defects, in fact only gp91^{phox} is very phagocyte specific, while the other autosomal components are expressed elsewhere as well (76). Membrane bound (cytochrome b558, comprised of gp91^{phox} and p22^{phox}) and cytosolic (p47^{phox}, p67^{phox}, and p40^{phox}) components are the structural genes of the oxidase. Since p22^{phox} is expressed in other tissues and gp91^{phox} is not, there must be other partners for p22^{phox} and the other members of the NADPH oxidase in other tissues. In general, these are the *Nox* family of proteins (77). Therefore, as patients with CGD recover more often from their infections, we are noticing other subtle abnormalities, such as high rates of vascular disease and diabetes in p47^{phox} deficient CGD patients (78) or exclusively inflammatory bowel disease in p40^{phox} deficient CGD (73).

The monkey wrench in this overall elegant explanation is that MPO deficiency is very common, while CGD is not (79). Therefore, while MPO is important in microbial killing, it is not essential,

suggesting that the final conversion of H_2O_2 to HOCl is not the critical step in microbial control. While the metabolites of superoxide themselves can contribute to bacterial killing, the generation of superoxide has broader implications (80). With the generation of superoxide, a charge is imparted to the phagolysosome that is rectified by the rapid influx of K^+ (81). This potassium influx leads to activation of the now intraphagosomal peptides, which mediate microbial killing. Therefore, under this hypothesis, reactive oxidants are working more as intracellular signaling molecules, leading to activation of other non-oxidative pathways, in addition to causing killing directly. Therefore, CGD has allowed us to dissect out the spectrum of mammalian microbicidal activity, using NADPH oxidase to regulate complex peptide activity, a process more nuanced than simply oxidative and non-oxidative pathways.

The NADPH oxidase has been implicated in the activation of neutrophil extracellular traps (NETs), extracellular assemblies of DNA and antimicrobial peptides released from apoptotic neutrophils (82). Gene therapy repair of the NADPH oxidase in a patient with X-CGD led to reconstitution of NET function and clearing of infection, suggesting that NET formation is impaired in CGD, dependent on NADPH function and important in infection control (83).

EPIDEMIOLOGY

A large, voluntary, retrospective study in the USA and Europe found rates of CGD of around 1:200 000–1:250 000 live births (84, 85). Incidences vary depending on ethnic practices and degrees of intermarriage: Sweden 1/450 000; Japan 1/300 000; Israeli Jews 1/218 000; Israeli Arabs 1/111 000 (86). However, the relative rates of X-linked compared to recessive CGD are directly tied to the rates of consanguineous marriage. For instance, in Iran, the recognized rate of recessive CGD is over 85%, with X-linked CGD being only 12.5% (87).

RESIDUAL SUPEROXIDE

Mechanistically, the level of residual neutrophil superoxide production determines survival, at least in X-linked CGD (86). Long-term survival is higher in X-CGD patients with higher residual superoxide production than those with lower rates. Mutations that lead to no protein production (nonsense mutations, deletions, certain splice defects) support no residual superoxide production and have the lowest level of survival. However, those that permit protein production and superoxide generation (missense mutations before amino acid 310 except histidine 222) are associated with higher survival rates. Surprisingly, mutations at or beyond amino acid 310 do not support superoxide production, suggesting strict gp91^{phox} structural requirements for the intracellular binding of NADPH and FAD (88).

INTERFERON GAMMA

The idea that superoxide formation was the central defect in CGD and that its regulation could alter the course of disease was one of the major hypotheses that drove the investigation and development of interferon gamma (IFN-γ). With the seminal demonstrations by Murray et al. (45) and Nathan et al. (89) in 1983, showing that IFN-γ was the cytokine that induced macrophage production of superoxide, the race was on to see whether IFN-γ could replace the defective superoxide production in CGD. Of course, the bigger picture for IFN-γ was thought to be cancer therapy, but that is another story. However, CGD offered a clean, essentially monogenic human disease in which the restoration of superoxide production might be curative. And it worked, to some extent. That is, it was clear already from the demonstration that IFN-γ upregulated superoxide and hydrogen peroxide production that IFN-γ had some sort of effect on NADPH oxidase activity. The demonstration

of exactly what that effect was awaited the cloning of the genes to look at transcription, translation and trafficking (51).

In 1987, Ezekowitz et al. (90) showed *in vitro* that macrophages that had some residual superoxide production (variant CGD) increased their superoxide production significantly in response to IFN-γ, while those who produced no superoxide (classic CGD), did not respond. These authors went on to show that IFN-γ was upregulating the expression of gp91^{phox} mRNA (91). Sechler et al. (92) found more IFN-γ responsiveness than Ezekowitz et al. (91), and showed that the same effect could be achieved by *in vivo* administration of IFN-γ. Importantly, not only did the IFN-γ work to increase superoxide production as measured by NBT reduction, the effect was also durable over several days. This was confirmed by another study by Ezekowitz et al. (93). Subsequently, explaining the long duration of IFN-γ action, Ezekowitz et al. showed that the effect was at the level of a hematopoietic progenitor cell (94).

Taking all these provocative data that IFN-γ improved cellular function in CGD together, investigators led by John Gallin at the NIH and sponsored by Genentech conducted a multi-site, multinational, placebo-controlled, double blind trial involving 128 CGD patients (95). This study was set up to measure both laboratory and clinical endpoints. The laboratory endpoints were along the lines of the previous studies, looking at NBT, superoxide and staphylococcal killing. The clinical endpoints were time to first infection and granulomatous complications. The effect on time to infection was clear and dramatic, with an ≈70% reduction in development of infections in the IFN-γ recipients compared to those who received placebo. This effect was most pronounced in the cohort aged under ten years who had X-linked disease (Fig. 13.2). Surprisingly, none of the laboratory components that had been so compelling in the previously published papers (superoxide production, NBT staining, increased

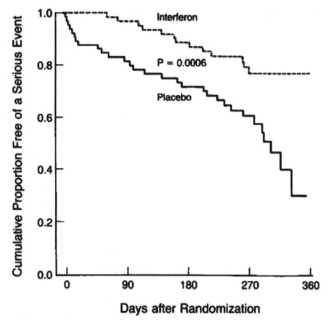

FIGURE 13.2   The effect of IFNg on development of first infection after randomization. By permission of the *New England Journal of Medicine* (93).

cytochrome levels) were found. This was a confusing result.

On the one hand, the clinical data were quite robust; on the other, the proposed mechanism, which had seemed so solid, was inapparent. What could explain this? It had long been apparent that there were two distinct sets of patients in the IFN-γ reconstitution trials, those who responded, who usually had some residual function in the first place, and those who did not. Could the data be explained by only those with residual function responding? This conundrum has continued to reverberate down the years, resulting in a split between those who think that there is very little if any role for IFN-γ in CGD prophylaxis now that there are good antifungals and those who think that the data are compelling even if the mechanism is not clear. If nothing else, the persistent dichotomy over this subject highlights the importance of mechanistic studies, which provide fundamental explanations and help justify clinical applications.

SURVIVAL

It was in part to understand the possible role of the genetics of CGD in the IFN-γ response that Kuhns et al. (88) undertook the systematic review of residual superoxide production and its relationship to overall survival. Examining 287 patients from 244 kindreds, they showed a clear correlation between neutrophil residual superoxide production and overall survival. The demonstration that superoxide production was a continuous variable, that is, the less you had the worse your survival chances, solidifies the case for superoxide deficiency as the major determinant of fatality in CGD. However, this study was surprising in that the mortality curves between high and low superoxide producers only diverged after age 20. What does it mean that infections begin early in life, and mortality diverges later in life? To understand this we need to inquire further about what causes death in CGD.

The causes of and ages at death in CGD have changed over the years. In the 1950s and 1960s, most of the deaths were from infection and most of them occurred in childhood. However, with the introduction of prophylactic antibiotics in the 1970s and 1980s (42) and prophylactic antifungals in the 1990s (96), the ages and causes of death have changed. Theo Heller and colleagues, investigating the gastrointestinal manifestations of CGD, found that liver function abnormalities were common, as were hepatic granulomata (75%) and lobular hepatitis (90%) (97). They also noted extensive portal and central vein venopathy, which suggests that the splenomegaly in CGD is secondary to liver involvement and portal hypertension. While collecting these data, Heller noticed that liver disease was actually a strong predictor of mortality. The independent variables of alkaline phosphatase elevations, decline in platelet count (a marker of splenomegaly and portal hypertension), and liver abscess, together contributed to overall mortality in CGD (Fig. 13.3) (98). Alkaline phosphatase elevations, liver abscess and low residual superoxide

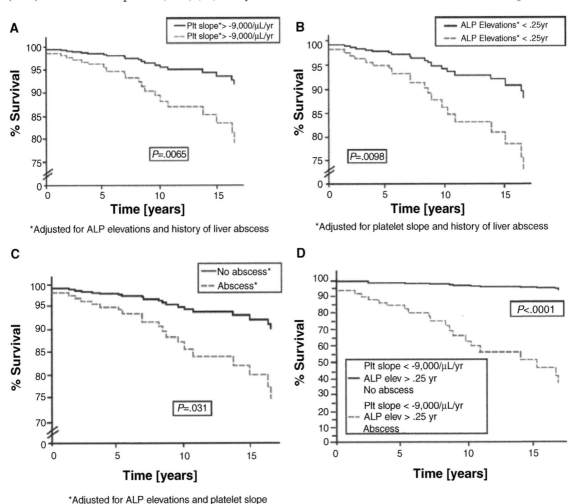

FIGURE 13.3 The effects of declining platelet slope, alkaline phosphatase elevations, history of liver abscess and their conjunction on mortality in CGD. By permission of *Gastroenterology* (98).

production were all independent variables, whereas platelet reduction was a reflection of portal hypertension which was much more common with low residual superoxide production (88). Interestingly, the colitis that is so common in CGD had no correlation with residual superoxide production, no bearing on mortality and no bearing on the frequency or severity of liver disease (88). Therefore, what is going on at the level of the intestinal epithelium is not especially related to what is going on in the liver, not determined by residual function, and while morbid is not mortal.

BACTERIAL INFECTIONS

CGD was first noted for its recurrent infections despite hypergammaglobulinemia. However, what has become so abundantly clear over the decades is that CGD is remarkable for its very narrow spectrum of profound infection susceptibility. In the USA, only about five organisms account for most infections: *Staphylococcus aureus* (lymphadenitis, liver abscess, cellulitis, osteomyelitis), *Serratia marcescens* (cellulitis, pneumonia), *Burkholderia cepacia* complex (pneumonia, sepsis), *Nocardia* species (pneumonia, dissemination) and *Aspergillus* species (pneumonia, osteomyelitis) (75). The properties that unify these organisms, which span the distance from prokaryotes to eukaryotes, remains unclear.

The *Burkholderia cepacia* complex is quite large, with numerous species, a surprisingly large number of which occur in CGD (99). Unlike in cystic fibrosis, there are not epidemic strains in CGD, and patients tend to develop infections with different strains over time. The closely related *Burkholderia gladioli*, the cause of onion rot, also infects patients with CGD (100). *Chromobacterium violaceum* is a Gram-negative rod found in brackish waters, such as those around the Gulf of Mexico in the USA, and causes severe sepsis in CGD (101). *Francisella philomiragia* is also a

Gram-negative rod found in brackish waters, such as the Chesapeake Bay, Long Island Sound and around Nova Scotia, and it also causes sepsis in CGD (102).

Granulibacter bethesdensis is a novel Gram-negative rod that was discovered from the necrotic lymph nodes of a man with CGD who had been sick for months with fever and weight loss despite numerous courses of antibiotics (103). After the organism was identified and treated in this patient it recurred numerous times, with prolonged asymptomatic periods in between episodes, suggesting that *Granulibacter bethesdensis* can have latent and active phases, similar to tuberculosis. The rate of seropositivity for this organism is about 50% in CGD patients, the majority of whom have not had recognized infections. However, the rate of seropositivity in normals is around 25%, indicating broad exposure. Perhaps there is a clinical syndrome of *Granulibacter bethesdensis* infection in normal individuals yet to be identified (104). Regardless of its effects in normals, in CGD, it is associated with necrotic lymphadenopathy that can be quite prolonged. There have been fatal cases of *Granulibacter bethesdensis* sepsis reported from Europe, suggesting that there may be differences between US and other strains of *Granulibacter bethesdensis* (105). Recently, Chu et al., (106) have shown that *Granulibacter bethesdensis* can persist for days within CGD macrophages, with neither host nor pathogen viability affected.

FUNGAL INFECTIONS

CGD patients get some fungal infections that are rare in any other hosts: *Aspergillus nidulans*, *Paecilomyces variotti*, *Paecilomyces lilacinus* and *Neosartorya udagawae*. These organisms are highly pathogenic in CGD but not in any other patient group, including transplant recipients (107). Mucormycosis occurs in CGD but mostly in the setting of significant immunosuppression (108). In contrast to these filamentous molds that are highly virulent in CGD, the endemic dimorphic mold infections histoplasmosis, blastomycosis, and coccidioidomycosis do not occur in CGD, nor does cryptococcosis.

Elegant work in mice by the Dinauer lab in 1997 (109) showed that while CGD alveolar macrophage killing of conidia was normal *in vitro*, pulmonary fungal infections were eventually fatal, even at very low doses of *Aspergillus fumigatus*. During these studies, they noticed that even sterilized fungi elicited an exuberant and durable inflammatory response in the lung, with granulomata and inflammation. This confirmed that at least some of the uncontrolled inflammation in CGD was occurring without persistent infections but could in fact be due to an over-aggressive response. A human correlate to the severe inflammatory response seen with sterile hyphae occurs in the syndrome of "mulch pneumonitis", an acute syndrome caused by inhalation of aerosolized mold, such as found in mulch, hay or dead leaves (42). The typical clinical presentation is that a previously well CGD patient spreads mulch, turns compost or clears moldy leaves, inhaling numerous fungal spores and hyphae; 1–10 days later a syndrome similar to hypersensitivity pneumonitis begins with fever and dyspnea; chest radiographs show diffuse interstitial infiltrates; bronchoscopy is usually uninformative but may yield *Aspergillus*; lung biopsy shows acute inflammation with necrotizing granulomata and fungi. The most successful treatments of this syndrome have included steroids for the inflammation coupled with antifungals. A possible mechanistic explanation for this phenomenon has been provided by the Romani group, who found high levels of IL-17-producing T-cells in CGD mice. They linked the IL-17 inflammatory response to adverse reactions to fungi (110). It remains to be clarified whether this is the same process in humans.

INFLAMMATION

Inflammation in CGD was recognized early on as being prominent in the gastrointestinal and genitourinary tracts. Esophageal, jejunal, ileal, cecal, rectal and perirectal involvement with granulomata mimicking Crohn's disease has been described (25, 111, 112). Perirectal disease is especially common. The rates of inflammatory bowel disease in CGD reach close to 50% for those with the X-linked form, and less for p47^{phox} deficiency (111). Recent review of our aging patients suggests that those with p47^{phox} deficiency develop more inflammatory bowel disease with age, with rates approaching those in X-linked disease. Therefore, CGD offers a remarkable opportunity to study the risks for and causes of inflammatory bowel disease (IBD). While TNF-alpha blocking agents are highly effective and rapidly suppress IBD, they carry a very high risk of infection and death in CGD (113).

Liver abscesses occur in around 35% of patients and were typically treated with surgery (114). Whether only certain CGD patients are predisposed to liver abscesses, or whether having had one liver abscess alters hepatic architecture and blood flow, making subsequent infection more likely is unclear. However, rates of liver abscess are identical between X-linked and p47^{phox} deficient CGD, suggesting that residual superoxide formation is not the determinant. Regardless of cause, treatment presents problems. Similar to the case in the lung with mulch pneumonitis, addressing the inflammation directly appears to be very helpful. This was first pointed out for liver abscesses by Yamazaki-Nakashimada et al. in 2006 (115), who successfully treated two refractory liver abscesses with steroids in addition to antibiotics. Leiding et al. (2012) followed the same path in the treatment of nine CGD patients with liver abscess using only steroids in addition to their antibiotics and avoiding surgery (116). Therefore, addressing the exuberant inflammation as well as the immediate infectious cause may be important in CGD.

LYONIZATION

Female carriers of X-linked CGD have two populations of phagocytes: one that produces superoxide and one that does not, giving carriers a characteristic mosaic pattern on oxidative testing (Fig. 13.4). Infections are infrequent unless the normal neutrophils are <10%, and even then uncommon until the percent age of normal cells falls below 5%. Severe skewing of X-inactivation creates a disease that is indistinguishable from X-linked CGD (117). The balance of wild-type to mutant cells may vary over time in the same woman, but we do not know whether this is directional, moving toward the CGD or the normal population (118). Discoid lupus erythematosus-like lesions, aphthous ulcers and photosensitive rashes have been seen in gp91^{phox} carriers. The rates of unsuspected CGD carriers are high

among patients with discoid lupus erythematosus (38, 39, 119). Finally, severe skewing can also occur following somatic mutation in one copy of CYBB, leading to late onset of disease (120).

BONE MARROW TRANSPLANTATION

Bone marrow transplantation cuts across many fields, from cancer to immunodeficiency to metabolic disease. In CGD, there have been several trials attempting to find the right regimen that is matched to the changing risk of CGD. That is, the survival of CGD patients is improving, but the complications of CGD, such as inflammatory bowel disease, growth retardation and infections continue. As is true for some other immunodeficiencies and distinct from cancer,

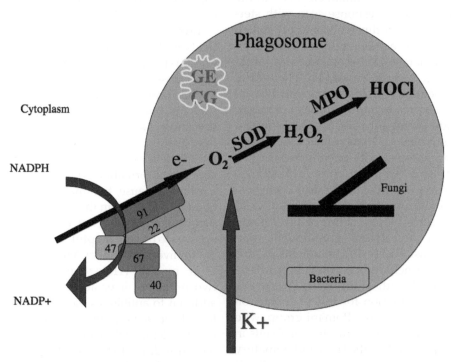

FIGURE 13.4 The assembled NADPH oxidase on the wall of the phagolysosome. The generation of superoxide, hydrogen peroxide, and hypohalous acid are shown. GE is granulocyte elastase and CG is cathepsin G, both enzymes that may be activated by the processes set in motion after superoxide generation. This figure is reproduced in color in the color section.

complete replacement of recipient cells is not required, it is sufficient to supply enough fully functional cells to prevent infections, and likely prevent the other complications as well. One of the special aspects of bone marrow transplantation in CGD is the frequency of fungal infections. On the one hand, fungal infections are a relative contraindication to transplantation; on the other hand, fungal infections are a major and critical reason to move forward with transplantation, especially in the setting of severe refractory molds (121). Matched sibling and unrelated donor bone marrow transplants from Newcastle on Tyne (122) showed an overall long-term success rate of 90% with a striking reconstitution of impaired pretransplant growth. Therefore, much of the growth retardation and delay seen in CGD must be due to various aspects of hematopoietic dysfunction, ranging from colitis to chronic inapparent infection. The 2013 publication by Gungor et al. (123) of a multicenter study using fludaribine, serotherapy and targeted busulfan reported a 96% overall survival rate with very modest toxicities. These developments promise further improvements in transplantation for CGD, with survival rates that are clearly better than untransplanted and amelioration of the other morbidities of the disease. However, since not all patients have access to or are eligible for transplantation, medical management will continue to be important.

GENE THERAPY

CGD is a single gene defect that can be reconstituted *in vitro* and does not require complete correction to be effective. Unfortunately, unlike severe combined immune deficiency (SCID) or adenosine deaminase (ADA) deficiency, corrected CGD cells do not have a growth or survival advantage in the marrow or in the tissue. Therefore, selection and augmentation of those cells is an area of effort. Currently, mildly to moderately myeloablative approaches that make room for the corrected cells are being pursued (124).

OTHER ROLES FOR NADPH OXIDASE

NADPH oxidase does many things outside the neutrophil. It is involved in signaling for NFκB activation in the setting of liver activation, either by ethanol or by carcinogens (125, 126). CGD mice appear to be protected from fatty deposits in the aorta when crossed with the ApoE $-/-$ hypercholesterolemic mouse model (127), whereas coronary arteries are not (128), suggesting that the factors that regulate NADPH activity in the coronaries and the aorta are distinct. p47^{phox} also appears to be critical for the coordinated response of the coronary muscle to biomechanical stress (129). Hippocampal long-term potentiation, the *in vitro* brain slice equivalent to memory, is also NADPH dependent, and CGD mice have trouble with certain learning tasks (130). CGD patients had an elevated rate of lower IQ and learning issues compared to disease controls (131). NADPH oxidase somatic and hematopoietic activity is involved in strokes (132) and in pulmonary vascular permeability (133). Therefore, it is clear that this enzyme is active in many more sites than just phagocytes, suggesting that CGD is more complex and has more to teach than about infections and bone marrow transplants alone.

CONCLUSION

Over the 60 years since its discovery, CGD has evolved dramatically. New understandings, new drugs and new approaches keep the diagnosis and management of CGD at the forefront of medicine and science.

Acknowledgment

This work supported by the Division of Intramural Research, NIAID, NIH.

References

1. Janeway CA, Craig J, Davidson M, Downey W, Gitlin D, Sullivan JC. Hypergammaglobulinemia associated with severe, recurrent and chronic non-specific infection. *Am J Dis Child* 1954;**88**:388–92.

2. Bruton OC. Agammaglobulinemia. *Pediatrics* 1952;**9**:722–8.

3. Berendes H, Bridges RA, Good RA. A fatal granulomatous disease of childhood: The clinical study of a new syndrome. *Minn Med* 1957;**40**:309.

4. Bridges RA, Berendes H, Good RA. A fatal granulomatous disease of childhood. *Am J Dis Child* 1959;**97**:387.

5. Landing BH, Shirkey HS. A syndrome of recurrent infection and infiltration of viscera by pigmented lipid histiocytes. *Pediatrics* 1957;**20**:431–8.

6. Holmes B, Quie PG, Windhorst DB, Good RA. Fatal granulomatous disease of childhood. An inborn abnormality of phagocytic function. *Lancet* 1966;**1**:1225–8.

7. Quie PG, White JG, Holmes B, Good RA. In vitro bactericidal capacity of human polymorphonuclear leukocytes: diminished activity in chronic granulomatous disease of childhood. *J Clin Invest* 1967;**46**:668–79.

8. Baldrige CW, Gerard RW. The extra respiration of phagocytosis. *Am J Physiol* 1933;**103**:235–6.

9. Baehner RL, Nathan DG. Leukocyte oxidase: defective activity in chronic granulomatous disease. *Science* 1967;**155**:835–6.

10. Holmes B, Page AR, Good RA. Studies of the metabolic activity of leukocytes from patients with a genetic abnormality of phagocytic function. *J Clin Invest* 1967;**46**:1422–32.

11. Carson MJ, Chadwick DL, Brubaker CA, Cleland RS, Landing BH. Thirteen boys with progressive septic granulomatosis. *Pediatrics* 1965;**35**:405–12.

12. Windhorst DB, Page AR, Holmes B, Quie PG, Good RA. The pattern of genetic transmission of the leukocyte defect in fatal granulomatous disease of childhood. *J Clin Invest* 1968;**47**:1026–34.

13. Quie PG, Kaplan EL, Page AR, Gruskay FL, Malawista SE. Defective polymorphonuclear-leukocyte function and chronic granulomatous disease in two female children. *N Engl J Med* 1968;**278**:976–80.

14. Baehner RL, Nathan DG. Quantitative nitroblue tetrazolium test in chronic granulomatous disease. *N Engl J Med* 1968;**278**:971–6.

15. Azimi PH, Bodenbender JG, Hintz RL, Kontras SB. Chronic granulomatous disease in three female siblings. *J Am Med Assoc* 1968;**206**:2865–70.

16. Baehner RL, Karnovsky ML. Deficiency of reduced nicotinamide-adenine dinucleotide oxidase in chronic granulomatous disease. *Science* 1968;**162**:1277–9.

17. Klebanoff SJ, White LR. Iodination defect in the leukocytes of a patient with chronic granulomatous disease of childhood. *N Engl J Med* 1969;**280**:460–6.

18. Chang YC, Segal BH, Holland SM, Miller GF, Kwon-Chung KJ. Virulence of catalase-deficient aspergillus nidulans in p47(phox)−/− mice. Implications for fungal pathogenicity and host defense in chronic granulomatous disease. *J Clin Invest* 1998;**101**:1843–50.

19. Messina CG, Reeves EP, Roes J, Segal AW. Catalase negative Staphylococcus aureus retain virulence in mouse model of chronic granulomatous disease. *FEBS Lett* 2002;**518**:107–10.

20. Johnston Jr RB, Baehner RL. Improvement of leukocyte bactericidal activity in chronic granulomatous disease. *Blood* 1970;**35**:350–5.

21. Baehner RL, Nathan DG, Karnovsky ML. Correction of metabolic deficiencies in the leukocytes of patients with chronic granulomatous disease. *J Clin Invest* 1970;**49**:865–70.

22. Giblett ER, Klebanoff SJ, Pincus SH. Kell phenotypes in chronic granulomatous disease: a potential transfusion hazard. *Lancet* 1971;**1**:1235–6.

23. Francke U, Ochs HD, de Martinville B, et al. Minor Xp21 chromosome deletion in a male associated with expression of Duchenne muscular dystrophy, chronic granulomatous disease, retinitis pigmentosa, and McLeod syndrome. *Am J Hum Genet* 1985;**37**:250–67.

24. Martyn LJ, Lischner HW, Pileggi AJ, Harley RD. Chorioretinal lesions in familial chronic granulomatous disease of childhood. *Am J Ophthalmol* 1972;**73**:403–18.

25. Ament ME, Ochs HD. Gastrointestinal manifestations of chronic granulomatous disease. *N Engl J Med* 1973;**288**:382–7.

26. Cyr WL, Johnson H, Balfour J. Granulomatous cystitis as a manifestation of chronic granulomatous disease of childhood. *J Urol* 1973;**110**:357–9.

27. Raubitschek AA, Levin AS, Stites DP, Shaw EB, Fudenberg HH. Normal granulocyte infusion therapy for aspergillosis in chronic granulomatous disease. *Pediatrics* 1973;**51**:230–3.

28. Curnutte JT, Whitten DM, Babior BM. Defective superoxide production by granulocytes from patients with chronic granulomatous disease. *N Engl J Med* 1974;**290**:593–7.

29. Hohn DC, Lehrer RI. NADPH oxidase deficiency in X-linked chronic granulomatous disease. *J Clin Invest* 1975;**55**:707–13.

30. Curnutte JT, Kipnes RS, Babior BM. Defect in pyridine nucleotide dependent superoxide production by a particulate fraction from the cranulocytes of patients with chronic granulomatous disease. *N Engl J Med* 1975;**293**:628–32.

31. Babior BM, Curnutte JT, Kipnes BS. Pyridine nucleotide-dependent superoxide production by a cell-free system from human granulocytes. *J Clin Invest* 1975;**56**:1035–42.

32. Segal AW, Peters TJ. Characterisation of the enzyme defect in chronic granulomatous disease. *Lancet* 1976;**1**:1363–5.

33. Segal AW, Jones OT, Webster D, Allison AC. Absence of a newly described cytochrome b from neutrophils of patients with chronic granulomatous disease. *Lancet* 1978;**2**:446–9.

34. Segal AW, Jones OT. Novel cytochrome b system in phagocytic vacuoles of human granulocytes. *Nature* 1978;**276**:515–7.

35. Bass DA, O'Flaherty JT, Szejda P, DeChatelet LR, McCall CE. Role of arachidonic acid in stimulation of hexose transport by human polymorphonuclear leukocytes. *Proc Natl Acad Sci USA* 1980;**77**:5125–9.

36. Newburger PE, Cohen HJ, Rothchild SB, Hobbins JC, Malawista SE, Mahoney MJ. Prenatal diagnosis of chronic granulomatous disease. *N Engl J Med* 1979;**300**:178–81.

37. Chang KP. Leishmanicidal mechanisms of human polymorphonuclear phagocytes. *Am J Trop Med Hyg* 1981;**30**:322–33.

38. Schaller J. Illness resembling lupus erythematosus in mothers of boys with chronic granulomatous disease. *Ann Intern Med* 1972;**76**:747–50.

39. Brandrup F, Koch C, Petri M, Schiodt M, Johansen KS. Discoid lupus erythematosus-like lesions and stomatitis in female carriers of X-linked chronic granulomatous disease. *Br J Dermatol* 1981;**104**:495–505.

40. Macher AM, Casale TB, Fauci AS. Chronic granulomatous disease of childhood and Chromobacterium violaceum infections in the southeastern United States. *Ann Intern Med* 1982;**97**:51–5.

41. Chusid MJ, Gelfand JA, Nutter C, Fauci AS. Letter: Pulmonary aspergillosis, inhalation of contaminated marijuana smoke, chronic granulomatous disease. *Ann Intern Med* 1975;**82**:682–3.

42. Siddiqui S, Anderson VL, Hilligoss DM, et al. Fulminant mulch pneumonitis: an emergency presentation of chronic granulomatous disease. *Clin Infect Dis* 2007;**45**:673–81.

43. Weening RS, Kabel P, Pijman P, Roos D. Continuous therapy with sulfamethoxazole-trimethoprim in patients with chronic granulomatous disease. *J Pediatr* 1983;**103**:127–30.

44. Jacobs RF, Wilson CB. Activity of antibiotics in chronic granulomatous disease leukocytes. *Pediatr Res* 1983;**17**:916–9.

45. Murray HW, Rubin BY, Rothermel CD. Killing of intracellular Leishmania donovani by lymphokine-stimulated human mononuclear phagocytes. Evidence that interferon-gamma is the activating lymphokine. *J Clin Invest* 1983;**72**:1506–10.

46. Henderson WR, Klebanoff SJ. Leukotriene production and inactivation by normal, chronic granulomatous disease and myeloperoxidase-deficient neutrophils. *J Biol Chem* 1983;**258**:13522–7.

47. Hamers MN, de Boer M, Meerhof LJ, Weening RS, Roos D. Complementation in monocyte hybrids revealing genetic heterogeneity in chronic granulomatous disease. *Nature* 1984;**307**:553–5.

48. Weening RS, Corbeel L, de Boer M, et al. Cytochrome b deficiency in an autosomal form of chronic granulomatous disease. A third form of chronic granulomatous disease recognized by monocyte hybridization. *J Clin Invest* 1985;**75**:915–20.

49. Volkman DJ, Buescher ES, Gallin JI, Fauci AS. B cell lines as models for inherited phagocytic diseases: abnormal superoxide generation in chronic granulomatous disease and giant granules in Chediak-Higashi syndrome. *J Immunol* 1984;**133**:3006–9.

50. Heyneman RA, Vercauteren RE. Activation of a NADPH oxidase from horse polymorphonuclear leukocytes in a cell-free system. *J Leukoc Biol* 1984;**36**:751–9.

51. Royer-Pokora B, Kunkel LM, Monaco AP, et al. Cloning the gene for an inherited human disorder – chronic granulomatous disease – on the basis of its chromosomal location. *Nature* 1986;**322**:32–8.

52. Baehner RL, Kunkel LM, Monaco AP, et al. DNA linkage analysis of X chromosome-linked chronic granulomatous disease. *Proc Natl Acad Sci USA* 1986;**83**:3398–401.

53. Kousseff B. Linkage between chronic granulomatous disease and Duchenne's muscular dystrophy? *Am J Dis Child* 1981;**135**:1149.

54. Dinauer MC, Orkin SH, Brown R, Jesaitis AJ, Parkos CA. The glycoprotein encoded by the X-linked chronic granulomatous disease locus is a component of the neutrophil cytochrome b complex. *Nature* 1987;**327**:717–20.

55. Teahan C, Rowe P, Parker P, Totty N, Segal AW. The X-linked chronic granulomatous disease gene codes for the beta-chain of cytochrome b-245. *Nature* 1987;**327**:720–1.

56. Segal AW. Absence of both cytochrome b-245 subunits from neutrophils in X-linked chronic granulomatous disease. *Nature* 1987;**326**:88–91.

57. Segal AW, Heyworth PG, Cockcroft S, Barrowman MM. Stimulated neutrophils from patients with autosomal recessive chronic granulomatous disease fail to phosphorylate a Mr-44,000 protein. *Nature* 1985;**316**:547–9.

58. Heyworth PG, Segal AW. Further evidence for the involvement of a phosphoprotein in the respiratory burst oxidase of human neutrophils. *Biochem J* 1986;**239**:723–31.

59. Hayakawa T, Suzuki K, Suzuki S, Andrews PC, Babior BM. A possible role for protein phosphorylation in the activation of the respiratory burst in human neutrophils. Evidence from studies with cells from patients with chronic granulomatous disease. *J Biol Chem* 1986;**261**:9109–15.

60. Curnutte JT, Kuver R, Scott PJ. Activation of neutrophil NADPH oxidase in a cell-free system. Partial purification of components and characterization of the activation process. *J Biol Chem* 1987;**262**:5563–9.

61. Curnutte JT, Berkow RL, Roberts RL, Shurin SB, Scott PJ. Chronic granulomatous disease due to a defect in the cytosolic factor required for nicotinamide adenine dinucleotide phosphate oxidase activation. *J Clin Invest* 1988;**81**:606–10.

62. Volpp BD, Nauseef WM, Clark RA. Two cytosolic neutrophil oxidase components absent in autosomal chronic granulomatous disease. *Science* 1988;**242**:1295–7.

63. Nunoi H, Rotrosen D, Gallin JI, Malech HL. Two forms of autosomal chronic granulomatous disease lack distinct neutrophil cytosol factors. *Science* 1988;**242**:1298–301.

64. Clark RA, Malech HL, Gallin JI, et al. Genetic variants of chronic granulomatous disease: prevalence of deficiencies of two cytosolic components of the NADPH oxidase system. *N Engl J Med* 1989;**321**:647–52.

65. Lomax KJ, Leto TL, Nunoi H, Gallin JI, Malech HL. Recombinant 47-kilodalton cytosol factor restores NADPH oxidase in chronic granulomatous disease. *Science* 1989;**245**:409–12.

66. Volpp BD, Nauseef WM, Donelson JE, Moser DR, Clark RA. Cloning of the cDNA and functional expression of the 47-kilodalton cytosolic component of human neutrophil respiratory burst oxidase. *Proc Natl Acad Sci USA* 1989;**86**:7195–9.

67. Leto TL, Lomax KJ, Volpp BD, et al. Cloning of a 67-kD neutrophil oxidase factor with similarity to a noncatalytic region of p60c-src. *Science* 1990;**248**:727–30.

68. Parkos CA, Dinauer MC, Jesaitis AJ, Orkin SH, Curnutte JT. Absence of both the 91kD and 22kD subunits of human neutrophil cytochrome b in two genetic forms of chronic granulomatous disease. *Blood* 1989;**73**:1416–20.

69. Dinauer MC, Pierce EA, Bruns GA, Curnutte JT, Orkin SH. Human neutrophil cytochrome b light chain (p22-phox). Gene structure, chromosomal location, and mutations in cytochrome-negative autosomal recessive chronic granulomatous disease. *J Clin Invest* 1990;**86**:1729–37.

70. Okamura N, Babior BM, Mayo LA, Peveri P, Smith RM, Curnutte JT. The p67-phox cytosolic peptide of the respiratory burst oxidase from human neutrophils. Functional aspects. *J Clin Invest* 1990;**85**:1583–7.

71. Rotrosen D, Leto TL. Phosphorylation of neutrophil 47-kDa cytosolic oxidase factor. Translocation to membrane is associated with distinct phosphorylation events. *J Biol Chem* 1990;**265**:19910–5.

72. Heyworth PG, Curnutte JT, Nauseef WM, Volpp BD, Pearson DW, Rosen H, et al. Neutrophil nicotinamide adenine dinucleotide phosphate oxidase assembly. Translocation of p47-phox and p67-phox requires interaction between p47-phox and cytochrome b558. *J Clin Invest* 1991;**87**:352–6.

73. Matute JD, Arias AA, Wright NA, et al. A new genetic subgroup of chronic granulomatous disease with autosomal recessive mutations in p40 phox and selective defects in neutrophil NADPH oxidase activity. *Blood* 2009;**114**:3309–15.

74. Marciano BE, Wesley R, De Carlo ES, et al. Long-term interferon-gamma therapy for patients with chronic granulomatous disease. *Clin Infect Dis* 2004;**39**:692–9.

75. Segal BH, Leto TL, Gallin JI, Malech HL, Holland SM. Genetic, biochemical, and clinical features of chronic granulomatous disease. *Medicine (Baltimore)* 2000;**79**:170–200.

76. Ushio-Fukai M. Localizing NADPH oxidase-derived ROS. *Sci STKE* 2006;**2006**:re8.

77. Nauseef WM. Detection of superoxide anion and hydrogen peroxide production by cellular NADPH oxidases. *Biochim Biophys Acta* 2014;**1840**:757–67.

78. Leiding JW, Marciano BE, Zerbe CS, Deravin SS, Malech HL, Holland SM. Diabetes, renal and cardiovascular disease in p47 phox−/− chronic granulomatous disease. *J Clin Immunol* 2013;**33**:725–30.

79. Klebanoff SJ, Kettle AJ, Rosen H, Winterbourn CC, Nauseef WM. Myeloperoxidase: a front-line defender against phagocytosed microorganisms. *J Leukoc Biol* 2013;**93**:185–98.

80. Tkalcevic J, Novelli M, Phylactides M, Iredale JP, Segal AW, Roes J. Impaired immunity and enhanced resistance to endotoxin in the absence of neutrophil elastase and cathepsin G. *Immunity* 2000;**12**:201–10.

81. Reeves EP, Lu H, Jacobs HL, et al. Killing activity of neutrophils is mediated through activation of proteases by K+ flux. *Nature* 2002;**416**:291–7.

82. Fuchs TA, Abed U, Goosmann C, et al. Novel cell death program leads to neutrophil extracellular traps. *J Cell Biol* 2007;**176**:231–41.

83. Bianchi M, Hakkim A, Brinkmann V, et al. Restoration of NET formation by gene therapy in CGD controls aspergillosis. *Blood* 2009;**114**:2619–22.

84. Winkelstein JA, Marino MC, Johnston Jr RB, et al. Chronic granulomatous disease. Report on a national registry of 368 patients. *Medicine (Baltimore)* 2000;**79**:155–69.

85. van den Berg JM, van Koppen E, Ahlin A, et al. Chronic granulomatous disease: the European experience. *PLoS One* 2009;**4**:e5234.

86. Wolach B, Gavrieli R, de Boer M, et al. Chronic granulomatous disease in Israel: clinical, functional and molecular studies of 38 patients. *Clin Immunol* 2008;**129**:103–14.

87. Fattahi F, Badalzadeh M, Sedighipour L, et al. Inheritance pattern and clinical aspects of 93 Iranian patients with chronic granulomatous disease. *J Clin Immunol* 2011;**31**:792–801.

88. Kuhns DB, Alvord WG, Heller T, et al. Residual NADPH oxidase and survival in chronic granulomatous disease. *N Engl J Med* 2010;**363**:2600–10.

89. Nathan CF, Murray HW, Wiebe ME, Rubin BY. Identification of interferon-gamma as the lymphokine that activates human macrophage oxidative metabolism and antimicrobial activity. *J Exp Med* 1983;**158**:670–89.

90. Ezekowitz RA, Orkin SH, Newburger PE. Recombinant interferon gamma augments phagocyte superoxide production and X-chronic granulomatous disease gene expression in X-linked variant chronic granulomatous disease. *J Clin Invest* 1987;**80**:1009–16.

91. Newburger PE, Ezekowitz RA, Whitney C, Wright J, Orkin SH. Induction of phagocyte cytochrome b heavy chain gene expression by interferon gamma. *Proc Natl Acad Sci USA* 1988;**85**:5215–9.

92. Sechler JM, Malech HL, White CJ, Gallin JI. Recombinant human interferon-gamma reconstitutes defective phagocyte function in patients with chronic granulomatous disease of childhood. *Proc Natl Acad Sci USA* 1988;**85**:4874–8.

93. Ezekowitz RA, Dinauer MC, Jaffe HS, Orkin SH, Newburger PE. Partial correction of the phagocyte defect in patients with X-linked chronic granulomatous disease by subcutaneous interferon gamma. *N Engl J Med* 1988;**319**:146–1451.

94. Ezekowitz RA, Sieff CA, Dinauer MC, Nathan DG, Orkin SH, Newburger PE. Restoration of phagocyte function by interferon-gamma in X-linked chronic granulomatous disease occurs at the level of a progenitor cell. *Blood* 1990;**76**:2443–8.

95. The International Chronic Granulomatous Disease Cooperative Study Group. A controlled trial of interferon gamma to prevent infection in chronic granulomatous disease. *N Engl J Med* 1991;**324**:509–16.

96. Gallin JI, Alling DW, Malech HL, et al. Itraconazole to prevent fungal infections in chronic granulomatous disease. *N Engl J Med* 2003;**348**:2416–22.

97. Hussain N, Feld JJ, Kleiner DE, et al. Hepatic abnormalities in patients with chronic granulomatous disease. *Hepatology* 2007;**45**:675–83.

98. Feld JJ, Hussain N, Wright EC, et al. Hepatic involvement and portal hypertension predict mortality in chronic granulomatous disease. *Gastroenterology* 2008;**134**:1917–26.

99. Greenberg DE, Goldberg JB, Stock F, Murray PR, Holland SM, Lipuma JJ. Recurrent Burkholderia infection in patients with chronic granulomatous disease: 11-year experience at a large referral center. *Clin Infect Dis* 2009;**48**:1577–9.

100. Ross JP, Holland SM, Gill VJ, DeCarlo ES, Gallin JI. Severe Burkholderia (Pseudomonas) gladioli infection in chronic granulomatous disease: report of two successfully treated cases. *Clin Infect Dis* 1995;**21**:1291–3.

101. Sirinavin S, Techasaensiri C, Benjaponpitak S, Pornkul R, Vorachit M. Invasive Chromobacterium violaceum infection in children: case report and review. *Pediatr Infect Dis J* 2005;**24**:559–61.

102. Mailman TL, Schmidt MH. Francisella philomiragia adenitis and pulmonary nodules in a child with chronic granulomatous disease. *Can J Infect Dis Med Microbiol* 2005;**16**:245–8.

103. Greenberg DE, Shoffner AR, Zelazny AM, et al. Recurrent Granulibacter bethesdensis infections and chronic granulomatous disease. *Emerg Infect Dis* 2010;**16**:1341–8.

104. Greenberg DE, Shoffner AR, Marshall-Batty KR, et al. Serologic reactivity to the emerging pathogen Granulibacter bethesdensis. *J Infect Dis* 2012;**206**:943–51.

105. Lopez FC, de Luna FF, Delgado MC, et al. Granulibacter bethesdensis isolated in a child patient with chronic granulomatous disease. *J Infect* 2008;**57**:275–7.

106. Chu J, Song HH, Zarember KA, Mills TA, Gallin JI. Persistence of the bacterial pathogen Granulibacter bethesdensis in chronic granulomatous disease monocytes and macrophages lacking a functional NADPH oxidase. *J Immunol* 2013;**191**:3297–307.

107. Falcone EL, Holland SM. Invasive fungal infection in chronic granulomatous disease: insights into pathogenesis and management. *Curr Opin Infect Dis* 2012;**25**:658–69.

108. Vinh DC, Freeman AF, Shea YR, et al. Mucormycosis in chronic granulomatous disease: association with iatrogenic immunosuppression. *J Allergy Clin Immunol* 2009;**123**:1411–3.

109. Morgenstern DE, Gifford MA, Li LL, Doerschuk CM, Dinauer MC. Absence of respiratory burst in X-linked chronic granulomatous disease mice leads to abnormalities in both host defense and inflammatory response to Aspergillus fumigatus. *J Exp Med* 1997;**185**:207–18.

110. Romani L, Fallarino F, De Luca A, et al. Defective tryptophan catabolism underlies inflammation in mouse chronic granulomatous disease. *Nature* 2008;**451**:211–5.

111. Marciano BE, Rosenzweig SD, Kleiner DE, et al. Gastrointestinal involvement in chronic granulomatous disease. *Pediatrics* 2004;**114**:462–8.

112. Marks DJ, Miyagi K, Rahman FZ, Novelli M, Bloom SL, Segal AW. Inflammatory bowel disease in CGD reproduces the clinicopathological features of Crohn's disease. *Am J Gastroenterol* 2009;**104**:117–24.

113. Uzel G, Orange JS, Poliak N, Marciano BE, Heller T, Holland SM. Complications of tumor necrosis factor-alpha blockade in chronic granulomatous disease-related colitis. *Clin Infect Dis* 2010;**51**:1429–34.

114. Lublin M, Bartlett DL, Danforth DN, et al. Hepatic abscess in patients with chronic granulomatous disease. *Ann Surg* 2002;**235**:383–91.

115. Yamazaki-Nakashimada MA, Stiehm ER, Pietropaolo-Cienfuegos D, Hernandez-Bautista V, Espinosa-Rosales F. Corticosteroid therapy for refractory infections in chronic granulomatous disease: case reports and review of the literature. *Ann Allergy Asthma Immunol* 2006;**97**:257–61.

116. Leiding JW, Freeman AF, Marciano BE, et al. Corticosteroid therapy for liver abscess in chronic granulomatous disease. *Clin Infect Dis* 2012;**54**:694–700.

117. Anderson-Cohen M, Holland SM, Kuhns DB, et al. Severe phenotype of chronic granulomatous disease presenting in a female with a de novo mutation in gp91-phox and a non familial, extremely skewed X chromosome inactivation. *Clin Immunol* 2003;**109**:308–17.

118. Rosen-Wolff A, Soldan W, Heyne K, Bickhardt J, Gahr M, Roesler J. Increased susceptibility of a carrier of X-linked chronic granulomatous disease (CGD) to Aspergillus fumigatus infection associated with age-related skewing of lyonization. *Ann Hematol* 2001;**80**:113–5.

119. Rupec RA, Petropoulou T, Belohradsky BH, et al. Lupus erythematosus tumidus and chronic discoid lupus erythematosus in carriers of X-linked chronic granulomatous disease. *Eur J Dermatol* 2000;**10**:184–9.

120. Wolach B, Scharf Y, Gavrieli R, de Boer M, Roos D. Unusual late presentation of X-linked chronic granulomatous disease in an adult female with a somatic mosaic for a novel mutation in CYBB. *Blood* 2005;**105**:61–6.

121. Kang EM, Marciano BE, DeRavin S, Zarember KA, Holland SM, Malech HL. Chronic granulomatous disease: overview and hematopoietic stem cell transplantation. *J Allergy Clin Immunol* 2011;**127**:1319–26 quiz 27-28.

122. Soncini E, Slatter MA, Jones LB, et al. Unrelated donor and HLA-identical sibling haematopoietic stem cell transplantation cure chronic granulomatous disease with good long-term outcome and growth. *Br J Haematol* 2009;**145**:73–83.

123. Gungor T, Teira P, Slatter M, et al. Reduced-intensity conditioning and HLA-matched haemopoietic stem-cell transplantation in patients with chronic granulomatous disease: a prospective multicentre study. *Lancet* 2013 Oct 22.

124. Kang EM, Malech HL. Gene therapy for chronic granulomatous disease. *Methods Enzymol* 2012;**507**:125–54.

125. Kono H, Rusyn I, Yin M, et al. NADPH oxidase-derived free radicals are key oxidants in alcohol-induced liver disease. *J Clin Invest* 2000;**106**:867–72.

126. Rusyn I, Kadiiska MB, Dikalova A, et al. Phthalates rapidly increase production of reactive oxygen species in vivo: role of Kupffer cells. *Mol Pharmacol* 2001;**59**: 744–50.

127. Barry-Lane PA, Patterson C, van der Merwe M, et al. p47phox is required for atherosclerotic lesion progression in ApoE(−/−) mice. *J Clin Invest* 2001;**108**: 1513–22.

128. Hsich E, Segal BH, Pagano PJ, et al. Vascular effects following homozygous disruption of p47(phox) : An essential component of NADPH oxidase. *Circulation* 2000;**101**:1234–6.

129. Patel VB, Wang Z, Fan D, et al. Loss of p47phox subunit enhances susceptibility to biomechanical stress and heart failure because of dysregulation of cortactin and actin filaments. *Circ Res* 2013;**112**:1542–56.

130. Kishida KT, Hoeffer CA, Hu D, Pao M, Holland SM, Klann E. Synaptic plasticity deficits and mild memory impairments in mouse models of chronic granulomatous disease. *Mol Cell Biol* 2006;**26**:5908–20.

131. Pao M, Wiggs EA, Anastacio MM, et al. Cognitive function in patients with chronic granulomatous disease: a preliminary report. *Psychosomatics* 2004;**45**:230–4.

132. Walder CE, Green SP, Darbonne WC, et al. Ischemic stroke injury is reduced in mice lacking a functional NADPH oxidase. *Stroke* 1997;**28**:2252–8.

133. Gao XP, Standiford TJ, Rahman A, et al. Role of NADPH oxidase in the mechanism of lung neutrophil sequestration and microvessel injury induced by Gram-negative sepsis: studies in p47phox−/− and gp91phox−/− mice. *J Immunol* 2002;**168**:3974–82.

Severe Combined Immunodeficiency – from Discovery to Newborn Screening

Jennifer M. Puck[1], Robert Currier[2]

[1]Division of Allergy, Immunology and Blood and Bone Marrow Transplantation, Department of Pediatrics, University of California San Francisco and UCSF Benioff Children's Hospital, San Francisco, CA, USA

[2]Genetic Disease Screening Program, California Department of Public Health, Richmond, CA, USA

HISTORY OF SCID AND ITS TREATMENT

All infants affected with severe combined immunodeficiency (SCID) have absent or extremely low production of T-cells in the thymus, affecting both T-cell number and diversity. Depending on genotype, B-cells may be absent, defective or present but, even in genotypes without an intrinsic defect of B-cells, specific antibody production is severely impaired due to the lack of T-helper cells. The combined defects of T and B-cells, plus absent natural killer (NK) cells in some forms of SCID, severely compromise an infant's ability to resist infections.

Primary Immunodeficiency Disorders: A Historic and Scientific Perspective

The first descriptions of a fatal congenital deficiency of lymphocytes date from the 1950s and were made in Switzerland, when pathologists Glanzmann and Riniker (1) described an idiopathic wasting syndrome with fatal *Candida albicans* infection. In 1958, Hitzig et al. (2) and Tobler and Cottier (3), five years after the discovery of agammaglobulinemia by Bruton (4), reported familial alymphocytosis combined with agammaglobulinemia with a fatal outcome in infancy. The term "Swiss-type agammaglobulinemia" was originally used to distinguish infants with fungal infections, lymphopenia and early death from the less severely affected children with isolated agammaglobulinemia who came to medical attention somewhat later in life. However, the term was a source of confusion, in part because the "Swiss-type" label was then applied to kindreds in which SCID was autosomal recessive. Nezeloff syndrome was another term originally describing cases of thymus dysplasia with preservation of immunoglobulins (5), but this designation is no longer used, having been abandoned after it became clear that B-cell immunity could not develop in the absence of T-cells even though B-cells were sometimes present and that maternally-derived IgG could be found in the blood of young infants affected with SCID. In subsequent decades, SCID designations have become more precise thanks to improved tools to identify T-cell, B-cell and NK-cell subsets, plus the identification of mutations in immunodeficiency-causing genes.

SCID is rare and has been difficult to recognize clinically as the underlying cause in infants presenting with recurrent diarrhea, pneumonia, septicemia, fungal infections or failure to thrive, partly because so many different infectious scenarios can occur. In kindreds in which affected infants died in earlier generations and even today, many individuals with SCID were mistakenly thought to have dietary intolerances or cystic fibrosis because of diarrhea, poor weight gain and pulmonary infections. Some were given diagnoses such as scarlet fever because of the rash observed when maternally-derived lymphocytes caused graft-versus-host disease (GVHD).

Complications following live vaccines may signify an underlying SCID diagnosis. In countries where newborns are routinely vaccinated against tuberculosis with the live attenuated mycobacterial organism *Bacillus* Calmette–Guérin (BCG), infants with SCID may develop fatal disseminated BCG infection. Administration of attenuated live poliovirus vaccine has caused paralytic polio in infants with SCID, and has also resulted in prolonged carriage of the virus in the gastrointestinal tract. Most recently, live attenuated rotavirus vaccine inadvertently given to infants who later proved to have SCID has caused severe diarrheal disease (6). Unfortunately, SCID is generally not suspected upon presentation to primary physicians or even referral centers.

Prior to 1968, patients with SCID died in infancy. However, in that year, Richard Gatti, Robert Good and colleagues reported the pioneering immune reconstitution of a case of SCID following bone marrow transplantation (BMT) from an HLA identical sibling (7). A few years later the unavailability of a histocompatible sibling donor led to the use of isolation in a germ-free environment for "David the Bubble Boy", a famous Texas patient who came to represent SCID to the general public (see Chapter 25). Although David succumbed to complications of newly recognized Epstein–Barr virus lymphoproliferative disease after a bone marrow transplant was finally performed at age 12 (8), techniques for transplantation with T-cell-depleted haploidentical bone marrow from a parent (9) and with unrelated adult or cord blood hematopoietic cells from well-matched donors have made SCID treatable for all patients (10–13).

Adenosine deaminase (ADA) deficient SCID was the first type of SCID for which the underlying cause was discovered, as reported by Eloise Giblett in 1972 (14, 15). After linkage mapping and discovery that the gene for the common gamma chain of cytokine receptors is defective in X-linked SCID in 1993 (16, 17), over a dozen more genes have been reported in which deleterious mutations lead to SCID (10, 13, 18). There are still many cases of SCID for which no gene defects have been discovered, indicating further SCID genes are likely to be found in the future.

In addition to treatment by replacing the hematopoietic stem cells that give rise to lymphoid lineages, enzyme replacement therapy has also been developed for ADA-deficient SCID, after it was recognized that transfused erythrocytes from healthy individuals could detoxify the purine metabolites found in excess in all blood and tissues of ADA-deficient patients (15, 19). Furthermore, ADA and X-linked SCID were the first human diseases successfully treated by gene therapy (20). With these advances in treatment, healthy survival and long life of patients affected with SCID has become possible.

WHY SCREEN FOR SCID?

From the earliest reports and in all standard medical textbooks, SCID has been described in terms of how the combined absence of T-cell and antibody immunity causes affected infants to come to medical attention with serious infections that lead to diarrhea and failure to thrive. The diagnosis and, indeed, the definition of SCID revolved around the infections experienced by the SCID infant, infections often caused by weakly pathogenic, opportunistic organisms. Prior to 1968, when the first successful bone marrow transplant was performed, SCID was always fatal, but now it can be treated by transplantation of bone marrow stem cells from a healthy donor, or by enzyme replacement or even gene therapy, *provided infections can be controlled.* Knowing how SCID is inherited has permitted some families, often following a tragic loss of an affected infant due to infection, to learn the diagnosis in subsequent affected children at birth, or even before birth. In these fortunate circumstances, early treatment of healthy infants with SCID who have avoided infections affords a very high likelihood of survival free of complications. Therefore, population-based newborn screening (NBS) for SCID was suggested as early as 1997, based on data showing that pre-symptomatic identification and treatment before 3.5 months of age were associated with improved survival for infants born with SCID (21–24). Because most cases of SCID are sporadic, universal screening would be required to give all affected infants the same survival odds as those with a known family history that allowed diagnosis prior to onset of infections.

NEWBORN SCREENING

The concept of NBS is to identify rare and otherwise unsuspected genetic defects that are treatable and that cause serious illness if infants remain undetected and untreated. NBS began in the USA in the 1960s, after Robert Guthrie developed a test for phenylketonuria (PKU), an inborn error of phenylalanine metabolism. In children who lack the enzyme phenylalanine hydroxylase and who are given a normal diet, severe mental retardation develops due to the neurotoxicity of high levels of phenylalanine (25). Guthrie's simple, sensitive test for PKU required absorbing a drop of an infant's blood on a piece of filter paper that was allowed to dry and then used in a bacterial inhibition assay for phenylalanine hydroxylase activity. Early detection of PKU allowed dietary modification by protein restriction for affected infants, such that intellectual disability could be avoided (26).

Following the successful institution of the PKU test, the dried blood spot became the universal sample on which many further screening tests could be conducted. Criteria for adding a test to a population screening program have been under discussion in the public health and genetics communities since 1968 (27).

Screening has been considered justifiable only if the targeted condition is an important public health problem, the natural history of which is well understood and whose symptoms are alleviated by early identification and treatment. In addition to PKU, other diseases detectable by performing assays on dried blood spots and for which early intervention improves outcome were added to newborn screening programs over time, including galactosemia, maple syrup urine disease and homocystinuria in the 1960s, congenital hypothyroidisim in the 1970s, and sickle cell disease and other hemoglobinopathies in the 1980s. Hearing screening was also instituted as a point of care test performed in nurseries (not a dried blood spot test), and congenital heart disease screening by nursery pulse oximetry has recently been recommended. Each state in the USA has developed its own newborn screening program independently, and these programs are still directed and financed by state departments of public health that control their own tests.

An interesting and illustrative instance of newborn screening in the context of inherited immune disorders was an early trial of dried blood spot analysis for SCID due to ADA deficiency. From 1970 until 1982, New York State conducted newborn screening for ADA-deficient SCID using a colorimetric ADA enzyme assay (28, 29). Unfortunately, after screening 2.5 million newborns, no ADA SCID cases were detected, but during that time two cases not detected by screening due to non-enrollment were documented. Furthermore, 13 infants, including 11 descendents of Kung tribes people from Africa and the Caribbean basin, were found who lacked ADA in red blood cells due to a genetic variant of the enzyme with decreased stability, but who had normal immune function due to adequate ADA activity in their other tissues. The kinetics of variant ADA enzymes was elegantly worked out by Rochelle Hirschhorn and colleagues and published in 1979 (29). These results led to the discontinuation of ADA

screening in New York; the disorder was felt to be so rare, and the false-positive results of the test so troublesome, that it was deemed not cost effective.

With tremendous advances in rare disease discovery, many more conditions have become amenable to screening using newborn dried blood spots. A USA Department of Health and Human Services Secretary's Advisory Committee on Heritable Disorders in Newborns and Children (SACHDNC) initially met in 2004 and was established as an official entity under the Public Health Service Act in 2008 to perform evidence-based reviews of new disorders nominated to be included in a uniform newborn screening panel (30, 31). The Committee's criteria were as follows: (1) the disorder has to be considered medically serious; (2) there should be prospective pilot data from population-based screening; (3) the spectrum of the disorder should be well described in the medical literature; (4) the screening test should be inexpensive and have very low false-negative results and as low as possible false-positive results; (5) if the spectrum of the disorder is broad, those most likely to benefit from early treatment should be identifiable; and (6) an effective treatment is available, but for best outcome must be given before the infant becomes symptomatic. A limited number of false-positive results has always been viewed as a necessary price to pay to avoid missing affected infants. It is inevitable that some healthy individuals will initially screen positive for any given disease; therefore, screening programs typically require follow-up testing after the initial non-normal result. As of 2009, the Secretary's Committee had endorsed a uniform panel of 29 conditions, many of which were detected by a single mass spectrometry method (32). SCID was the first condition to be recommended for addition to the screening panel in 2010 (see below), and further conditions continue to be added. However, to this day, it remains up to individual states to decide upon implementation and financing of the recommended testing.

THE T-CELL RECEPTOR EXCISION CIRCLE (TREC) SCREENING TEST FOR SCID

The first suggestion that all newborns be screened for SCID grew from recognizing that the majority of cases could be identified by a complete blood count and differential to determine the absolute number of lymphocytes in relation to age-adjusted normal values. Lymphocytes in newborns are normally present in twice the numbers found in older children and adults, T-cells are approximately 70% of lymphocytes in healthy infants, and absence of T-cells causes the total lymphocyte count of most infants with SCID to be quite low (21, 31). However, some forms of SCID are associated with the presence of B lymphocytes, and maternal cells are also present in the blood of some infants with SCID. Therefore, total lymphocyte counts, though simple to perform, would not capture all SCID cases. T-lymphocyte counts would be more specific, but these cannot be measured in dried blood spots.

Therefore, the TREC test was developed for dried blood spots by Kee Chan and Jennifer Puck (32, 33) and validated by other groups (34). Recombination of the T-cell receptor (TCR) genes in the thymus is the process whereby a diverse repertoire of T-cells is generated from alternate variable (V), diversity (D) and joining (J) segments to synthesize a unique rearrangement in each cell. TRECs are circular DNA molecules formed from the leftover fragments generated from this rearrangement. TREC DNA circles are measured by a quantitative polymerase chain reaction (PCR) across the joint of the circular DNA (35, 36). Normal infant blood samples have one TREC per 10 T-cells (35), reflecting the high rate of new T-cell generation early in life. Infants with SCID lack TRECs (32, 33).

Occasional dried blood spots fail to amplify TREC DNA adequately for technical reasons; such samples need to have repeat determinations made along with a control DNA segment. If both TREC and control PCR fail, a new blood spot must be requested from the infant. Repeated unsatisfactory tests where PCR fails, as well as tests indicating low or absent TRECs need to be followed up with a liquid blood sample from the infant that is tested for total lymphocyte numbers and subsets of T, B, and NK cells as well as naïve and memory cells by flow cytometry.

With this test, Dr Puck, along with patient advocacy groups, including the Jeffrey Modell Foundation and the Immune Deficiency Foundation, applied to the SACHDNC committee in 2008 for consideration of including this testing for SCID in the recommended universal screening panel. The matter was referred to an evidence review committee, which reported back in 2009 that there was insufficient data from pilot screening programs to prove that the test could be implemented in population-based state programs (37). However, by that time, Wisconsin, under the leadership of Ronald Laessig, Mei Baker and John Routes, had undertaken statewide pilot screening with a scaled-up version of the dried blood spot TREC test (38). In 2010, upon reconsideration of the SCID application with additional information from the Wisconsin pilot, the SACHDNC committee recommended and DHHS Secretary Sibelius endorsed inclusion of SCID screening in the newborn screening panels of all states (31).

Based on these recommendations, infants with abnormally low numbers of T-cells would need to be referred promptly to an expert in pediatric immunology to determine whether the infant has SCID. In addition to SCID, other conditions in which T-cell numbers are low, known as secondary targets, could be flagged by TREC testing and promptly treated.

The TREC method was first adapted to statewide testing in Wisconsin in 2008 (38–40), followed by Massachusetts in 2009 (41, 42), and

California (43) and New York in 2010 (44). Now, many more states are conducting TREC screening for SCID, and still more are in the planning stages of offering this testing. As of this writing over half of infants born in the USA are being screened for SCID with the TREC test. Pilot SCID screening is also under way in Canada and several European countries. The Jeffrey Modell Foundation, which provides support and advocacy for primary immunodeficiencies, issued a Berlin Declaration in 2013 signed by a large number of international primary immunodeficiency experts, endorsing adoption of SCID newborn screening and early treatment of affected infants along with continued efforts for outreach and education in the field (Fred and Vicki Modell, personal communication).

Because SCID was originally described in terms of its infectious complications, which are precisely the features that are *not* present in very young affected infants, new definitions have been needed to categorize patients with inadequate T-cell immunity based on laboratory tests of blood lymphocyte numbers, function and genotype. These may need modification as we observe the full spectrum of patients detected by newborn screening. Table 14.1 shows the working definitions developed by experts in pediatric immunology in the US Rare Disease Network established in 2009, the Primary Immune Deficiency Treatment Consortium (PIDTC) (33, 44, 45).

RESULTS TO DATE OF TREC NEWBORN SCREENING FOR SCID

As of the fall of 2013, several newborn screening programs have successfully identified pre-symptomatic infants with SCID and other conditions characterized by low T-cell counts, allowing these infants to receive prompt treatment without the burden of devastating infections. Wisconsin (38), Massachusetts (39, 40) and California (41) have published SCID cases discovered by screening. An important feature of the Californian

TABLE 14.1   Laboratory definition of SCID and T-cell lymphopenia

TYPICAL SCID

<300 CD3 T-cells/microliter of peripheral blood
Absent or <10% of control T-cell proliferation
    to phytohemagglutinin (PHA)
OR
T-cells of maternal origin present

LEAKY SCID

Low number of CD3 T-cells (other than expanded
    oligoclonal T-cells)

    under 2 years of age <1000/microliter
    between 2 and 4 years <800/microliter
    over 4 years <600/microliter
No maternal T-cell engraftment
Reduced T-cell proliferation to PHA
May have immune dysregulation or autoimmunity

OMENN SYNDROME

Skin rash with generalized erythroderma
No maternal T-cell engraftment
Detectable CD3 T-cells, but with a low proportion of naïve
    versus memory phenotype (low CD45RA and high
    CD45RO)
Absent or decreased T-cell proliferation to antigens to
    which the patient has been exposed
At least 4 of the following supportive criteria:
    Hepatomegaly and/or splenomegaly
    Lymphadenopathy
    Elevated IgE
    Elevated absolute eosinophil count
    Reduced proliferative response in mixed leukocyte
        reaction, <30% of lower limit of normal
    Mutation(s) in a SCID-causing gene that are consistent
        with some preservation of function in the gene product
    Abnormal antibody responses

program is integration of T-cell enumeration by flow cytometry on a new liquid blood sample as a standardized second test integrated into the newborn screening program. This allows all infants with non-normal TRECs to receive immediate testing, including a complete and differential white blood cell count and enumeration of T, B and NK cells along with T-cell subsets and ascertainment of naïve versus memory phenotype T-cells. This information along with a clinical summary obtained by screening coordinator staff is sufficient for the program immunology experts to direct referrals for appropriate further diagnosis and treatment.

As with all newborn screening tests that have been instituted over the years, screening with TRECs has revealed additional infants affected with conditions other than the primary target of screening, which is SCID. Table 14.2 shows the range of conditions that have been found to be associated with low TRECs and T-cell lymphopenia, but this list is likely to grow with further screening experience (33, 38, 40, 43).

Congenital syndromes most commonly associated with T-cell lymphopenia are DiGeorge syndrome, most often associated with interstitial deletion of chromosome 22q11.2, trisomy 21 and CHARGE (Coloboma of the eye, Heart defects, Atresia of the choanae, Retardation of growth and development, Genital and urinary abnormalities, and Ear abnormalities and deafness) syndrome, most often due to defects in the CHD7 gene.

Of particular interest are cases referred to as variant SCID (see Table 14.2), newborns with persistently low T-cell counts and no established syndromic diagnosis who do not have mutations found in known SCID genes. These infants may have as yet unknown disorders or pre-symptomatic presentations of disorders that have not been appreciated to have any neonatal abnormalities. Deep sequencing, including whole-exome or whole-genome sequencing may reveal underlying causes for variant SCID. As an example, in two instances in California, clinically healthy infants with T lymphopenia detected by screening were found to have ataxia telangiectasia (46). With the diagnosis established by gene sequencing, rising levels of serum alpha fetoprotein and eventual neurological symptoms, these cases were moved from the variant SCID category to the category of congenital syndromes.

Infants with secondary T lymphopenia due to causes listed in Table 14.2 will achieve normal T-cell numbers after treatment and resolution

TABLE 14.2   Conditions found by screening for low or absent TRECs

---

PRIMARY TARGET OF SCREENING: TYPICAL SCID, LEAKY SCID AND OMENN SYNDROME

Underlying gene defects including *IL2RG* (X-linked SCID), *ADA, IL7R, JAK3, RAG1, RAG2, DCLRE1C (Artemis), TCRD, TCRE, TCRZ* and *CD45*; additional as yet unknown genotypes

PREVIOUSLY UNSUSPECTED T-CELL LYMPHOPENIC DISORDERS

Variant SCID, representing a group of infants with persistently low T-cells but no defect found in any of the above known SCID genes

CONGENITAL SYNDROMES WITH VARIABLE DEGREES OF IMPAIRED T-CELL PRODUCTION

Complete DiGeorge syndrome or partial DiGeorge syndrome with low T-cells

Trisomy 21

CHARGE syndrome

Ataxia telangiectasia

Cartilage hair hypoplasia

Jacobsen syndrome

RAC2 dominant interfering mutation with neutropenia

DOCK8 deficient hyper-IgE syndrome

INFANTILE DISEASES WITH CONSEQUENTIAL INCREASED DESTRUCTION OR IMPAIRED PRODUCTION OF T-CELLS

Neonatal hydrops

Neonatal cardiac defects requiring immediate surgical correction (with thymectomy in many instances)

Chylothorax

Gastrointestinal malformations (such as gastroschisis, intestinal atresia)

Neonatal leukemia

Maternal treatment with immunosuppressive medications during pregnancy

(Possibly severe prenatal HIV infection, but no cases described to date)

PREMATURITY ALONE (T-CELL LYMPHOPENIA RESOLVES TO NORMAL OVER TIME)

---

of the underlying condition. A special category of self-resolving T lymphopenia is preterm birth (see Table 14.2). Most infants born prematurely have normal T-cell numbers, but a small proportion are initially T lymphopenic. Regardless of the cause of low T-cell count, identification by TREC screening has given medical providers important information regarding potential heightened risks of infection in these infants.

Although composite data from all states conducting SCID screening are not yet available, California, the most populous state with great ethnic diversity and one eighth of all births in the USA, has screened nearly 1.5 million infants in its first three years of TREC screening and serves as a microcosm of the effectiveness of these programs (41; J Puck and R Currier, unpublished data). As shown in Table 14.3, during this time, 26 infants with SCID were found and promptly referred to specialized centers for treatment by allogeneic hematopoietic cell transplantation, enzyme replacement therapy (for ADA-deficient SCID) or experimental gene therapy (for ADA-deficient and X-linked SCID). The false-positive rates of testing were very low, and the positive predictive value of non-normal TREC results of 44% is remarkably good compared to other newborn screening tests.

Now that TREC screening has become available and its effectiveness has been shown, spreading its implementation to all states and countries outside the USA is important. As screening becomes widespread, the true incidence and proportions of each type of SCID in different populations can be documented.

TABLE 14.3 California experience after 3 years of newborn screening for SCID

Infants screened	1 495 177
Infants with abnormal TREC tests who required flow cytometry to determine T-cell number	234 (1 in 6400 births, 95% CI 1 in 5600–1 in 7400)
Infants proven to have T-cell lymphopenia	104 (1 in 14 000 births, 95% CI 1 in 12 000–1 in 17 000)
Typical and leaky SCID and Omenn syndrome	26 (1 in 58 000 births, 95% CI 1 in 41 000–1 in 94 000)

TREC NEWBORN SCREENING – EXCELLENT, BUT NOT PERFECT FOR IDENTIFYING PRIMARY IMMUNODEFICIENCIES

TREC newborn screening followed by lymphocyte subset measurement has been proven to have clinical utility, as described above. As more experience accumulates and more states add newborn TREC screening, it will be important to document the outcomes of infants identified by screening programs. Not only the total incidence but also the severity spectrum, relative incidence by genotype, subpopulations with particular risks and best treatments for these rare conditions remain to be defined.

We will also learn which T-cell deficiency diseases are not detected by the TREC test. Diseases expected to be missed are those in which T-cells can develop in the thymus to the point of T-cell receptor recombination and TREC formation, but in which function of mature T-cells is impaired. For example, newborns with T-cell receptor ζ chain associated protein 70 (ZAP70) deficiency, MHC class II deficiency and nuclear factor kappa B (BNF-κB) essential modulator deficiency (NEMO deficiency) have been shown to have normal TRECs in the neonatal period, as has one patient with delayed-onset ADA deficiency.

All primary immunodeficiencies, and not just SCID, stand to benefit from early diagnosis. Continued advances in molecular and genomic technology may soon allow screening for lack of B-cells, by testing for B-cell κ chain excision circles (KRECs) (47–49). Moreover, it is possible that future newborns will have extensive testing for DNA variations, or even sequencing of their entire genome, from which a blueprint of risks for a great variety of conditions affecting health can be ascertained (50, 51). Even predisposition to the more common multifactorial immune disorders with later onset may become possible through deep sequence analysis of DNA from newborns. However, since genotypes do not fully predict phenotype for these conditions, much more needs to be learned about the true predictive value of each proposed type of screening. Thus the history of newborn screening for immune disorders has just begun to be written.

Acknowledgments

Support for the initial phase of CA SCID screening was provided by contract HHSN267200603430C to New York State from the Eunice Kennedy Shriver Institute of Child Health and Development, The Jeffrey Modell Foundation, and Perkin Elmer Genetics. Analysis of California SCID cases was supported by the Primary Immune Deficiency Treatment Consortium, NIH AI U54 082973, with case development criteria facilitated by NIH R13 AI094943 from the NIH Office of Rare Disease Research of the National Center for Advancing Translational Sciences (NCATS). JMP received support from the UCSF CTSI, NIH NCATS 1UL1 RR024131, RO3 HD 060311, RO1 AI 078248 and the Jeffrey Modell Foundation.

References

1. Glanzmann E, Riniker P. Essentielle lymphocytophtose. Ein neues Krankeitsbild aus der Säuglingspathologie. *Ann Paediat* 1950;**174**:1–5.
2. Hitzig WH, Biro Z, Bosch H, Huser HJ. Agammaglobulinemia & alymphocytosis with atrophy of lymphatic tissue. *Helvet Paediatr Acta* 1958;**13**:551–85.
3. Tobler R, Cottier H. Familial lymphopenia with agammaglobulinemia & severe moniliasis: the essential lymphocytophthisis as a special form of early childhood agammaglobulinemia. *Helvet Paediatr Acta* 1958;**13**:313–38.
4. Bruton OC. Agammaglobulinemia. *Pediatrics* 1952;**9**: 722–7.
5. Nezelof C. Thymic dysplasia with normal immunoglobulins and immunologic deficiency: Pure alymphocytosis. *Birth Defects Orig Art Ser* 1968;**4**:4–112.
6. Bakare N, Menschik D, Tiernan R, Hua W, Martin D. Severe combined immunodeficiency (SCID) and rotavirus vaccination: reports to the Vaccine Adverse Events Reporting System (VAERS). *Vaccine* 2010;**28**:6609–12.
7. Gatti RA, Meuwissen HJ, Allen HD, et al. Immunological reconstitution of sex-linked lymphopenic immunological deficiency. *Lancet* 1968;**2**:1366.
8. Shearer WT, Ritz J, Finegold MJ, et al. Epstein-Barr virus-associated B-cell proliferations of diverse clonal origins after bone marrow transplantation in a 12-year-old patient with severe combined immunodeficiency. *N Engl J Med* 1985;**312**:1151–9.
9. Bortin MM, Rimm AA. Severe combined immunodeficiency disease: characterization of the disease and results of transplantation. *J Am Med Assoc* 1977;**238**:591–600.
10. Buckley RH. Molecular defects in human severe combined immunodeficiency and approaches to reconstitution. *Annu Rev Immunol* 2004;**22**:625–55.
11. O'Reilly RJ, Schiff RI, Small TN. Transplantation approaches for severe combined immunodeficiency disease, Wiskott-Aldrich syndrome, and other lethal genetic combined immunodeficiency disorders. In: Forman SJ, Blume KG, Thomas ED, editors. *Bone Marrow Transplantation*. Boston: Blackwell Scientific Publications; 1994. p. 849–67.
12. Cavazzana-Calvo M, Carlier F, Le Deist F, et al. Long-term T-cell reconstitution after hematopoietic stem-cell transplantation in primary T-cell-immunodeficient patients is associated with myeloid chimerism and possibly the primary disease phenotype. *Blood* 2007;**109**:4575–81.
13. Buckley RH. Transplantation of hematopoietic stem cells in human severe combined immunodeficiency: longterm outcomes. *Immunol Res* 2011;**49**:25–43.
14. Giblett ER, Anderson JE, Cohen F, Pollara B, Meuwissen HJ. Adenosine-deaminase deficiency in two patients with severely impaired cellular immunity. *Lancet* 1972;**2**:1067–9.
15. Giblett ER. ADA and PNP deficiencies: how it all began. *Ann NY Acad Sci* 1985;**451**:1–8.
16. Noguchi M, Yi H, Rosenblatt HM, et al. Interleukin-2 receptor gamma chain mutation results in X-linked severe combined immunodeficiency in humans. *Cell* 1993;**73**:147–57.
17. Puck JM, Deschênes SM, Porter JC, et al. The interleukin-2 receptor gamma chain maps to Xq13.1 and is mutated in X-linked severe combined immunodeficiency, SCIDX1. *Hum Mol Genet* 1993;**8**:1099–104.
18. Al-Herz W, Bousfiha A, Casanova JL, et al. Primary immunodeficiency diseases: an update on the classification from the international union of immunological societies expert committee for primary immunodeficiency. *Front Immunol* 2011;**2**:1–26.
19. Polmar SH. Enzyme replacement and other biochemical approaches to the therapy of adenosine deaminase deficiency. *Ciba Found Symp* 1978;**68**:213–30.
20. Cavazzana-Calvo M, Fischer A, Hacein-Bey-Abina S, Aiuti A. Gene therapy for primary immunodeficiencies: Part 1. *Curr Opin Immunol* 2012;**24**:580–4.
21. Buckley RH, Schiff RI, Schiff SE, et al. Human severe combined immunodeficiency: genetic phenotypic, and functional diversity in one hundred eight infants. *J Pediatr* 1997;**130**:378–87.
22. Myers LA, Patel DD, Puck JM, Buckley RH. Hematopoietic stem cell transplantation for severe combined immunodeficiency in the neonatal period leads to superior thymic output and improved survival. *Blood* 2002;**99**: 872–8.

23. Brown L, Xu-Bayford J, Allwood Z, et al. Neonatal diagnosis of severe combined immunodeficiency leads to significantly improved survival outcome: the case for newborn screening. *Blood* 2011;**117**:3243–6.

24. Chan A, Scalchunes C, Boyle M, Puck JM. Early vs. delayed diagnosis of severe combined immunodeficiency: a family perspective survey. *Clin Immunol* 2011;**138**:3–8.

25. Guthrie R, Susi A. A simple phenylalanine method for detecting phenylketonuria in large populations of newborn infants. *Pediatrics* 1963;**32**:338–43.

26. Santos LL, Magalhaes Mde C, Januario JN, et al. The time has come: a new scene for PKU treatment. *Genet Mol Res* 2006;**5**:33–44.

27. Wilson JM, Jungner YG. [Principles and practice of mass screening for disease]. *Bol Oficina Sanit Panam* 1968;**65**:281–93.

28. Moore EC, Meuwissen HJ. Screening for ADA deficiency. *J Pediatr* 1974;**85**:802–4.

29. Hirschhorn R, Roegner V, Jenkins T, Seaman C, Piomelli S, Borkowsky W. Erythrocyte adenosine deaminase deficiency without immunodeficiency. Evidence for an unstable mutant enzyme. *J Clin Invest* 1979;**64**:1130–9.

30. on behalf of the American College of Medical Genetics' Newborn Screening Expert Group. Newborn screening: toward a uniform screening panel and system. Watson MS, Mann MY, Lloyd-Puryear MA, Rinaldo P, Howell R, *Genet Med* 2006;**8**(Suppl):1S–252S.

31. Buckley RH. The long quest for neonatal screening for severe combined immunodeficiency. *J Allergy Clin Immunol* 2012;**129**:597–604.

32. Chan K, Puck JM. Development of population-based newborn screening for severe combined immunodeficiency. *J Allergy Clin Immunol* 2005;**115**:391.

33. Puck JM. Laboratory technology for population-based screening for severe combined immunodeficiency in neonates: The winner is T-cell receptor excision circles. *J Allergy Clin Immunol* 2012;**129**:607–16.

34. Morinishi Y, Imai K, Nakagawa N, et al. Identification of severe combined immunodeficiency by T-cell receptor excision circles quantification using neonatal guthrie cards. *J Pediatr* 2009;**155**:829–33.

35. Douek DC, Koup RA, McFarland RD, Sullivan JL, Luzuriaga K. Effect of HIV on thymic function before and after antiretroviral therapy in children. *J Infect Dis* 2000;**181**:1479–82.

36. Hazenberg MD, Verschuren MC, Hamann D, Miedema F, van Dongen JJ. T-cell receptor excision circles as markers for recent thymic emigrants: basic aspects, technical approach, and guidelines for interpretation. *J Mol Med* 2001;**79**:631–40.

37. Lipstein EA, Vorono S, Browning MF, et al. Systematic evidence review of newborn screening and treatment of severe combined immunodeficiency. *Pediatrics* 2010;**125**:e1226–1235.

38. Routes JM, Grossman WJ, Verbsky J, et al. Statewide newborn screening for severe T-cell lymphopenia. *J Am Med Assoc* 2009;**302**:2465–70.

39. Chase NM, Verbsky JW, Routes JM. Newborn screening for SCID: three years of experience. *Ann NY Acad Sci* 2011;**1238**:99–106.

40. Verbsky J, Thakar M, Routes J. The Wisconsin approach to newborn screening for severe combined immunodeficiency. *J Allergy Clin Immunol* 2012;**129**:662–7.

41. Hale JE, Bonilla FA, Pai SY, et al. Identification of an infant with severe combined immunodeficiency by newborn screening. *J Allergy Clin Immunol* 2010;**126** 1073–4.

42. Comeau AM, Hale JE, Pai SY, et al. Guidelines for implementation of population-based newborn screening for severe combined immunodeficiency. *J Inherit Metab Dis* 2010;**33**(Suppl 2):S273–281.

43. Kwan A, Church JA, Cowan MJ, et al. Newborn screening for severe combined immunodeficiency and T-cell lymphopenia in California: results of the first 2 years. *J Allergy Clin Immunol* 2013;**132**:140–50.

44. Clinical and Laboratory Standards Institute (CLSI). Newborn blood spot screening for severe combined immunodeficiency by measurement of T-cell receptor excision circles; Approved guidline. 2013;**33**:1–70.

45. Shearer WT, Dunn E, Notarangelo LD, et al. Establishing diagnostic criteria for severe combined immunodeficiency disease (SCID), leaky SCID, and Omenn syndrome: the Primary Immune Deficiency Treatment Consortium experience. *J Allergy Clin Immunol* 2014;**133**(4):1092–8.

46. Mallott J, Kwan A, Church J, et al. Newborn screening for SCID identifies patients with ataxia telangiectasia. *J Clin Immunol* 2013;**33**:540–9.

47. Sottini A, Ghidini C, Zanotti C, et al. Simultaneous quantification of recent thymic T-cell and bone marrow B-cell emigrants in patients with primary immunodeficiency undergone to stem cell transplantation. *Clin Immunol* 2010;**136**:217–27.

48. Nakagawa N, Imai K, Kanegane H, et al. Quantification of kappa-deleting recombination excision circles in Guthrie cards for the identification of early B-cell maturation defects. *J Allergy Clin Immunol* 2011;**128**:223–5.

49. Borte S, von Dobeln U, Fasth A, et al. Neonatal screening for severe primary immunodeficiency diseases using high-throughput triplex real-time PCR. *Blood* 2012;**119**: 2552–5.

50. Goldenberg AJ, Sharp RR. The ethical hazards and programmatic challenges of genomic newborn screening. *J Am Med Assoc* 2012;**307**:461–2.

51. Dondorp WJ, de Wert GM, Niermeijer MF. Genomic sequencing in newborn screening programs. 2012;**307**:2146; author reply 2147.

Severe Combined Immunodeficiency as Diseases of Defective Cytokine Signaling

Warren J. Leonard

Laboratory of Molecular Immunology and The Immunology Center, National Heart, Lung, and Blood Institute, National Institutes of Health, Bethesda, MD, USA

OUTLINE

(Continued)

Primary Immunodeficiency Disorders: A Historic and Scientific Perspective
2014 Published by Elsevier Inc.

INTRODUCTION

Severe combined immunodeficiency disease (SCID) represents a broad range of inherited genetic diseases that collectively comprise the most severe forms of primary immunodeficiency disease (PIDD) (1). Although the basic clinical manifestations are similar in many forms of SCID, with affected individuals exhibiting defective T- and B-cell function, these diseases can be subdivided according to the presence or absence of T-cells, B-cells, and natural killer (NK) cells. The most common form of SCID is X-linked (X) SCID, which accounts for approximately half of all cases of SCID and occurs in approximately 1 in every 80 000 live births. Affected boys have profoundly diminished numbers of T-cells and NK cells, and B-cells, while present in normal numbers, are non-functional. XSCID is often known as the "Bubble Boy Disease", named for David Vetter (see http://en.wikipedia.org/wiki/David_Vetter and Chapter 24), a young boy with XSCID who lived in a protective "bubble" in Houston, Texas, for many years prior to his receiving an unsuccessful bone marrow transplant from his sister and eventually succumbing to an Epstein–Barr (EBV)-mediated lymphoma.

As of the early 1990s, there was extensive knowledge of the clinical features and immunological phenotype of XSCID, and the gene responsible for this disease had been putatively mapped to Xq13 (2), but the identity of the gene was unknown. Interestingly, X-inactivation studies in heterozygous carrier females revealed that X-inactivation was random in fibroblasts but strictly non-random in T-cells (3) and NK cells (4), which indicated that only those cells within these lineages that have activated the wild-type X chromosome could mature, explaining the absence of T- and NK cells in boys with XSCID. Additionally, X-inactivation was shown to be random in immature B-cells but it was strictly non-random in more mature B-cells (5), indicating an essential role for the wild-type gene not only for the development of T-cells and NK cells but also for the maturation of B-cells. Efforts to clone the gene whose mutation resulted in XSCID by positional cloning would likely have identified the gene eventually. However, it was basic science studies, focused on hormones of the immune system known as cytokines, which resulted in discovery of the molecular basis of XSCID. This was found to result from mutations in the IL2RG gene (6), thereby demonstrating that XSCID was a disease that resulted from defective cytokine signaling.

IL-2 AND IL-2 RECEPTORS

IL-2 was discovered in 1976 as a T-cell growth promoting activity (7) that was produced by T lymphocytes following antigen activation. IL-2 is structurally a four α-helical bundle type 1 cytokine that is historically particularly important, in that it was the first type 1 cytokine that was cloned, and the first type 1 cytokine for which a receptor component was cloned (8). In addition to its actions as a T-cell growth factor, IL-2 was subsequently recognized as a pleiotropic cytokine that exhibits a broad range of biological actions on multiple lineages. As of the early 1990s, IL-2 had also been shown to be capable of augmenting the cytolytic activity of lymphokine-activated killer cells, natural killer (NK) cells and cytotoxic T-cells and, moreover, was recognized to be capable of augmenting immunoglobulin production by B-cells (9). Subsequent to these studies, IL-2 was shown to have a range of other important actions as well, including the ability to drive the differentiation of regulatory T-cells (Treg cells) (10), to promote activation-induced cell death (11) and to promote the differentiation of a range of T-helper cell (Th) populations, including Th1 and Th2 cells, while inhibiting the differentiation of Th17 cells (12–14). IL-2 was known to signal via three classes of IL-2 receptors that could bind IL-2 with low, intermediate or high affinity (Fig. 15.1). The low-affinity IL-2 receptor consists of the IL-2 receptor α chain (IL-2Rα), the intermediate-affinity IL-2 receptor contains IL-2Rβ and IL-2Rγ, and the high-affinity IL-2 receptor contains all three chains (8, 15, 16). Intermediate-affinity receptors are expressed on NK cells and some other resting lymphocytes, including resting CD8$^+$ T-cells. Expression of IL-2Rα is induced by TCR stimulation, converting the intermediate-affinity receptors to high-affinity receptors, and additionally resulting in the expression of low-affinity receptors in great abundance, so that peripheral blood human T-cells have a ratio of approximately ten low-affinity receptors for every

high-affinity receptor. Expression of IL-2Rα can be further increased by cellular exposure to a number of cytokines, including IL-2 itself (8, 17).

At the time of these basic studies, although IL-2 was known to exhibit actions on T, B and NK cells, lineages that were all affected in XSCID, a relationship of IL-2 with XSCID seemed extremely unlikely given that both humans lacking IL-2 (18, 19) and *Il2*-deficient mice (20) were known to have normal T-cell and NK-cell development even though they had severe combined immunodeficiency, findings that seemed to minimize the possibility that mutation in the genes encoding either IL-2 or a component of the IL-2 receptor would cause XSCID (6). In addition, whereas XSCID was obviously an X-linked disease, the *IL2* gene was known to be located at human chromosome 4q26-27 (21), the *IL2RA* gene at 10p14-15 (22), and the *IL2RB* gene at 22q11-12 (23). However, following the cDNA cloning of *IL2RG* in the laboratory of Dr Kazuo Sugamura in 1992 (16), Masayuki Noguchi in my laboratory at NIH began studying the gene. We additionally collaborated with Dr William S. Modi in the NCI to map the chromosomal location by *in situ* hybridization and then with Drs Huafang Yi and Dr O. Wesley McBride, also in the NCI, to perform somatic cell hybrid analysis experiments, with fine mapping using Centre d'Etude du Polymorphisme

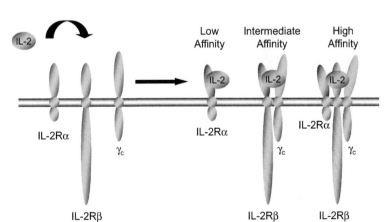

FIGURE 15.1 Three classes of IL-2 receptors. IL-2 binds to three classes of IL-2 receptors. The low-affinity receptor (Kd ≈10^{-8} M) consists of IL-2Rα, the intermediate-affinity receptor (Kd ≈10^{-9} M) consists of IL-2Rβ and γ$_c$, and the high-affinity receptor (K$_d$ ≈10^{-11} M) consists of IL-2Rα, IL-2Rβ and γ$_c$. The intermediate- and high-affinity forms of the receptor are functional. (*Figure is a modification of one prepared by Dr Jian-Xin Lin, NHLBI.*)

Humain (CEPH) pedigrees and single stranded conformation polymorphism (SSCP) analyses, which collectively mapped the *IL2RG* gene to human Xq13 (6) at a position indistinguishable from that previously determined to be the locus for XSCID (2). We then sequenced DNA isolated from EBV-transformed B-cell lines derived from three patients with classic clinical and immunological manifestations of XSCID, including DNA from David the "Bubble Boy", and discovered inactivating premature stop codons in all three individuals. The identification of such mutations in three unrelated individuals with XSCID established that XSCID indeed resulted from mutations in *IL2RG* (6) (Fig. 15.2). Drs Masayuki Noguchi, O. Wesley McBride,

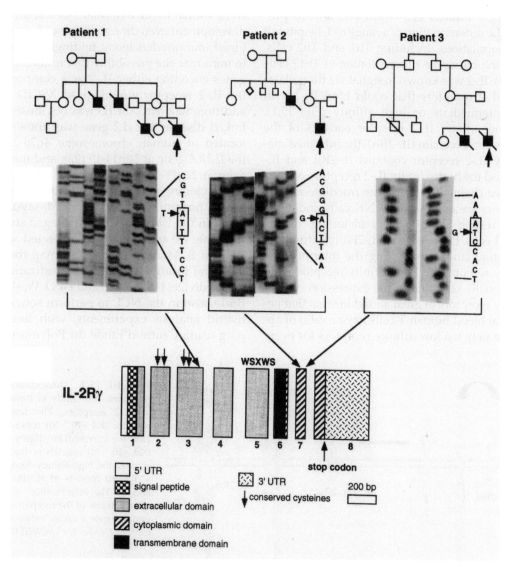

FIGURE 15.2 The first three XSCID patients whose mutations were identified (6) *(From Noguchi et al., 1993).*

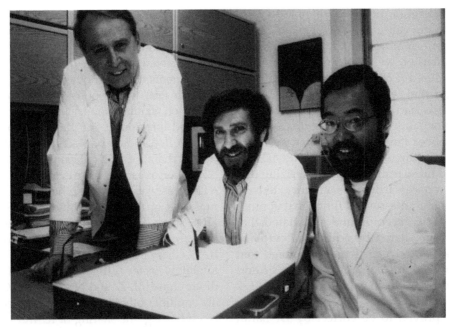

FIGURE 15.3   Drs Masayuki Noguchi (right), O. Wesley McBride (left) and Warren J. Leonard (center).

and Warren J. Leonard were the key investigators in to the identification of the defect in XSCID (Fig. 15.3). Subsequently, a large number of mutations in *IL2RG* were rapidly identified in patients with XSCID (24), including mutations that essentially "destroy" the protein by frameshifts resulting from nucleotide insertions or deletions, premature stop codons or mutation of splice acceptor or donor sites. Additionally, many single amino changes have been found to be inactivating and to result in XSCID; these can either interfere with cytokine binding (e.g. if located in the extracellular domain) or the coupling to key signaling pathways (if located in the cytoplasmic domain) (24). Interestingly, affected males in a family with a moderate form of X-linked combined immunodeficiency known as XCID were also found to result from mutation of *IL2RG*, which results in a single amino acid change, L271Q, with attenuation rather than abrogation of γc function (25).

THE CONUNDRUM AND THE SPECULATION THAT IL-2Rγ WAS SHARED BY MULTIPLE CYTOKINES, LEADING TO THE DESIGNATION OF THIS GROUP OF CYTOKINES AS THE γC FAMILY OF CYTOKINES

The identification of *IL2RG* as the gene that was mutated in XSCID was an extremely exciting and provocative observation, but it also was most puzzling. Simply stated, the conundrum was: how could mutations in a component of the IL-2 receptor result in XSCID even though neither defective expression of IL-2 in humans (18, 19) nor deletion of the *Il2* gene in mice (20) resulted in developmental abnormalities in lymphoid lineages (6)? Although not then known, subsequent studies would reveal that mutations in the *IL2RA* gene in humans (26) or targeted deletion of *Il2ra* in mice (27), or targeted deletion of the *Il2rb* gene in mice (28) also did not cause defective lineage developmental abnormalities. In any case, *a priori*, as mentioned earlier, we would never have seriously considered *IL2RG* as a rational candidate for the gene that was mutated in XSCID, but yet we already had experimental proof that *IL-2RG* mutations indeed caused the disease, given the presence of inactivating mutations in the *IL2RG* gene in three patients with XSCID. We rationalized these results by hypothesizing that IL-2Rγ must necessarily have a more extensive role than being an essential functional component of the IL-2 receptor, and speculated that IL-2Rγ may be shared by additional cytokine receptors as well. In particular, we noted that only single receptor components, IL-4Rα and IL-7Rα, were known to exist for IL-4 and IL-7 and therefore may utilize IL-2Rγ. This speculation was proven to be correct by Dr Sarah Russell (now Professor at the Peter McCallum Cancer Centre and the Centre for MicroPhotonics at Swinburne University of Technology in Melbourne) and Dr Masayuki Noguchi (now Professor, Hokkaido University Graduate School of Medicine) in my lab, and independently by Dr Motonari Kondo in the laboratory of Dr Kazuo Sugamura, who demonstrated in a series of papers published in 1993 in *Science* that both IL-4 (29, 30) and IL-7 (31, 32) shared IL-2Rγ as a critical receptor component. Our decision to investigate IL-2Rγ as a component of the IL-7 receptor was additionally motivated by the known importance of IL-7 for T-cell development, as discussed below. The observation that IL-2Rγ was shared by additional cytokine receptors led my laboratory to propose that IL-2Rγ should be renamed as the common cytokine receptor γ chain, γ_c (31). Subsequent studies then revealed that that γ_c was a component of the IL-9 receptor (25, 33), IL-15 (34) and IL-21 (35) receptors as well. Thus, γ_c is shared by the receptors for a total of six cytokines (Fig. 15.4), each of which is inactivated

in XSCID, underscoring that XSCID is indeed a disease of defective cytokine signaling, as has been previously reviewed (24).

It is noteworthy that the sharing of receptor components occurs in other cytokine systems as well. In fact, we proposed the γ_c nomenclature based on similar nomenclature that had been adopted previously for the hematopoietic cytokines, IL-3, IL-5 and GM-CSF, which share a common β chain, β_c, as a critical receptor component (37), whereas each of these cytokines has a specific α chain as well. Moreover, the signal transducing molecule, gp130 (encoded by the *IL6ST* gene, for "*IL-6 Signal Transducer*") is a component of multiple receptors (38, 39), including those for IL-6, IL-11, ciliary neurotropic factor (CNTF), leukemia inhibitory factor (LIF), oncostatin M (OSM), cardiotrophin-1, IL-27 and NNT/BSF-3/CLC. There are multiple other examples of shared cytokine receptors, e.g. the sharing of IL-4Rα by IL-4 and IL-13 receptors, and of IL-7Rα by IL-7 and thymic stromal lymphopoietin (TSLP) receptors (40, 41).

CRITICAL ROLE FOR HETERODIMERIZATION OF IL-2Rβ AND γ_C IN IL-2 SIGNALING: THE ASSOCIATION OF JAK1 WITH IL-2Rβ AND JAK3 WITH γ_C

JAK-STAT signaling represents a rapid cytoplasmic to nuclear signaling pathway that was originally discovered in the context of the type 1 interferons (IFN-α/β), but subsequently was recognized to also be utilized by type 2 interferon (IFN-γ) and type I four α-helical bundle cytokines, as well as by the IL-10 family of type II cytokines, which are more closely related to the IFNs (42). JAK kinases are also known as Janus family tyrosine kinases. There are four of these; JAK1, JAK2, TYK2, and JAK3 (42). Janus was the Roman mythological god with two faces, and these correspond

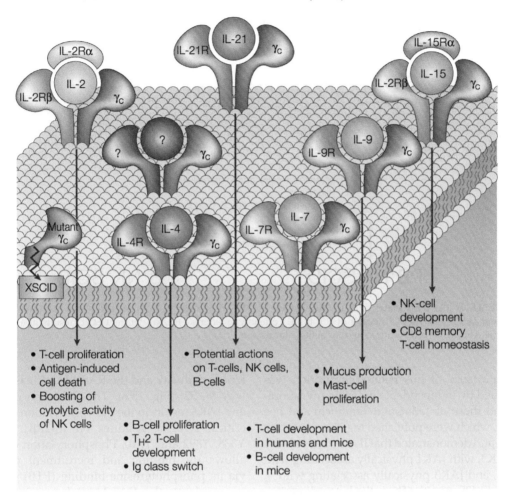

FIGURE 15.4 The γ_c family of cytokines. Originally discovered as IL-2Rγ, γ_c is now known to be an essential component of the receptors for IL-2, IL-4, IL-7, IL-9, IL-15 and IL-21. *(From Leonard et al. 2001) (36).* This figure is reproduced in color in the color section.

to the presence in each of these kinases of a pseudokinase domain in addition to a functional kinase domain that is catalytically active. JAK1, JAK2, and TYK2 are ubiquitously expressed and relatively constitutive in their expression, whereas JAK3 is lympho-hematopoietic restricted, and its expression is more inducible, for example after TCR activation (42). JAK3 is the most recently identified,

being co-discovered in the laboratories of John O'Shea, James Ihle, and Leslie Berg (43–45).

An analysis of the role for JAK kinases in IL-2 signaling was motivated in part by studies from my lab and that of Phil Greenberg published in *Nature* in 1994 which used different chimeric receptor approaches to demonstrate that heterodimerization of IL-2Rβ and IL-2Rγ is required for IL-2 signaling (46, 47). The basis for

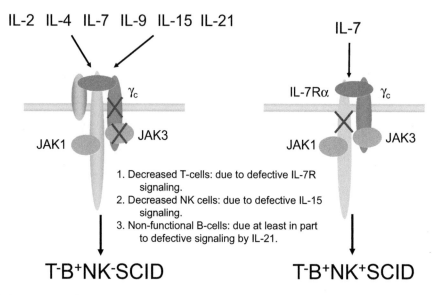

FIGURE 15.5 JAK1 associates with IL-2Rβ, IL-4Rα, IL-7Rα, IL-9R and IL-21R, whereas JAK3 associates with γ_c. XSCID and JAK3-deficient SCID result in a T⁻B⁺NK⁻ phenotype whereas IL-7Rα-deficient SCID results in a T⁻B⁺NK⁺ phenotype. This figure is reproduced in color in the color section.

the importance of this heterodimerization was elucidated by independent studies in my laboratory and those of Tadatsugu Taniguchi and Lee Nadler, which were published together in *Science* in 1994 and demonstrated that IL-2 activates JAK1 and JAK3, with JAK1 physically associating with IL-2Rβ, and JAK3 physically associating with γ_c (Fig. 15.5) (25, 48, 49). Thus, the heterodimerization of IL-2Rβ and γ_c results in the juxtapositioning of subsequent activation of JAK1 and JAK3. Further studies revealed that the activation of these two kinases by IL-2 primarily results in the activation of STAT5A and STAT5B, with weaker activation of STAT1 and STAT3 (50, 51).

IL-2 ALSO ACTIVATES PI 3-KINASE AND MAP-KINASE COUPLED PATHWAYS

In addition to its activation of the JAK-STAT pathway, IL-2 signaling also activates the phosphoinositol 3-kinase (PI 3-kinase)-AKT-p70 S6 kinase pathway and the RAS-MAP kinase pathway (8, 52) (Fig. 15.6). The activation of JAK1 and JAK3 results in the phosphorylation of three critical tyrosine residues on IL-2Rβ, namely Y338, Y392 and Y510. Phosphorylation of Y338 allows the binding and recruitment of SHC via its phosphotyrosine binding (PTB) domain to couple to the RAS-MAP kinase pathway, whereas the phosphorylation of Y392 and Y510 mediates the docking of STAT5A and STAT5B to facilitate their own tyrosine phosphorylation, subsequent dimerization and nuclear translocation (53). PI 3-kinase recruitment does not involve tyrosine phosphorylation of IL-2Rβ but, interestingly, recruitment of the p85 subunit of PI 3-kinase is dependent on a membrane proximal region of IL-2Rβ as well as on JAK1, which associates with the membrane proximal region of IL-2Rβ, and JAK1 mediates the tyrosine phosphorylation of p85 (54). Thus, three major signaling pathways are activated by IL-2; the JAK-STAT, PI 3-kinase-AKT-p70 S6 kinase and RAS-MAP kinase pathways.

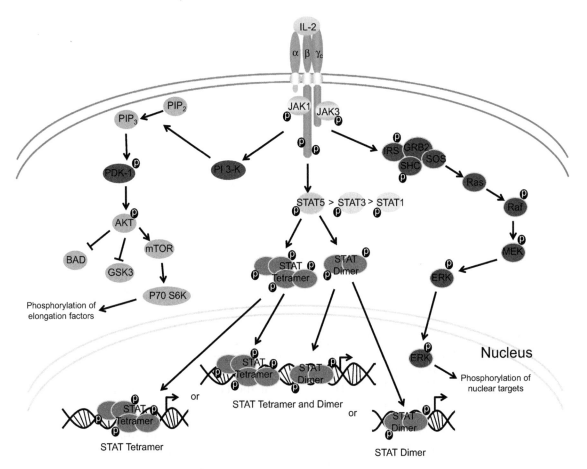

FIGURE 15.6 Signaling pathways utilized by IL-2. IL-2 signals via JAK-STAT, PI 3-kinase/Akt and RAS-MAP kinase pathways. STAT5A and STAT5B are the principal STAT proteins activated by IL-2, with STAT1 and STAT3 activated to a lesser degree. *(Modified from a figure in (8).)* This figure is reproduced in color in the color section.

THE DISCOVERY OF JAK3-DEFICIENT HUMAN SCID AS ANOTHER FORM OF SCID RESULTING FROM DEFECTIVE CYTOKINE SIGNALING AND THE DEVELOPMENT OF JAK3 INHIBITORS AS POTENT IMMUNOSUPPRESSANTS

The observation that JAK3 associated with and thus was immediately downstream of γ_c and its contributions to IL-2 signaling led us to speculate that mutations in *JAK3* would cause a form of SCID that would be clinically and immunologically indistinguishable from XSCID, except that it would presumably occur with equal frequency in boys and girls as a homozygous recessive disease, given that the *JAK3* gene is located on chromosome 19p13.1 rather than on the X chromosome. Studying a patient of Dr Rebecca Buckley at Duke University, we found that this was indeed the case, with similar findings reported by the group of Dr Luigi Notarangelo (55, 56). The fact that XSCID and JAK3-deficient SCID represent indistinguishable

$T^-B^+NK^-$ phenotypes underscores the special relationship between γ_c and JAK3, with γ_c-dependent signaling requiring JAK3, and JAK3 not having a major biological role except for its mediation of γ_c-dependent signaling. The discovery of JAK3-deficient SCID underscored the critical role of JAK3 in immune function, and we hypothesized that "the current study further suggests that any agents that inactivate Jak3 function may be potent immunosuppressants". This provided the motivation for the development of JAK3 inhibitors (57, 58). Indeed, such an agent, Tofacitinib, was approved by the Food and Drug Administration (FDA) in 2012 for the treatment of rheumatoid arthritis, with thousands of patients having been treated and further investigation related to this and other related compounds in progress.

DISCOVERY OF IL-7Rα-DEFICIENT SCID: ESTABLISHING THAT DEFECTIVE IL-7Rα-DEPENDENT SIGNALING EXPLAINS THE DEFECTIVE T-CELL DEVELOPMENT IN XSCID AND JAK3-DEFICIENT SCID

Given that patients with XSCID and JAK3 deficiency each exhibit greatly decreased numbers of T and NK cells and normal numbers of B-cells that are non-functional, how can these defects be explained? Based on the defective T-cell development in mice lacking expression of IL-7 or IL-7Rα, we speculated that defective IL-7 signaling was responsible for the T-cell defect; in fact, as noted above, the lack of T-cells in humans with XSCID was one of the original motivations for investigating whether IL-2Rγ was also part of the IL-7 receptor. We noted with great interest that in a large study of SCID patients published by Rebecca Buckley's group in the *Journal of Pediatrics*, two patients lacked T-cells but had normal to elevated levels

of NK cells (59). We speculated that these individuals had defects related specifically to IL-7 signaling and therefore initiated another collaboration with Dr Buckley. When we analyzed EBV-induced B-cell lines established from these individuals, we discovered that both had defective expression of IL-7Rα and inactivating mutations in the *IL7R* gene, whereas their *IL7* genes were intact (60, 61). This identification of the first IL-7Rα-deficient SCID patients partially elucidated the basis for the XSCID issue, given that IL-7Rα-deficient individuals have a $T^-B^+NK^+$ phenotype. In other words, defective expression of IL-7Rα indeed resulted in a selective loss of T-cells, indicating that defective IL-7Rα-mediated signaling causes the T-cell defect in XSCID. Surprisingly, no one has reported patients with IL-7-deficient SCID, even though it is reasonable to hypothesize that such individuals would be similar phenotypically to IL-7Rα-deficient SCID patients, although IL-7-deficient SCID may simply be a very rare cause of SCID. Alternatively, it is conceivable that the lack of IL-7 expression either results in a milder than anticipated phenotype so that such individuals do not have SCID and thus do not come to medical attention, or that the lack of IL-7 is much more severe than assumed, resulting in embryonic lethality. Because a second cytokine, thymic stromal lymphopoietin (TSLP) also shares IL-7Rα as a receptor component (62, 63), it should be noted that IL-7Rα-deficient SCID patients cannot signal in response to either IL-7 or TSLP, whereas IL-7-deficiency will only result in a loss of IL-7 signaling. Additional investigation is required to determine the ramifications of defective expression of IL-7, as opposed to IL-7Rα, in humans *in vivo*. Despite the lack of identification of IL-7-deficient patients, based on our results from IL7Rα-deficient SCID patients as well as data from mouse models, we speculate that defective IL-7 signaling likely causes the T-cell defect in XSCID and JAK3-deficient SCID.

DEFECTIVE IL-15-DEPENDENT SIGNALING EXPLAINS THE NK-CELL DEVELOPMENTAL DEFECT IN PATIENTS WITH XSCID AND JAK3 DEFICIENCY

Of the cytokines known to share γ_c, IL-15 was the only one known to affect NK-cell development. In fact, both IL-15 (64) and IL-15Rα-deficient mice (65) lack NK cells. Thus, the defective NK-cell development in XSCID and JAK3 deficiency undoubtedly results from the lack of IL-15 signaling.

THE DISCOVERY OF IL-21 AND FINDING THAT THE B-CELL DEFECT IN XSCID RESULTS AT LEAST IN PART FROM DEFECTIVE SIGNALING BY IL-21

The cause of defective B-cell function in XSCID and JAK3-deficient SCID was less evident. Whereas mice lacking IL-7 or IL-7Rα exhibit defective B-cell development as well as T-cell development, human B-cell development is normal in the absence of IL-7Rα, as well as in the setting of JAK3 or IL-2Rγ deficiency, underscoring a fundamental difference in human and mouse B-cell development. Thus, either IL-7 signaling is not required for human B-cell development or, alternatively, in the absence of IL-7 signaling, signaling by another cytokine or by a cytokine-independent mechanism complements the absence of IL-7.

The key to the nature of the B-cell defect in XSCID and JAK3-deficient SCID came with the discovery of IL-21. The receptor for IL-21 was discovered as an orphan cytokine receptor based on large-scale sequencing projects. Strikingly, this orphan receptor was very similar to IL-2Rβ in its sequence and was encoded by a gene that is adjacent to the gene encoding IL-4Rα in both humans and mice. The gene encoding the ligand, IL-21, is located adjacent to the gene encoding IL-2. IL-21 has a broad range of important actions on multiple lymphoid lineages. It can cooperate with IL-7 or IL-15 to drive the expansion of CD8+ T-cells, it can activate NK cells and it has a range of B-cell actions. Not only is IL-21 pro-apoptotic for incompletely activated B-cells (66–68), perhaps serving a role analogous to that of IL-2 in the process of activation-induced cell death and potentially preventing autoimmune responses (69), but it also drives terminal differentiation of B-cells to plasma cells when they are activated during an effective immune response (68, 70). Moreover, based on studies in human cells as well as in IL-21R-deficient mice, IL-21 is required for normal production of certain classes of IgG (particularly IgG1 and IgG3) (71). Strikingly, mice lacking IL-21R also have elevated levels of IgE (72, 73), which appears to result at least in part from the ability of IL-21 to inhibit Cε transcription (74). Because class switch to IgG1- and IgE-producing cells is known to be dependent on IL-4, *Il21r-Il4* double knockout mice were generated to determine whether this elevated production of IgE was in fact dependent on IL-4 (73). Indeed, mice lacking expression of both IL-21R and IL-4 no longer produced IgE but, remarkably, they exhibited a pan-hypogammaglobulinemia, with defective production not only of IgE but also of all IgG subclasses, although significant production of IgM was retained (73). This immunoglobulin profile essentially mimicked the B-cell defect in humans with XSCID. Subsequent studies with human cells have corroborated a critical role for IL-21 for normal B-cell function (75).

Thus, whereas defects in IL-7 and IL-15, respectively, can explain the defective T- and NK-cell development in XSCID, the inactivation of signaling by IL-21 contributes to the non-functional B-cells in XSCID (see Fig. 15.5). Because IL-2 is also known to be capable of promoting immunoglobulin production by B-cells, it is conceivable that defective signaling by IL-2 might additionally contribute to the B-cell defect in XSCID. Presumably defective IL-7, IL-15 and IL-4/IL-21 signaling also explains the defective T-cell and NK-cell development and B-cell function in JAK3-deficent SCID as well.

In addition to this important role related to B-cell function, it is striking that IL-21 has important anti-tumor actions (69), and is currently in phase 2 clinical trials as an anti-cancer agent. Additionally, it contributes critically to the development of a range of autoimmune diseases in animal models (76–78), with inhibitors of IL-21 now in early clinical trials as medicines for rheumatoid arthritis. Given that JAK3 inhibitors inhibit the action of IL-21, the inactivation of this cytokine provides the basis for some of the actions of these inhibitors.

THE EFFICACY OF BONE MARROW TRANSPLANTATION FOR CELLULAR RECONSTITUTION AND B-CELL FUNCTION IN XSCID, JAK3-DEFICIENT SCID AND IL-7Rα-DEFICIENT SCID

Without treatment, individuals with XSCID, JAK3 deficiency or IL-7Rα deficiency typically die during the first year of life, after the protective effect of maternal antibodies has dissipated. Bone marrow transplantation is the treatment of choice, with highly successful engraftment of T-cells in all three diseases and NK cells in XSCID and JAK3 deficiency (79, 80). One of the reasons for the success of bone marrow transplantation in these diseases is that engrafted cells have a relative growth advantage in terms of their ability to expand in response to the relevant cytokines, so that successful engraftment of even a limited number of hematopoietic stem cells may be sufficient. Similar properties make these diseases ideal candidates for successful gene therapy (see below). A major issue in the treatment of XSCID and JAK3 deficiency, however, relates to the fact that normalization of B-cell function does not always occur. Whereas

donor T and NK cells typically engraft, successful engraftment is more variable with donor B-cells, given that the patients have normal numbers of non-functional "host" B-cells. Without successful B-cell engraftment, chronic intravenous immunoglobulin (IVIG) may be required, even after successful bone marrow transplantation for T-cells and NK cells. The situation for IL-7Rα-deficient SCID is distinctive, however. In this disease, not only are NK cells present, but B-cells can respond normally to IL-21 and IL-4 and thus they are not intrinsically defective, as they are in XSCID and JAK3-deficient SCID. Thus, once normal T-cells are reconstituted by bone marrow transplantation with the ability to produce cytokines including IL-4 and IL-21, one would predict that the B lymphocytes can signal properly. Indeed, reconstitution of B-cell function is more common in IL7Rα-deficient SCID, without the need for chronic IVIG administration (79).

Following successful gene therapy in animal models of XSCID (81), gene therapy was also performed for humans with XSCID. Whereas retroviral-based gene therapy successfully reconstituted T and NK cells with normalization of immune function (82), the therapy was unfortunately associated with serious adverse events, with leukemia resulting from retrovirally-mediated activation of the *LMO2* gene (83, 84). Newer approaches to gene therapy that can hopefully avoid this serious problem are under development (85, 86).

DOES STAT DEFICIENCY RESULT IN SCID?

Because IL-2, IL-7, IL-9 and IL-15 primarily utilize STAT5A and STAT5B as their major STAT proteins, one might predict that the inactivation of these STAT5 proteins might result in defective IL-7 and IL-15 signaling and thus in defective T-cell and NK-cell development. Indeed, in mouse models, deletion of either the *Stat5a* (87) or *Stat5b* (88) gene resulted in partial defects related to T-cell development and more significant defects, particularly in the case of *Stat5b*, related to NK-cell development and function. Additionally, the simultaneous deletion of both *Stat5a* and *Stat5b* genes results in essential abrogation of both T-cell and NK-cell development (89). However, because STAT5A and STAT5B are encoded by distinct, albeit adjacent genes, a large deletion might be required to simultaneously inactivate both genes. Interestingly, however, humans with mutations in the *STAT5B* gene have been reported and these patients have defects associated with growth hormone insensitivity, IGF-1 deficiency and T-cell lymphopenia, particularly of T-cells as well as of NK cells (90–92). Thus, although these individuals do not have SCID, their T- and NK-cell phenotype suggests that a complete lack of STAT5 proteins might indeed result in a profound decrease in the number of T- and NK cells, consistent with what has been described in corresponding knockout mice.

IL-21R-DEFICIENT PATIENTS

Patients lacking expression of IL-21R have recently been identified. Although these patients do not have SCID, they exhibit a primary immunodeficiency disease characterized by impaired immunoglobulin class switching of B-cells, diminished NK cytolytic function and defective production of cytokines by T-cells, associated with cryptosporida infection and liver disease (93). These manifestations are consistent with defective signaling by IL-21.

CONCLUSIONS

SCID represents a spectrum of disorders with many causes. Of these, at least three forms are diseases of defective cytokine signaling, including XSCID, JAK3-deficient SCID and IL-7Rα-deficient SCID. In addition, mutations in *IL2RA*, *IL2RB*, *IL21R* and *STAT5B* are also established causes of human diseases (Table 15.1). The careful

TABLE 15.1 Causes of human inherited immunodeficiency disorders associated with the γ_c family of cytokines

XSCID	T⁻B⁺NK⁻ SCID
JAK3-deficient SCID	T⁻B⁺NK⁻ SCID
IL-7Rα-deficient SCID	T⁻B⁺NK⁺ SCID
IL-2Rα-deficiency	Decreased peripheral T-cell numbers with abnormal proliferation and extensive infiltration of tissues
IL-2Rβ-deficiency	Absent NK cells
IL-21R-deficiency	Defective Ig class switch of B-cells, decreased NK cytolytic function, defective cytokine production by T-cells, cryptosporida infection and liver disease
STAT5B-deficiency	Growth hormone insensitivity, IGF-1 deficiency, decreased T- and NK cells

analysis of the molecular basis of XSCID has helped to establish the existence of the γ_c-family of cytokines and has led to our understanding of their actions. Moreover, the identification of the molecular causes of these forms of SCID has provided the basis for better prenatal and postnatal diagnosis, and paved the way for the potential of successful gene therapy.

Acknowledgments

Supported by the Division of Intramural Research, National Heart, Lung, and Blood Institute, National Institutes of Health. I thank Drs Jian-Xin Lin and Rosanne Spolski for critical comments.

References

1. Buckley RH. Molecular defects in human severe combined immunodeficiency and approaches to immune reconstitution. *Annu Rev Immunol* 2004;**22**:625–55.
2. de Saint Basile G, Arveiler B, Oberle I, et al. Close linkage of the locus for X chromosome-linked severe combined

immunodeficiency to polymorphic DNA markers in Xq11-q13. *Proc Natl Ac Sci USA* 1987;**84**:7576–9.

3. Puck JM, Nussbaum RL, Conley ME. Carrier detection in X-linked severe combined immunodeficiency based on patterns of X chromosome inactivation. *J Clin Invest* 1987;**79**:1395–400.

4. Wengler GS, Allen RC, Parolini O, Smith H, Conley ME. Nonrandom X chromosome inactivation in natural killer cells from obligate carriers of X-linked severe combined immunodeficiency. *J Immunol* 1993;**150**:700–4.

5. Conley ME, Lavoie A, Briggs C, Brown P, Guerra C, Puck JM. Nonrandom X chromosome inactivation in B-cells from carriers of X chromosome-linked severe combined immunodeficiency. *Proc Natl Acad Sci USA* 1988;**85**: 3090–4.

6. Noguchi M, Yi H, Rosenblatt HM, et al. Interleukin-2 receptor gamma chain mutation results in X-linked severe combined immunodeficiency in humans. *Cell* 1993;**73**:147–57.

7. Morgan DA, Ruscetti FW, Gallo R. Selective in vitro growth of T lymphocytes from normal human bone marrows. *Science* 1976;**193**:1007–8.

8. Liao W, Lin JX, Leonard WJ. Interleukin-2 at the crossroads of effector responses, tolerance, and immunotherapy. *Immunity* 2013;**38**:13–25.

9. Waldmann TA. The multi-subunit interleukin-2 receptor. *Annu Rev Biochem* 1989;**58**:875–911.

10. Malek TR, Yu A, Vincek V, Scibelli P, Kong L. CD4 regulatory T-cells prevent lethal autoimmunity in IL-2Rbeta-deficient mice. Implications for the nonredundant function of IL-2. *Immunity* 2002;**17**:167–78.

11. Lenardo MJ. Interleukin-2 programs mouse alpha beta T lymphocytes for apoptosis. *Nature* 1991;**353**:858–61.

12. Liao W, Schones DE, Oh J, et al. Priming for T helper type 2 differentiation by interleukin 2-mediated induction of interleukin 4 receptor alpha-chain expression. *Nat Immunol* 2008;**9**:1288–96.

13. Liao W, Lin JX, Wang L, Li P, Leonard WJ. Modulation of cytokine receptors by IL-2 broadly regulates differentiation into helper T-cell lineages. *Nat Immunol*. 2011;**12**: 551–9.

14. Laurence A, Tato CM, Davidson TS, et al. Interleukin-2 signaling via STAT5 constrains T helper 17 cell generation. *Immunity* 2007;**26**:371–81.

15. Wang X, Rickert M, Garcia KC. Structure of the quaternary complex of interleukin-2 with its alpha, beta, and gammac receptors. *Science* 2005;**310**:1159–63.

16. Takeshita T, Asao H, Ohtani K, et al. Cloning of the gamma chain of the human IL-2 receptor. *Science* 1992;**257**:379–82.

17. Depper JM, Leonard WJ, Drogula C, Kronke M, Waldmann TA, Greene WC. Interleukin 2 (IL-2) augments transcription of the IL-2 receptor gene. *Proc Natl Acad Sci USA* 1985;**82**:4230–4.

18. Pahwa R, Chatila T, Pahwa S, et al. Recombinant interleukin 2 therapy in severe combined immunodeficiency disease. *Proc Natl Acad Sci USA* 1989;**86**:5069–73.

19. Weinberg K, Parkman R. Severe combined immunodeficiency due to a specific defect in the production of interleukin-2. *N Engl J Med* 1990;**322**:1718–23.

20. Schorle H, Holtschke T, Hunig T, Schimpl A, Horak I. Development and function of T-cells in mice rendered interleukin-2 deficient by gene targeting. *Nature* 1991;**352**: 621–4.

21. Seigel LJ, Harper ME, Wong-Staal F, Gallo RC, Nash WG, O'Brien SJ. Gene for T-cell growth factor: location on human chromosome 4q and feline chromosome B1. *Science* 1984;**223**:175–8.

22. Leonard WJ, Donlon TA, Lebo RV, Greene WC. Localization of the gene encoding the human interleukin-2 receptor on chromosome 10. *Science* 1985;**228**:1547–9.

23. Gnarra JR, Otani H, Wang MG, McBride OW, Sharon M, Leonard WJ. Human interleukin 2 receptor beta-chain gene: chromosomal localization and identification of 5' regulatory sequences. *Proc Natl Acad Sci USA* 1990;**87**:3440–4.

24. Leonard WJ. The molecular basis of X-linked severe combined immunodeficiency: defective cytokine receptor signaling. *Annu Rev Med* 1996;**47**:229–39.

25. Russell SM, Johnston JA, Noguchi M, et al. Interaction of IL-2R beta and gamma c chains with Jak1 and Jak3: implications for XSCID and XCID. *Science* 1994;**266**:1042–5.

26. Sharfe N, Dadi HK, Shahar M, Roifman CM. Human immune disorder arising from mutation of the alpha chain of the interleukin-2 receptor. *Proc Natl Acad Sci USA* 1997;**94**:3168–71.

27. Willerford DM, Chen J, Ferry JA, Davidson L, Ma A, Alt FW. Interleukin-2 receptor alpha chain regulates the size and content of the peripheral lymphoid compartment. *Immunity* 1995;**3**:521–30.

28. Suzuki H, Kundig TM, Furlonger C, et al. Deregulated T-cell activation and autoimmunity in mice lacking interleukin-2 receptor beta. *Science* 1995;**268**:1472–6.

29. Russell SM, Keegan AD, Harada N, et al. Interleukin-2 receptor gamma chain: a functional component of the interleukin-4 receptor. *Science* 1993;**262**:1880–3.

30. Kondo M, Takeshita T, Ishii N, et al. Sharing of the interleukin-2 (IL-2) receptor gamma chain between receptors for IL-2 and IL-4. *Science* 1993;**262**:1874–7.

31. Noguchi M, Nakamura Y, Russell SM, et al. Interleukin-2 receptor gamma chain: a functional component of the interleukin-7 receptor. *Science* 1993;**262**:1877–80.

32. Kondo M, Takeshita T, Higuchi M, et al. Functional participation of the IL-2 receptor gamma chain in IL-7 receptor complexes. *Science* 1994;**263**:1453–4.

33. Kimura Y, Takeshita T, Kondo M, et al. Sharing of the IL-2 receptor gamma chain with the functional IL-9 receptor complex. *Internatl Immunol* 1995;**7**:115–20.

34. Giri JG, Ahdieh M, Eisenman J, et al. Utilization of the beta and gamma chains of the IL-2 receptor by the novel cytokine IL-15. *EMBO J* 1994;**13**:2822–30.

35. Asao H, Okuyama C, Kumaki S, et al. Cutting edge: the common gamma-chain is an indispensable subunit of the IL-21 receptor complex. *J Immunol* 2001;**167**:1–5.

36. Leonard WJ. Cytokines and immunodeficiency diseases. *Nat Rev Immunol* 2001;**1**:200–8.

37. Kitamura T, Sato N, Arai K, Miyajima A. Expression cloning of the human IL-3 receptor cDNA reveals a shared beta subunit for the human IL-3 and GM-CSF receptors. *Cell* 1991;**66**:1165–74.

38. Gearing DP, Comeau MR, Friend DJ, et al. The IL-6 signal transducer, gp130: an oncostatin M receptor and affinity converter for the LIF receptor. *Science* 1992;**255**:1434–7.

39. Taga T, Narazaki M, Yasukawa K, et al. Functional inhibition of hematopoietic and neurotrophic cytokines by blocking the interleukin 6 signal transducer gp130. *Proc Natl Acad Sci USA* 1992;**89**:10998–1001.

40. Ozaki K, Leonard WJ. Cytokine and cytokine receptor pleiotropy and redundancy. *J Biol Chem* 2002;**277**:29355–8.

41. Leonard WJ. *Type I cytokines and interferons and their receptors.* 6th edn. Philadelphia: Wolters Kluwer/Lippincott Williams & Wilkins; 2008.

42. Leonard WJ, O'Shea JJ. Jaks and STATs: biological implications. *Annu Rev Immunol* 1998;**16**:293–322.

43. Kawamura M, McVicar DW, Johnston JA, et al. Molecular cloning of L-JAK, a Janus family protein-tyrosine kinase expressed in natural killer cells and activated leukocytes. *Proc Natl Acad Sci USA* 1994;**91**:6374–8.

44. Witthuhn BA, Silvennoinen O, Miura O, et al. Involvement of the Jak-3 Janus kinase in signalling by interleukins 2 and 4 in lymphoid and myeloid cells. *Nature* 1994;**370**:153–7.

45. Thomis DC, Gurniak CB, Tivol E, Sharpe AH, Berg LJ. Defects in B lymphocyte maturation and T lymphocyte activation in mice lacking Jak3. *Science* 1995;**270**:794–7.

46. Nakamura Y, Russell SM, Mess SA, et al. Heterodimerization of the IL-2 receptor beta- and gamma-chain cytoplasmic domains is required for signalling. *Nature* 1994;**369**:330–3.

47. Nelson BH, Lord JD, Greenberg PD. Cytoplasmic domains of the interleukin-2 receptor beta and gamma chains mediate the signal for T-cell proliferation. *Nature* 1994;**369**:333–6.

48. Boussiotis VA, Barber DL, Nakarai T, et al. Prevention of T-cell anergy by signaling through the gamma c chain of the IL-2 receptor. *Science* 1994;**266**:1039–42.

49. Miyazaki T, Kawahara A, Fujii H, et al. Functional activation of Jak1 and Jak3 by selective association with IL-2 receptor subunits. *Science* 1994;**266**:1045–7.

50. Lin JX, Migone TS, Tsang M, et al. The role of shared receptor motifs and common Stat proteins in the generation of cytokine pleiotropy and redundancy by IL-2, IL-4, IL-7, IL-13, and IL-15. *Immunity* 1995;**2**:331–9.

51. Lin JX, Leonard WJ. The role of Stat5a and Stat5b in signaling by IL-2 family cytokines. *Oncogene* 2000;**19**:2566–76.

52. Lin JX, Leonard WJ. Signaling from the IL-2 receptor to the nucleus. *Cytokine Growth Factor Rev* 1997;**8**:313–32.

53. Friedmann MC, Migone TS, Russell SM, Leonard WJ. Different interleukin 2 receptor beta-chain tyrosines couple to at least two signaling pathways and synergistically mediate interleukin 2-induced proliferation. *Proc Natl Acad Sci USA* 1996;**93**:2077–82.

54. Migone TS, Rodig S, Cacalano NA, Berg M, Schreiber RD, Leonard WJ. Functional cooperation of the interleukin-2 receptor beta chain and Jak1 in phosphatidylinositol 3-kinase recruitment and phosphorylation. *Mol Cell Biol* 1998;**18**:6416–22.

55. Macchi P, Villa A, Giliani S, et al. Mutations of Jak-3 gene in patients with autosomal severe combined immune deficiency (SCID). *Nature* 1995;**377**:65–8.

56. Russell SM, Tayebi N, Nakajima H, et al. Mutation of Jak3 in a patient with SCID: essential role of Jak3 in lymphoid development. *Science* 1995;**270**:797–800.

57. Changelian PS, Flanagan ME, Ball DJ, et al. Prevention of organ allograft rejection by a specific Janus kinase 3 inhibitor. *Science* 2003;**302**:875–8.

58. Kontzias A, Kotlyar A, Laurence A, Changelian P, O'Shea JJ. Jakinibs: a new class of kinase inhibitors in cancer and autoimmune disease. *Curr Opin Pharmacol* 2012;**12**:464–70.

59. Buckley RH, Schiff RI, Schiff SE, et al. Human severe combined immunodeficiency: genetic, phenotypic, and functional diversity in one hundred eight infants. *J Pediatr* 1997;**130**:378–87.

60. Puel A, Ziegler SF, Buckley RH, Leonard WJ. Defective IL7R expression in T(-)B(+)NK(+) severe combined immunodeficiency. *Nat Genet* 1998;**20**:394–7.

61. Puel A, Leonard WJ. Mutations in the gene for the IL-7 receptor result in T(-)B(+)NK(+) severe combined immunodeficiency disease. *Curr Opin Immunol* 2000;**12**:468–73.

62. Liu YJ, Soumelis V, Watanabe N, et al. TSLP: an epithelial cell cytokine that regulates T-cell differentiation by conditioning dendritic cell maturation. *Annu Rev Immunol* 2007;**25**:193–219.

63. Rochman Y, Spolski R, Leonard WJ. New insights into the regulation of T-cells by gamma(c) family cytokines. *Nat Rev Immunol* 2009;**9**:480–90.

64. Kennedy MK, Glaccum M, Brown SN, et al. Reversible defects in natural killer and memory CD8 T-cell lineages in interleukin 15-deficient mice. *J Exp Med* 2000;**191**:771–80.

65. Lodolce JP, Boone DL, Chai S, et al. IL-15 receptor maintains lymphoid homeostasis by supporting lymphocyte homing and proliferation. *Immunity* 1998;**9**:669–76.

66. Mehta DS, Wurster AL, Whitters MJ, Young DA, Collins M, Grusby MJ. IL-21 induces the apoptosis of resting and activated primary B-cells. *J Immunol* 2003;**170**:4111–8.

67. Jin H, Carrio R, Yu A, Malek TR. Distinct activation signals determine whether IL-21 induces B-cell costimulation,

growth arrest, or Bim-dependent apoptosis. *J Immunol* 2004;**173**:657–65.

68. Ozaki K, Spolski R, Ettinger R, et al. Regulation of B-cell differentiation and plasma cell generation by IL-21, a novel inducer of Blimp-1 and Bcl-6. *J Immunol* 2004;**173**: 5361–71.

69. Spolski R, Leonard WJ. Interleukin-21: basic biology and implications for cancer and autoimmunity. *Annu Rev Immunol* 2008;**26**:57–79.

70. Ettinger R, Sims GP, Fairhurst AM, et al. IL-21 induces differentiation of human naive and memory B-cells into antibody-secreting plasma cells. *J Immunol* 2005;**175**: 7867–79.

71. Pene J, Gauchat JF, Lecart S, et al. Cutting edge: IL-21 is a switch factor for the production of IgG1 and IgG3 by human B-cells. *J Immunol* 2004;**172**:5154–7.

72. Kasaian MT, Whitters MJ, Carter LL, et al. IL-21 limits NK cell responses and promotes antigen-specific T-cell activation: a mediator of the transition from innate to adaptive immunity. *Immunity* 2002;**16**:559–69.

73. Ozaki K, Spolski R, Feng CG, et al. A critical role for IL-21 in regulating immunoglobulin production. *Science* 2002;**298**:1630–4.

74. Suto A, Nakajima H, Hirose K, et al. Interleukin 21 prevents antigen-induced IgE production by inhibiting germ line C(epsilon) transcription of IL-4-stimulated B-cells. *Blood* 2002;**100**:4565–73.

75. Recher M, Berglund LJ, Avery DT, et al. IL-21 is the primary human common gamma chain-binding cytokine required for human B-cell differentiation *in vivo. Blood* 2011;**118**:6824–35.

76. Spolski R, Kashyap M, Robinson C, Yu Z, Leonard WJ. IL-21 signaling is critical for the development of type I diabetes in the NOD mouse. *Proc Natl Acad Sci USA* 2008;**105**:14028–33.

77. Bubier JA, Sproule TJ, Foreman O, et al. A critical role for IL-21 receptor signaling in the pathogenesis of systemic lupus erythematosus in BXSB-Yaa mice. *Proc Natl Acad Sci USA* 2009;**106**:1518–23.

78. Wang L, Yu CR, Kim HP, et al. Key role for IL-21 in experimental autoimmune uveitis. *Proc Natl Acad Sci USA.* 2011;**108**:9542–7.

79. Buckley RH, Win CM, Moser BK, Parrott RE, Sajaroff E, Sarzotti-Kelsoe M. Post-transplantation B-cell function in different molecular types of SCID. *J Clin Immunol* 2013;**33**:96–110.

80. Cavazzana-Calvo M, Andre-Schmutz I, Fischer A. Haematopoietic stem cell transplantation for SCID patients: where do we stand? *Br J Haematol* 2013;**160**:146–52.

81. Lo M, Bloom ML, Imada K, et al. Restoration of lymphoid populations in a murine model of X-linked severe combined immunodeficiency by a gene-therapy approach. *Blood* 1999;**94**:3027–36.

82. Cavazzana-Calvo M, Hacein-Bey S, de Saint Basile G, et al. Gene therapy of human severe combined immunodeficiency (SCID)-X1 disease. *Science* 2000;**288**:669–72.

83. Hacein-Bey-Abina S, Von Kalle C, Schmidt M, et al. LMO2-associated clonal T-cell proliferation in two patients after gene therapy for SCID-X1. *Science* 2003;**302**:415–9.

84. Hacein-Bey-Abina S, Garrigue A, Wang GP, et al. Insertional oncogenesis in 4 patients after retrovirus-mediated gene therapy of SCID-X1. *J Clin Invest* 2008;**118**:3132–42.

85. Wang GP, Berry CC, Malani N, et al. Dynamics of gene-modified progenitor cells analyzed by tracking retroviral integration sites in a human SCID-X1 gene therapy trial. *Blood* 2010;**115**:4356–66.

86. Fischer A, Hacein-Bey-Abina S, Cavazzana-Calvo M. Strategies for retrovirus-based correction of severe, combined immunodeficiency (SCID). *Meth Enzymol* 2012;**507**:15–27.

87. Nakajima H, Liu XW, Wynshaw-Boris A, et al. An indirect effect of Stat5a in IL-2-induced proliferation: a critical role for Stat5a in IL-2-mediated IL-2 receptor alpha chain induction. *Immunity* 1997;**7**:691–701.

88. Imada K, Bloom ET, Nakajima H, et al. Stat5b is essential for natural killer cell-mediated proliferation and cytolytic activity. *J Exp Med* 1998;**188**:2067–74.

89. Yao Z, Cui Y, Watford WT, et al. Stat5a/b are essential for normal lymphoid development and differentiation. *Proc Natl Acad Sci USA* 2006;**103**:1000–10005.

90. Cohen AC, Nadeau KC, Tu W, et al. Cutting edge: Decreased accumulation and regulatory function of CD4+ CD25(high) T-cells in human STAT5b deficiency. *J Immunol* 2006;**177**:2770–4.

91. Nadeau K, Hwa V, Rosenfeld RG. STAT5b deficiency: an unsuspected cause of growth failure, immunodeficiency, and severe pulmonary disease. *J Pediatr* 2011;**158**:701–8.

92. Jenks JA, Seki S, Kanai T, et al. Differentiating the roles of STAT5B and STAT5A in human CD4 T-cells. *Clin Immunol* 2013;**148**:227–36.

93. Kotlarz D, Zietara N, Uzel G, et al. Loss-of-function mutations in the IL-21 receptor gene cause a primary immunodeficiency syndrome. *J Exp Med* 2013;**210**:433–43.

The Hyper IgM Syndromes – a Long List of Genes and Years of Discovery

Anne Durandy[1,2,3], Sven Kracker[1,2]

[1]Institut National de la Santé et de la Recherche Médicale U768,
Hôpital Necker Enfants Malades, Paris, France
[2]Faculté de Médecine, Descartes-Sorbonne Paris Cité University of Paris,
Imagine Institute, France
[3]Centre d'étude des déficits immunitaires, Hôpital Necker Enfants Malades,
Paris, France

OUTLINE

Primary Immunodeficiency Disorders: A Historic and Scientific Perspective

INTRODUCTION

Immunoglobulin (Ig) class switch recombination deficiencies (CSR-Ds, which were previously named "dysgammaglobulinemia" and then "hyper-IgM syndromes") are characterized by elevated (or sometimes normal) serum IgM levels and a considerable decrease in (or the absence of) IgG, IgA and IgE – suggesting defective CSR. The production of switched isotypes is essential for optimizing the humoral response against pathogens, since the different Ig isotypes vary in activity (their half-life, binding to different Fc receptors, ability to activate the complement system, etc.) and tissue location (since IgA can be secreted by plasma cells in mucosal membranes). From a clinical standpoint, CSR-Ds are associated with increased susceptibility to bacterial infections. Depending on the exact nature of the molecular defect, the CSR-D may be associated with a defect in the second event in B-cell maturation, i.e. the introduction of mutations into the variable (V) region of the Ig locus. This somatic hypermutation (SHM) process is required for the selection of B-cells expressing a B-cell receptor (BCR) with a high affinity for antigen.

Class switch recombination deficiencies were first reported decades ago (even before the description of the different Ig isotypes), when a French group described a case of "agammaglobulinemia with increased β2-macroglobulinemia" (Fig. 16.1) (1). Since the age of three years, the patient (a teenage boy at the time of diagnosis) had presented with recurrent bacterial infections of the respiratory tract, bronchiectasis and several episodes of meningitis. These symptoms had responded effectively to treatment with antibiotics and thermal spas. No viral infections or neutropenia were reported. An episode of cervical lymphadenopathy had been cured by treatment with a combination of antibiotics and corticosteroids. Laboratory tests revealed elevated β2-macroglobulin but the absence of γ-globulin. The patient was prescribed intermittent, intramuscular injections of γ-globulin. Isohemagglutinins were present, prompting the investigators to suggest a correlation between the respective levels of these natural antibodies and β2-macroglobulins. The authors concluded their report by suggesting that recurrent infections could be secondary to atypical agammaglobulinemia with qualitative (not only quantitative) abnormalities in protein levels. One year later, Fred Rosen described two boys affected with recurrent bacterial infections and "dysgammaglobulinemia". One of the boys was also suffering from a *Mycobacterium tuberculosis* infection, together with severe anemia and thrombocytopenia that led to splenectomy. Histological examination of the spleen and a lymph node revealed the complete absence of germinal centers. The second boy did not present with lymphadenopathy but the results of a lymph node biopsy also suggested the absence of germinal centers; in addition, he suffered from severe, persistent neutropenia. In both patients, ultracentrifugation and immunoelectrophoresis of plasma proteins revealed a lack of 7S globulins and marked elevation of 19S γ-globulins. Following immunization with typhoid vaccine, one patient failed to produce antibody, but the second patient produced high titers of anti-O antibody known to be of the 19S (IgM) isotype, but not anti-H antibody (7S). Both patients were found to have isohemagglutinins and Forssman antibody known to be of the IgM isotype. Both patients were treated with intramuscular immunoglobulin infusions (2).

Of note, these initial clinical observations were made even before the World Health Organisation in 1964 designated the 19S γ-globulin as IgM (M standing for "macroglobulin"). The term "dysgammaglobulinemia" (3–6) was used until 1973, when the Second International Workshop on Primary Immunodeficiency Diseases in Man, held in St Petersberg, Florida, adopted the name "immunodeficiency with increased IgM" (7, 8). Interestingly, only the X-linked form

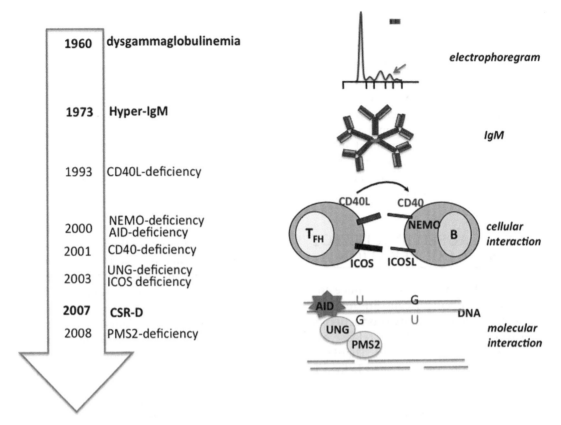

FIGURE 16.1 Milestones in the characterization of CSR-Ds. An electropherogram reveals dysgammaglobulinemia (elevated 19S globulins and a lack of γ-globulins). Pentameric IgM ("M" stands for "macroglobulinemia"). Cellular interactions: T follicular helper (T_FH) cells cross-talk with B-cells through interaction between CD40L/CD40 and – to a lesser extent – ICOS/ICOSL. CD40 signaling in B-cells involves the NF-κB pathway via activation of NF-κB essential modulator (NEMO). Molecular interactions: Activation induced cytidine deaminase (AID) introduces cytidine > uridine DNA lesions within S and V regions. The U:G base pairs that are misintegrated into the DNA are mostly processed by uracil-N glycosylase (UNG), leading to double-strand DNA breaks. PMS2 endonuclease activity contributes to DNA breakage downstream from UNG. The names of the disease entities described at the different time periods are indicated in bold letters. This figure is reproduced in color in the color section.

was reported in this workshop (see Fig. 16.1). Given the defective production of IgG and IgA and contrastingly high IgM levels, a defect in CSR was suspected in "hyper-IgM immunodeficiency". The first reports suggested an intrinsic B-cell defect because the patients' B-cells were shown only to express IgM and IgD and were unable to produce IgG and IgA *in vitro* (9, 10). However, the activators used in these experiments were inducers of switched B-cell proliferation rather than of actual CSR in naïve B-cells.

In 1985, evidence of a causative T-cell defect in this condition was provided by the observation that Sezary T-cells are capable of inducing patients' B-cells to undergo CSR *in vitro* (11). Since *in vivo* CSR is defective in all patients and IgM levels can be in the normal range in some individuals (especially after Ig substitution), we suggested that the term "CSR deficiency" was more appropriate (12).

Interestingly, almost all patients initially reported in the literature were male – suggesting

X-linked transmission (1–5). In 1963, Rosen described a female adult patient as being affected by "acquired dysgammaglobulinemia" (13). Autosomal recessive (AR) inheritance was suggested by the first description of dysgammaglobulinemia in a female infant, in 1965. In contrast to what had been observed in the previous publications, germinal centers were present (14). Autosomal dominant inheritance has been reported in a later publication (15), adding genetic heterogeneity to the previously observed phenotypic diversity.

CSR-D CAUSED BY A DEFECT IN T:B COOPERATION

It has long been known that in mice and humans, antigen-specific T-cells are required for thymus-dependent antibody production *in vivo* (either directly or through the release of non-specific soluble factors) (16–19). Thanks to the cognate interaction with activated helper T-cells and cytokine production, mature (but still naïve) IgM$^+$IgD$^+$ B-cells in the secondary lymphoid organs (spleen, lymph nodes, tonsils) are rescued from apoptosis when they encounter an antigen that is specifically recognized by their B-cell receptor (BCR). These B-cells proliferate vigorously and give rise to a unique lymphoid structure: the germinal center (GC) (20–25). Here, the B-cells undergo the two major events in maturation: CSR (to produce IgG, IgA and IgE) and SHM (to generate antibodies with high affinity for antigen) (25). In 1996, researchers found evidence to suggest that a particular population of T-cells (T-follicular helper (T$_{FH}$) cells) interact with B-cells in the GC during Ig production (26). The T$_{FH}$ cells have subsequently been found to express the CXC chemokine receptor 5 and activation markers and to produce large amounts of cytokines (including IL-21) (27, 28). Therefore, the discovery of a CSR-D caused by defective T:B cooperation was not unexpected.

X-LINKED CSR-D DUE TO CD40L DEFICIENCY

In view of the published data on phenotypes and, when available, pedigrees, it is very likely that a majority of the first patients reported suffered from an X-linked CSR-D, which accounts for approximately 50% of all CSR-Ds and is by far the most frequent condition of this type. The clinical features include both humoral and cellular immune defects, with early onset and a poor prognosis. Impaired production of IgG and IgA is responsible for specific susceptibility to recurrent bacterial infections. Although no antibodies of the IgG isotype against infectious agents or vaccines are produced, isohemagglutinins and antipolysaccharide IgM antibodies can generally be detected. Recently, anti-non-typable *Haemophilus influenzae* IgM antibodies found in patients' serum and saliva were shown to be microbiologically and clinically protective (29). Although patient B-cells cannot form germinal centers in secondary lymphoid organs *in vivo*, they are intrinsically normal, since they can be induced to proliferate and undergo CSR *in vitro* upon activation by CD40 agonists and appropriate cytokines (30). Most (but not all) patients present low counts of CD27$^+$ "memory" IgM$^-$IgD$^-$ (switched) B-cells and a low frequency of SHM (31). In addition to this humoral defect, affected patients suffer from an abnormal cellular immune response, which leads to marked susceptibility to opportunistic infections. This susceptibility cannot be controlled by Ig replacement therapy. Liver disease is very common; sclerosing cholangitis (often associated with *Cryptosporidium*) is particularly severe and may lead to terminal liver failure. Intermittent or chronic neutropenia is also a common feature (32). Other less frequent complications, such as autoimmune manifestations or cancer, have been reported in some cases (Tables 16.1, 16.2).

In 1987, two research groups reported that the gene involved in this defect was located

TABLE 16.1   The main clinical features of CSR-Ds

Gene defect	Inheritance	Infections	Lymphadenopathies	Autoimmunity/ inflammation	Cancers	Other features
CD40L	X-linked	Bacterial and opportunistic	Sometimes present	+	Sometimes present	Neutropenia
CD40	AR	Bacterial and opportunistic	Sometimes present	Not reported	Nnot reported	Neutropenia
NEMO	X-linked	Bacterial and opportunistic Mycobacteria	Sometimes present	Not reported	Not reported	AED
ICOS	AR	Almost bacterial	–	+	Not reported	
AID	AR AD	Bacterial	+++ (giant GCs) +	++ Not reported	Sometimes present Not reported	
UNG	AR	Bacterial	++	++	Risk of lymphoma?	
PMS2	AR	Bacterial	Not reported	Not reported	+++	Café au lait skin spots
ATM	AR	Bacterial	Sometimes present	Not reported	+++	Ataxia telangiectasia

DNA repair deficiencies caused by a lack of PMS2and ATM are also shown here because some affected patients develop humoral immunodeficiencies as the first symptom. AR: autosomal recessive; AD: autosomal dominant; AED: anhidrotic ectodermal dysplasia; GC: germinal center.

TABLE 16.2   The main biological features of CSR-Ds

Gene defect	*In vivo* CSR	*In vitro* CD40L-induced CSR	CD27+ B-cell count	SHM
CD40L	Generally absent But IgA may be present	Normal	Low	Absent or low
CD40	Absent	Absent	Low	Absent or low
NEMO	Variable, defective anti-polysaccharide response	Variable	Low	Variable
ICOS	Variable	Normal	Low	Low
AID-AR	Absent	Absent	Normal	Absent
AID-AD	Variable for IgG, IgA always absent	Absent	Normal	Normal
UNG	Absent	Absent	Normal	Present but pattern is abnormal
PMS2	Low(mostly affecting IgA and IgG2/IgG4 subclasses)	Low	Low	Normal
ATM	Variable, sometimes drastic	Absent	Normal	Normal

DNA repair deficiencies caused by a lack of PMS2 and ATM are also shown here because some affected patients develop humoral immunodeficiencies as the first symptom.

on the long arm of the X chromosome (at Xq 21.3-22) (33, 34). The CD40-ligand gene (*CD40L*) was cloned in 1992 (35) (see Fig. 16.1). CD40L (also known as CD154) is an activation molecule that is transiently expressed on T helper CD4$^+$ cells (including T$_{FH}$ cells). It interacts with the CD40 receptor, which is constitutively expressed on B-cells and monocytes, dendritic cells and other non-hematopoietic cells (36). As *CD40L* had been cloned in a genetic region (35) previously reported as linked to the defect (32, 34), and as patients' B-cells were normally able to undergo CSR upon CD40L activation (30, 37), several groups simultaneously investigated the possibility of a CD40L defect as responsible for this X-linked, puzzlingly severe CSR-D. Deleterious mutations were identified in *CD40L* gene, resulting in impaired CD40L protein expression (37–41). The finding of markedly low or absent membrane CD40L expression by *in vitro* activated CD4$^+$ T-cells made the diagnosis of this syndrome simple and straightforward. The resulting CD40 trans-activation defect of B-cells results in an antibody deficiency, since CSR and SHM are initiated in germinal centers when T$_{FH}$ and B-cells cooperate through a CD40L/CD40 interaction (32). Defective interaction between T-cells and monocytes/dendritic cells impairs the maturation of dendritic cells. The lack of IL-12 production by dendritic cells and macrophages and the absence of T-cell priming result in an abnormal cellular immune response (42). Given that myeloid progenitors express CD40 molecules, the neutropenia reported in about half of all patients (and which is not linked to infectious episodes) may result from a defective "stress"-induced, CD40-dependent granulopoiesis. However, the details of this latter process are less clear (43).

Although mutations affect the entire *CD40L* gene, they are irregularly distributed; the majority affect exon 5, which contains most of the tissue necrosis factor homology domain (44). A strict relationship between genotype and phenotype has not been established. Although the CD40L gene is located on the X chromosome, female patients can sometimes be affected because of a skewed X inactivation pattern (45) or a chromosomal translocation (46).

The detection of serum IgA and the presence of SHM in some patients suggests that there are alternative diversification pathways for production of IgA (upon CpG- or proliferating inducible ligand (APRIL)-induced activation of B-cells in the gut lamina propria) (47) and for SHM in a T-cell-independent manner – possibly as an innate defence mechanism in the splenic marginal zone (48, 49).

Treatment of CD40L deficiency is based on intravenous or subcutaneous Ig infusion, antibiotic prophylaxis and (in cases of severe neutropenia) administration of G-CSF. However, even under adequate treatment, the prognosis remains poor: in a large study including 56 patients, 13 patients were shown to die either from severe infection, liver damage or cancer (32). Thus, hematopoietic stem cell transplantation should be considered if a human leukocyte antigen-identical donor is available. Subcutaneous administration of rCD40L is unable to trigger B-cell responses but can restore T$_{H1}$-cell function and thus can be considered for severe, opportunistic, infectious episodes (50).

AUTOSOMAL RECESSIVE CSR-D DUE TO CD40 DEFICIENCY

CD40 deficiency was recognized in 2001 as an AR inherited disease in a small cohort of patients on the basis of a lack of CD40 expression on the surface of B lymphocytes and monocytes (40) (see Fig. 16.1). However, more recent studies provided evidence that some mutations not affecting CD40 expression can also be pathogenic and are expected to result in defective intracellular signalling (51). The clinical and immunological characteristics of CD40-deficient patients are identical to those reported in CD40L deficiency

(40, 51, 52), except that CD40-deficient B-cells are unable to proliferate and undergo *in vitro* CSR upon activation with CD40 agonists and cytokines (40).

X-LINKED CSR-D DUE TO DEFECTIVE NUCLEAR FACTOR KAPPA B (NF-κB) ACTIVATION

Cross-linking of CD40 results in activation of the canonical NF-κB signaling pathway which is critical in CSR, as shown by patients affected with ectodermal dysplasia associated with immunodeficiency (EDA-ID) (53–57). Although this syndrome is phenotypically diverse, patients are characterized by normal-to-elevated IgM levels, low levels of serum IgG and IgA, impaired antibody responses to polysaccharide antigens and susceptibility to mycobacterial infections. EDA-ID is inherited as an X-linked trait caused by hypomorphic mutations that are commonly found in the zinc-finger domain of the NF-κB essential modulator (NEMO, also known as IKKγ). NEMO acts as a scaffolding protein that binds to the IKKα and IKKβ kinases required for NF-κB activation and nuclear translocation. *In vitro* CSR and SHM can be either normal or defective, depending on the type and location of the NEMO mutation. Because NF-κB nuclear translocation is required for many signaling pathways (including those involving the T- and B-cell antigen receptors), the defect is not restricted to CD40 activation. Ectodermal dysplasia is a key feature of this syndrome and results from NEMO deficiency, since the ectodysplasin receptor (expressed on tissues derived from the ectoderm) activates NF-κB via the IKKα/β NEMO complex. Recently, a similar clinical phenotype was described in a handful of patients with AD-EDA-ID due to heterozygous gain-of-function mutations in the IKBA gene, encoding for the NF-κB inhibitor IκB-α (58–62).

AUTOSOMAL RECESSIVE CSR-D DUE TO ICOS-DEFICIENCY

Although inducible co-stimulatory molecule (ICOS) deficiency was first described as a common variable immunodeficiency (63), it generally leads to a CSR-D, as shown by the elevated or normal IgM levels and low IgG and IgA levels observed in these patients (64). ICOS is expressed on activated T-cells including the T_{FH} and is involved in the generation and function (i.e. cytokine production) of T_{FH} in germinal centers (65). Furthermore, ICOS must interact with its cognate ligand, ICOS-L, which is constitutively expressed on B-cells, for full antibody production. The SHM process is impaired in $CD27^+$ B-cells and the corresponding cell count is very low. A cell-mediated immune defect has been reported in some (but not all) patients (66). ICOS-deficiency is a rare condition; only two different *ICOS* gene mutations have been reported to date (63, 66).

The molecular characterization of CSR-Ds due to defective T:B cooperation highlighted the essential role of CD40L–CD40 interactions and, to a lesser extent, of ICOS–ICOS-L interactions, in mature antibody production within the germinal center. A defect in T_{FH} generation/function or a defective follicular B-cell response may underlie these conditions.

CSR-Ds WITH NORMAL *IN VITRO* B-CELL RESPONSES TO CSR ACTIVATION

Interestingly, several patients with CSR-D display normal *in vitro* CSR but are not affected either by CD40L or ICOS deficiencies. The phenotype of this group of patients differs significantly from that observed in CD40L-deficient patients, since the former group is not overly susceptible to opportunistic infections and tends to present only with bacterial infections that can be controlled by Ig replacement therapy.

Lymphadenopathies with enlarged germinal centers are characteristic findings. There is phenotypic heterogeneity, since the frequencies of SHM can be normal or low. A number of possible causes of *in vivo* CSR-D have been excluded, e.g. congenital rubella and major histocompatibility complex class II deficiency, respectively, which are associated with diminished expression of CD40L by activated CD4$^+$ T-cells (67, 68). A defect in T_{FH} generation, T_{FH} activation and/or the interaction between T_{FH} and follicular B-cells has been suspected in these rare CSR-Ds.

CSR-Ds CAUSED BY AN INTRINSIC B-CELL DEFECT

This category of CSR-Ds is caused by an intrinsic B-cell defect interfering with the CSR machinery itself. B-cells of affected patients are unable to undergo *in vitro* CSR upon appropriate stimulation. Somatic hypermutation may be normal or defective, depending on the molecular basis of the defect. Characterization of these conditions has greatly contributed to our understanding of the mechanisms involved in antibody maturation. The major events in antibody maturation, first described in 1978 (69), include CSR that involves recombination between two different switch (S) DNA regions located upstream of the constant (C) regions, and deletion of the intervening DNA by formation of an excision circle (70–74). In CSR, it is thought that a DNA lesion is introduced during transcription by a putative "switchase" or "recombinase" (75). This lesion leads to the generation of DNA double-strand breaks (DSBs) that are required for the recombination process. As shown by experiments involving corresponding knockout mice or isolated B-cells, most of the DSBs are repaired through the conventional non-homologous end-joining (c-NHEJ) pathway (76–78). However, an alternative end-joining process can also promote CSR – especially if the NHEJ is defective (79, 80). Mismatch repair (MMR) enzymes also

have a role in CSR – probably through DSB induction and processing (81–86). The complex mechanism underlying CSR ultimately leads to replacement of the Cμ region by a downstream Cx region from another class of Ig, resulting in the production of antibodies of different isotypes (IgG, IgA, and IgE) with the same variable (V) region and thus the same antigen specificity and affinity.

The other process of antibody maturation, SHM, was first described in 1970 (87). This process introduces mutations (and, less frequently, deletions or insertions) into the V regions of immunoglobulins (88–93). At the outset, SHM is triggered by activation of the BCR and CD40/CD40L interaction (94, 95). The mutations occur at a high frequency in the V regions and their proximal flanks (1×10^{-3} bases/generation). As is the case for CSR, it has been hypothesized that a DNA lesion in V regions can be introduced during transcription by a "putative mutator". This type of mutation does not involve DNA double strand breaks and, as shown by experiments in knockout mice, is repaired by error-prone polymerases and the MMR complex (96–99). The mutated B-cells undergo a final selection process within the germinal centers, with proliferation of clonal B-cells expressing a high-affinity BCR (100, 101). Although CSR and SHM both occur within the GC, it has been postulated that the two processes are mediated by different molecular mechanisms because the respective DNA targets and induced DNA modifications are quite distinct (102). The mutations induced by these mechanisms affect the generation of IgM, IgG and IgA (103).

AUTOSOMAL RECESSIVE ACTIVATION-INDUCED CYTIDINE DEAMINASE DEFICIENCY

By 1997, we had identified a cohort of CSR-D patients with a phenotype that differed markedly from that observed in CD40L-deficient patients. Consanguinity was found in 71% of cases. The equal frequencies of female and male patients strongly suggested an AR inherited disease. The onset of symptoms generally occurred during the first decade of life, although a few patients were diagnosed after reaching adulthood. The clinical symptoms included recurrent bacterial infections of the respiratory and digestive tracts but not opportunistic infections. Neutropenia had not been observed. Lymphoid hyperplasia was a prominent feature of this entity and affected 75% of patients. While there was massive enlargement of the germinal centers, which were accordingly described as "giant", the mantle zone was almost completely absent. The germinal centers were filled with actively proliferating B-cells that co-express CD38, surface IgM and sIgD, all of which are markers of germinal center founder cells. Autoimmunity caused by autoantibodies of the IgM isotype resulting in hemolytic anemia, thrombocytopenia, hepatitis and systemic lupus erythematosus had been found more frequently in this group of patients than in CD40L deficiency, and affected almost 30% of the patients (104). With regular infusions of Ig, the patients' prognosis has been excellent. However, IgG replacement therapy did not control lymphoid hyperplasia nor the autoimmune manifestations (see Table 16.1).

CSR was drastically impaired in these patients, while serum IgG, IgA and IgE were never detected, IgM levels can be very high. *In vitro*, patient B-cells proliferated, produced large amounts of IgM and upregulated the CD23 activation marker when cultured in the presence of sCD40L and cytokines. However, under these conditions, no IgG, IgA or IgE was generated via CSR (105). The proportion of peripheral blood B-cells carrying the CD27 marker was normal (20–50%), although all B-cells were IgM$^+$ and IgD$^+$. Somatic hypermutation was either absent or occurred at very low levels in the memory CD27$^+$ B-cell subset (see Table 16.2). Overall, this condition was the second most

frequent molecularly defined CSR-D (accounting for 10% of patients) and is therefore rarer than CD40L deficiency.

Homozygosity mapping performed in 1999 in consanguineous or multiplex families allowed us to define a small (4.5 cM) region on chromosome 12 (12p13) that segregated with the disease. One year later, Honjo et al. cloned the gene encoding activation-induced cytidine deaminase AID (*AICDA*) located exactly within this genetic locus. At the same time, Honjo's group had found that AID was strongly expressed in germinal centers (106). At that point, a collaboration was established between our two laboratories, leading to the identification of bi-allelic, harmful mutations in all patients with this unique phenotype (107) (see Fig. 16.1). Finally, the phenotype observed in humans was confirmed by the simultaneously generated AID-knockout mouse (108). AID appeared to be both the putative "switchase" and the "mutator", the existence of which had been postulated for a long time. Thus AID turned out to be a key molecule for both CSR and SHM and constitutes the first link between these two events in antibody maturation.

The AID protein's mode of action in CSR and SHM is still subject to debate. As a cytidine deaminase, AID deaminates cytidine to uridine (106). It acts upstream of the DNA DSBs required for S region recombination and probably induces DNA lesions in the S and V regions (109). Given that the gene sequence of AID is similar to that of the RNA-editing enzyme APOBEC-1, it had originally been suggested that AID edits an mRNA coding for a substrate that is involved in both CSR and SHM such as an endonuclease (108, 110). The finding that AID is mostly (if not totally) located in the cytoplasm constituted an argument in favor of this hypothesis (111). However, data from subsequent experiments performed *in vitro* or after transfection into *Escherichia coli* strongly suggested that AID exerts its cytidine deaminase activity directly on DNA (112–116).

In the most current model, phosphorylated AID acts directly on single-strand DNA within stalled transcription bubbles. This model integrates (1) the requirement for S region DNA transcription in CSR and (2) AID's association with RNA-polymerase II (117) via the transcription elongation factor SPT5 (118). The RNA exosome is involved in targeting AID- and transcription-dependent DNA deamination on both DNA strands (119). It is thought that AID shuttles between the cytoplasm and the nucleus (where it edits DNA) as a result of the balance between active transport into the nucleus (which depends on the nuclear import signal located within the N-terminal part of AID), nuclear export (which depends on the nuclear export domain located at the C-terminal end of AID) and cytoplasmic retention (which is also a function of AID's C-terminal domain) (120). Cytoplasmic retention may be important for AID's stability, since the protein is rapidly degraded by the nuclear proteasome (121).

The mutations identified in patients are scattered throughout the *AICDA* gene without hotspots and lead to a defect in both CSR and SHM (122). However, mutations located in the C-terminal (C^{ter}) part of *AICDA*, abrogating the protein's nuclear export signal (NES), result in a complete lack of CSR but do not affect SHM (123). Interestingly, normal germinal centers were observed in a cervical lymph node from one patient. It has been found that the C^{ter}-mutated enzyme retains normal cytidine deaminase activity *in vitro* and, if transfected into *E. coli*, is translocated into the nucleus (123, 124). Although CSR is defective in C^{ter}-mutated patients, CSR-induced mutations and DSBs have been found in the Sµ regions of murine B-cells expressing nuclear export signal (NES)-deleted AID (125). Taken together, these observations suggest that AID has another activity in addition to its deamination of cytidine – probably as a docking protein (120, 124, 126, 127).

Another unexpected finding (128) is that heterozygous, nonsense mutations in the C-terminal domain (resulting in loss of the last eleven (V186X) or nine (R190X) amino acids of the NES ($AID^{\Delta NES}$)) lead to a variable, autosomal dominant inherited CSR-D, in which serum IgG levels are low and IgA is absent (see Tables 16.1, 16.2). Haplo-insufficiency is highly unlikely (despite the fact that it has been reported in mice, with weak effects on Ig levels) (129) because all human heterozygotes for AID deficiency have normal Ig levels. The mutated allele's dominant negative effect is expected if AID participates in a homomeric complex; however, this hypothesis is still subject to debate (120, 126).

The discovery of AID has profoundly modified our concept of antibody maturation and has opened new avenues for better characterizing the molecular events that play major roles in the maturation process.

URACIL-N GLYCOSYLASE (UNG) DEFICIENCY

The observation that uracil-N glycosylase (UNG) is involved in antibody maturation provided strong evidence of AID's DNA-editing activity, since UNG is able to recognize and remove uracils that have been misintegrated into DNA. The first demonstration of UNG's role was provided in 2002, when Neuberger et al. reported that *ung*-deficient mice have partial CSR-D and a skewed SHM pattern (130). This report reminded us that one of our CSR-D patients presented with the very same SHM abnormality. We sequenced the *UNG* gene in this patient and indeed found an UNG deficiency (see Fig. 16.1). When we discussed our discovery with colleagues from Tokyo and Seattle, two additional cases of UNG deficiency were identified and all three published together in 2003 (131). We were indeed lucky, since this form of CSR-D appears to be very rare; as of this date only one additional patient with UNG deficiency has been identified (132). All three patients from the original report have a history of frequent bacterial infections of the respiratory tract, which are easily controlled by regular intravenous Ig infusions. Two of the three patients presented with lymphadenopathy, including transient enlargement of mediastinal or cervical lymph nodes. No giant germinal centers were found in a biopsy sample of a cervical lymph node. One of the patients developed autoimmune manifestations, hemolytic anemia and Sjögren syndrome during adulthood (see Table 16.1) (131). The three first described patients were shown to have a severe *in vivo* and *in vitro* CSR defect (131, 132). Interestingly, SHM is normal in terms of frequency but is characterized by a skewed nucleotide substitution pattern, with an abnormally high proportion of transitions on C:G residues (see Table 16.2). In the absence of uracil removal, transitions probably arise from the replication of AID-induced uracil:guanosine lesions (C > T, G > A). There is also the possibility that MMR enzymes recognize and repair these mismatches by introducing mutations on neighboring nucleotides that result in both transitions and transversions on A:T residues (133).

Five different mutations in the UNG catalytic domain have been found in the four patients identified to date (131, 132). Three patients have small deletions that create a premature stop codon (a homozygous mutation in two patients with consanguineous parents and two heterozygous mutations in the other). The fourth patient bears a homozygous missense mutation (131). The fact that UNG expression and function were entirely absent in Epstein–Barr virus (EBV)-transformed patient B-cell lines provided evidence for the lack of any compensatory UNG-DNA glycosylase activity – at least in B-cells. Accordingly, it has been shown that expression of SMUG1 (another uracil-N glycosylase) is downregulated during B-cell activation (134).

Since UNG is part of the DNA base excision pathway involved in the repair of spontaneously occurring lesions, it constitutes an antimutagenic defense strategy. Indeed, elderly UNG-deficient mice develop B-cell lymphomas (135). It is therefore possible, although it has not yet been observed, that human UNG deficiency predisposes to this type of tumor in adulthood. Another consequence of UNG deficiency also reported in *ung*-deficient mice, is post-ischemic brain injury which is much more severe in deficient than in control animals; this is probably a consequence of defective mitochondrial DNA repair (136).

The investigation of UNG deficiency greatly strengthened the hypothesis that AID is a DNA-editing enzyme (112). The AID-induced uracil is removed by UNG, leading to an abasic site which is either repaired by error-prone translesion synthesis polymerases during SHM or cleaved by an apurinic/apyrimidinic endonuclease (APE, defined as APEX 1 and APEX2 in mice) during CSR (137). The subsequent nicks are further processed into the DSB required for inter-switch region recombination.

Ig CSR-DEFICIENCIES WITH UNKNOWN MOLECULAR DEFECT(S)

At least half of Ig CSR-deficiencies due to an intrinsic B-cell defect are not related to AID or UNG deficiencies. Although most of the reported cases are sporadic, the mode of inheritance observed in a few multiplex or consanguineous families is suggestive of an AR pattern. The clinical phenotype is very similar to that seen in AID deficiency and includes increased susceptibility to bacterial infections of the respiratory and gastrointestinal tracts. Lymphoid hyperplasia is milder and less frequent (in 50% of patients) and consists of moderate follicular hyperplasia in the absence of the giant germinal centers that typify AID deficiency. Autoimmune manifestations have been reported (12). The *in vivo* CSR defect appears to be milder than in AID and UNG deficiencies, since residual serum levels of IgG can be detected in some patients. However, switched B-cells are barely detectable in peripheral blood and *in vitro* CSR is always defective. The defect is restricted to CSR, since SHM is normal in terms of both frequency and pattern in the CD27+ B-cell subset which numerically is also normal.

Although the underlying molecular defect(s) in this condition has not yet been identified, the phenotype is very similar to that observed in AID C^ter mutations, suggesting direct or indirect impairment of the ability of AID to shuttle between the cytoplasm and nucleus and/or target S regions and/or recruitment of factors involved in DNA repair or recombination.

Ig CSR-Ds as Part of DNA Repair Deficiencies

CSR-D can occur in patients with a defined DNA repair deficiency. Although other complications are by far the pre-eminent symptoms of these conditions, in some patients, the CSR-D may be the first and only phenotypic feature for several years. Hence, DNA repair deficiencies should always be investigated whenever a molecularly defined "classic" CSR-D cannot be found.

AUTOSOMAL RECESSIVE MISMATCH REPAIR DEFICIENCY AND CSR-D

Mono-allelic mutations in genes coding for the MMR enzymes MutSα (mostly an MSH2/6 complex) and MutLα (mostly a PMS2/MLH1 complex) lead to a hereditary predisposition to the development of non-polyposis colon carcinoma (also known as Lynch syndrome) (138) in adulthood. Furthermore, bi-allelic mutations in either MutSα or MutLα lead to a variety of cancers with onset in the first few years of life (139). IgA deficiency (combined with an IgG2 deficiency in one case) has been reported as a minor finding in four MutSα–deficient patients

(140–142). We observed that the CSR-D can be a prominent symptom (or, indeed, the sole symptom for several years) in PMS2 deficiency (81). Indeed, four of the nine PMS2-deficient patients that we studied were diagnosed with a primary, humoral immunodeficiency because of recurrent bacterial infections; Ig replacement therapy was provided to three individuals (see Table 16.1). In all nine patients, serum IgM levels were found to be normal or elevated, whereas serum IgG2 and IgG4 levels were low. Total IgG and/or IgA levels were low in the four symptomatic patients. The IgG and IgA levels rose with age – probably as a result of the accumulation of long-lived plasma cells. *In vitro*, B-cells were unable to undergo CSR following activation with CD40L and appropriate cytokines. In terms of SHM, the frequency and nucleotide substitution pattern were normal but the peripheral blood CD27+ B-cell count was always low (see Table 16.2).

We had the opportunity to study two male patients with bi-allelic hypomorphic mutations in MLH1, the partner of PMS2. Although serum Ig levels were normal (other than a lack of IgG4), the number of circulating switched B-cells was low in both patients and *in vitro* CSR was weak but still detectable (our unpublished data). The mildness of this effect may be related to residual MLH1 expression and activity, which was nevertheless insufficient to protect the affected patients from cancer. Eight patients with mutations that completely abrogated the expression of MSH6 were also investigated; although *in vivo* susceptibility to infection was not reported, low serum IgG2 and IgG4 levels, low circulating switched B-cell counts and weak *in vitro* CSR were observed in all cases. Moreover, SHM exhibited an abnormal substitution pattern, with almost no mutations on A:T residues (which is reminiscent of data published in mice) (86, 98). Patients with MSH2 deficiency were not available for assessment – probably because of the very early onset and severity of cancer in this context (143).

As has been reported in mice (82, 84, 85, 133, 144), the MMR pathway has a role in antibody maturation in humans as well (81, 86). The MMR enzymes recognize mismatched nucleotides, provided that a nick on the same DNA strand is already present (145). One can thus hypothesize that the MMR enzymes act downstream of UNG-APE and repair the remaining U:G mismatches not processed by UNG. It appears that in the absence of a nick close to the mismatch site, PMS2 is able to create a nick through its endonuclease activity (146). Thus, the MMR enzymes in general and PMS2 in particular seem to play a specific role in the CSR-induced generation of DNA breaks in S regions (147). Our observation of a biased pattern of SHM in cases of MSH6 deficiency emphasizes the role of MutSα (but not MutLα) in this process.

The most prominent symptom of MMR deficiency is the occurrence of cancers in childhood. Nevertheless, the CSR-D (which appears to be present in all the patients studied to date – at least *in vitro*) can lead to infectious complications during chemotherapy. The presence of a CSR-D should be considered prior to the initiation of chemotherapy in patients with a diagnosed MMR deficiency. Conversely, MMR deficiency should be considered in patients with a CSR-D that lacks a molecular definition. A non-specific but suggestive symptom is the presence of cutaneous café au lait spots.

CSR-D IN ATAXIA TELANGIECTASIA

CSR-D as a complication of ataxia telangiectasia (AT) was first suspected several decades ago (6). Ataxia telangiectasia is caused by bi-allelic mutations in the *ATM* gene. A devastating disease, it combines progressive neurodegeneration (ataxia), cutaneous abnormalities (telangiectasia), a predisposition to malignancy and immunodeficiency (see Table 16.1 and Chapter 8). The

immunodeficiency may be the sole symptom for several years and is characterized by susceptibility to bacterial infections (especially of the respiratory tract) requiring Ig replacement therapy. In addition to a progressive T-cell defect, AT patients may present a CSR-D, as suggested by elevated IgM and very low IgG and IgA levels (148) and B-cells that are unable to undergo CSR *in vitro*. Analysis of the recombined switch junctions in Ig gene loci indicates a failure of DNA repair during CSR and suggests that the *ATM* gene product has a role in CSR-induced DSB repair in S regions (149). In contrast, the presence of normal SHM in the CD27+ B-cell subpopulation (which is normal in number) indicates that ATM is not essential for DNA repair in V regions (see Table 16.2).

The Ig CSR deficiency observed in AT is very similar to that caused by mutations located in the C terminal part of AID and/or to the defect observed in molecularly undefined CSR-D linked to an intrinsic B-cell defect (see above). Thus, it is imperative to screen for AT (with the α-fetoprotein assay) in molecularly undefined CSR-Ds with a normal CD27+ cell count and normal SHM.

OTHER Ig CSR-DEFICIENCIES ASSOCIATED WITH A DNA REPAIR DEFECT

As is the case for ATM, the MRE11/RAD50/NBS1 complex is involved in the repair of CSR-induced DSBs. Hence, a CSR-D can be observed in the "AT-like" disease due to *MRE11* mutations and Nijmegen breakage syndrome (due to *NBS1* mutations). A CSR-D associated with elevated radiosensitivity of fibroblasts has been reported in RIDDLE syndrome and is caused by mutations in the gene for ubiquitin ligase RNF168 (150). However, the CSR-D is rarely the pre-eminent symptom in these conditions. An Ig CSR-D phenotype has also been observed as part of a combined immunodeficiency related to leaky mutations in genes encoding NHEJ factors (such as Cernunnos, DNA ligase IV and Artemis) (151–154).

Strikingly, some patients with molecularly undefined CSR-Ds (seven cases in our cohort of 152 patients) developed non-EBV-induced B lymphomas; this observation suggests that these individuals may be affected by an as yet undetermined DNA repair defect (155). The *in vivo* CSR defect in these seven patients is severe, with variable *in vitro* CD40-induced B-cell responses and a persistently low circulating CD27$^+$ B-cell count. B-cell function could not be investigated further since the patients received chemotherapy. This observation should remind physicians that some CSR-Ds can be complicated by malignant manifestations and thus require careful follow up.

CONCLUSION AND PERSPECTIVES

The clinical description and molecular analysis of a large cohort of CSR-Ds carried out during the last 50 years indicate that these conditions are much more heterogeneous than was first expected. The clinical phenotype varies from very severe – as observed in T:B cell interaction deficiencies, to mild – as in cases of intrinsic B-cell defects with the caveat that autoimmunity and cancers can occur in some forms. Although several gene defects have been identified, a large proportion of CSR-Ds have yet to be defined in molecular terms. One can expect that the rapid development of new molecular techniques (such as whole-exome and whole-genome sequencing) will greatly accelerate the characterization of these latter conditions. However, gene analysis in CSR-Ds is even more complicated than anticipated. For example, mutations in the *AICDA* gene lead to different phenotypes and even different patterns of inheritance, depending on their location. While most CSR-Ds are monogenic, it is possible that some conditions have a more complex genetic background – as

observed in common variable immunodeficiency, a subset of which overlaps with CSR-Ds. Molecular characterization is essential for (1) establishing a firm diagnosis, (2) offering appropriate genetic counseling and accurate follow up and (3) opening up the way to specific therapeutic approaches such as the administration of recombinant soluble CD40L, which has already been safely performed in a few patients. Lastly, the detailed analysis of these conditions is essential for gaining a better understanding of antibody production and maturation.

References

1. Israel-Asselain R, Burtin P, Chebat J. [A new biological disorder: agammaglobulinemia with beta2-macroglobulinemia (a case)]. *Bull Mem Soc Med Hop Paris* 1960;**76**:519–23.
2. Rosen FS, Kevy SV, Merler E, Janeway CA, Gitlin D. Recurrent bacterial infections and dysgammaglobulinemia: deficiency of 7S gamma-globulins in the presence of elevated 19S gamma-globulins. Report of two cases. *Pediatrics* 1961;**28**:182–95.
3. Hong R, Schubert WK, Perrin EV, West CD. Antibody deficiency syndrome associated with beta-2 macroglobulinemia. *J Pediatr* 1962;**61**:831–42.
4. Ackerman BD. Dysgammaglobulinemia: Report of a case with a family history of a congenital gamma globulin disorder. *Pediatrics* 1964;**34**:211–9.
5. Stiehm ER, Fudenberg HH. Clinical and immunologic features of dysgammaglobulinemia type I. Report of a case diagnosed in the first year of life. *Am J Med* 1966;**40**: 805–15.
6. Goldman AS, Ritzmann SE, Houston EW, Sidwell S, Bratcher R, Levin WC. Dysgammaglobulinemic antibody deficiency syndrome. *J Pediatr* 1967;**70**:16–27.
7. Cooper MD, Faulk WP, Fudenberg HH, et al. Meeting report of the Second International Workshop on Primary Immunodeficiency Disease in Man held in St Petersburg, Florida, February, 1973. *Clin Immunol Immunopathol* 1974;**2**:416–45.
8. Cooper MD, Faulk WP, Fudenberg HH, et al. Classification of primary immunodeficiencies. *N Engl J Med* 1973;**288**:966–7.
9. Geha RS, Hyslop N, Alami S, Farah F, Schneeberger EE, Rosen FS. Hyper immunoglobulin M immunodeficiency. (Dysgammaglobulinemia). Presence of immunoglobulin M-secreting plasmacytoid cells in peripheral blood and failure of immunoglobulin M-immunoglobulin G switch in B-cell differentiation. *J Clin Invest* 1979;**64**:385–91.

10. Levitt D, Haber P, Rich K, Cooper MD. Hyper IgM immunodeficiency. A primary dysfunction of B lymphocyte isotype switching. *J Clin Invest* 1983;**72**:1650–7.

11. Mayer L, Posnett DN, Kunkel HG. Human malignant T-cells capable of inducing an immunoglobulin class switch. *J Exp Med* 1985;**161**:134–44.

12. Durandy A, Taubenheim N, Peron S, Fischer A. Pathophysiology of B-cell intrinsic immunoglobulin class switch recombination deficiencies. *Adv Immunol* 2007;**94**:275–306.

13. Rosen FS, Bougas JA. Acquired dysgammaglobulinemia: Elevation of the 19s gamma globulin and deficiency of the 7s gamma globulin in a woman with chronic progressive bronchiectasis. *N Engl J Med* 1963;**269**:1336–40.

14. Gleich GJ, Condemi JJ, Vaughan JH. Dysgammaglobulinemia in the presence of plasma cells. *N Engl J Med* 1965;**272**:331–40.

15. Brahmi Z, Lazarus KH, Hodes ME, Baehner RL. Immunologic studies of three family members with the immunodeficiency with hyper-IgM syndrome. *J Clin Immunol* 1983;**3**:127–34.

16. Bluestein HG, Pierce CW. Cellular requirements for development of primary anti-hapten antibody responses *in vitro*. *J Immunol* 1973;**111**:137–42.

17. Kapp JA, Pierce CW, Benacerraf B. Genetic control of immune responses *in vitro*. II. Cellular requirements for the development of primary plaque-forming cell responses to the random terpolymer 1-glutamic acid 60-1-alanine30-1-tyrosine10 (GAT) by mouse spleen cells *in vitro*. *J Exp Med* 1973;**138**:1121–32.

18. Mudawwar FB, Yunis EJ, Geha RS. Antigen-specific helper factor in man. *J Exp Med* 1978;**148**:1032–43.

19. Marrack PC, Kappler JW. Antigen-specific and non-specific mediators of T-cell/B-cell cooperation. I. Evidence for their production by different T-cells. *J Immunol* 1975;**114**:1116–25.

20. Ledbetter JA, Clark EA. Surface phenotype and function of tonsillar germinal center and mantle zone B-cell subsets. *Hum Immunol* 1986;**15**:30–43.

21. Rousset F, Garcia E, Banchereau J. Cytokine-induced proliferation and immunoglobulin production of human B lymphocytes triggered through their CD40 antigen. *J Exp Med* 1991;**173**:705–10.

22. Lederman S, Yellin MJ, Krichevsky A, Belko J, Lee JJ, Chess L. Identification of a novel surface protein on activated CD4+ T-cells that induces contact-dependent B-cell differentiation (help). *J Exp Med* 1992;**175**:1091–101.

23. Spriggs MK, Armitage RJ, Strockbine L, et al. Recombinant human CD40 ligand stimulates B-cell proliferation and immunoglobulin E secretion. *J Exp Med* 1992;**176**:1543–50.

24. Zan H, Cerutti A, Dramitinos P, Schaffer A, Li Z, Casali P. Induction of Ig somatic hypermutation and class switching in a human monoclonal IgM+ IgD+ B-cell line *in vitro*: definition of the requirements and modalities of hypermutation. *J Immunol* 1999;**162**:3437–47.

25. Butch AW, Chung GH, Hoffmann JW, Nahm MH. Cytokine expression by germinal center cells. *J Immunol* 1993;**150**:39–47.

26. Zheng B, Han S, Kelsoe G. T helper cells in murine germinal centers are antigen-specific emigrants that down-regulate Thy-1. *J Exp Med* 1996;**184**:1083–91.

27. Breitfeld D, Ohl L, Kremmer E, et al. Follicular B helper T-cells express CXC chemokine receptor 5, localize to B-cell follicles, and support immunoglobulin production. *J Exp Med* 2000;**192**:1545–52.

28. Ma CS, Deenick EK, Batten M, Tangye SG. The origins, function, and regulation of T follicular helper cells. *J Exp Med* 2012;**209**:1241–53.

29. Micol R, Kayal S, Mahlaoui N, et al. Protective effect of IgM against colonization of the respiratory tract by nontypeable *Haemophilus influenzae* in patients with hypogammaglobulinemia. *J Allergy Clin Immunol* 2012;**129**:770–7.

30. Durandy A, Schiff C, Bonnefoy JY, et al. Induction by anti-CD40 antibody or soluble CD40 ligand and cytokines of IgG, IgA and IgE production by B-cells from patients with X-linked hyper IgM syndrome. *Eur J Immunol* 1993;**23**:2294–9.

31. Agematsu K, Nagumo H, Shinozaki K, et al. Absence of IgD-CD27(+) memory B-cell population in X-linked hyper-IgM syndrome. *J Clin Invest* 1998;**102**:853–60.

32. Levy J, Espanol-Boren T, Thomas C, et al. Clinical spectrum of X-linked hyper-IgM syndrome. *J Pediatr* 1997;**131**:47–54.

33. Mensink EJ, Thompson A, Sandkuyl LA, et al. X-linked immunodeficiency with hyperimmunoglobulinemia M appears to be linked to the DXS42 restriction fragment length polymorphism locus. *Hum Genet* 1987;**76**:96–9.

34. Malcolm S, de Saint Basile G, Arveiler B, et al. Close linkage of random DNA fragments from Xq 21.3-22 to X-linked agammaglobulinaemia (XLA). *Hum Genet* 1987;**77**:172–4.

35. Graf D, Korthauer U, Mages HW, Senger G, Kroczek RA. Cloning of TRAP, a ligand for CD40 on human T-cells. *Eur J Immunol* 1992;**22**:3191–4.

36. Revy P, Geissmann F, Debre M, Fischer A, Durandy A. Normal CD40-mediated activation of monocytes and dendritic cells from patients with hyper-IgM syndrome due to a CD40 pathway defect in B-cells. *Eur J Immunol* 1998;**28**:3648–54.

37. Allen RC, Armitage RJ, Conley ME, et al. CD40 ligand gene defects responsible for X-linked hyper-IgM syndrome. *Science* 1993;**259**:990–3.

38. Korthauer U, Graf D, Mages HW, et al. Defective expression of T-cell CD40 ligand causes X-linked immunodeficiency with hyper-IgM. *Nature* 1993;**361**:539–41.

39. DiSanto JP, Bonnefoy JY, Gauchat JF, Fischer A, de Saint Basile G. CD40 ligand mutations in X-linked immunodeficiency with hyper-IgM. *Nature* 1993;**361**:541–3.

40. Ferrari S, Giliani S, Insalaco A, et al. Mutations of CD40 gene cause an autosomal recessive form of immunodeficiency with hyper IgM. *Proc Natl Acad Sci USA* 2001;**98**:12614–9.

41. Aruffo A, Farrington M, Hollenbaugh D, et al. The CD40 ligand, gp39, is defective in activated T-cells from patients with X-linked hyper-IgM syndrome. *Cell* 1993;**72**:291–300.

42. Lougaris V, Badolato R, Ferrari S, Plebani A. Hyper immunoglobulin M syndrome due to CD40 deficiency: clinical, molecular, and immunological features. *Immunol Rev* 2005;**203**:48–66.

43. Banchereau J, Bazan F, Blanchard D, et al. The CD40 antigen and its ligand. *Annu Rev Immunol* 1994;**12**:881–922.

44. Hollenbaugh D, Grosmaire LS, Kullas CD, et al. The human T-cell antigen gp39, a member of the TNF gene family, is a ligand for the CD40 receptor: expression of a soluble form of gp39 with B-cell co-stimulatory activity. *EMBO J* 1992;**11**:4313–21.

45. de Saint Basile G, Tabone MD, Durandy A, Phan F, Fischer A, Le Deist F. CD40 ligand expression deficiency in a female carrier of the X-linked hyper-IgM syndrome as a result of X chromosome lyonization. *Eur J Immunol* 1999;**29**:367–73.

46. Imai K, Shimadzu M, Kubota T, et al. Female hyper IgM syndrome type 1 with a chromosomal translocation disrupting CD40LG. *Biochim Biophys Acta* 2006;**1762**:335–40.

47. He B, Xu W, Santini PA, et al. Intestinal bacteria trigger T-cell-independent immunoglobulin A(2) class switching by inducing epithelial-cell secretion of the cytokine APRIL. *Immunity* 2007;**26**:812–26.

48. Weller S, Faili A, Aoufouchi S, et al. Hypermutation in human B-cells *in vivo* and *in vitro*. *Ann NY Acad Sci* 2003;**987**:158–65.

49. Scheeren FA, Nagasawa M, Weijer K, et al. T-cell-independent development and induction of somatic hypermutation in human IgM+ IgD+ CD27+ B-cells. *J Exp Med* 2008;**205**:2033–42.

50. Jain A, Kovacs JA, Nelson DL, et al. Partial immune reconstitution of X-linked hyper IgM syndrome with recombinant CD40 ligand. *Blood* 2011;**118**:3811–7.

51. Karaca NE, Forveille M, Aksu G, Durandy A, Kutukculer N. Hyper-immunoglobulin M syndrome type 3 with normal CD40 cell surface expression. *Scand J Immunol* 2012;**76**:21–5.

52. Al-Saud BK, Al-Sum Z, Alassiri H, et al. Clinical, immunological, and molecular characterization of hyper-IgM syndrome due to CD40 deficiency in eleven patients. *J Clin Immunol* 2013. DOI: 10.1007/s10875-013-9951-9.

53. Zonana J, Elder ME, Schneider LC, et al. A novel X-linked disorder of immune deficiency and hypohidrotic ectodermal dysplasia is allelic to incontinentia pigmenti and due to mutations in IKK-gamma (NEMO). *Am J Hum Genet* 2000;**67**:6.

54. Doffinger R, Smahi A, Bessia C, et al. X-linked anhidrotic ectodermal dysplasia with immunodeficiency is caused by impaired NF-kappaB signaling. *Nat Genet* 2001;**27**:277–85.

55. Jain A, Ma CA, Liu S, Brown M, Cohen J, Strober W. Specific missense mutations in NEMO result in hyper-IgM syndrome with hypohydrotic ectodermal dysplasia. *Nat Immunol* 2001;**2**:223–8.

56. Hanson EP, Monaco-Shawver L, Solt LA, et al. Hypomorphic nuclear factor-kappaB essential modulator mutation database and reconstitution system identifies phenotypic and immunologic diversity. *J Allergy Clin Immunol* 2008;**122**:1169–77.

57. Jain A, Ma CA, Lopez-Granados E, et al. Specific NEMO mutations impair CD40-mediated c-Rel activation and B-cell terminal differentiation. *J Clin Invest* 2004;**114**:1593–602.

58. Courtois G, Smahi A, Reichenbach J, et al. A hypermorphic IkappaBalpha mutation is associated with autosomal dominant anhidrotic ectodermal dysplasia and T-cell immunodeficiency. *J Clin Invest* 2003;**112**:1108–15.

59. Lopez-Granados E, Keenan JE, Kinney MC, et al. A novel mutation in NFKBIA/IKBA results in a degradation-resistant N-truncated protein and is associated with ectodermal dysplasia with immunodeficiency. *Hum Mutat* 2008;**29**:861–8.

60. McDonald DR, Mooster JL, Reddy M, Bawle E, Secord E, Geha RS. Heterozygous N-terminal deletion of IkappaBalpha results in functional nuclear factor kappaB haploinsufficiency, ectodermal dysplasia, and immune deficiency. *J Allergy Clin Immunol* 2007;**120**:900–7.

61. Ohnishi H, Miyata R, Suzuki T, et al. A rapid screening method to detect autosomal-dominant ectodermal dysplasia with immune deficiency syndrome. *J Allergy Clin Immunol* 2012;**129**:578–5780.

62. Schimke LF, Rieber N, Rylaarsdam S, et al. A novel gain-of-function IKBA mutation underlies ectodermal dysplasia with immunodeficiency and polyendocrinopathy. *J Clin Immunol* 2013;**33**:1088–99.

63. Grimbacher B, Hutloff A, Schlesier M, et al. Homozygous loss of ICOS is associated with adult-onset common variable immunodeficiency. *Nat Immunol* 2003;**4**:261–8.

64. Warnatz K, Bossaller L, Salzer U, et al. Human ICOS deficiency abrogates the germinal center reaction and provides a monogenic model for common variable immunodeficiency. *Blood* 2006;**107**:3045–52.

65. Bossaller L, Burger J, Draeger R, et al. ICOS deficiency is associated with a severe reduction of CXCR5+CD4 germinal center Th cells. *J Immunol* 2006;**177**:4927–32.

66. Takahashi N, Matsumoto K, Saito H, et al. Impaired CD4 and CD8 effector function and decreased memory T-cell populations in ICOS-deficient patients. *J Immunol* 2009;**182**:5515–27.

67. Kawamura N, Okamura A, Furuta H, et al. Improved dysgammaglobulinaemia in congenital rubella syndrome after immunoglobulin therapy: correlation with CD154 expression. *Eur J Pediatr* 2000;**159**:764–6.

68. Nonoyama S, Etzioni A, Toru H, et al. Diminished expression of CD40 ligand may contribute to the defective humoral immunity in patients with MHC class II deficiency. *Eur J Immunol* 1998;**28**:589–98.

69. Honjo T, Kataoka T. Organization of immunoglobulin heavy chain genes and allelic deletion model. *Proc Natl Acad Sci USA* 1978;**75**:2140–4.

70. Iwasato T, Shimizu A, Honjo T, Yamagishi H. Circular DNA is excised by immunoglobulin class switch recombination. *Cell* 1990;**62**:143–9.

71. Matsuoka M, Yoshida K, Maeda T, Usuda S, Sakano H. Switch circular DNA formed in cytokine-treated mouse splenocytes: evidence for intramolecular DNA deletion in immunoglobulin class switching. *Cell* 1990;**62**:135–42.

72. von Schwedler U, Jack HM, Wabl M. Circular DNA is a product of the immunoglobulin class switch rearrangement. *Nature* 1990;**345**:452–6.

73. Kinoshita K, Honjo T. Unique and unprecedented recombination mechanisms in class switching. *Curr Opin Immunol* 2000;**12**:195–8.

74. Manis JP, Tian M, Alt FW. Mechanism and control of class-switch recombination. *Trends Immunol* 2002;**23**:31–9.

75. Nikaido T, Yamawaki-Kataoka Y, Honjo T. Nucleotide sequences of switch regions of immunoglobulin C epsilon and C gamma genes and their comparison. *J Biol Chem* 1982;**257**:7322–9.

76. Casellas R, Nussenzweig A, Wuerffel R, et al. Ku80 is required for immunoglobulin isotype switching. *EMBO J* 1998;**17**:2404–11.

77. Manis JP, Dudley D, Kaylor L, Alt FW. IgH class switch recombination to IgG1 in DNA-PKcs-deficient B-cells. *Immunity* 2002;**16**:607–17.

78. Manis JP, Gu Y, Lansford R, et al. Ku70 is required for late B-cell development and immunoglobulin heavy chain class switching. *J Exp Med* 1998;**187**:2081–9.

79. Yan CT, Boboila C, Souza EK, et al. IgH class switching and translocations use a robust non-classical end-joining pathway. *Nature* 2007;**449**:478–82.

80. Boboila C, Yan C, Wesemann DR, et al. Alternative end-joining catalyzes class switch recombination in the absence of both Ku70 and DNA ligase 4. *J Exp Med* 2010;**207**:417–27.

81. Peron S, Metin A, Gardes P, et al. Human PMS2 deficiency is associated with impaired immunoglobulin class switch recombination. *J Exp Med* 2008;**205**:2465–72.

82. Schrader CE, Vardo J, Stavnezer J. Role for mismatch repair proteins Msh2, Mlh1, and Pms2 in immunoglobulin class switching shown by sequence analysis of recombination junctions. *J Exp Med* 2002;**195**:367–73.

83. Schrader CE, Edelmann W, Kucherlapati R, Stavnezer J. Reduced isotype switching in splenic B-cells from mice deficient in mismatch repair enzymes. *J Exp Med* 1999;**190**:323–30.

84. Ehrenstein MR, Rada C, Jones AM, Milstein C, Neuberger MS. Switch junction sequences in PMS2-deficient mice reveal a microhomology-mediated mechanism of Ig class switch recombination. *Proc Natl Acad Sci USA* 2001;**98**:14553–8.

85. Ehrenstein MR, Neuberger MS. Deficiency in Msh2 affects the efficiency and local sequence specificity of immunoglobulin class-switch recombination: parallels with somatic hypermutation. *EMBO J* 1999;**18**:3484–90.

86. Gardes P, Forveille M, Alyanakian MA, et al. Human MSH6 deficiency is associated with impaired antibody maturation. *J Immunol* 2012;**188**:2023–9.

87. Weigert MG, Cesari IM, Yonkovich SJ, Cohn M. Variability in the lambda light chain sequences of mouse antibody. *Nature* 1970;**228**:1045–7.

88. Bothwell AL, Paskind M, Reth M, Imanishi-Kari T, Rajewsky K, Baltimore D. Heavy chain variable region contribution to the NPb family of antibodies: somatic mutation evident in a gamma 2a variable region. *Cell* 1981;**24**:625–37.

89. Rogerson B, Hackett Jr J, Peters A, Haasch D, Storb U. Mutation pattern of immunoglobulin transgenes is compatible with a model of somatic hypermutation in which targeting of the mutator is linked to the direction of DNA replication. *EMBO J* 1991;**10**:4331–41.

90. Bernard O, Hozumi N, Tonegawa S. Sequences of mouse immunoglobulin light chain genes before and after somatic changes. *Cell* 1978;**15**:1133–44.

91. Bothwell AL, Paskind M, Reth M, Imanishi-Kari T, Rajewsky K, Baltimore D. Somatic variants of murine immunoglobulin lambda light chains. *Nature* 1982;**298**:380–2.

92. Liu YJ, Joshua DE, Williams GT, Smith CA, Gordon J, MacLennan IC. Mechanism of antigen-driven selection in germinal centres. *Nature* 1989;**342**:929–31.

93. Gearhart PJ, Johnson ND, Douglas R, Hood L. IgG antibodies to phosphorylcholine exhibit more diversity than their IgM counterparts. *Nature* 1981;**291**:29–34.

94. Storb U, Peters A, Klotz E, et al. Cis-acting sequences that affect somatic hypermutation of Ig genes. *Immunol Rev* 1998;**162**:153–60.

95. Jacobs H, Bross L. Towards an understanding of somatic hypermutation. *Curr Opin Immunol* 2001;**13**:208–18.

96. Delbos F, Aoufouchi S, Faili A, Weill JC, Reynaud CA. DNA polymerase eta is the sole contributor of A/T modifications during immunoglobulin gene hypermutation in the mouse. *J Exp Med* 2007;**204**:17–23.

97. Wiesendanger M, Kneitz B, Edelmann W, Scharff MD. Somatic hypermutation in MutS homologue (MSH)3-, MSH6-, and MSH3/MSH6-deficient mice reveals a role for the MSH2-MSH6 heterodimer in modulating the base substitution pattern. *J Exp Med* 2000;**191**:579–84.

98. Martomo SA, Yang WW, Gearhart PJ. A role for Msh6 but not Msh3 in somatic hypermutation and class switch recombination. *J Exp Med* 2004;**200**:61–8.

99. Rada C, Ehrenstein MR, Neuberger MS, Milstein C. Hot spot focusing of somatic hypermutation in MSH2-deficient mice suggests two stages of mutational targeting. *Immunity* 1998;**9**:135–41.

100. Rajewsky K. Clonal selection and learning in the antibody system. *Nature* 1996;**381**:751–8.

101. Frazer JK, LeGros J, de Bouteiller O, et al. Identification and cloning of genes expressed by human tonsillar B lymphocyte subsets. *Ann NY Acad Sci* 1997;**815**:316–8.

102. Honjo T, Kinoshita K, Muramatsu M. Molecular mechanism of class switch recombination: linkage with somatic hypermutation. *Annu Rev Immunol* 2002;**20**:165–96.

103. Kaartinen M, Griffiths GM, Markham AF, Milstein C. mRNA sequences define an unusually restricted IgG response to 2-phenyloxazolone and its early diversification. *Nature* 1983;**304**:320–4.

104. Quartier P, Bustamante J, Sanal O, et al. Clinical, immunologic and genetic analysis of 29 patients with autosomal recessive hyper-IgM syndrome due to activation-induced cytidine deaminase deficiency. *Clin Immunol* 2004;**110**:22–9.

105. Durandy A, Hivroz C, Mazerolles F, et al. Abnormal CD40-mediated activation pathway in B lymphocytes from patients with hyper-IgM syndrome and normal CD40 ligand expression. *J Immunol* 1997;**158**:2576–84.

106. Muramatsu M, Sankaranand VS, Anant S, et al. Specific expression of activation-induced cytidine deaminase (AID), a novel member of the RNA-editing deaminase family in germinal center B-cells. *J Biol Chem* 1999;**274**:18470–6.

107. Revy P, Muto T, Levy Y, et al. Activation-induced cytidine deaminase (AID) deficiency causes the autosomal recessive form of the hyper-IgM syndrome (HIGM2). *Cell* 2000;**102**:565–75.

108. Muramatsu M, Kinoshita K, Fagarasan S, Yamada S, Shinkai Y, Honjo T. Class switch recombination and hypermutation require activation-induced cytidine deaminase (AID), a potential RNA editing enzyme. *Cell* 2000;**102**:553–63.

109. Catalan N, Selz F, Imai K, Revy P, Fischer A, Durandy A. The block in immunoglobulin class switch recombination caused by activation-induced cytidine deaminase deficiency occurs prior to the generation of DNA double strand breaks in switch mu region. *J Immunol* 2003;**171**:2504–9.

110. Kinoshita K, Honjo T. Linking class-switch recombination with somatic hypermutation. *Nat Rev Mol Cell Biol* 2001;**2**:493–503.

111. Rada C, Jarvis JM, Milstein C. AID-GFP chimeric protein increases hypermutation of Ig genes with no evidence of nuclear localization. *Proc Natl Acad Sci USA* 2002;**99**:7003–8.

112. Petersen-Mahrt SK, Harris RS, Neuberger MS. AID mutates E. coli suggesting a DNA deamination mechanism for antibody diversification. *Nature* 2002;**418**:99–104.

113. Bransteitter R, Pham P, Scharff MD, Goodman MF. Activation-induced cytidine deaminase deaminates deoxycytidine on single-stranded DNA but requires the action of RNase. *Proc Natl Acad Sci USA* 2003;**100**:4102–7.

114. Chaudhuri J, Tian M, Khuong C, Chua K, Pinaud E, Alt FW. Transcription-targeted DNA deamination by the AID antibody diversification enzyme. *Nature* 2003;**422**:726–30.

115. Dickerson SK, Market E, Besmer E, Papavasiliou FN. AID mediates hypermutation by deaminating single stranded DNA. *J Exp Med* 2003;**197**:1291–6.

116. Ramiro AR, Stavropoulos P, Jankovic M, Nussenzweig MC. Transcription enhances AID-mediated cytidine deamination by exposing single-stranded DNA on the nontemplate strand. *Nat Immunol* 2003;**4**:452–6.

117. Nambu Y, Sugai M, Gonda H, et al. Transcription-coupled events associating with immunoglobulin switch region chromatin. *Science* 2003;**302**:2137–40.

118. Pavri R, Gazumyan A, Jankovic M, et al. Activation-induced cytidine deaminase targets DNA at sites of RNA polymerase II stalling by interaction with Spt5. *Cell* 2010;**143**:122–33.

119. Basu U, Meng FL, Keim C, et al. The RNA exosome targets the AID cytidine deaminase to both strands of transcribed duplex DNA substrates. *Cell* 2011;**144**:353–63.

120. Patenaude AM, Orthwein A, Hu Y, et al. Active nuclear import and cytoplasmic retention of activation-induced deaminase. *Nat Struct Mol Biol* 2009;**16**:517–27.

121. Aoufouchi S, Faili A, Zober C, et al. Proteasomal degradation restricts the nuclear lifespan of AID. *J Exp Med* 2008;**205**:1357–68.

122. Durandy A, Peron S, Taubenheim N, Fischer A. Activation-induced cytidine deaminase: structure-function relationship as based on the study of mutants. *Hum Mutat* 2006;**27**:1185–91.

123. Ta VT, Nagaoka H, Catalan N, et al. AID mutant analyses indicate requirement for class-switch-specific cofactors. *Nat Immunol* 2003;**4**:843–8.

124. Geisberger R, Rada C, Neuberger MS. The stability of AID and its function in class-switching are critically sensitive to the identity of its nuclear-export sequence. *Proc Natl Acad Sci USA* 2009;**106**:6736–41.

125. Barreto V, Reina-San-Martin B, Ramiro AR, McBride KM, Nussenzweig MC. C-terminal deletion of AID uncouples class switch recombination from somatic hypermutation and gene conversion. *Mol Cell* 2003;**12**:501–8.

126. Doi T, Kato L, Ito S, et al. The C-terminal region of activation-induced cytidine deaminase is responsible for a recombination function other than DNA cleavage in class switch recombination. *Proc Natl Acad Sci USA* 2009;**106**:2758–63.

127. Kracker S, Imai K, Gardes P, Ochs HD, Fischer A, Durandy AH. Impaired induction of DNA lesions during immunoglobulin class-switch recombination in humans influences end-joining repair. *Proc Natl Acad Sci USA* 2010;**107**:22225–30.

128. Imai K, Zhu Y, Revy P, et al. Analysis of class switch recombination and somatic hypermutation in patients affected with autosomal dominant hyper-IgM syndrome type 2. *Clin Immunol* 2005;**115**:277–85.

129. Takizawa M, Tolarova H, Li Z, et al. AID expression levels determine the extent of cMyc oncogenic translocations and the incidence of B-cell tumor development. *J Exp Med* 2008;**205**:1949–57.

130. Rada C, Williams GT, Nilsen H, Barnes DE, Lindahl T, Neuberger MS. Immunoglobulin isotype switching is inhibited and somatic hypermutation perturbed in UNG-deficient mice. *Curr Biol* 2002;**12**:1748–55.

131. Imai K, Slupphaug G, Lee WI, et al. Human uracil-DNA glycosylase deficiency associated with profoundly impaired immunoglobulin class-switch recombination. *Nat Immunol* 2003;**4**:1023–8.

132. aan de Kerk DJ, Jansen MH, ten Berge IJ, van Leeuwen EM, Kuijpers TW. Identification of B-cell defects using age-defined reference ranges for *in vivo* and *in vitro* B-cell differentiation. *J Immunol* 2013;**190**:5012–9.

133. Rada C, Di Noia JM, Neuberger MS. Mismatch recognition and uracil excision provide complementary paths to both Ig switching and the A/T-focused phase of somatic mutation. *Mol Cell* 2004;**16**:163–71.

134. Di Noia JM, Rada C, Neuberger MS. SMUG1 is able to excise uracil from immunoglobulin genes: insight into mutation versus repair. *EMBO J* 2006;**25**:585–95.

135. Nilsen H, Stamp G, Andersen S, et al. Gene-targeted mice lacking the Ung uracil-DNA glycosylase develop B-cell lymphomas. *Oncogene* 2003;**22**:5381–6.

136. Endres M, Biniszkiewicz D, Sobol RW, et al. Increased postischemic brain injury in mice deficient in uracil-DNA glycosylase. *J Clin Invest* 2004;**113**:1711–21.

137. Guikema JE, Linehan EK, Tsuchimoto D, et al. APE1- and APE2-dependent DNA breaks in immunoglobulin class switch recombination. *J Exp Med* 2007;**204**:3017–26.

138. Niessen RC, Berends MJ, Wu Y, et al. Identification of mismatch repair gene mutations in young patients with colorectal cancer and in patients with multiple tumours associated with hereditary non-polyposis colorectal cancer. *Gut* 2006;**55**:1781–8.

139. Nakagawa H, Lockman JC, Frankel WL, et al. Mismatch repair gene PMS2: disease-causing germline mutations are frequent in patients whose tumors stain negative for PMS2 protein, but paralogous genes obscure mutation detection and interpretation. *Cancer Res* 2004;**64**:4721–7.

140. Ostergaard JR, Sunde L, Okkels H. Neurofibromatosis von Recklinghausen type I phenotype and early onset of cancers in siblings compound heterozygous for mutations in MSH6. *Am J Med Genet A* 2005;**139**:96–105 discussion 96.

141. Whiteside D, McLeod R, Graham G, et al. A homozygous germ-line mutation in the human MSH2 gene predisposes to hematological malignancy and multiple cafe-au-lait spots. *Cancer Res* 2002;**62**:359–62.

142. Scott RH, Mansour S, Pritchard-Jones K, Kumar D, MacSweeney F, Rahman N. Medulloblastoma, acute myelocytic leukemia and colonic carcinomas in a child with biallelic MSH6 mutations. *Nat Clin Pract Oncol* 2007; **4**:130–4.

143. Bougeard G, Charbonnier F, Moerman A, et al. Early onset brain tumor and lymphoma in MSH2-deficient children. *Am J Hum Genet* 2003;**72**:213–6.

144. Schrader CE, Vardo J, Stavnezer J. Mlh1 can function in antibody class switch recombination independently of Msh2. *J Exp Med* 2003;**197**:1377–83.

145. Kadyrov FA, Dzantiev L, Constantin N, Modrich P. Endonucleolytic function of MutLalpha in human mismatch repair. *Cell* 2006;**126**:297–308.

146. van Oers JM, Roa S, Werling U, et al. PMS2 endonuclease activity has distinct biological functions and is essential for genome maintenance. *Proc Natl Acad Sci USA* 2010;**107**(30):13384–9.

147. Stavnezer J, Schrader CE. Mismatch repair converts AID-instigated nicks to double-strand breaks for antibody class-switch recombination. *Trends Genet* 2006;**22**:23–8.

148. Etzioni A, Ben-Barak A, Peron S, Durandy A. Ataxia-telangiectasia in twins presenting as autosomal recessive hyper-immunoglobulin M syndrome. *Isr Med Assoc J* 2007;**9**:406–7.

149. Meyts I, Weemaes C, De Wolf-Peeters C, et al. Unusual and severe disease course in a child with ataxia-telangiectasia. *Pediatr Allergy Immunol* 2003;**14**:330–3.

150. Soresina A, Meini A, Lougaris V, et al. Different clinical and immunological presentation of ataxia-telangiectasia within the same family. *Neuropediatrics* 2008;**39**:43–5.

151. Buck D, Malivert L, de Chasseval R, et al. Cernunnos, a novel nonhomologous end-joining factor, is mutated in human immunodeficiency with microcephaly. *Cell* 2006;**124**:287–99.

152. Pan-Hammarstrom Q, Jones AM, Lahdesmaki A, et al. Impact of DNA ligase IV on nonhomologous end joining pathways during class switch recombination in human cells. *J Exp Med* 2005;**201**:189–94.

153. Du L, van der Burg M, Popov SW, et al. Involvement of Artemis in nonhomologous end-joining during immunoglobulin class switch recombination. *J Exp Med* 2008;**205**:3031–40.

154. O'Driscoll M, Cerosaletti KM, Girard PM, et al. DNA ligase IV mutations identified in patients exhibiting developmental delay and immunodeficiency. *Mol Cell* 2001;**8**:1175–85.

155. Peron S, Pan-Hammarstrom Q, Imai K, et al. A primary immunodeficiency characterized by defective immunoglobulin class switch recombination and impaired DNA repair. *J Exp Med* 2007;**204**:1207–16.

Unraveling the Complement System and its Mechanism of Action

Michael M. Frank

Duke University Medical Center, Durham, NC, USA

OUTLINE

INTRODUCTION

The progress we have made in understanding the complement system parallels the history of immunology and its origins in the field of microbiology. Observations made in the latter half of

the 19th century clarified the role of microbes in causing human infectious diseases. Vaccination had been known to be effective since the late 18th century, and immunization appeared to be a likely method of protecting against infectious disease.

Primary Immunodeficiency Disorders: A Historic and Scientific Perspective

The concept of antimicrobials had not yet been developed, but it was clear that individuals exposed to attenuated microorganisms often became immune to that organism. It was natural to ask the question: "What is responsible for that protection?" The work of Ehrlich, Metchnikoff and a large number of others had led to an understanding that both antibodies and cells were important in such protection (see Chapter 1). There was, however, controversy as to which of these provided the most important element of host defense.

EARLY DISCOVERY: HEAT LABILE ALEXIN AND HEAT STABLE ANTIBODY

As early as 1874, Traube and Gscheidlen had shown that microorganisms injected into the circulation are rapidly destroyed and the bloodstream maintains its sterility (1). The concept of bloodstream clearance of organisms was developed by these investigators and others (2). By 1884, Grohmann had noted that cell free serum *in vitro* is capable of killing microorganisms (3). Nuttall, in 1888, also inoculated defribinated blood with bacteria and showed that outside of the body, serum retained its bactericidal activity, although activity was lost after a period of storage and the activity was heat labile (4). Buchner, in 1889, is generally credited with the first expression of the humoral theory of immunity: that a principal in fresh blood that he termed alexin (protective substance) was capable of killing bacteria. He thought that the alexin acted as an enzyme in destroying the bacteria, but he did not differentiate antibody from alexin (5). It is interesting that Baumgartner presented the theory, not taken seriously, that alexin damaged the membrane of organisms, causing an osmotic death (6).

In 1894, Pfeiffer began a series of experiments that revolutionized our understanding of the bactericidal action of serum (7). He reported that

injection of cholera vibrios into the peritoneum of immunized guinea pigs led to rapid dissolution of the organisms, which was not observed when injected into non-immunized control animals. In controls, the organisms multiplied and killed the guinea pig, but serum from an immunized animal, if injected into controls, conferred protection. Heated serum also conferred protection, although it had lost all alexin activity. It was quickly shown by Bordet and Metschnikoff that these phenomena could be duplicated *in vitro*, and Bordet reported that cells were not required for the bactericidal effect (8,9). Bordet extended the concept of alexin by clearly identifying the heat-stable, specific antibody that recognizes the antigen and the heat labile, non-specific activity, present in all normal sera, that had no activity by itself, but was lytic in the presence of antibody (10). Bordet also noted that a similar set of phenomena govern the lysis of red blood cells and introduced the concept of sensitization of the cells by immune serum. He noted that sensitized red cells were lysed by fresh serum and realized that this represented the release of hemoglobin and the formation of erythrocyte ghosts that could be recovered from the reaction mixture. Others made the same observations independently. Bordet was awarded the Nobel prize for this series of observations.

HEMOLYTIC COMPLEMENT

Ehrlich and Morgenroth extended these studies and presented their now famous side chain theory as an explanation (11). They believed that immune cells had on their surface specific receptors that recognize antigen. On further immunization with antigen, more of these specific receptors (side chains) were formed and were shed from the cells to circulate in the blood. These receptors had a specific antigen-recognition site (haptophore group) that bound to antigen. They also had a specific complement-binding site (complementophore group) that activated complement after the receptors had bound to antigen. Ehrlich conceived of complement acting like a bacterial toxin to damage the cells. He used the term complement rather than alexin to emphasize the fact that the fresh serum factor "complemented" the specific receptors. He used the term amboceptor to emphasize the two types of recognition sites on the antibody ("side chain") molecule.

There was much confusion over the fact that complement appeared more or less lytic in different systems, and the complement from different species differed in its ability to lyse various targets. Ehrlich postulated that each specific amboceptor had its own complement, and Bordet believed that there was only one type of complement. As early as 1909, Muir observed that complement of a given species tended to be ineffective in lysing the cells of that species but were lytic for cells of other species (12). At that time, the reason was unclear. We now know that there are species-specific complement regulatory molecules that downregulate the lysis of one's own cells. Nevertheless, by 1910, it was known that alexin or complement activity depended on specific antibody and immunologically non-specific protein factors. The concept that antibody, together with complement, promoted the killing of these organisms was established, and the first studies of complement chemistry were yielding results.

In the early 1900s, the first attempts at protein separation revealed that complement was not a single substance but consisted of multiple proteins. Buchner separated serum into water-soluble pseudoglobulin and water-insoluble euglobulin. He showed that the pseudoglobulin and euglobulin fractions each no longer had bactericidal activity (13). Ferrata confirmed that neither the pseudoglobulin nor euglobulin fractions of serum maintained hemolytic activity (14). When these fractions were mixed together, complement activity was restored. It was possible to examine the action of these fractions sequentially and classify these fractions into "mid-piece" and "end-piece", fractions that reacted sequentially to complete the lytic reaction.

Ehrlich's laboratory focused on the lysis of sheep erythrocytes sensitized with rabbit antibody in the presence of guinea pig serum. The reason for choosing this strange mixture of reagents is apparent. His laboratory had noted empirically that sheep erythrocytes are exquisitely sensitive to lysis by rabbit anti-sheep erythrocyte antibody and complement. We now know that sheep erythrocytes have on their surface a lipopolysaccharide antigen, the Forssman antigen, to which rabbits make large amounts of IgM and IgG hemolytic antibody. Sheep erythrocytes differ from those of most other species in that they are smaller, more spherical, and therefore highly sensitive to lysis by antibody and complement. The hemolytic titer of fresh guinea pig serum, as reflected in the ability to lyse antibody sensitized sheep erythrocytes, is higher than that of most other species. Thus, it was possible, using reagents easily available in the laboratory at the turn of the century, to develop a highly sensitive system for analyzing complement.

Brand, examining red cell lysis, recognized that complement action could be defined by a series of cellular intermediates important in hemolysis (15). By 1919, it was believed that complement consisted of at least three factors. This belief depended upon the earlier observation in 1903 by Flexner and Noguchi that cobra venom could destroy complement activity (16). Both mid-piece and end-piece contained factors that could restore this activity. By 1913, Weil had prepared an erythrocyte–antibody–mid-piece intermediate that could be hemolyzed by the addition of heated serum (17). This finding was confirmed by Nathan, who showed that as these cells were hemolyzed, the third component was consumed (18). Ueno, in 1938, working in isolation in Japan, described a cellular intermediate product consisting of erythrocytes, antibody, C1, C4 and end-piece. His experiments established the order of the complement cascade (19). As early as 1914, Coca reported that what was then called the third component of complement was capable of combining with yeast, and that yeast could be used to remove the third component from serum (20). This early work by Coca allowed for the experiments performed in the 1950s by Pillemer that led to the discovery of the alternative pathway (21).

Investigators at that time were interested primarily in how human beings protect themselves against infection. In retrospect, it is clear that the system chosen for these studies, consisting of sheep erythrocytes, rabbit antibody and guinea pig serum, differed considerably from the systems that they were interested in understanding. However, the assumption was made that discoveries in this model would be immediately applicable to the systems of particular interest. It was much easier to study the lysis of sheep erythrocytes since, at the end of the reaction time, the cells could be sedimented and the hemoglobin in the supernatant easily determined. If one studied the lysis of bacteria, one had to perform colony counts at the end of the reaction to determine whether the bacteria had been killed, which is a more laborious process. Therefore, they missed the existence of a second or alternative pathway. Moreover, investigators focused on what bound to the sheep erythrocyte or bacteria, neglecting what was left behind in the supernatant. They missed the important fact that peptides left behind following complement activation have important inflammatory properties.

THE ISOLATION OF COMPLEMENT COMPONENTS

The serum fraction termed mid-piece contained a protein that could interact with erythrocyte-bound hemolytic antibody. This protein was termed C1. Lepow's group demonstrated that C1 was a three-part molecule, consisting of C1q, C1r and C1s held together in the presence of ionic calcium (22). Both by electron microscopy and by direct structural examination, C1q

was found to resemble a group of six tulips held together at the base by a core protein. Each tulip head was capable of binding to the Fc portion of an IgG or IgM molecule (23). In 1926, Gordon et al. reported the existence of the fourth component of complement, showing that ammonia destroyed a heat-stable factor in serum that was distinct from the third component (24). By the 1950s, it was believed that there were four components of complement but, in the late 1950s, M. M. Mayer's group developed a mathematical model of complement action that suggested that the last acting complement protein, termed C3, might actually be a complex of multiple protein components (25). This proved to be the case (26). Shortly after, again using the approach of mathematical modeling, Borsos and Rapp performed elegant experiments that indicated that a single molecule of IgM antibody was sufficient to activate the classical complement pathway to destroy one sheep erythrocyte, but a doublet of complement activating IgG antibody was required to provide the same activity (27).

Nelson et al., using laborious hemolytic assays and techniques of simple column chromatography, reported the existence of at least six separate factors that together made up C3 (28). Muller-Eberhard and his colleagues, using more sophisticated techniques for protein isolation, began the arduous task of separating these proteins and characterizing them individually. His laboratory isolated many of the human proteins, studied their levels in various diseases and examined their physiological function (29). This led to a massive expansion in our understanding of these proteins in disease. As early biochemical methods became available for the separation of proteins, they were applied to complement-system proteins, and the proteins were separated. Mathematical models were applied in the 1950s to analyze the protein reactions and to clarify the steps in the sequence. With the advent of modern molecular biology, the focus turned to sequencing the proteins, determining their chromosomal location and examining their structure.

Humphrey and Dourmashkin, in the 1960s, using electron microscopy, first demonstrated complement lesions in red cells (30). Ultimately, this finding confirmed the prediction of Mayer and his colleagues that a single hit, comprised of antibody and all of the relevant proteins of the complement sequence, was sufficient to produce a lesion in the membrane of a sheep erythrocyte (2). The lesion had a hydrophobic outer surface and a hydrophilic core that allowed the free transport of ions across the cell membrane. This hole or donut-like lesion in the cell surface destabilized the cell's osmotic equilibrium and induced lysis.

THE ALTERNATIVE COMPLEMENT PATHWAY

In 1954, Pillemer and his colleagues reported the results of a startling series of experiments (21). In studying the ability of a Baker's yeast fraction, zymosan, to interact with serum, they noted that this insoluble polysaccharide interacted with the late-acting complement proteins without appearing to utilize the earlier proteins in the sequence. The zymosan appeared to interact with a previously undescribed serum protein that they termed properdin. Properdin-mediated C3 activation occurred in the absence of antibody, but did require the presence of a series of cofactors that they termed properdin factor A and properdin factor B. Factor A was ammonia and hydrazine sensitive and heat stable. Factor B was stable to ammonia, but was heat labile. Pillemer believed that these properdin factors were different from the previously described proteins of the classical complement pathway.

Pillemer's group found that at a temperature of 17°C, properdin could interact with zymosan without consuming complement (21). The mixture of zymosan and serum could be centrifuged, leading to properdin-depleted serum in which complement was no longer activated by zymosan. However, properdin could be eluted from the zymosan, reintroduced into the serum and, at 37°C, zymosan could again activate the complement. Thus, Pillemer believed that he had discovered a new complement pathway, not requiring antibody, but requiring properdin and the late-acting proteins of complement.

It was now possible to study the ability of various microorganisms to be damaged by serum from which properdin had been removed by absorption. A series of papers followed that suggested that there was a wide variety of microorganisms whose destruction depended on the so-called properdin pathway. These included bacteria, viruses such as Newcastle disease virus, and even erythrocytes from patients with the disease paroxysmal nocturnal hemoglobinuria (31). Properdin was thought to play an important role in non-specific immunity against infection.

In 1968, Nelson suggested a second interpretation of Pillemer's data (32). He suggested that properdin was nothing more than IgM antibody to zymosan and that this antibody was required for zymosan to interact with complement. If the IgM antibody to zymosan were removed by absorption, zymosan would no longer interact with complement. If the antibody were restored, the interaction would again take place. The only additional concept needed was that a wide variety of organisms might have antigens on their surface that cross-reacted with the polysaccharides important in the antigenicity of zymosan. Thus, absorption might remove an IgM antibody that cross-reacted with a variety of microbial antigens and that was required for complement activation. The arguments seemed compelling, since IgM was only then being identified, and the properdin theory went into eclipse for over a decade. However, several of Pillemer's students, including Lepow and Pensky, continued to follow up Pillemer's original observations, identifying properdin as a protein and showing that it was unique and not identical to IgM (33).

The later history of the alternative pathway followed other paths of development. In 1963,

Schur and Becker suggested that F(ab)'2 fragments of rabbit antibody in the form of antigen–antibody complexes could fix or activate complement (34). Since C1 binding requires Fc fragment of antibody, this was a surprising observation. Moreover, the F(ab)'2 antigen–antibody complexes could not bind all of the complement proteins. The investigators came to the conclusion that there were two types of complement. One type could be absorbed with the F(ab)'2-containing fragments. The other only interacted with complement containing undigested antibody. This work did not receive wide recognition.

By the end of the 1960s, it was also shown that guinea pig IgG antibodies could be separated into $\gamma 1$ and $\gamma 2$ fractions. $\gamma 2$ antibodies behaved as expected, fixing or activating all of the components of complement after interacting with antigen. $\gamma 1$ antibodies also activated or fixed complement proteins when immune complexes were formed; however, only the late-acting proteins appeared to be activated and the early proteins were spared (35).

At about that time, Gewurz and colleagues noted that bacterial endotoxins consumed the late-acting complement proteins but appeared to spare the early-acting proteins (36). This occurred even in agammaglobulinemic serum with no detectable antibody. The difficulty with interpretation of these studies was that no one knew whether the observed variability reflected differences in antibody function, assay variability, or the presence of confounding regulatory molecules. Shortly thereafter, however, a strain of guinea pigs was identified that totally lacked the early-acting protein of the classical pathway, C4 (37). The serum of these guinea pigs had no ability to lyse sheep erythrocytes because of the failure of classical pathway activity. However, antigen–antibody complexes could activate the late-acting proteins in this serum. Thus, there was now no question that there was an alternative pathway of complement activation.

With many lines of evidence in multiple laboratories suggesting the presence of an alternative pathway of complement activation, there was a resurgence of interest in Pillemer's earlier observations. The laboratory of Muller-Eberhard turned to the analysis of an old observation (38). There is a protein in cobra venom capable of activating complement (16). Study suggested that C1, C4 and C2 of the classical pathway were spared by cobra venom factor, but C3 and the later proteins were activated and destroyed. Further analysis suggested that co-factors were required, and the co-factors had the same properties as those originally described by Pillemer: co-factor A and co-factor B. Extensive analysis in several laboratories ultimately showed that cobra venom protein is a cobra analog of human C3b (38). It binds to factor B of the alternative pathway, and in the presence of magnesium ion, factor D of the alternative pathway cleaves the factor B to form a C3 convertase. The search then began for the cobra venom protein analog in human serum. Ultimately, it was found that hydrolyzed C3 in the presence of factors B and D formed a C3 convertase (38). The search was made easier by the critical finding that C3 structure was stabilized by an internal thioester bond and that activation of this molecule was accompanied by cleavage of the thioester (39). This convertase containing C3 and factor B was stabilized by properdin. Native C3 did not form a convertase, but only C3 with the hydrolyzed thioester that had undergone this conformational change or C3b itself served this function. Thus, the original hypothesis of Pillemer was confirmed.

COMPLEMENT FRAGMENTS AND REGULATORY PROTEINS

Research performed during the late 1970s and 1980s failed to discover new classical or alternative pathway complement proteins. These, in general, had been discovered earlier. As these proteins were purified, it became clear that most were activated by specific proteolytic cleavages. Often the small cleavage fragment had biological activity. For example, the small fragments of C3, C4 and C5 – C3a, C4a, and C5a – acted as anaphylatoxins, causing the degranulation of mast cells (40). C5a was the principal complement-derived chemotactic factor in serum. Now the emphasis was on characterizing the protein fragments and in defining regulatory steps in the complement cascade. An astonishing number of regulatory proteins were discovered, and the reasons for their presence became clear (41). There are relatively few individuals missing any complement proteins, but complement is frequently involved in the causation of human disease. In these cases, it is almost always true that the complement is functioning in a physiologically normal fashion, but it is damaging host tissues rather than damaging invading microorganisms or foreign cells. The many regulatory molecules control activation in the host in an attempt to regulate possible damage to host tissue. The 1970s and 1980s focused on defining these regulatory molecules, their mechanism of action, whether they were free in plasma or bound to cell surfaces, and how they function. Moreover, the critical role of complement in controlling cell function became the focus of study. Surprisingly, although individual complement proteins in general are rarely missing, abnormalities of regulatory molecules often contribute to disease. For example, factor H abnormalities are now associated with atypical hemolytic uremic syndrome and macular degeneration in the elderly (42). This factor regulates the degradation of C3, and it is believed that failure to appropriately regulate C3 degradation is responsible for the development of disease. Paroxysmal nocturnal hemoglobinuria was found to be due to a defect in the synthesis of phosphatidylinositol-linked cell membrane proteins (43). Here the defect is in CD59, a protein responsible for downregulating the ability of the late-acting complement proteins to lyse erythrocytes. The disease has been treated with monoclonal antibody to C5 (eculizumab), leading to dramatic decrease in hemolysis (44).

MANNOSE-BINDING PROTEINS

The history of the discovery of the lectin pathway is a much more recent event. The classical pathway is so named because it was the first to be discovered, at the turn of the 20th century. The presence of an alternative pathway was recognized in the 1950s but it was not until the 1970s and 80s that it was carefully explored and its existence confirmed. In the 1980s, it was recognized that animals had both cell-bound and circulating mannose-binding proteins. In 1988, Ezekowitz and colleagues isolated a similar protein from human plasma (45). This protein, termed mannose-binding lectin, binds mannose and a series of related sugars. It became clear in the 1990s and 2000s that this protein was part of the complement system and was analogous to C1q of the classical pathway. It had a similar structure to C1q and, because of its lectin-binding groups, could bind to the surface of a microorganism (46, 47). It circulated with associated proteins termed MASPs, mannose-binding protein associated serum proteases. Three MASPs and some related proteins were identified. The principal pathway through which mannose-binding protein activates complement appears to be via MASP2 which, in turn, like C1q, C1r and C1s, binds and cleaves C4 to enter the classical pathway. However, it appears that by activating MASP3, the pathway may cleave and activate C3 directly. Like the alternative pathway, the lectin pathway is ancient in evolutionary terms and does not require antibody for activation. A protein related to mannose-binding lectin, ficolin, appears to recognize acetylated sugars, but otherwise seems to resemble mannose-binding lectin in its function (46, 47). The three complement activation pathways, the classical pathway, the alternative pathway and the lectin pathway, all proceed to the formation of a C-3 convertase which leads to the activation of C3. They all proceed together to activation of C9, the last protein in the lytic sequence. All complement activation pathways can lyse susceptible cells or microbes. All can cause C3 to bind to the surface of microbes leading to opsonization. This occurs because C3 fragments can bind to specific receptors on phagocytes supporting the phagocytic process. All pathways can generate the biologically active fragments of complement proteins that support inflammation.

DISCOVERY THAT COMPLEMENT RECEPTORS REMOVE IMMUNE COMPLEXES

From the turn of the last century, it was known that bacteria injected into the bloodstream of animals might be found in phagocytic cells and also might be detected on the surface of cellular elements in the bloodstream. Exudates derived from phagocytic cells were more lytic for bacteria than was serum or plasma. In severe cases of hemolytic anemia, it was noted that erythrocytes might be bound to the surface of monocytes or neutrophils or might be present in phagocytic cells. Nelson and his colleagues extended these studies in the 1950s to develop the concept of immune adherence, the coating of a particle by opsonic complement fragments that mediated adherence to phagocytic cells (48). Specific receptors for these complement fragments were identified and, ultimately, Fearon and colleagues isolated the first of these complement receptors (CD 35) (49). Specific receptors for C4b, C3b, iC3b, C3dg, C1q, C5a, C3a and others were found during subsequent decades. Their importance in the biological activity of complement was established, and the focus was redirected to determining how they trigger alterations in cellular function. Their importance

in the removal of immune complexes from the circulation and in the phagocytosis of opsonized particles became apparent. Their function during the induction phase of the immune response is just starting to become clear.

THE HISTORY OF COMPLEMENT-COMPONENT DEFICIENCIES

The stage is set for a more detailed understanding of the role of normal complement function in human disease. Part of the reason that this area has proceeded slowly is the absence of patients in large numbers missing these complement factors, and the limited studies of immune function in these rare individuals. Individuals missing most or all of the complement proteins have been described (50–52). Complement deficiency individuals often have autoimmune disease or propensity for severe infection, particularly with *Neisseria* organisms. However, the number of individuals has been few, and animal models of complement deficiency have been relied on for the available knowledge.

ANIMAL MODELS OF COMPLEMENT DEFICIENCY

In the 1920s, a guinea pig strain lacking C3 was reported by Hyde and colleagues (53). At that time, it was not known that what was then termed "C3" consisted of multiple proteins. The strain was lost over the years, and we do not know which specific late protein was missing in these guinea pigs. The development of C2- and C4-deficient guinea pigs missing classical pathway activity and C5-deficient mice, missing the activity of one of the later components, made study of these portions of the complement cascade easier (37). To explore the contribution of the classic complement pathway to antibody responses *in vivo*, C2- and C4-deficient guinea pigs were immunized with the T-cell-dependent neoantigen bacteriophage φX174. These complement-deficient animals produced less antibody than control guinea pigs, were unable to maintain measurable antibody levels, and failed to develop amplification and isotype switching from IgM to IgG antibody following booster immunization. This defect could be overcome by increasing the antigen dose or by injecting normal guinea pig serum at the time of the primary immunization. The likely explanation for this observation is the requirement of C3 activation and the formation of antigen–antibody–complement complex that leads to antigen accumulation in lymphoid organs and entrapment of antigen-specific B cells (54, 55). C6-deficient rabbits were described, but these animals were difficult to breed and few were available for only a limited number of studies (56). C6-deficient rats, C3-deficient dogs and factor H-deficient pigs have been described, but these animals are not generally available (56).

LESSONS LEARNED: THE NATURE OF SCIENTIFIC ADVANCE

The fact that the number of investigators working for more than a century on understanding of complement and its function is small allows the historian to examine how science develops and progresses, and how scientists form opinions, interpret/misinterpret data and eventually overcome the challenge.

One striking finding is that one finds what one looks for. Early on, a series of assumptions were made that influenced the design of the experiments that were selected and thus the conclusions reached. While the early investigators were interested in the complement system and its function, the experiments they performed did not clearly raise that question. The experiments they designed asked how does complement kill bacteria or, even more specifically, what binds to a bacterial surface that ultimately leads to the death of the microorganism?

Interestingly, the fact that complement generates fragments during activation, and that these split products mediate many features of inflammation, was not considered. Those fragments were not discovered until some hundred years later. When it was found that the optimal conditions for the lysis of a bacterium are similar to the conditions necessary for the lysis of a red cell, the decision was made to study the lysis of red cells, mainly because the lysis of erythrocytes is much easier to study. The fact that the proteins required in these two different systems are not always the same was overlooked, and the presence of an alternative pathway of complement activation was only recognized some 80 years later. It took 100 years of complement-related research to discover a third activation, the lectin pathway, although elements of this pathway had been clearly observed in the previous decades.

The detailed study of the lysis of sheep erythrocytes generated a series of questions that were not easily answered. If one adds to a series of test tubes a large but equal number of antibody-coated sheep erythrocytes, and studies their lysis by adding an increasing amount of fresh serum to each tube, one generates a lysis curve: by adding very small amounts of fresh serum no lysis is observed but lysis occurs with increasing amounts of serum, following an S-shaped curve. The early investigators argued that if one adds 1 billion erythrocytes to each tube and a small amount of serum, only a few erythrocytes are expected to undergo lysis, while adding a larger amount of serum, more lysis should occur. However, the curve generated under those conditions should be a straight line with lysis increasing proportionally with the increasing amount of serum added. To explain the S-shaped curve, the complement pioneers postulated that complement produced holes in the red cell surface that caused lysis. If single holes in the cells were not sufficient for lysis, it might be that multiple holes were required before the cells started to lyse. Therefore, a certain amount of fresh serum would have to be added to the tube before there were enough holes in the erythrocyte membrane for lysis to begin. To prove this theory, mathematical modeling was done, demonstrating that the S-shaped curve could be exactly fitted if one assumed that it took six lesions on a red cell surface to lyse the cell. As we now know, a single lesion which passes through the cell membrane is sufficient to lyse a red cell. The reason for the S-shaped curve is the fact that multiple protein steps are required before that lesion forms, and that a certain amount of fresh serum is required before one gets through all of the steps needed for the formation of a lytic lesion. The lesson learned is that it is possible to model mathematically a not yet fully understood event and that, even if the model accurately follows the data, it does not at all explain how the system really functions.

The single-minded attention paid to understanding guinea pig complement in the early years delayed studies of the human complement system, which in turn led to a delay in our understanding of human disease. Furthermore, the emphasis on studies of the proteins required to lyse a red cell or bacterium led to a delay in discovering the control proteins that regulate the complement system which, as has been recently shown, play a major role in complement activity. It is now clear that polymorphism of some of those proteins that increase or decrease activity has an important role in human disease. It is only now that we have come to recognize that many renal diseases as well as acquired diseases, such as macular degeneration, are due to abnormalities in complement regulation rather then in the complement steps required for the lysis of a bacterium or red cell.

Looking back at the 140 year history of the complement field, it is obvious that the nature of questions thought to be important at the time profoundly influence the type of experiments that are selected. Those in turn influence the type of data generated, their interpretation, and the understanding of human disease. We believe that we have come to a new era in complement exploration, but who knows: perhaps in 100 years people will marvel at how naïve we have been in the types of questions we are asking at this moment of time and reflection.

References

1. Traube M, Gscheidlen R. Uber Faeulniss und den Widerstand der lebenden Organismen gegendieselbe. Zweiundfuenfzigster Jahres Bericht der Schlesischen Gesellschaft fur vaterlaendische Cultur 1874; p. 179.
2. Mayer MM. Complement. Historical perspectives and some current issues. *Complement* 1984;**1**:2–26.
3. Grohmann W. *Uber die Einwirkung des zellenfreien Blutplasma auf einigepflanzliche Microorganismen (Schimmel, Spross, pathogene und nichtpathogene Spaltpilze)*. C. Mattiesen: Dorpat; 1884 p. 34.
4. Nuttall G. Experimente ueber die bacterienfeindlichen Einfluesse des thierischen Koerpers. *Z HygInfektionskr* 1888;**4**:353.
5. Buchner H. Ueber die bakterien toedtendeWirkung des zellfreien Blutserums. *Zbl Bakt (Naturwiss)* 1889;**5**:817.
6. Baumgartner. Lehrbuch der pathogenen Mikroorg. Hirzel: Leipzig, 1911. Quoted in Zinsser H. Infection and Resistance. Macmillan: New York, 1923; p. 154.
7. Pfeiffer R, Issaeff R. Ueber die specifische Bedeutung der Cholera immunitaet. *Z HygInfektionskr* 1894;**17**:355.
8. Bordet J. Sur l'agglutination et la dissolution des globules rouges par le serum d'animaux injectes de sang defibrine. *Ann Inst Pasteur (Paris)* 1898;**12**:688.
9. Metschnikoff E. Sur la lutte des cellules de l'organisme contre l'invasion des microbes. *Ann Inst Pasteur (Paris)* 1887;**1**:321.
10. Bordet J. Résumé of immunity. *Studies in Immunity*. New York: J. Wiley and Sons; 1909.
11. Ehrlich P. *Studies in Immunity*. New York: J. Wiley and Sons; 1906.
12. Muir R. On the relationships between the complement and the immune bodies of different animals. *J Pathol Bacteriol* 1911;**16**:523.
13. Buchner H. Ueber die naehere Natur der bakterien toedtenden Substanz in Blutserum. *Zbl Bakt (Naturwiss)* 1889;**6**:561.
14. Ferrata A. Die Unwirksamkeit der komplexen Haemolysine in salzfreien Loesungen und ihre Ursache. *Berlin Klin Wochenschr* 1907;**44**:366.
15. Brand E. Ueber das Verhalten der Komplementebei der Dialyse. *Berlin Klin Wochenschr* 1907;**44**:1075.
16. Flexner S, Noguchi H. Snake venom in relation to haemolysis, bacteriolysis, and toxicity. *J Exp Med* 1903;**6**:277.

17. Weil E. Ueber die Wirkungsweise des Komplements bei der Haemolyse. *Biochem Z* 1913;**48**:347.

18. Nathan E. Ueber die Beziehungen der Komponenten des Komplements zu den ambozeptor beladenen Blutkoeperchen. *Z Immunitaetsforsch* 1914;**21**:259.

19. Ueno S. Studienueber die Komponenten des Komplementes. *Jap J Med Sci Trans Astr VII* 1938;**2**:201–36.

20. Coca AF. A study of the anticomplementary action of yeast of certain bacteria and of cobra venom. *Z Immunitaetsforsch* 1914;**21**:604.

21. Pillemer L, Blum L, Lepow IH, Ross OA, Todd EW, Wardlaw AC. The properdin system and immunity. I. Demonstration and isolation of a new serum protein, properdin, and its role in immune phenomena. *Science* 1954;**120**:279–85.

22. Naff GB, Pensky J, Lepow IH. The macromolecular nature of the first component of human complement. *J Exp Med* 1964;**119**:593–613.

23. Reid KBM. C1q and mannose-binding lectin. In: Frank MM, Volanakis JE, editors. *The Human Complement System in Health and Disease.* Marcel Dekker, Inc.; 1998. p. 33–48.

24. Gordon J, Whitehead HR, Wormall A. The action of ammonia on complement. The fourth component. *J Biochem* 1926;**20**:1028.

25. Mayer MM, Levine L, Rapp HJ, Marucci AA. Kinetic studies on immune hemolysis. VII. Decay of EAC'1, 4, 2, fixation of C'3, and other factors influencing the hemolytic action of complement. *J Immunol* 1954;**73**:443–54.

26. Rapp HJ. Mechanism of immune hemolysis: recognition of two steps in the conversion of EAC'1,4,2 to E*. *Science* 1958;**127**:234.

27. Borsos T, Rapp HJ. Complement fixation on cell surfaces by 19S and 7S antibodies. *Science* 1965;**150**:505–6.

28. Nelson RA, Jensen J, Gigli I, Tamura N. Methods for the separation, purification and measurement of nine components of hemolytic complement in guinea pig serum. *Immunochemistry* 1966;**3**:111.

29. Muller-Eberhard HJ. Complement. *Annu Rev Biochem* 1975;**44**:697–724.

30. Humphrey JH, Dourmashkin RR. The lesions in cell membranes caused by complement. *Adv Immunol* 1969;**11**:75–115.

31. May JE, Rosse W, Frank MM. Paroxysmal nocturnal hemoglobinuria – alternate-complement-pathway-mediated lysis induced by magnesium. *N Engl J Med* 1973;**289**:705–9.

32. Nelson RA. An alternative mechanism for the properdin system. *J Exp Med* 1968;**108**:515–35.

33. Lepow IH. Presidential address to American Association of Immunologists Louis Pillemer, properdin, and scientific controversy. *J Immunol* 1980;**125**:471–5.

34. Schur PH, Becker EL. Pepsin digestion of rabbit and sheep antibodies. The effect on complement fixation. *J Exp Med* 1963;**118**:891–904.

35. Sandberg AL, Osler AG. Dual pathways of complement interaction with guinea pig immunoglobulins. *J Immunol* 1971;**107**:1268–73.

36. Gewurz H, Shin HS, Mergenhagen SE. Interactions of the complement system with endotoxic lipopolysaccharide: consumption of each of the six terminal complement components. *J Exp Med* 1968;**128**:1049–57.

37. Ellman L, Frank MM, Green I. Genetically controlled total deficiency of the fourth component of complement in the guinea pig. *Science* 1970;**170**:74–5.

38. Gotze O, Muller-Eberhard HJ. The C3 activation system: an alternate pathway of complement activation. *J Exp Med* 1971;**134**:90s–108s.

39. Levine RP, Dodds AW. The thioester bond of C3. *Curr Top Microbiol Immunol* 1990;**153**:73–82.

40. Abe M, et al. Contribution of anaphylatoxins to allergic inflammation in human lungs. *Microbiol Immunol* 2005;**49**:981–6.

41. Liszewski MK, Atkinson JP. Regulatory proteins of complement. *The Human Complement System in Health and Disease.* Marcel Dekker, Inc.; 1998. p. 149–67.

42. Park HJ, Atkinson JP. Autoimmunity: homeostasis of innate immunity gone awry. *J Clin Immunol* 2012;**32**:1148–52.

43. Luzzatto L. Paroxysmal nocturnal hemoglobinuria: an acquired X-linked genetic disease with somatic-cell mosaicism. *Curr Opin Genet Devel* 2006;**16**:317–22.

44. Hillmen P, et al. The complement inhibitor eculizumab in paroxysmal nocturnal hemoglobinuria. *N Engl J Med* 2006;**355**:1233–43.

45. Ezekowitz RA, Day LE, Herman GA. A human mannose-binding protein is an acute-phase reactant that shares sequence homology with other vertebrate lectins. *J Exp Med* 1988;**167**:1034–46.

46. Heitzeneder S, et al. Mannan-binding lectin deficiency – Good news, bad news, doesn't matter? *Clin Immunol* 2012;**143**:22–38.

47. Degn SE, Thiel S, Jensenius JC. New perspectives on mannan-binding lectin-mediated complement activation. *Immunobiology* 2007;**212**:301–11.

48. Nelson RA. The immune-adherence phenomenon. A hypothetical role of erythrocytes in defence against bacteria and viruses. *Proc R Soc Med* 1956;**49**:55.

49. Fearon DT. Identification of the membrane glycoprotein that is the C3b receptor of the human erythrocyte, polymorphonuclear leukocyte, B lymphocyte and monocyte. *J Exp Med* 1980;**152**:20.

50. Figueroa J, Densen P. Infectious diseases associated with complement deficiencies. *Clin Microbiol Rev* 1991;**4**:369–95.

51. Ross SC, Densen P. Complement deficiency states and infection: epidemiology, pathogenesis and consequences of neisserial and other infections in an immune deficiency. *Medicine (Baltimore)* 1984;**63**:243–73.

52. Sullivan KE, Winkelstein JA. Genetically determined deficiencies of the complement system. In: Ochs HD, Smith CIE, Puck JM, editors. *Primary Immunodeficiency Diseases, A Molecular and Genetic Approach*. 2nd edn. Oxford University Press: New York; 2007. p. 589–608.

53. Hyde RR. Complement-deficient guinea-pig serum. *J Immun* 1923;**8**:167.

54. Ochs HD, Wedgwood RJ, Frank MM, Heller SR, Hosea SW. The role of complement in the induction of antibody responses. *Clin Exp Immunol* 1983;**53**:208–16.

55. Ochs HD, Wedgwood RJ, Heller SR, Beatty PG. Complement, membrane glycoproteins and complement receptors: their role in regulation of the immune response. *Clin Immunol Immunopathol* 1986;**40**:94–104.

56. Frank MM. Animal models for complement deficiencies. *J Clin Immunol* 1995;**15**:113S–21S.

18

DiGeorge Syndrome: A Serendipitous Discovery

Kathleen E. Sullivan[1], Donna M. McDonald-McGinn[2]

[1]The Division of Allergy Immunology, Philadelphia, PA, USA
[2]The Division of Human Genetics The Children's Hospital of Philadelphia,
Philadelphia, PA, USA

ANCIENT HISTORY

The history of DiGeorge syndrome parallels our understanding of the immune system. Its recognition as an immune deficiency occurred at the dawn of our understanding of T-cells and the thymus. In antiquity, at a time when most people understood human disease as an imbalance in humors or a religious judgment, Thucydides, a Greek historian recording the plague that hit Athens in 430 stated:

> But those that were recovered had much compassion both on them that died and on them that lay sick, as having both known the misery themselves and now no more subject to the danger. For this disease never took any man the second time so as to be mortal.

This represents one of the first descriptions of immunity, however, for over 2000 years, there was no understanding of the mechanism by which people became immune. Indeed, for the most part, disease continued to be perceived as an imbalance in bodily humors. As mysterious as the concept of immunity was for much of human history, the thymus was yet more mysterious. The name thymus is thought to come from

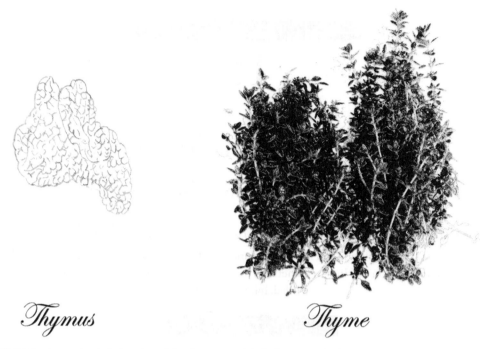

FIGURE 18.1   An anatomical drawing of the thymus modeled after an early Gray's Anatomy. Adjacent is an enhanced photograph of two bunches of thyme.

the Latin derivation of the Greek *thymos* meaning warty excretion, due to its resemblance to the flowers of the thyme plant (Fig. 18.1) (1). The homonym *thymos* translates as soul or spirit and, for this reason, the thymus was represented as the seat of the soul by the early Greeks. A Greek anatomist, Rufus of Ephesus, lived in approximately 100 AD and trained in Alexandria, where he was thought to have been taught about the thymus. A translation of his work (2) states that:

> ... amongst the glands there is one called the thymus, situated under the head of the heart, oriented towards the seventh vertebra of the neck and towards the end of the trachea that touches the lung; it is not to be found in all animals.

This is the first recognized recording of the thymus. Galen is perhaps the most famous physician of all antiquity and he is credited with being the first to note that the thymus is larger in young animals and dwindles in size with age. This is a critical observation that continues to inspire research even today. Galen's other important contribution to the study of the thymus was to refer to it as an "organ of mystery", a term that remained prevalent for almost 2000 years.

During the Renaissance, there were many dissection studies of the thymus and it was often felt to represent a cushion or a regulatory organ for the lungs (3, 4). In 1777, it was finally recognized as part of the lymphatic organs by William Hewson, however, this was a hollow advance since the role of the lymphatic organs was not yet understood (5). In 1846, with improved microscopy, Hassall and Vanarsdale identified the famous Hassall's corpuscles (6). Unfortunately, that did nothing to elucidate further the function of the thymus and, for the next 100 years, it was widely misunderstood by clinicians. In fact, it has been said that the thymus is one of the biggest medical scapegoats of all history. Throughout the 1800s, thymic asthma and status thymicolymphaticus, both terms for a large thymus, were seen as causes of abrupt death in early childhood (7). The thymus gland was a victim of an increasing requirement that a specific cause of death be identified as opposed to "a visitation from God" to explain early childhood deaths. It is now easily understood that children who died rapidly would have had a large thymus compared to those who had a more lingering death and had stress-induced involution of the thymus, however, at that time the large thymus was felt to be pathological. For this reason, there were a number of decades where thymectomy and thymic irradiation were prescribed to shockingly large numbers of children (8, 9). It was only with our increased understanding of the immune system and the role of the thymus that these practices came to a stop.

THE THYMUS AS AN ORGAN OF CELLULAR IMMUNITY

In the early 1900s, tentative steps in understanding the role of the small lymphocyte were taken by James Murphy (10, 11). He determined that these cells were responsible for allograft rejection, however, there were still many decades before a more complete understanding of the lymphocyte evolved. As late as 1951, Arnold Rich stated that "nothing of importance is known regarding the potentialities of small lymphocytes". He found that to be a "disgraceful gap in all medical knowledge" (12). The pioneering work of Max Cooper and Robert Good, published in 1965, was pivotal in our understanding of immune deficiencies by defining the disparate roles of two types of lymphocytes (13) (see Chapter 2). They identified two distinct types of lymphocytes in the chicken. By surgically excising either the thymus or the bursa of Fabricius in chickens, Cooper was able to demonstrate that plasma

cells and immunoglobulins were dependent on the bursa; however, cellular immunity, as defined by graft rejection and delayed-type hypersensitivity, was unaffected by bursectomy. In contrast, thymectomy reduced the population of lymphocytes in the blood and functional cellular immunity, while leaving immunoglobulin production intact. His seminal recognition of these two independent sites required for the differentiation of two distinct types of lymphocytes laid the foundation for our understanding of DiGeorge syndrome. Shortly after Cooper's publication, he moderated a session at the Society for Pediatric Research meeting where these findings in chickens were applied to human disease by Angelo DiGeorge (14).

It is important to place these findings in the greater historical context. It was only in 1952 that Colonel Bruton described agammaglobulinemia (15). It was only in 1961 that the thymus gland was identified as an immunologically important organ (16). A graph demonstrating the number of publications related to the thymus demonstrates rare descriptions until the 1970s, when the impact of its role was explored (Fig. 18.2). During the centuries when medicine was viewed through the prism of anatomy, two studies stand out: one describing absence of the thymus gland and the other describing absence of the parathyroids (17, 18). At the time of the 1965 meeting moderated by Max Cooper, primary immune deficiencies involving lymphocytes were just becoming recognized. The Swiss-type agammaglobulinemia (now known to have represented severe combined immune deficiency [SCID]), ataxia telangiectasia and Bruton's agammaglobulinemia, were all well described clinical phenotypes (15, 19–21). It was already appreciated that immune deficiencies could affect antibody production alone. Swiss-type agammaglobulinemia affected both cellular and humoral immunity. At nearly the same time, the great pathologist, Christian Nezelof, described a pure cellular immunity defect with thymus tissue that resembled that seen in

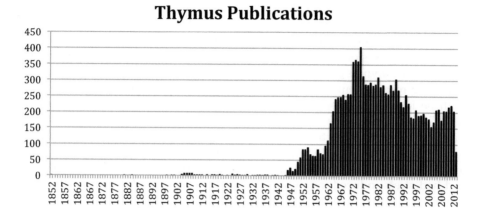

FIGURE 18.2   The graph demonstrates the number of publications with the word thymus in the title each year since 1850.

Swiss-type agammaglobulinemia. The concept of purely humoral, purely cellular and mixed-type immune deficiencies is useful even today. The clinical framework was formally introduced by Drs Hitzig and Soothill and was accepted by the World Health Organization in 1970 (21–24). The great insight of Angelo DiGeorge from St Christopher's Hospital for Children in Philadelphia was the recognition that thymic aplasia leads to a specific absence in cellular immunity with preservation (in some cases) of humoral immunity. He was able to connect his findings conceptually to those seen in Cooper's chicken model when he removed the thymus gland. DiGeorge's patients had congenital hypoparathyroidism and absence of cellular immunity, and he recognized the embryological connection between the two organs. In three of the patients, an autopsy was performed at which point congenital absence of the thymus gland was identified. DiGeorge clearly understood that this represented a pure thymic defect in humans and was the root cause of the infectious susceptibility. He went on to publish the critical inception series in 1965 where he focused on the thymus and the parathyroid glands (25). With Harold Lischner, he published further immunological studies in the 1960s and, in the

1970s, the broad spectrum of the severity of the immune deficiency was emphasized (26, 27). Some patients had complete thymic aplasia and generally had impaired antibody production, while others had thymic hypoplasia with generally preserved antibody production.

ADDITIONAL PHENOTYPIC FEATURES

Since the profound insight of DiGeorge, various individuals have described cohorts of patients with similar findings. In many cases, the associated clinical features rather than the immune deficiency drove the description, and it was in these latter papers that the full phenotypic spectrum came to be appreciated. In 1968, William Strong from Cleveland reported an association of cardiac abnormalities and developmental delay, while others focused on facial features (28–31). Palatal features were highlighted in an early report from Czechoslovakia (32). In time, these clinical cohorts were recognized to have significant overlap, however, in the absence of an identified gene defect it was difficult to know if these syndromes represented a spectrum of a single genetic defect or distinct disorders.

THE GENETIC BASIS OF DIGEORGE SYNDROME

In the early 1980s, several patients with DiGeorge syndrome were found to have a microscopic rearrangement of chromosome 22 (33, 34). This key finding opened the door for additional studies to identify the gene defect. Peter Scambler developed a fluorescent *in situ* hybridization approach to identify patients with the microdeletions within chromosome 22 and, with this new tool in hand, it was determined that the majority of patients with DiGeorge syndrome had a deletion within chromosome 22 (35). Similarly, the majority of patients with velocardiofacial syndrome, conotruncal anomaly face syndrome and some patients with autosomal dominant Opitz G/BBB syndrome and Caylor cardiofacial syndrome were ultimately found to have the deletion (36–40). Having said that, there remain a small number of patients with DiGeorge syndrome who do not have the deletion, and it is important to recognize that DiGeorge syndrome can be due to intrauterine exposures such as diabetes and isotretinoin. Because of this, the term DiGeorge syndrome and all of the other syndromic names should be reserved for those patients who do not have a genetic diagnosis. In all patients with a chromosome 22q11.2 deletion, their diagnostic descriptor should be simply chromosome 22q11.2 deletion to minimize confusion.

In terms of the genetic diagnosis, there was one more key finding to be made, and that was the identification of rare individual patients with TBX1 mutations with the same phenotype (41). This was critical, because within the commonly deleted region it was known that over 30 genes were affected and for many years it was not clear whether the clinical phenotype was due to haploinsufficiency of a single gene, combinations of genes, or impacted by the gene background. Studies which focused on the function of TBX1 have demonstrated that haploinsufficiency for this transcription factor is responsible for most of the phenotypic manifestations of the deletion in mice (42–44). These pioneering studies initially generated deletions in mice that mimicked those in humans. By tightening the window of the deletion, the critical region was refined and TBX1 was identified. TBX1 is specifically responsible for the thymic hypoplasia, although background genes contribute (44).

CLINICAL AND LABORATORY FEATURES OF DIGEORGE SYNDROME

Our understanding of the immune deficiency in DiGeorge syndrome has matured in parallel with our understanding of the etiology. Early studies by Angelo DiGeorge and Harold Lischner emphasized a high mortality rate and the occurrence of frequent infections (25–27). One of the first studies to characterize the immune deficiency using newly developed laboratory methods was that of Mary Ellen Conley in 1979 (45). Her study of 25 patients with DiGeorge syndrome also exhibited an astonishingly high mortality rate of 96%, although it is important to remember that cardiac repair surgery was not available at that time, and that there was a selection bias, since the study cohort consisted of autopsy cases collected by the Seattle pathologist Bruce Beckwith. Conley carefully characterized the infections and identified many of the now well-known syndromic features such as cardiac anomalies. Her study used microscopic delineation of the immune deficiency and found that both the T-cell and B-cell dependent areas of lymphoid tissue were depleted. The infections that she identified in this series were not often the cause of death and consisted of *Candida*, diarrhea, and rhinitis.

Two studies from Germany that appeared in 1988 and 1989 were the first true effort to utilize modern laboratory methods to define the immune deficiency (46, 47). At that point, T-cells were identified by sheep erythrocyte rosetting, and the patients were found to have a broad range of T-cell counts in the peripheral blood. Immunoglobulin levels were found to be largely normal, although several patients were clearly hypogammaglobulinemic. These two papers also defined T-cell function using mitogen proliferation and interleukin 2 (IL-2) responses. As we know today, the patients largely had normal proliferative responses with only the three most severely affected babies having absent proliferation. The relatively newly developed flow cytometric methodology was also applied to the study population, and it was found that T-cells were relatively selectively affected in the syndrome. In the following year, a study appeared which demonstrated that a low CD4 T-cell count as well as poor responses to phytohemagglutinin (PHA) defined a subset of patients who seemed to have more clinical evidence of immune deficiency (48). These latter three papers clearly demonstrated that there is a spectrum of thymic hypoplasia. In both the German papers and the US study, there were subjects who had normal T-cell counts and normal T-cell function as well as those who were profoundly immune deficient. The study by John Bastian was important because it demonstrated that the immune deficiency was stable over time. In the ensuing years, there have been a number of efforts to define more carefully the immune deficiency and today we now understand several key features related to the decrement in T-cell production.

Modern technologies and systematic study have continued to advance the understanding of the immune deficiency. We have learned that among patients with chromosome 22q11.2 deletion syndrome, most infants will have low T-cell counts, although in adulthood the majority will have normal T-cell counts. While the age-dependent changes in the immune system have been well elucidated for decades, the effect in chromosome 22q11.2 deletion syndrome has been identified in only a few studies (49–51). We also understand that the repertoire and the quality of the T-cell compartment are impacted by thymic hypoplasia, even in adults whose T-cell count is normal. Studies have demonstrated that the T-cell repertoire can be compromised in chromosome 22q11.2 deletion syndrome, but it is not necessarily affected (52–55). The heterogeneity of the thymic hypoplasia was highlighted in the early studies of Angelo DiGeorge and his colleague, Harold Lischner (25–27). Additional studies of the anatomy of the thymus in this syndrome revealed that small thymic remnants can occur at anatomically anomalous locations (27, 56). In spite of their anomalous location, these remnants of thymic tissue have been shown to support T-cell production. Another important finding from the detailed study of T-cells in chromosome 22q11.2 deletion syndrome was that the adults with normal CD3 T-cell counts can have compromised function because the subset distribution of the T-cells is aberrant or because qualitatively the T-cells are skewed towards a Th2 phenotype (49).

In spite of the abundant data about the T-cells in this syndrome, there have been remarkably few studies of the clinical aspects related to immune deficiency. Early studies of DiGeorge syndrome documented a high frequency of infection (45). Several early studies emphasized graft-versus-host disease and severe infections, commensurate with ascertainment of the most severely affected children (57–65). In fact, there are only a handful of modern studies that have evaluated infection, a significant medical knowledge gap. Infections are known to occur commonly in the affected children (66, 67). A study of adults by Andrew Gennery found ongoing infections in adulthood, suggesting this will be a fruitful area for continued investigation (68).

In contrast to the lack of data identifying the frequency of infections in patients, there have been more studies regarding the frequency of autoimmune disease. These landmark clinical studies defined the strong association of chromosome 22q11.2 deletion syndrome with a variety of autoimmune diseases. In adults, autoimmune thyroid disease was found to be the most common, although the definition of autoimmune thyroid disease has been difficult because the thyroid can be anatomically involved in the syndrome (69–71). Nevertheless, a genuine increase in autoimmune thyroiditis has been described in patients with chromosome 22q11.2 deletion syndrome. Other autoimmune diseases that have been found with increased frequency are juvenile idiopathic arthritis, autoimmune hemolytic anemia, idiopathic thrombocytopenic purpura and inflammatory bowel disease (72–80). The reason for the increased frequency of autoimmune disease has been examined in several studies. These patients were found to have lower numbers of regulatory T-cells in the peripheral blood which may drive increased susceptibility to autoimmunity (81). Although the absolute regulatory T-cell counts were found to be lower in the syndrome, it did not appear that the ratio of regulatory to effector T-cells was decreased. One other hypothesis related to the autoimmune disease was that it flows from the homeostatic proliferation that occurs secondary to the diminished thymic output (51). It had been demonstrated in laboratory models that homeostatic proliferation selects for self-reactive T-cells and, in extreme cases, will lead to autoreactivity (82).

TREATMENT OF THE IMMUNE DEFICIENCY

Patients with either DiGeorge syndrome or chromosome 22q11.2 deletion syndrome have sometimes been said to have "complete DiGeorge syndrome". The implication of the word "complete" is that thymic aplasia is complete. Because this ignores the other organ systems involved, it has led to some confusion among patients and their families, however, this terminology has persisted. Regardless of the terminology, patients with DiGeorge syndrome or

chromosome 22q11.2 deletion syndrome who have no T-cells will die without intervention. There are several early papers attempting to use either extracts of thymus, extracts of leukocytes, or direct implantation of fetal thymic tissue as a curative strategy (83–86). Charles August and Humphrey Kay developed thymic transplantation as a cure for the thymic deficiency in DiGeorge syndrome in 1968 (83, 87). This represented a heroic intervention for a fatal disease. Improvement after thymic transplantation was highly variable and there was little understanding of graft rejection at the time. Graft-versus-host disease was a recognized threat. Furthermore, access to fetal thymic tissue was becoming more difficult in the later 1970s. To circumvent the technical and logistical issues associated with thymic transplants, efforts were made to develop a soluble extract which could be provided to induce T-cell development. Joseph Bellanti and Peralta Serrano both creatively implanted a thymus encased in a Millipore chamber to prevent graft-versus-host diseases (88, 89). Both children exhibited significant improvement, supporting a move to enhance cellular immunity via injectable therapeutics. Various extracts and dialysates were developed, with thymosin ultimately achieving prominence. These efforts may seem extraordinarily refracted through a modern lens. Joe Bellanti and Diane Wara, early pioneers in the treatment of T-cell defects, remember the time as being one of great optimism and energy. Recognition of adenosine deaminase (ADA) and purine nucleoside phosphorylase (PNP) deficiencies in the 1970s led to the conceptual approach of replacing soluble enzymes and factors to treat immune deficiencies. From within this framework, the use of soluble thymic extracts was logical.

Thymosin was originally purified and its potential recognized by Alan Goldstein (90). It was demonstrated to enhance cell-mediated immunity in mice and had significant benefit on human cells *in vitro* (91). It was natural for immunologists to attempt to harness its power therapeutically. Morton Cowan's group in San Francisco treated five patients with thymosin as a bridge while awaiting thymus transplantation. The treatment was well tolerated and appeared to benefit a subset of patients. There would ultimately be over 1500 papers published on the use of thymosin and, today, there are still those who feel it was clinically beneficial. Nevertheless, the lack of controlled trials and a failure of further purified components to reproduce the effect led to its gradual disuse (92).

In the intervening years, bone marrow transplantation came into its own, and fully matched sibling transplants were found to be curative of the immune deficiency in the syndrome (93, 94). In spite of the fact that there is no thymic tissue for the stem cells to mature in, the transplanted mature T-cells provided sufficient repertoire diversity that the patients were able to defend adequately against infection. In fact, long-term outcomes of some of the early patients have demonstrated an excellent quality of life and reasonably good health (95). Nevertheless, the unavailability of full sibling matches for many patients has limited the usefulness of the strategy.

Frustrated with the lack of treatment options for these patients, Louise Markert began to explore possible postnatal thymic transplantation as a curative strategy (96). This represented a dramatic change in the approach to the treatment of complete DiGeorge syndrome. Her incredible efforts over the past 20 years have led to the treatment of now nearly 100 children, all of whom had very serious immune deficiency and likely would not have survived (97, 98). With her revolutionary thymic transplantation method, the vast majority of these patients have survived and she has elegantly demonstrated the functionality of their new immune systems. These landmark studies were revolutionary because she did not match the thymic tissue to the host. Based on the work of Zinkernagel, it was believed that a non-matched thymic transplant would lead to intolerable graft-versus-host

disease as a result of stem cells being educated on foreign major histocompatibility complex (MHC) (92). However, Dr Markert has elegantly demonstrated that T-cells do undergo thymic education and do become tolerant to the host MHC (99).

SECONDARY HUMORAL DEFECTS IN CHROMOSOME 22Q11.2 DELETION SYNDROME

The second critical finding for the management of patients with chromosome 22q11.2 deletion syndrome has been the identification of secondary B-cell defects. Variable humoral defects were described in the initial series of patients described at Saint Christopher's Hospital for Children (26, 27). Sporadic reports continued to appear on small numbers of patients with defective antibody production (100–103). Anne Junker reported on 13 patients using carefully controlled studies and found that humoral immunity was largely intact (104). Andrew Gennery in the UK was the first to highlight the association of immunoglobulin abnormalities with autoimmunity and recurrent infections (68). This has now been followed by several studies that have attempted to identify the mechanism by which patients with chromosome 22q11.2 deletion syndrome have B-cell functional deficits. These studies have demonstrated an impaired ability to undergo class switching, which suggests that the B-cell deficit is secondary to a lack of T-cell help, as anticipated (49). In addition, the frequency of hypogammaglobulinemia has now been identified and appears to be present in roughly 5–6% of the patient population (105). The frequency of specific antibody deficiency is not known, although it is likely that specific antibody deficiency is at least as common as hypogammaglobulinemia. It is presumed, but not yet demonstrated, that replacement of immunoglobulin in these patients would be of clinical benefit.

SUMMARY

DiGeorge syndrome and chromosome 22q11.2 deletion syndrome represent one of the great treatment success stories in clinical immunology. Prior to the early 1980s, when cardiac bypass became widely available, the majority of babies with DiGeorge syndrome (and likely chromosome 22q11.2 deletion syndrome) died as a result of their cardiac anomaly. That single innovation with its unlikely dependence on the tubing used to dispense beer has saved more children with DiGeorge syndrome than any other. We now know that approximately 20% of infants with chromosome 22q11.2 deletion syndrome have no cardiac disease and, while these patients presumably survived, they were likely not diagnosed. Once cardiac repair surgery became routinely available in the early 1980s, the mortality rate in chromosome 22q11.2 deletion syndrome fell dramatically. Today, mortality associated with chromosome 22q11.2 deletion syndrome is exceedingly low and the majority of deaths are due to cardiac disease (106). Having said that, there is a small subset of patients who require treatment for immune deficiency. The pivotal work on thymus transplantation by Louise Markert has ensured that the majority of those patients will also reach their full potential.

References

1. Skinner H. *Origin of medical terms*. 2nd edn. Baltimore: Williams and Wilkins; 1961.
2. May M. *Galen on the usefulness of the parts of the body*. Ithaca: Cornell University Press; 1968.
3. Crotty A. *Thyroid and thymus*. Philadelphia: Lea and Febiger; 1922.
4. Wright HG. Use of the thymus gland: an original theory, with explanatory remarks. *Lond J Med* 1852;**4**:443–50.
5. Hewson W. Experimental inquiries III. In Cadell, T., (ed.), *Experimental inquiries into the properties of the blood*. London: T. Cadell, 1777; pp. 1–223.
6. Hassall A, Vanarsdale H. Illustrations of the microscopic anatomy of the human body in health and disease. In: Hassall A, editor. *Miscroscopic anatomy of the human body in health and disease*. London: Wood; 1846. p. 1–79.
7. Jacobs MT, Frush DP, Donnelly LF. The right place at the wrong time: historical perspective of the relation of the thymus gland and pediatric radiology. *Radiology* 1999;**210**:11–6.
8. McCardle W. Status lymphaticus in relation to general anesthesia. *Lancet* 1908;**1908**:196–202.
9. Moncrieff A. Enlargement of the thymus in infants with special reference to clinical evidence of so called status thymolymphaticus. *Proc R Soc Med* 1937;**31**:537–44.
10. Murphy JB. Studies in tissue specificity : II. the ultimate fate of mammalian tissue implanted in the chick embryo. *J Exp Med* 1914;**19**:181–6.
11. Murphy JB. Factors of resistance to heteroplastic tissue-grafting : studies in tissue specificity. III. *J Exp Med* 1914;**19**:513–22.
12. Rich A. *The pathogenesis of tuberculosis*. Springfield: C Thomas; 1951.
13. Cooper MD, Peterson RD, Good RA. Delineation of the thymic and bursal lymphoid systems in the chicken. *Nature* 1965;**205**:143–6.
14. Cooper M, Peterson R, Good R. A new concept of the cellular basis of immunity. *J Pediatr* 1965;**67**:907–8.
15. Bruton OC. Agammaglobulinemia. *Pediatrics* 1952;**9**: 722–8.
16. Miller J. Immunological function of the thymus. *Lancet* 1961;**1961**:748–9.
17. Harington LH. Absence of the thymus gland. *Lond Med Gaz* 1829;**3**:314–20.
18. Lobdell DH. Congenital absence of the parathyroid gland. *Arch Pathol* 1959;**67**:412–8.
19. Wells CE, Shy GM. Progressive familial choreoathetosis with cutaneous telangiectasia. *J Neurol Neurosurg Psychiatr* 1957;**20**:98–104.
20. Louis-Bar D. Sur un syndrome progressif comprenant des telangiectasies capillaires cutanees et conjonctivales symetriques, a disposition naevoide et des troubles cerebelleux. *Confin Neurol* 1941;**4**:32–42.
21. Hitzig WH, Biro Z, Bosch H, Huser HJ. [Agammaglobulinemia & alymphocytosis with atrophy of lymphatic tissue]. *Helv Paediatr Acta* 1958;**13**:551–85.
22. Soothill JF. Classification of immunological deficiency diseases. *Proc R Soc Med* 1968;**61**:881–3.
23. Fudenberg HH, Good RA, Hitzig W, et al. Classification of the primary immune deficiencies: WHO recommendation. *N Engl J Med* 1970;**283**:656–7.
24. Hitzig WH. Combined cellular and humoral deficiency states. *Proc R Soc Med* 1968;**61**:887–9.
25. DiGeorge AM. Congenital absence of the thymus and its immunological consequences: concurrance with congenital hypothyroidism. *Birth Defects* 1968;**4**:116–21.
26. Lischner HW. DiGeorge syndrome(s). *J Pediatr* 1972;**81**: 1042–4.
27. Lischner HW, Huff DS. T-cell deficiency in DiGeorge syndrome. *Birth Defects* 1975;**11**:16–21.

28. Strong WB. Familial syndrome of right-sided aortic arch, mental deficiency, and facial dysmorphism. *J Pediatr* 1968;**73**:882–8.

29. Shprintzen RJ, Goldberg RB, Lewin ML, et al. A new syndrome involving cleft palate, cardiac anomalies, typical facies, and learning disabilities: velo-cardio-facial syndrome. *Cleft Palate J* 1978;**15**:56–62.

30. Caylor G. Cardiofacial syndrome: Congenital heart disease and facial weakness, a hitherto unrecognized association. *Arch Dis Child* 1969;**44**:69–75.

31. Kinouchi A, Mori K, Ando M, Takao A. Facial appearance of patients with conotruncal anomalies. *Pediatr Japan* 1976;**17**:84–7.

32. Sedlackova E. The syndrome of the congenitally shortening of the soft palate. *Cas Lek Ces* 1955;**94**:1304–7.

33. de la Chapelle A, Herva R, Koivisto M, Aula P. A deletion in chromosome 22 can cause DiGeorge syndrome. *Hum Genet* 1981;**57**:253–6.

34. Kelley RI, Zackai EH, Emanuel BS, Kistenmacher M, Greenberg F, Punnett HH. The association of the DiGeorge anomalad with partial monosomy of chromosome 22. *J Pediatr* 1982;**101**:197–200.

35. Scambler PJ, Carey AH, Wyse RK, et al. Microdeletions within 22q11 associated with sporadic and familial DiGeorge syndrome. *Genomics* 1991;**10**:201–6.

36. Driscoll DA, Salvin J, Sellinger B, et al. Prevalence of 22q11 microdeletions in DiGeorge and velocardiofacial syndromes: implications for genetic counselling and prenatal diagnosis. *J Med Genet* 1993;**30**:813–7.

37. Burn J, Takao A, Wilson D, et al. Conotruncal anomaly face syndrome is associated with a deletion within chromosome 22q11. *J Med Genet* 1993;**30**:822–4.

38. Matsuoka R, Takao A, Kimura M, et al. Confirmation that the conotruncal anomaly face syndrome is associated with a deletion within 22q11.2. *Am J Med Genet* 1994;**53**:285–9.

39. McDonald-McGinn D, Driscoll D, Bason L, et al. Autosomal dominant "Opitz" GBBB due to a 22q11.2 deletion. *Am J Med Genet* 1995;**59**:103–13.

40. Giannotti A, Digilio MC, Marino B, Mingarelli R, Dallapiccola B. Cayler cardiofacial syndrome and del 22q11: part of the CATCH22 phenotype. *Am J Med Genet* 1994;**53**:303–4.

41. Yagi H, Furutani Y, Hamada H, et al. Role of TBX1 in human del22q11.2 syndrome. *Lancet* 2003;**362**:1366–73.

42. Lindsay EA, Vitelli F, Su H, et al. Tbx1 haploinsufficieny in the DiGeorge syndrome region causes aortic arch defects in mice. *Nature* 2001;**410**:97–101.

43. Merscher S, Funke B, Epstein JA, et al. TBX1 is responsible for cardiovascular defects in velo-cardio- facial/DiGeorge syndrome. *Cell* 2001;**104**:619–29.

44. Jerome LA, Papaioannou VE. DiGeorge syndrome phenotype in mice mutant for the T-box gene, Tbx1. *Nat Genet* 2001;**27**:286–91.

45. Conley ME, Beckwith JB, Mancer JF, Tenckhoff L. The spectrum of the DiGeorge syndrome. *J Pediatr* 1979;**94**:883–90.

46. Muller W, Peter HH, Wilken M, et al. The DiGeorge syndrome. I. Clinical evaluation and course of partial and complete forms of the syndrome. *Eur J Pediatr* 1988;**147**:496–502.

47. Muller W, Peter HH, Kallfelz HC, Franz A, Rieger CH. The DiGeorge sequence. II. Immunologic findings in partial and complete forms of the disorder. *Eur J Pediatr* 1989;**149**:96–103.

48. Bastian J, Law S, Vogler L, et al. Prediction of persistent immunodeficiency in the DiGeorge anomaly. *J Pediatr* 1989;**115**:391–6.

49. Jawad AF, Luning Prak E, Boyer J, et al. A prospective study of influenza vaccination and a comparison of immunologic parameters in children and adults with chromosome 22q11.2 deletion syndrome (DiGeorge syndrome/velocardiofacial syndrome). *J Clin Immunol* 2011;**31**:927–35.

50. Jawad AF, McDonald-McGinn DM, Zackai E, Sullivan KE. Immunologic features of chromosome 22q11.2 deletion syndrome (DiGeorge syndrome/velocardiofacial syndrome). *J Pediatr* 2001;**139**:715–23.

51. Piliero LM, Sanford AN, McDonald-McGinn DM, Zackai EH, Sullivan KE. T-cell homeostasis in humans with thymic hypoplasia due to chromosome 22q11.2 deletion syndrome. *Blood* 2004;**103**:1020–5.

52. Pierdominici M, Marziali M, Giovannetti A, et al. T-cell receptor repertoire and function in patients with DiGeorge syndrome and velocardiofacial syndrome. *Clin Exp Immuno* 2000;**121**:127–32.

53. Pierdominici M, Mazzetta F, Caprini E, et al. Biased T-cell receptor repertoires in patients with chromosome 22q11.2 deletion syndrome (DiGeorge syndrome/velocardiofacial syndrome). *Clin Exp Immunol* 2003;**132**:323–31.

54. Haire RN, Buell RD, Litman RT, et al. Diversification, not use, of the immunoglobulin VH gene repertoire is restricted in DiGeorge syndrome. *J Exp Med* 1993;**178**:825–34.

55. Cancrini C, Romiti ML, Finocchi A, et al. Post-natal ontogenesis of the T-cell receptor CD4 and CD8 Vbeta repertoire and immune function in children with DiGeorge syndrome. *J Clin Immunol* 2005;**25**:265–74.

56. Bale PM, Sotelo-Avila C. Maldescent of the thymus: 34 necropsy and 10 surgical cases, including 7 thymuses medial to the mandible. *Pediatr Pathol* 1993;**13**:181–90.

57. Ocejo-Vinyals JG, Lozano MJ, Sanchez-Velasco P, Escribano de Diego J, Paz-Miguel JE, Leyva-Cobian F. An unusual concurrence of graft versus host disease caused by engraftment of maternal lymphocytes with DiGeorge anomaly. *Arch Dis Child* 2000;**83**:165–9.

58. Deerojanawong J, Chang AB, Eng PA, Robertson CF, Kemp AS. Pulmonary diseases in children with severe combined immune deficiency and DiGeorge syndrome. *Pediatr Pulmonol* 1997;**24**:324–30.

59. Brouard J, Morin M, Borel B, et al. Syndrome de DiGeorge compliqué d'une reaction du greffon contre l'hote. *Arch Fr Pediatr* 1985;**42**:853–5.

60. Wintergerst U, Meyer U, Remberger K, Belohradsky BH. Graft versus host reaction in an infant with DiGeorge syndrome. *Monats Kinderheilk* 1989;**137**:345–7.

61. Washington K, Gossage DL, Gottfried MR. Pathology of the liver in severe combined immunodeficiency and DiGeorge syndrome. *Pediatr Pathol* 1993;**13**:485–504.

62. Washington K, Gossage DL, Gottfried MR. Pathology of the pancreas in severe combined immunodeficiency and DiGeorge syndrome: acute graft-versus-host disease and unusual viral infections. *Hum Pathol* 1994;**25**: 908–14.

63. Tuvia J, Weisselberg B, Shif I, Keren G. Aplastic anaemia complicating adenovirus infection in DiGeorge syndrome. *Eur J Pediatr* 1988;**147**:643–4.

64. Wood DJ, David TJ, Chrystie IL, Totterdell B. Chronic enteric virus infection in two T-cell immunodeficient children. *J Med Virol* 1988;**24**:435–44.

65. Gilger MA, Matson DO, Conner ME, Rosenblatt HM, Finegold MJ, Estes MK. Extraintestinal rotavirus infections in children with immunodeficiency. *J Pediatr* 1992;**120**:912–7.

66. Sullivan KE, McDonald-McGinn D, Driscoll D, Emanuel BS, Zackai EH, Jawad AF. Longitudinal analysis of lymphocyte function and numbers in the first year of life in chromosome 22q11.2 deletion syndrome (DiGeorge syndrome/velocardiofacial syndrome). *Clin Diagn Lab Immunol* 1999;**6**:906–11.

67. Sullivan KE, Jawad AF, Randall P, et al. Lack of correlation between impaired T-cell production, immunodeficiency and other phenotypic features in chromosome 22q11.2 deletions syndrome (DiGeorge syndrome/velocardiofacial syndrome). *Clin Immunol Immunopathol* 1998;**84**:141–6.

68. Gennery AR, Barge D, O'Sullivan JJ, Flood TJ, Abinun M, Cant AJ. Antibody deficiency and autoimmunity in 22q11.2 deletion syndrome. *Arch Dis Child* 2002;**86**: 422–5.

69. Zimmerman D, Lteif AN. Thyrotoxicosis in children. *Endocrinol Metab Clin North Am* 1998;**27**:109–26.

70. Ham Pong AJ, Cavallo A, Holman GH, Goldman AS. DiGeorge syndrome: Long term survival complicated by Graves disease. *J Pediatr* 1985;**106**:619–20.

71. Kawame H, Adachi M, Tachibana K, et al. Graves' disease in patients with 22q11.2 deletion. *J Pediatr* 2001;**139**:892–5.

72. Pinchas-Hamiel O, Engelberg S, Mandel M, Passwell JH. Immune hemolytic anemia, thrombocytopenia and liver disease in a patient with DiGeorge syndrome. *Israel J Med Sci* 1994;**30**:530–2.

73. DePiero AD, Lourie EM, Berman BW, Robin NH, Zinn AB, Hostoffer RW. Recurrent immune cytopenias in two patients with DiGeorge/velocardiofacial syndorme. *J Pediatr* 1997;**131**:484–6.

74. Levy A, Michel G, Lemerrer M, Philip N. Idiopathic thrombocytopenic purpura in two mothers of children with DiGeorge sequence: a new component manifestation of deletion 22q11? *Am J Med Genet* 1997;**69**: 356–9.

75. Sullivan KE, McDonald-McGinn DM, Driscoll DA, et al. Juvenile rheumatoid arthritis-like polyarthritis in chromosome 22q11.2 deletion syndrome (DiGeorge anomalad/velocardiofacial syndrome/conotruncal anomaly face syndrome). *Arthritis Rheum* 1997;**40**:430–6.

76. Lawrence S, McDonald-McGinn DM, Zackai E, Sullivan KE. Thrombocytopenia in patients with chromosome 22q11.2 deletion syndrome. *J Pediatr* 2003;**143**:277–8.

77. Etzioni A, Pollack S. Autoimmune phenomena in DiGeorge syndrome. *Israel J Med Sci* 1994;**30**:853.

78. Berglund J, Christensen SB, Hallengren B. Total and age-specific incidence of Graves' thyrotoxicosis, toxic nodular goitre and solitary toxic adenoma in Malmo 1970–74. *J Intern Med* 1990;**227**:137–41.

79. Lima K, Abrahamsen TG, Wolff AB, et al. Hypoparathyroidism and autoimmunity in the 22q11.2 deletion syndrome. *Eur J Endocrinol* 2011;**165**:345–52.

80. McLean-Tooke A, Spickett GP, Gennery AR. Immunodeficiency and autoimmunity in 22q11.2 deletion syndrome. *Scand J Immunol* 2007;**66**:1–7.

81. Sullivan KE, McDonald-McGinn D, Zackai EH. CD4(+) CD25(+) T-cell production in healthy humans and in patients with thymic hypoplasia. *Clin Diagn Lab Immunol* 2002;**9**:1129–31.

82. King C, Ilic A, Koelsch K, Sarvetnick N. Homeostatic expansion of T-cells during immune insufficiency generates autoimmunity. *Cell* 2004;**117**:265–77.

83. August CS, Filler RM, Janeway CA, Markowski B, Kay HEM. Implanation of a fetal thymus restoring immunological competence in a patient with thymic aplasia (DiGeorge's syndrome). *Lancet* 1968;**2**:1210–1.

84. Barrett DJ, Wara DW, Ammann AJ, Cowan MJ. Thymosin therapy in the DiGeorge syndrome. *J Pediatr* 1980;**97**:66–71.

85. Pahwa S, Pahwa R, Incefy G, et al. Failure of immunologic reconstitution in a patient with the DiGeorge syndrome after fetal thymus transplantation. *Clin Immunol Immunopathol* 1979;**14**:96–106.

86. Smith J, Hsieh H. Treatment of DiGeorge syndrome with dialyzable leukocyte extract. *Clin Res* 1979;**27**:A588.

87. Cleveland WW, Fogel BJ, Brown WT, Kay HE. Foetal thymic transplant in a case of Digeorge's syndrome. *Lancet* 1968;**2**:1211–4.

88. Steele RW, Limas C, Thurman GB, Schuelein M, Bauer H, Bellanti JA. Familial thymic aplasia. Attempted reconstitution with fetal thymus in a Millipore diffusion chamber. *N Engl J Med* 1972;**287**:787–91.

89. Cruz Garcia Rodriguez M, Gracia Bouthelier R, Fontan Casariego G, et al. Implante de timo en un caso de sindrome de DiGeorge. *Ann Esp Pediatr* 1978;**11**:771–6.

90. Goldstein AL, Guha A, Zatz MM, Hardy MA, White A. Purification and biological activity of thymosin, a hormone of the thymus gland. *Proc Natl Acad Sci USA* 1972;**69**:1800–3.

91. Hardy MA, Quint J, Goldstein AL, State D, White A. Effect of thymosin and an antithymosin serum on allograft survival in mice. *Proc Natl Acad Sci USA* 1968;**61**:875–82.

92. Zinkernagel RM, Callahan GN, Klein J, Dennert G. Cytotoxic T-cells learn specificity for self H-2 during differentiation in the thymus. *Nature* 1978;**271**:251–3.

93. Borzy MS, Ridgway D, Noya FJ, Shearer WT. Successful bone marrow transplantation with split lymphoid chimerism in DiGeorge syndrome. *J Clin Immunol* 1989;**9**:386–92.

94. Goldsobel AB, Haas A, Stiehm ER. Bone marrow transplantation in DiGeorge syndrome. *J Pediatr* 1987;**111**:40–4.

95. Land MH, Garcia-Lloret MI, Borzy MS, et al. Long-term results of bone marrow transplantation in complete DiGeorge syndrome. *J Allergy Clin Immunol* 2007;**120**:908–15.

96. Markert ML, Kostyu DD, Ward FE, et al. Successful formation of a chimeric human thymus allograft following transplantation of cultured postnatal human thymus. *J Immunol* 1997;**158**:998–1005.

97. Markert ML, Sarzotti M, Ozaki DA, et al. Thymus transplantation in complete DiGeorge syndrome: immunologic and safety evaluations in 12 patients. *Blood* 2003;**102**:1121–30.

98. Markert ML, Devlin BH, Chinn IK, McCarthy EA. Thymus transplantation in complete DiGeorge anomaly. *Immunol Res* 2009;**44**:61–70.

99. Chinn IK, Markert ML. Induction of tolerance to parental parathyroid grafts using allogeneic thymus tissue in patients with DiGeorge anomaly. *J Allergy Clin Immunol* 2011;**127**:1351–5.

100. Mayumi M, Kimata H, Suehiro Y, et al. DiGeorge syndrome with hypogammaglobulinaemia: a patient with excess suppressor T-cell activity treated with fetal thymus transplantation. *Eur J Pediatr* 1989;**148**:518–22.

101. Garcia Miranda JL, Otero Gomez A, Varela Ansedes H, et al. Monosomy 22 with humoral immunodeficiency: is there an immunoglobulin chain deficit? *J Med Genet* 1983;**20**:69–72.

102. Schubert MS, Moss RB. Selective polysaccharide antibody deficiency in familial DiGeorge syndrome. *Ann Allergy* 1992;**69**:231–8.

103. Etzioni A, Pollack S. Hypogammaglobulinaemia in DiGeorge sequence. *Eur J Pediatr* 1989;**149**:143–4.

104. Junker AK, Driscoll DA. Humoral immunity in DiGeorge syndrome. *J Pediatr* 1995;**127**:231–7.

105. Patel K, Akhter J, Kobrynski LJ, Gathmann B, Davis O, Sullivan K. Immunoglobulin deficiencies: The B side of DiGeorge Syndrome. *J Pediatr* 2012;**161**:950–3.

106. Bassett AS, Chow EW, Husted J, et al. Premature death in adults with 22q11.2 deletion syndrome. *J Med Genet* 2009;**46**:324–30.

19

The Many Faces of the Hyper-IgE Syndrome

Karin R. Engelhardt, Bodo Grimbacher

Centre for Chronic Immunodeficiency, University Medical Centre, Freiburg, Germany

INTRODUCTION

Excessive immunoglobulin E (IgE) production is associated with increased susceptibility to infection in several immunodeficiency syndromes (1). Some of these patients have specific features that help the diagnostic work-up, such as a bleeding disorder in Wiskott–Aldrich syndrome, specific hair-shaft abnormalities in Comèl–Netherton syndrome, or abnormal T-cells and lack of B-cells in Omenn syndrome. Many individuals with immunodeficiency/ recurrent infections and elevated IgE levels but without such specific features or a molecular genetic diagnosis are often classified within the mixed bag of hyper-IgE syndromes (HIES).

HIES may be autosomal dominant (AD) (MIM #147060) or autosomal recessive (AR) (MIM #243700), but most patients are sporadic. HIES are characterized by the clinical triad of extreme elevations of serum IgE, recurrent (staphylococcal) skin abscesses, and recurrent pneumonia (2). Staphylococcal skin abscesses may be "cold" without signs of inflammation, but are typically filled with pus (3, 4). Eczema may already be present as a newborn rash and is usually driven by *Staphylococcus aureus*, improving with anti-staphylococcal antibiotics. Chronic candidiasis of mucosa and nail beds can occur.

Pneumonias are frequently complicated by parenchymal abnormalities of the lung, mainly bronchiectasis and, in individuals with AD-HIES, also pneumatoceles. These represent a risk for pulmonary fungal superinfection, which is a major source of morbidity and mortality (5). AD-HIES is distinguished by pronounced abnormalities of the connective tissue, skeleton and dentition (6). Typical findings include characteristic asymmetric facies, scoliosis and hyperextensible joints; recurrent fractures upon minimal trauma, especially in the long bones; and delayed shedding of three or more primary teeth, which is probably caused by the lack of resorption of primary tooth roots (7–9). Vascular abnormalities include tortuosity, dilation and aneurysms of the coronary arteries (10). Furthermore, focal hyperintensities in the white matter of the CNS were found, predominantly in adult patients (11).

In contrast, patients with AR-HIES have an increased susceptibility to severe and refractory cutaneous viral infections (molluscum contagiosum, herpes zoster, herpes simplex, human papillomavirus), food allergies, neurological symptoms, failure to thrive and a predisposition to malignancy (12, 13). Mortality occurs at a younger age, often in the first and second decades of life. (For recent reviews of clinical features see 14, 15.)

Three different molecular genetic defects in HIES have been published to date: heterozygous dominant-negative mutations in *STAT3*, accounting for approximately 75% of all autosomal-dominant and sporadic cases of HIES (16–21); homozygous and compound heterozygous loss-of-function mutations in *DOCK8*, accounting for approximately 80% of all autosomal-recessive cases with HIES (22–27); and a homozygous mutation in *TYK2* in one patient (28).

The aim of this review is to provide a historical perspective of the characterization of HIES and the discovery of the genetic defects causing it.

CHARACTERIZATION OF THE DISEASE – PHENOTYPES, INHERITANCE PATTERN AND FIRST LINKAGE ANALYSIS: 1966–2004

In 1966, Davis, Schaller and Wedgwood described two girls who had recurrent respiratory infections, severe eczema and recurrent staphylococcal skin infections that were notable for their lack of inflammation such as surrounding warmth, erythema or tenderness (3). These skin abscesses reminded the authors of the biblical figure of Job; thus, with the citation from Job II:7, "So went Satan forth from the presence of the Lord, and smote Job with sore boils from the sole of his foot unto his crown" they created the term "Job syndrome". In 1972, the term "Buckley syndrome" was derived from a report by Buckley et al. about two boys with similar infectious problems, severe dermatitis, distinctive facial features and elevated serum levels of IgE (29). Two years later, with the finding of elevated IgE levels in the two girls of the first report (30), it was realized that Job syndrome and Buckley syndrome represented the same disease, which is now mainly called "hyper-immunoglobulin E recurrent infection syndrome" or just "hyper-IgE syndrome" (HIES).

By 1999, more than 200 cases of HIES had been published, describing various defects in the immune system, such as abnormal granulocyte chemotaxis (4, 30–33), defective antibody production (34–38), altered B-cell maturation and increased plasma B-cell activating factor (BAFF) levels (38, 39), deficiency of "suppressor" T-cells (40), and abnormalities in the responsiveness to interleukin 4 (IL-4) and interferon γ (IFN-γ) or their production, respectively (41–44). In addition, several reports had indicated findings

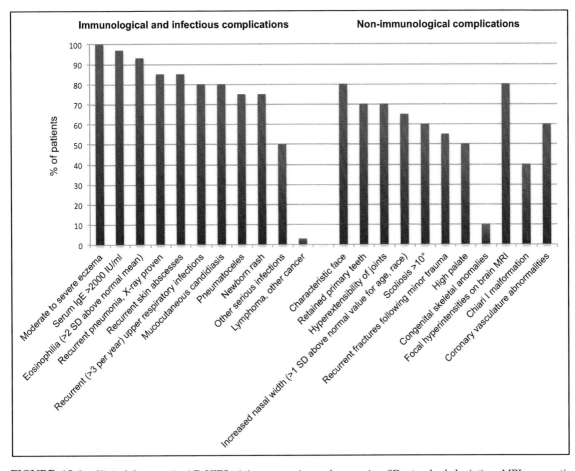

FIGURE 19.1 Clinical features in AD-HIES giving approximate frequencies. SD: standard deviation; MRI: magnetic resonance imaging.

unrelated to the immune system. Distinctive facial features, hyperextensibility of the joints, bone fractures and craniosynostosis occurred in many cases of HIES (3, 7, 8, 45–47).

When one of the authors of this review was a postdoctorate student in the group of Jennifer Puck at the US National Institutes of Health (NIH) in 1999, an investigation of non-immune features was performed in a cohort of 30 HIES patients collected by John Gallin, Harry Malech and Steve Holland. All subjects older than eight years had the following symptoms: scoliosis was present in 76%, hyperextensible joints in 67%, recurrent fractures upon minor

trauma in 57% and failure or delay of shedding of the primary teeth owing to lack of root resorption present in 72% (6). Thus, we showed that HIES is a multisystem disorder affecting the immune system, skeleton, dentition and the connective tissue (Fig. 19.1). Furthermore, we noted that the syndrome is inherited as a single-locus autosomal dominant trait with variable expressivity, even though most cases are sporadic (6).

Based on this extended phenotype of the disease, we developed a scoring system using clinical data and laboratory tests to assess patients presenting with symptoms of HIES (Table 19.1)

TABLE 19.1 NIH HIES score

Clinical findings	Points (circle appropriate box for each finding)									
	0	1	2	3	4	5	6	7	8	10
Highest IgE (IU/mL)	<200	200–500			501–1000				1001–2000	>2000
Skin abscesses (total #)	None		1–2		3–4				>4	
Pneumonias (X-ray proven, total #)	None		1		2		3		>3	
Parenchymal lung abnormalities	Absent						Bronchiectasis		Pneumatocele	
Other serious infection	None				Severe					
Fatal infection	Absent				Present					
Highest eosinophils/μL	<700			700-800			>800			
Newborn rash	Absent				Present					
Eczema (worst stage)	Absent	Mild	Moderate		Severe					
Sinusitis, otitis (# times in worst year)	1–2	3	4–6		>6					
Candidiasis	None	Oral, vaginal	Fingernail		Systemic					
Retained primary teeth	None	1	2		3					
Scoliosis, max curve	<10°		10–14°		15–20°				>20°	
Fractures with little trauma	None				1–2				>2	
Hyperextensibility	Absent				Present					
Characteristic face	Absent		Mild			Present				
Increased nose width (interalar distance)	<1 SD	1–2 SD		>2 SD						
High palate	Absent		Present							
Midline anomaly	Absent					Present				
Lymphoma	Absent				Present					
Young age add-on	>5 years			2–5 years		1–2 years		<1 year		

(48). The most points were given to findings specific to HIES, such as retention of primary teeth, serum-IgE levels >2000 IU/mL or pneumatocele formation, whereas features frequently also found in the general population, such as recurrent upper respiratory infections or eczema, received fewer points. The diagnosis of HIES is likely for patients with >40 points, possible for individuals with 20–40 points, and unlikely for individuals with <20 points. This score, known as the NIH HIES score, is still used today; often together with a modified score which predicts the likelihood of an individual with HIES to have a *STAT3* mutation (20).

Knowing the inheritance pattern, being able accurately to evaluate patients using the scoring system and having available multiplex kindred made genetic linkage analysis to find the cause(s) of HIES attractive. Prompted by the finding of a sporadic HIES patient with autism and mental retardation who had cytogenetic abnormalities on chromosome 4q (49), we performed a limited linkage analysis with microsatellite markers from this region of chromosome 4. We found linkage of our HIES families to the candidate region on chromosome 4 with a maximum multipoint logarithm of odds (LOD) score of >3.4, exceeding the standard genome-wide significance threshold of 3.0, but also with clear evidence of locus heterogeneity. Unfortunately, however, this linkage region turned out to be a "red herring" and it took eight more years to find the genetic defect of autosomal-dominant HIES (AD-HIES) on chromosome 17.

Yet before that, in 2004, a new form of HIES was described with autosomal-recessive inheritance and a slightly different phenotype. A total of 13 patients from six consanguineous families with autosomal recessive HIES (AR-HIES), all presenting with the immunological characteristics of HIES, such as highly elevated levels of serum IgE, recurrent staphylococcal skin abscesses, recurrent pneumonia, eczema, eosinophilia and candidiasis were found. However,

these patients did not develop pneumatoceles, and the non-immune features were absent. Instead, these AR-HIES patients experienced severe recurrent viral infections (molluscum contagiosum, herpes simplex, herpes zoster and human papillomavirus), autoimmunity, central nervous system abnormalities (hemiplegia, ischemic infarction and subarachnoid hemorrhages) and high mortality (12).

DISCOVERY OF THE FIRST TWO GENETIC DEFECTS – TYK2 AND STAT3: 2006 AND 2007

Curiously, it was the AR-HIES disease entity that brought forward the first publication of a genetic defect. In 2006, Minegishi et al. investigated a Japanese patient from consanguineous parents with eczema, recurrent sinopulmonary infections, staphylococcal skin abscesses, oral candidiasis, mild eosinophilia and high serum IgE (28). He had an NIH HIES score of 48, recurrent fungal (oral candidiasis) and viral infections (molluscum contagiosum, herpes simplex), but no skeletal abnormalities; hence, the diagnosis of AR-HIES was made. In addition, the patient showed an unusual susceptibility to mycobacteria and salmonella: he experienced BCG lymphadenitis and severe non-typhoid salmonella gastroenteritis leading to sepsis. A homozygous deletion of four nucleotides in the non-receptor tyrosine kinase *TYK2* was found, which resulted in a frameshift, leading to a premature stop codon and absence of full-length protein expression (28). However, no additional patient with findings of HIES was ever shown to have a *TYK2* mutation. We screened 15 autosomal-recessive families with HIES by genotyping on chromosome 19 and found the affected individuals in five families to be homozygous at a marker near the *TYK2* locus (50). We sequenced one patient from each family but did not find mutations in *TYK2*. We thus concluded that *TYK2* is not the

major "AR-HIES gene", but rather a gene causing a distinct disease entity associated with
susceptibility to intracellular bacteria. This
hypothesis was supported in 2011 and 2012 by
reports of a single patient with a homozygous
mutation in *TYK2* who had no signs of HIES
except for cutaneous viral infections (herpes
zoster), but with susceptibility to intracellular
bacterial infections, such as BCGosis and neu-
robrucellosis (51, 52).

In 2007, the long-awaited gene causing
AD-HIES was finally discovered, first by the
Japanese group led by Yoshi Minegishi. Spec-
ulating that a different molecule in the Janus
kinase (JAK)/STAT signaling pathway might
be implicated (TYK2 is a member of the JAK
family of tyrosine kinases), Minegishi's group
identified, through a combination of cytokine
stimulation assays and candidate gene sequencing, heterozygous dominant-negative
mutations in the signal transducer and activator of transcription 3 (*STAT3*) (16). Eight out of
15 unrelated sporadic HIES patients had five
different mutations in the DNA-binding domain. Patients' peripheral blood mononuclear
cells (PBMCs) showed defective responses to
the cytokines IL-6 and IL-10, as well as diminished ability of mutant STAT3 to bind to
DNA. Meanwhile, a group at NIH employed a
similar approach and published *STAT3* mutations in AD-HIES just a month later (17). Cytokine secretion assays and gene expression
arrays of resting and stimulated cells implicated *STAT3* as a candidate gene. Sequencing
of *STAT3* identified 18 different heterozygous
mutations in 50 familial and sporadic cases of
AD-HIES. Mutations were found not only in
the DNA-binding domain, but also in the Src
homology 2 (SH2) domain, which mediates
protein–protein interactions. Of interest, one
of the two original patients with Job syndrome
described by Davis et al., and two of her affected descendents, a son and a grandson, were
found to have a heterozygous missense mutation (p.R382W) in the DNA-binding domain of
STAT3 (53).

By 2010, >150 patients with dominant-
negative *STAT3* mutations had been described
(16–20, 54–57). The mutations, being mainly
missense or short in-frame deletions, never lead
to an absence of protein expression. Instead,
they impair the function of the mutated protein.
After cytokine binding to their receptors, Janus
kinases undergo activation and phosphorylate
STAT3, which then dimerizes and translocates to
the nucleus to control transcription of its target
genes (58). Mutated STAT3 confers a dominant-
negative effect on the homodimer formed with
a wild-type protein, thus preventing target gene
transcription (Fig. 19.2). Mutational hotspots are
found in the DNA-binding and SH2 domains of

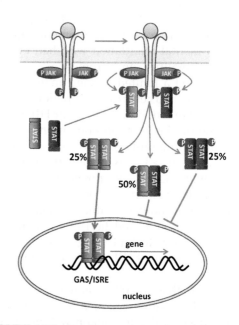

FIGURE 19.2 Dominant-negative effect of *STAT3* mutations on STAT3 signaling. Heterozygous *STAT3* mutations
result in half of the STAT3 proteins being non-functional, allowing only ≈25% intact STAT3 homodimers to form. These
are thought to be necessary for cell survival but seem to be
insufficient for the proper function of some subsets of cells,
such as leukocytes and cells that participate in bone formation. Mutated STAT3 is depicted red, wild-type STAT3 green.
(Figure courtesy of C. Woellner.) This figure is reproduced in
color in the color section.

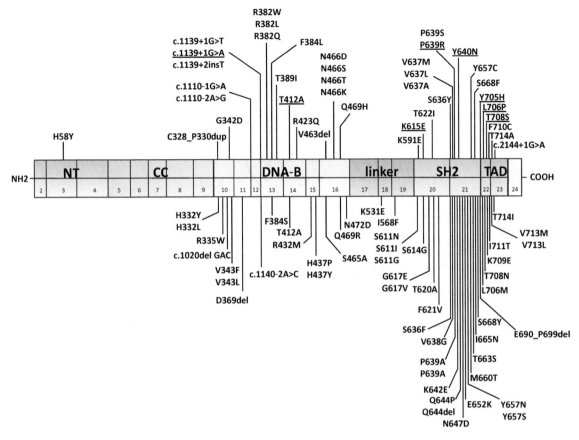

FIGURE 19.3 Schematic representation of *STAT3* showing mutational hotspots in the DNA-binding (DNA-B) and Src homology 2 (SH2) domain. Further mutations are located in the amino-terminal (NT), the linker and the transcriptional activation (TAD) domains. Mutations on the upper part of the cartoon were identified by Woellner et al., 2010. Underscored mutations have not yet been published in the literature. *(Figure modified after M.O. Chandesris et al., 2012 (59).)* This figure is reproduced in color in the color section.

STAT3 (Fig. 19.3). Despite functional differences in the two domains, a significant genotype–phenotype correlation could not be observed (20, 21).

The fact that patients with *STAT3* mutations have defective responses to both proinflammatory cytokines (IL-6) and anti-inflammatory cytokines (IL-10) might in part explain the dichotomy of AD-HIES being a disease of both too much and too little inflammation; while there is abundant inflammation in abscesses and pneumonia, systemic effects of inflammation such as fever and acute-phase proteins are impaired.

FINDINGS OF REDUCED NUMBERS OF TH17 CELLS AND DIMINISHED TH17 RESPONSES IN HIES PATIENTS: 2008

Another interesting discovery that helped to understand the pathogenesis of AD-HIES was the finding of reduced Th17 cells with diminished IL-17 production in patients with HIES. Milner et al. were the first to describe this deficiency in 2008 (54), followed three months later by reports from Australia, Seattle/Munich and Europe (19, 55, 60). *STAT3* is crucial for the

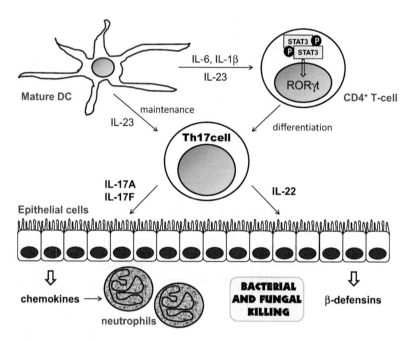

FIGURE 19.4   Role of *STAT3* in the differentiation of Th17 cells. Mature dendritic cells (DCs) secrete IL-1β, IL-6 and IL-23, which act on naïve CD4+ T-cells to activate STAT3-dependent expression of the Th17 lineage-determining transcription factor retinoic acid-related gamma t (RORγt). IL-23 is also important in the maintenance of differentiated Th17 cells. Th17 cells secrete IL-17A, IL17-F and IL-22, which act on epithelial cells triggering the production of chemokines and β-defensins, respectively. Chemokines recruit neutrophils, which kill pathogens by phagocytosis and neutrophil extracellular traps, whereas β-defensins are peptides with antimicrobial properties. Mutations in *STAT3* result in failure of Th17-cell differentiation which, in turn, leads to susceptibility to fungi and extracellular bacteria.

differentiation of Th17 cells by upregulating the Th17-specific transcription factor RORγt (61). Th17 cells are a T-helper cell subset that produces high amounts of IL-17A, IL-17F, IL-21 and IL-22, which are key cytokines in the production of antimicrobial peptides, including β-defensin (Fig. 19.4) (62). Antimicrobial peptides play important roles in the defense against extracellular bacteria and *Candida* species at mucosal surfaces (63); thus, lack of Th17 cells in HIES patients might account for the susceptibility to *Candida albicans* and staphylococcal skin infections. Indeed, keratinocytes and lung epithelial cells produce less β-defensin and neutrophil-recruiting chemokines when stimulated by T-cells from HIES patients (56). Keratinocytes and lung epithelial cells need the synergistic action of IL-17

and classical proinflammatory cytokines to produce antibacterial factors, whereas in other cell types, proinflammatory cytokines are sufficient. This might explain why in HIES, which is characterized by impaired IL-17 but normal proinflammatory cytokine production, susceptibility to bacterial infections occurs particularly in the skin and lung (6, 56).

DISCOVERY OF A GENETIC DEFECT FOR AR-HIES – DOCK8: 2009

At the end of 2009, with improved methods for gene identification in monogenetic diseases, two groups discovered, in parallel, homozygous and compound heterozygous deletions and mutations in *DOCK8* in patients with AR-HIES (22, 23). Zhang et al. published information about 11 patients with *DOCK8* mutations found by comparative genomic hybridization arrays and targeted gene sequencing (22). Using genome-wide single nucleotide polymorphism (SNP)-Chips and microsatellite markers to locate copy number variations and homozygous haplotypes, Engelhardt et al. found subtelomeric biallelic microdeletions including *DOCK8* in five patients, and a linkage region on chromosome 9p with 13 genes including *DOCK8* in others (23). In total, *DOCK8* was sequenced in 27 AR-HIES patients and mutations and deletions were found in 21. Most DOCK8-deficient patients had absent production of the full-length DOCK8 protein (23). At the same time, a *Dock8*-deficient mouse model that concentrated on the role of Dock8 on B-cell function was published (64). More reports of patients with *DOCK8* mutations followed, allowing a more detailed characterization of this deficiency (24–27, 65–70). In addition to the previously characterized features of AR-HIES such as elevated serum IgE, eosinophilia, eczema, staphylococcal skin infections, sinopulmonary infections, mucocutaneous candidiasis, viral skin infections and a predisposition to malignancy, DOCK8 deficiency is also associated with allergies, mainly severe food allergies, asthma and failure to thrive. Hallmarks of the disease are cutaneous viral infections, such as severe refractory molluscum contagiosum, warts, herpes zoster, and recurrent herpes simplex mucosal and skin infections. The connective tissue and skeletal manifestations characteristic of AD-HIES are infrequent in DOCK8 deficiency. Associated laboratory features are diminished serum IgM and abnormal T-lymphocyte proliferation.

DOCK8 is one of 11 members of the DOCK180 family of guanine nucleotide exchange factors (GEFs) that are thought to be involved in cytoskeletal rearrangements allowing cell migration, adhesion and growth (71). Several studies performed after 2009 investigated the role of DOCK8 in man and mice with a focus on B and T lymphocytes, dendritic cells (DCs), and natural killer (NK) cells (72–77).

Deficiency of DOCK8 results in an impaired ability to sustain circulating $CD27^+$ memory B-cells and long-lasting, affinity-matured antibody responses; a failure of memory $CD8^+$ T-cell persistence; a defective interstitial DC migration; and a reduced NK cell cytotoxicity. DOCK8 is crucial for the formation of a mature immunological synapse. In B-cells, accumulation of the integrin ligand ICAM-1 in the peripheral SMAC, needed for formation of a stable synapse that is able to deliver an integrin co-stimulus upon BRC activation, is dependent on DOCK8 (64). In $CD8^+$ T-cells, DOCK8 is required for the recruitment of the integrin LFA-1 to the $CD8^+$ T-cell synapse formed with DCs (64). Finally, the formation of a mature cytotoxic NK-cell synapse needs DOCK8 to polarize LFA-1, F-actin and cytotoxic granules towards the cell contact zone (76, 77) (Fig. 19.5). The lack of mature B- and $CD8^+$ T-cell synapses results in failure to generate long-lived humoral immunity and long-lived, recallable memory $CD8^+$ T-cells, whereas lack of a mature NK-cell synapse results in impaired NK-cell cytotoxicity.

In addition, other cellular pathways are impaired in DOCK8 deficiency. DOCK8 is critical for interstitial DC migration to the lymph node parenchyma and for T-cell priming during immune responses (75). Jabara et al. showed with B-cells from DOCK8-deficient patients that TLR9 activation depends on a DOCK8/MyD88/Pyk2 complex to drive activation of kinases Src/Lyn and Syk, followed by STAT3 phosphorylation, and ultimately leading to proliferation and differentiation of naïve B-cells (see Fig. 19.5) (72), suggesting that DOCK8 functions as an adaptor molecule in B-cells.

Lack of serological and cellular memory, lack of T-cell priming due to defective interstitial DC migration and lack of NK-cell cytotoxicity may explain the susceptibility to persistent viral and extracellular bacterial infections observed in DOCK8 deficiency.

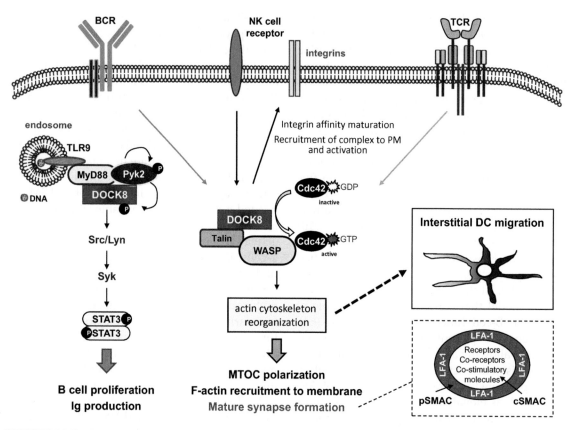

FIGURE 19.5 Role of DOCK8 in immune cells. In B-, T- and NK cells, DOCK8 is crucial for the formation of a stable immunological synapse with an integrin-containing pSMAC. It has been shown in NK cells that DOCK8 forms a complex with WASp and talin, leading to F-actin recruitment to the membrane, MTOC polarization and formation of the cytotoxic synapse. In this complex, DOCK8 might function as a GEF for Cdc42 which, in turn, activates WASp followed by actin cytoskeleton rearrangements. In DCs, DOCK8 plays a role in spatial activation of Cdc42 at the leading edge and interstitial DC migration. In B-cells, DOCK8 might have an additional role as an adaptor protein in a complex with MyD88 and Pyk2 downstream of TLR9 and upstream of STAT3, being important for B-cell proliferation and immunoglobulin production. BCR: B-cell receptor; DC: dendritic cell; GEF: guanine nucleotide exchange factor; MTOC: microtubule organizing center; NK cell: natural killer cell; PM: plasma membrane; pSMAC: peripheral supramolecular activation complex; TCR: T-cell receptor; WASp: Wiskott–Aldrich syndrome protein. This figure is reproduced in color in the color section.

HSCT EMERGES AS A SUCCESSFUL THERAPY FOR HIES: 1998–2012

An attempt to cure the severe immunodeficiency of HIES by hematopoietic stem cell transplantation (HSCT) was first made in two patients in 1998 and 2000, respectively (78, 79). Nester and colleagues reported a 46-year-old man with AD-HIES who had suffered from multiple episodes of pneumonia since early childhood. When he developed a B-cell lymphoma, he received bone marrow from an HLA-identical sibling. Following transplant, his serum IgE levels normalized and he did not experience any further severe infections. Yet, unfortunately, he died from interstitial pneumonitis six months after transplant (78). The second transplanted patient was a seven-year-old girl with severe AD-HIES due to a heterozygous mutation with dominant negative effect in *STAT3* who received bone marrow from a matched unrelated donor. She achieved full engraftment and, at first, her skin lesions cleared and serum IgE levels normalized. However, after four years, she again had high IgE levels and recurrent, albeit milder, infections. The authors suggested that susceptibility to infection may not have been confined to the hematopoietic system and did not regard HSCT as a curative therapy for HIES (79). However, long-term follow up of this patient by Andrew Cant and Andrew Gennery indicated considerable improvement and cessation of recurrent infections.

Additional transplants for AD-HIES appear to be more successful than initially thought. In 2010, the story of two unrelated boys with HIES due to mutations in *STAT3* was reported (80). They had received bone marrow from an HLA-identical sibling for treatment of high-grade non-Hodgkin's lymphoma 10 and 14 years earlier. Both patients achieved full donor chimerism and sustained immunological correction including normal IgE serum levels as well as absence of sinopulmonary and skin infections during the entire follow-up period. In one proband, non-immune manifestations like osteoporosis resolved. Furthermore, later-occurring complications such as coronary artery aneurysms, brain lesions, or degenerative joint disease did not develop after HSCT.

More recently, successful HSCT for AR-HIES patients with *DOCK8* mutations were reported by several groups. Bittner et al. described a DOCK8-deficient patient six years after a bone marrow transplant from her HLA-identical father (65). Partial myeloablative preconditioning was selected due to the severe pulmonary dysfunction, and resulted in mixed donor chimerism, but was sufficient to reduce her high IgE levels and improve her pulmonary function as well as to resolve her severe eczema, and severe infections such as molluscum contagiosum and candidiasis. However, she still suffers from multiple food allergies, which improved in severity but did not resolve completely. Gatz et al. reported two additional DOCK8-deficient

patients with classic phenotype who were able to lead a normal and healthy life after HSCT (67). Both received bone marrow from an HLA-matched unrelated donor and achieved full donor chimerism. One proband had a rather troublesome transplant course with brain abscesses and an acute Epstein–Barr virus infection. Finally, however, both had corrected immunological functions with complete clearance of molluscum contagiosum and absence of bacterial infections at last follow up two and four years after transplant, respectively. Additional successful HSCT of DOCK8-deficient patients followed these initial reports (66, 68, 70, 81–83), suggesting that transplantation is a curative treatment for DOCK8 deficiency, which is otherwise characterized by high morbidity and mortality, often within the first or second decade of life.

CONCLUDING REMARKS AND A GLIMPSE INTO THE FUTURE

The advances in whole-exome and genome sequencing will speed up the discovery of new genetic defects in HIES patients who have no molecular diagnosis yet. Characterization of new genetically defined HIES may identify unique features for each, making them distinct from the "classic" phenotypes of STAT3-AD-HIES and DOCK8-AR-HIES. With each discovery, the question will arise if it is still a form of HIES or rather a disease entity on its own. DOCK8 deficiency serves as an example: most individuals with a *DOCK8* mutation were originally diagnosed by their physicians as having AR-HIES; and some of the patients in the original publication describing AR-HIES as a new disease entity (12) were found in retrospect to have a mutation in *DOCK8*, making *DOCK8* a disease gene of AR-HIES. Despite that, due to a more pronounced T-cell defect than had originally been recognized, many investigators prefer to detach this entity from the term HIES and call it DOCK8 deficiency (26, 63, 83, 84) or DOCK8 immunodeficiency syndrome (85).

The same might happen to new and yet-to-be-identified forms of AD- and AR-HIES, including the glycosylation defect we found in some AR-HIES patients (86, 87).

Perhaps in the future, when the dust has settled, we will restrict the term HIES to the classic autosomal-dominant "Job syndrome", while the other forms will receive their own name based on the gene defect. At present, we recognize HIES as a disease with AD and AR traits, multiple genetic defects and considerable phenotypic variability, all sharing as common ground the classical clinical triad of HIES: high serum levels of IgE, recurrent pneumonias, and frequent staphylococcal lung and skin infections.

Acknowledgments

The authors are supported by the German Federal Ministry of Education and Research (BMBF 01 EO 0803).

References

1. Grimbacher B, Belohradsky BH, Holland SM. Immunoglobulin E in primary immunodeficiency diseases. *Allergy* 2002;**57**:995–1007.
2. Grimbacher B, Holland SM, Puck JM. Hyper-IgE syndromes. *Immunol Rev* 2005;**203**:244–50.
3. Davis SD, Schaller J, Wedgwood RJ. Job's syndrome. Recurrent, "cold", staphylococcal abscesses. *Lancet* 1966;**1**:1013–5.
4. Hill HR, Ochs HD, Quie PG, et al. Defect in neutrophil granulocyte chemotaxis in Job's syndrome of recurrent "cold" staphylococcal abscesses. *Lancet* 1974;**2**:617–9.
5. Freeman AF, Kleiner DE, Nadiminti H, et al. Causes of death in hyper-IgE syndrome. *J Allergy Clin Immunol* 2007;**119**:123412–40.
6. Grimbacher B, Holland SM, Gallin JI, et al. Hyper-IgE syndrome with recurrent infections – an autosomal dominant multisystem disorder. *N Engl J Med* 1999;**340**:692–702.
7. Borges WG, Hensley T, Carey JC, Petrak BA, Hill HR. The face of Job. *J Pediatr* 1998;**133**:303–5.
8. Kirchner SG, Sivit CJ, Wright PF. Hyperimmunoglobulinemia E syndrome: association with osteoporosis and recurrent fractures. *Radiology* 1985;**156**:362.
9. O'Connell AC, Puck JM, Grimbacher B, et al. Delayed eruption of permanent teeth in hyperimmunoglobulinemia E recurrent infection syndrome. *Oral Surg Oral Med Oral Pathol Oral Radiol Endod* 2000;**89**:177–85.
10. Freeman AF, Avila EM, Shaw PA, et al. Coronary artery abnormalities in Hyper-IgE syndrome. *J Clin Immunol* 2011;**31**:338–45.
11. Freeman AF, Collura-Burke CJ, Patronas NJ, et al. Brain abnormalities in patients with hyperimmunoglobulin E syndrome. *Pediatrics* 2007;**119**:e1121–1125.
12. Renner ED, Puck JM, Holland SM, et al. Autosomal recessive hyperimmunoglobulin E syndrome: a distinct disease entity. *J Pediatr* 2004;**144**:93–9.
13. Zhang Q, Su HC. Hyperimmunoglobulin E syndromes in pediatrics. *Curr Opin Pediatr* 2011;**23**:653–8.
14. Yong PF, Freeman AF, Engelhardt KR, Holland S, Puck JM, Grimbacher B. An update on the hyper-IgE syndromes. *Arthritis Res Ther* 2012;**14**:228.
15. Minegishi Y, Saito M. Cutaneous manifestations of Hyper IgE syndrome. *Allergol Int* 2012;**61**:191–6.
16. Minegishi Y, Saito M, Tsuchiya S, et al. Dominant-negative mutations in the DNA-binding domain of STAT3 cause hyper-IgE syndrome. *Nature* 2007;**448**:1058–62.
17. Holland SM, DeLeo FR, Elloumi HZ, et al. STAT3 mutations in the hyper-IgE syndrome. *N Engl J Med* 2007;**357**:1608–19.
18. Jiao H, Toth B, Erdos M, et al. Novel and recurrent STAT3 mutations in hyper-IgE syndrome patients from different ethnic groups. *Mol Immunol* 2008;**46**:202–6.
19. Renner ED, Rylaarsdam S, Anover-Sombke S, et al. Novel signal transducer and activator of transcription 3 (STAT3) mutations, reduced T(H)17 cell numbers, and variably defective STAT3 phosphorylation in hyper-IgE syndrome. *J Allergy Clin Immunol* 2008;**122**:181–7.
20. Woellner C, Gertz EM, Schaffer AA, et al. Mutations in STAT3 and diagnostic guidelines for hyper-IgE syndrome. *J Allergy Clin Immunol* 2010;**125**:424–32.
21. Heimall J, Davis J, Shaw PA, et al. Paucity of genotype-phenotype correlations in STAT3 mutation positive Hyper IgE Syndrome (HIES). *Clin Immunol* 2011;**139**:75–84.
22. Zhang Q, Davis JC, Lamborn IT, et al. Combined immunodeficiency associated with DOCK8 mutations. *N Engl J Med* 2009;**361**:2046–55.
23. Engelhardt KR, McGhee S, Winkler S, et al. Large deletions and point mutations involving the dedicator of cytokinesis 8 (DOCK8) in the autosomal-recessive form of hyper-IgE syndrome. *J Allergy Clin Immunol* 2009;**124**:1289–302.
24. Al-Herz W, Ragupathy R, Massaad MJ, et al. Clinical, immunologic and genetic profiles of DOCK8-deficient patients in Kuwait. *Clin Immunol* 2012;**143**:266–72.
25. Chu EY, Freeman AF, Jing H, et al. Cutaneous manifestations of DOCK8 deficiency syndrome. *Arch Dermatol* 2012;**148**:79–84.
26. Sanal O, Jing H, Ozgur T, et al. Additional diverse findings expand the clinical presentation of DOCK8 deficiency. *J Clin Immunol* 2012;**32**:698–708.
27. Alsum Z, Hawwari A, Alsmadi O, et al. Clinical, immunological and molecular characterization of DOCK8 and DOCK8-like deficient patients: single center experience of twenty-five patients. *J Clin Immunol* 2013;**33**:55–67.

28. Minegishi Y, Saito M, Morio T, et al. Human tyrosine kinase 2 deficiency reveals its requisite roles in multiple cytokine signals involved in innate and acquired immunity. *Immunity* 2006;**25**:745–55.

29. Buckley RH, Wray BB, Belmaker EZ. Extreme hyperimmunoglobulinemia E and undue susceptibility to infection. *Pediatrics* 1972;**49**:59–70.

30. Hill HR, Quie PG. Raised serum-IgE levels and defective neutrophil chemotaxis in three children with eczema and recurrent bacterial infections. *Lancet* 1974;**1**:183–7.

31. Van Scoy RE, Hill HR, Ritts RE, Quie PG. Familial neutrophil chemotaxis defect, recurrent bacterial infections, mucocutaneous candidiasis, and hyperimmunoglobulinemia E. *Ann Intern Med* 1975;**82**:766–71.

32. Van Epps DE, El-Naggar A, Ochs HD. Abnormalities of lymphocyte locomotion in immunodeficiency disease. *Clin Exp Immunol* 1983;**53**:679–88.

33. Gahr M, Muller W, Allgeier B, Speer CP. A boy with recurrent infections, impaired PMN-chemotaxis, increased IgE concentrations and cranial synostosis – a variant of the hyper-IgE syndrome? *Helv Paediatr Acta* 1987;**42**:185–90.

34. Buckley RH, Becker WG. Abnormalities in the regulation of human IgE synthesis. *Immunol Rev* 1978;**41**: 288–314.

35. Dreskin SC, Goldsmith PK, Gallin JI. Immunoglobulins in the hyperimmunoglobulin E and recurrent infection (Job's) syndrome. Deficiency of anti-Staphylococcus aureus immunoglobulin A. *J Clin Invest* 1985;**75**: 26–34.

36. Dreskin SC, Goldsmith PK, Strober W, Zech LA, Gallin JI. Metabolism of immunoglobulin E in patients with markedly elevated serum immunoglobulin E levels. *J Clin Invest* 1987;**79**:1764–72.

37. Vercelli D, Jabara HH, Cunningham-Rundles C, et al. Regulation of immunoglobulin (Ig)E synthesis in the hyper-IgE syndrome. *J Clin Invest* 1990;**85**:1666–71.

38. Meyer-Bahlburg A, Renner ED, Rylaarsdam S, et al. Heterozygous signal transducer and activator of transcription 3 mutations in hyper-IgE syndrome result in altered B-cell maturation. *J Allergy Clin Immunol* 2012;**129**: 559–62 62.

39. Speckmann C, Enders A, Woellner C, et al. Reduced memory B-cells in patients with hyper IgE syndrome. *Clin Immunol* 2008;**129**:448–54.

40. Geha RS, Reinherz E, Leung D, McKee Jr KT, Schlossman S, Rosen FS. Deficiency of suppressor T-cells in the hyperimmunoglobulin E syndrome. *J Clin Invest* 1981;**68**: 783–91.

41. King CL, Gallin JI, Malech HL, Abramson SL, Nutman TB. Regulation of immunoglobulin production in hyperimmunoglobulin E recurrent-infection syndrome by interferon gamma. *Proc Natl Acad Sci USA* 1989;**86**:10085–9.

42. Claasen JJ, Levine AD, Schiff SE, Buckley RH. Mononuclear cells from patients with the hyper-IgE syndrome produce little IgE when they are stimulated with recombinant human interleukin-4. *J Allergy Clin Immunol* 1991;**88**:713–21.

43. Paganelli R, Scala E, Capobianchi MR, et al. Selective deficiency of interferon-gamma production in the hyper-IgE syndrome. Relationship to in vitro IgE synthesis. *Clin Exp Immunol* 1991;**84**:28–33.

44. Rousset F, Robert J, Andary M, et al. Shifts in interleukin-4 and interferon-gamma production by T-cells of patients with elevated serum IgE levels and the modulatory effects of these lymphokines on spontaneous IgE synthesis. *J Allergy Clin Immunol* 1991;**87**:58–69.

45. Smithwick EM, Finelt M, Pahwa S, et al. Cranial synostosis in Job's syndrome. *Lancet* 1978;**1**:826.

46. Hoger PH, Boltshauser E, Hitzig WH. Craniosynostosis in hyper-IgE-syndrome. *Eur J Pediatr* 1985;**144**:414–7.

47. Holland SM, Gallin JI. Disorders of granulocytes and monocytes. In: Fauci AS, Braunwald E, Isselbacher KJ, editors. *Harrison's Principles of Internal Medicine*. 14th edn. New York: McGraw-Hill; 1998. p. 351–6.

48. Grimbacher B, Schaffer AA, Holland SM, et al. Genetic linkage of hyper-IgE syndrome to chromosome 4. *Am J Hum Genet* 1999;**65**:735–44.

49. Grimbacher B, Dutra AS, Holland SM, et al. Analphoid marker chromosome in a patient with hyper-IgE syndrome, autism, and mild mental retardation. *Genet Med* 1999;**1**:213–8.

50. Woellner C, Schaffer AA, Puck JM, et al. The hyper IgE syndrome and mutations in TYK2. *Immunity* 2007;**26**:535 author reply 6.

51. Grant AV, Boisson-Dupuis S, Herquelot E, et al. Accounting for genetic heterogeneity in homozygosity mapping: application to Mendelian susceptibility to mycobacterial disease. *J Med Genet* 2011;**48**:567–71.

52. Kilic SS, Hacimustafaoglu M, Boisson-Dupuis S, et al. A patient with tyrosine kinase 2 deficiency without hyper-IgE syndrome. *J Pediatr* 2012;**160**:1055–7.

53. Renner ED, Torgerson TR, Rylaarsdam S, et al. STAT3 mutation in the original patient with Job's syndrome. *N Engl J Med* 2007;**357**:1667–8.

54. Milner JD, Brenchley JM, Laurence A, et al. Impaired T(H)17 cell differentiation in subjects with autosomal dominant hyper-IgE syndrome. *Nature* 2008;**452**:773–6.

55. Ma CS, Chew GY, Simpson N, et al. Deficiency of Th17 cells in hyper IgE syndrome due to mutations in STAT3. *J Exp Med* 2008;**205**:1551–7.

56. Minegishi Y, Saito M, Nagasawa M, et al. Molecular explanation for the contradiction between systemic Th17 defect and localized bacterial infection in hyper-IgE syndrome. *J Exp Med* 2009;**206**:1291–301.

57. Al Khatib S, Keles S, Garcia-Lloret M, et al. Defects along the T(H)17 differentiation pathway underlie genetically distinct forms of the hyper IgE syndrome. *J Allergy Clin Immunol* 2009;**124**:342–8.

58. Levy DE, Lee CK. What does Stat3 do? *J Clin Invest* 2002;**109**:1143–8.

59. de Beaucoudrey L, Puel A, Filipe-Santos O, et al. Mutations in STAT3 and IL12RB1 impair the development of human IL-17-producing T-cells. *J Exp Med* 2008;**205**:1543–50.

60. McGeachy MJ, Cua DJ. Th17 cell differentiation: the long and winding road. *Immunity* 2008;**28**:445–53.

61. Ouyang W, Kolls JK, Zheng Y. The biological functions of T helper 17 cell effector cytokines in inflammation. *Immunity* 2008;**28**:454–67.

62. Aujla SJ, Chan YR, Zheng M, et al. IL-22 mediates mucosal host defense against Gram-negative bacterial pneumonia. *Nat Med* 2008;**14**:275–81.

63. Randall KL, Lambe T, Johnson AL, et al. Dock8 mutations cripple B-cell immunological synapses, germinal centers and long-lived antibody production. *Nat Immunol* 2009;**10**:1283–91.

64. Bittner TC, Pannicke U, Renner ED, et al. Successful long-term correction of autosomal recessive hyper-IgE syndrome due to DOCK8 deficiency by hematopoietic stem cell transplantation. *Klin Padiatr* 2010;**222**:351–5.

65. McDonald DR, Massaad MJ, Johnston A, et al. Successful engraftment of donor marrow after allogeneic hematopoietic cell transplantation in autosomal-recessive hyper-IgE syndrome caused by dedicator of cytokinesis 8 deficiency. *J Allergy Clin Immunol* 2010;**126**:1304–5.

66. Gatz SA, Benninghoff U, Schutz C, et al. Curative treatment of autosomal-recessive hyper-IgE syndrome by hematopoietic cell transplantation. *Bone Marrow Transplant* 2011;**46**:552–6.

67. Barlogis V, Galambrun C, Chambost H, et al. Successful allogeneic hematopoietic stem cell transplantation for DOCK8 deficiency. *J Allergy Clin Immunol* 2011;**128**:420–2.

68. Lee WI, Huang JL, Lin SJ, et al. Clinical, immunological and genetic features in Taiwanese patients with the phenotype of hyper-immunoglobulin E recurrent infection syndromes (HIES). *Immunobiology* 2011;**216**:909–17.

69. Metin A, Tavil B, Azik F, et al. Successful bone marrow transplantation for DOCK8 deficient hyper IgE syndrome. *Pediatr Transplant* 2012;**16**:398–9.

70. Ruusala A, Aspenstrom P. Isolation and characterisation of DOCK8, a member of the DOCK180-related regulators of cell morphology. *FEBS Lett* 2004;**572**:159–66.

71. Jabara HH, McDonald DR, Janssen E, et al. DOCK8 functions as an adaptor that links TLR-MyD88 signaling to B-cell activation. *Nat Immunol* 2012;**13**:612–20.

72. Lambe T, Crawford G, Johnson AL, et al. DOCK8 is essential for T-cell survival and the maintenance of CD8+ T-cell memory. *Eur J Immunol* 2011;**41**:3423–35.

73. Randall KL, Chan SS, Ma CS, et al. DOCK8 deficiency impairs CD8 T-cell survival and function in humans and mice. *J Exp Med* 2011;**208**:2305–20.

74. Harada Y, Tanaka Y, Terasawa M, et al. DOCK8 is a Cdc42 activator critical for interstitial dendritic cell migration during immune responses. *Blood* 2012;**119**:4451–61.

75. Mizesko MC, Banerjee PP, Monaco-Shawver L, et al. Defective actin accumulation impairs human natural killer cell function in patients with dedicator of cytokinesis 8 deficiency. *J Allergy Clin Immunol* 2013;**131**:840–8.

76. Ham H, Guerrier S, Kim J, et al. Dedicator of cytokinesis 8 interacts with talin and Wiskott-Aldrich syndrome protein to regulate NK cell cytotoxicity. *J Immunol* 2013;**190**:3661–9.

77. Nester TA, Wagnon AH, Reilly WF, Spitzer G, Kjeldsberg CR, Hill HR. Effects of allogeneic peripheral stem cell transplantation in a patient with Job syndrome of hyperimmunoglobulinemia E and recurrent infections. *Am J Med* 1998;**105**:162–4.

78. Gennery AR, Flood TJ, Abinun M, Cant AJ. Bone marrow transplantation does not correct the hyper IgE syndrome. *Bone Marrow Transplant* 2000;**25**:1303–5.

79. Goussetis E, Peristeri I, Kitra V, et al. Successful long-term immunologic reconstitution by allogeneic hematopoietic stem cell transplantation cures patients with autosomal dominant hyper-IgE syndrome. *J Allergy Clin Immunol* 2010;**126**:392–4.

80. Boztug H, Karitnig-Weiss C, Ausserer B, et al. Clinical and immunological correction of DOCK8 deficiency by allogeneic hematopoietic stem cell transplantation following a reduced toxicity conditioning regimen. *Pediatr Hematol Oncol* 2012;**29**:585–94.

81. Siah TW, Gennery A, Leech S, Taylor A. Gross generalized molluscum contagiosum in a patient with autosomal recessive hyper-IgE syndrome, which resolved spontaneously after haematopoietic stem-cell transplantation. *Clin Exp Dermatol* 2013;**38**:196–7.

82. Al-Mousa H, Hawwari A, Alsum Z. In DOCK8 deficiency donor cell engraftment post-genoidentical hematopoietic stem cell transplantation is possible without conditioning. *J Allergy Clin Immunol* 2013;**131**:1244–5.

83. Su HC. Dedicator of cytokinesis 8 (DOCK8) deficiency. *Curr Opin Allergy Clin Immunol* 2010;**10**:515–20.

84. Zhang Q, Davis JC, Dove CG, Su HC. Genetic, clinical, and laboratory markers for DOCK8 immunodeficiency syndrome. *Dis Markers* 2010;**29**:131–9.

85. Chandesris MO, Melki I, Natividad A, et al. Autosomal dominant STAT3 deficiency and hyper-IgE syndrome: molecular, cellular, and clinical features from a French national survey. *Medicine (Baltimore)* 2012;**91**:e1–19.

86. Zhang Y, Yu X, Ichikawa M, et al. Autosomal recessive phosphoglucomutase 3 (*PGM3*) mutations link glycosylation defects to atopy, immune deficiency, autoimmunity, and neurocognitive impairment. *J Allergy Clin Immunol* 2014;**133**:1400–9.

87. Sassi A, Lazaroski S, Wu G, et al. Hypomorphic homozygopus mutations in phosphoglucomutase 3 (*PGM3*) impair immunity and increase serum levels. *J Allergy Clin Immunol* 2014;**133**:1410–19.

ADA Deficiency – The First Described Genetic Defect Causing PID

Michael S. Hershfield[1], Hilaire J. Meuwissen[2], Rochelle Hirschhorn[3]

[1]Professor of Medicine and Biochemistry, Duke University School of Medicine, Durham, NC 27710

[2]Department of Pediatrics, Albany Medical Center, Albany, NY

[3]Professor Emerita of Medicine, Cell Biology and Pediatrics, Research Professor of Medicine, New York University Medical Center, NY

INTRODUCTION

The 1972 report by Giblett et al. (1) of the absence of adenosine deaminase (ADA) in two unrelated children with defective cellular immunity had a galvanizing effect on two disparate fields. At the time, clinical immunologists were categorizing inherited immune deficiencies based on mode of inheritance, clinical presentation, histopathology, immunoglobulin levels, and functional testing. Suddenly, a subset of patients with "Swiss type" lymphopenic immune deficiency (immunoglobulin deficiency combined with defects in lymphoid cells) had a metabolic disorder that could be diagnosed by a simple biochemical test. This new inborn error was just as surprising to physicians and biochemists working on "purine metabolic diseases". They were focused on gout and a few rare disorders that caused uric acid overproduction, the paradigm being the Lesch–Nyhan syndrome resulting from X-linked deficiency of

HPRT (hypoxanthine-guanine phosphoribosyl-transferase). ADA deficiency did not affect uric acid levels and had an equally devastating, but very different, phenotype.

Convinced of the importance of nucleotide/nucleoside metabolism to immune development (the connection was not obvious, to say the least), Giblett systematically screened other enzymes in patients with immune deficiency, reporting in 1975 the absence of purine nucleoside phosphorylase (PNP), one metabolic step removed from both ADA and HPRT (Fig. 20.1), in a girl with T-cell lymphopenia (2). In response to Giblett's discoveries, immunologists and purine biochemists were soon scrambling to make connections and share patient samples, to learn each other's language (if not their methods), and to help (and sometimes compete with) each other to gain an understanding of pathogenesis. Sadly, Eloise Giblett, known as Elo by her friends and colleagues, who died in 2009, is not here to

provide her own account of events that led to and followed her discovery of ADA deficiency.

The three authors of this chapter came from different disciplines and were at different stages in their careers in 1972. HJM was caring for patients with immune deficiency, RH had training and research interests in both immunology and genetics and MSH was acquiring postdoctoral training in molecular biology and biochemical genetics. Each will present their recollection of how they and their colleagues were involved in/reacted to, the discovery of ADA deficiency, and relate some observations of how, over the next few months and years, new knowledge emerged regarding mechanism, genetics, and therapy.

Although each of us regards their involvement in this story a bit differently: from providential to coincidental – we were all "at the right place at the right time", but with the medical and scientific training needed to play our part. For that we are grateful.

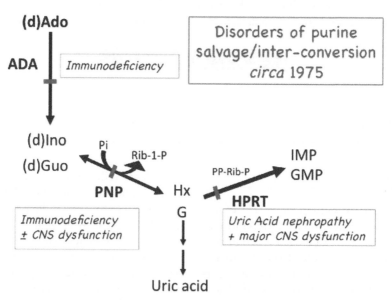

FIGURE 20.1 Major inherited disorders of purine metabolism *circa* 1975. ADA: adenosine deaminase; PNP: purine nucleoside phosphorylase; HPRT: hypoxanthine-guanine phosphoribosyltransferase; (d): deoxy; Ado: adenosine; Ino: inosine; Guo: guanosine; Hx: hypoxanthine; G: guanine; Pi: inorganic phosphate; Rib-1-P: ribose-1-phosphate; PP-rib-P: phosphoribosylpyrophosphate; CNS: central nervous system. This figure is reproduced in color in the color section.

Given the space allowance, our accounts are not comprehensive, and span only about two decades. We apologize to many of our colleagues whose contributions are not specifically mentioned.

HILAIRE J. MEUWISSEN

The First ADA Deficient Patient

I arrived in Dr Robert Good's lab in Minneapolis in 1965 to begin a research fellowship, and immediately became immersed in issues related to the thymus and bursa of Fabricius, antibody formation, immunodeficiency diseases and human histocompatibility antigens, to the extent that precise typing was then possible. In 1968, when a boy presented with severe combined immunodeficiency (SCID), a fatal condition, Dr Good decided to use a novel strategy: to transplant bone marrow from a sister compatible at the major histocompatibility locus. This resulted, for the first time, in a healthy baby who had a permanent graft that restored immune functions (3). Dr Good's success opened the way for a worldwide application of this new technique in patients with immunodeficiencies and other blood or stem cell diseases.

In 1970, I joined the NY State Birth Defects Institute at the Albany Medical Center to establish a marrow transplantation unit. My overriding interest was to show that Dr Good's strategy of transplanting marrow from a histocompatible sibling could be used for any patient with SCID. In 1972, a five-month-old girl with persistent mucocutaneous candidiasis, interstitial pneumonitis and significant T-cell and variable B-cell abnormalities was admitted to our marrow transplantation unit. We tissue typed the family but, unfortunately, she did not have an HLA compatible sibling and the findings were very confusing until we realized that the biological father and mother were consanguineous. To confirm this point I called around and heard of

Dr Eloise Giblett's reputation in red cell genetics. When I contacted her, she agreed to screen the red cells with a set of polymorphic proteins, and I sent blood samples from the family members to Seattle.

A few weeks later Dr Giblett called back and told me that the data analysis of the red cells was as confusing as that of the tissue typing. Then she added a surprising detail: our patient's red cells contained no ADA, while her biological father and the mother had approximately 50% of normal ADA activity, as measured by both starch gel electrophoresis (Fig. 20.2) and spectrophotometric assay. She had never come across ADA-negative red cells, and could recall no report of this abnormality; she did not quite know what to make of it, and considered it a red herring. I discussed this unusual coincidence with the experts at NY State health department, but no one could even imagine why the lack of an enzyme, hidden somewhere in nucleic acid metabolism, should or could interfere with cutaneous delayed hypersensitivity or antibody formation in lymph nodes or spleen. It made no sense to publish the case, even as a letter to the Editor, because we could not be sure of any connection between these two rare entities.

Several months later at the Children's Hospital of Michigan in Detroit, Dr Flossie Cohen, in despair about the poor condition of her little patient with severe immunodeficiency and pulmonary failure, decided to "pull all the stops" and prepare for bone marrow transplantation. That was when her friend, Elo Giblett, who had been thinking about the coincidence of SCID and lack of ADA in the Albany patient, called to ask if Flossie by chance had a SCID patient to test for red blood cell ADA activity. She had, and the result was that Dr Giblett encountered, twice in a very short time, a red cell abnormality she had never seen before. In Albany and Detroit, we were stunned to hear of a second patient with severe immune deficiencies compatible with SCID who also lacked ADA. I flew to Seattle, and discussed the implications with

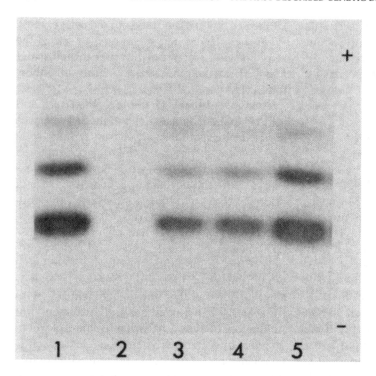

FIGURE 20.2 Starch-gel electrophoretic patterns of five hemolysates stained for ADA activity. Channels 1 and 5, from normal controls, and channels 3 and 4, from parents of patient, show the typical pattern of ADA-1. The hemolysate of the patient, in channel 2, contains no ADA activity. *(Adapted from (1).)*

Dr Giblett. She proposed that we write a short paper. I wrote the first draft, but she was the better writer and discussant, and she wrote most of the report that appeared in *The Lancet* (1).

As the Albany patient had no matched sibling available, we first transplanted, unsuccessfully, fetal thymus and liver. We then proceeded with a maternal bone marrow transplant, as the mother was non-reactive in the mixed leukocyte culture. The girl subsequently died of overwhelming cytomegalovirus (CMV) infection, possibly originating from the maternal graft.

A connection between ADA deficiency and SCID now appeared highly likely. Moreover, ADA-negative SCID was not as rare as we initially had thought, and pediatric immunologists were eager to identify these patients. Dr Ian Porter, head of the NY State Birth Defects Institute, encouraged us to organize a workshop on ADA–SCID, with the aim of defining the genetics, clinical characteristics and, if possible, the biochemical abnormalities. In preparation,

we contacted friends and colleagues working with patients suffering from SCID, and collected clinical, radiological and pathological data on a substantial number of patients from the USA, Canada, and Europe. We then asked two radiologists with extensive experience in pediatric immunodeficiency disease (J. Wolfson, V. Cross), and two similarly experienced pediatric pathologists (J. Huber, J. Kersey) to perform a blinded review of these data and specimens. The Workshop was convened in Albany NY in 1973 (4, 5). The participants provided expertise in genetics (Eloise Giblett, Rochelle Hirschhorn, Kurt Hirschhorn), clinical immunology (Flossie Cohen, Fred Rosen, Richard Pickering, Bernard Pollara, Ellen Moore, Richard Keightley, Arthur Amman, Diane Wara and myself), and therapy (Richard Hong).

The genetic data supported autosomal recessive inheritance, but it did not appear that a single deletion could eliminate both ADA and HLA antigen genes, as some had thought, as the ADA and HLA loci were on chromosomes 20 and 6, respectively. Instead, it seemed that the deficiency of ADA alone was responsible for the immune deficiency (confirmed later by restoration of immunocompetence by transfusing ADA-positive red blood cells and then PEG-ADA). Summarizing the clinical and immunological data for 22 patients with SCID and known ADA status, Dr Flossie Cohen concluded that no significant differences existed between ADA-positive and ADA-negative groups in respect to onset or severity of illness, or age at death. Residual T- and B-cell function was found in both groups. In the ADA-negative group, six of 12 patients had "severe SCID", while the others had some evidence of T- and/or B-cell functions.

The 37 SCID patients with known ADA status and available radiographs could easily be separated into ADA-positive or ADA-negative groups solely by blinded reading of X-rays. Abnormalities in severely affected ADA-negative patients involved the pelvis, spine, and ribs (concavity and flaring of anterior rib ends). These skeletal abnormalities were not observed in the ADA-positive group, or in ADA-negative patients with residual T- and/or B-cell function. Somewhat similar alterations of ribs and pelvis found in achondroplasia and other congenital abnormalities could be distinguished with careful examination from those associated with ADA-deficient SCID (5).

The pathology of tissues from four ADA-negative patients was also remarkable, primarily in respect to the thymus. Unlike the classical picture in SCID, all ADA-deficient children had evidence of preserved Hassall's corpuscles and differentiated thymic epithelium (5). A thymus shadow was seen on X-ray of one ADA-negative patient at seven months of age, who then developed severe immunodeficiency. It was suggested that during fetal development, the thymus may have been populated with lymphoid cells under the influence of maternal ADA crossing the placenta, with subsequent involution after birth.

Dr Richard Hong summarized treatment for SCID. With rare exceptions, administration of fetal liver and thymus transplants was unsuccessful, and transfer factor did not work. The best results for ADA-negative and ADA-positive SCID patients were obtained by transplantation of histocompatible stem cells. At this workshop, faced with the first single gene defect associated with a specific immunodeficiency, we considered the distinct possibility that immunodeficiencies might be corrected not only by transplantation, but also with gene therapy. I remember the excitement in the conference room when this prospect was raised, long before the first attempts at gene therapy were performed at the NIH in patients with ADA-negative SCID.

At the end of the workshop, Dr Hong saluted the "opening of the study of immunological deficiency to the precise tools of the biochemist" and Dr Kurt Hirschhorn concluded that "we have again succeeded in coordinating three different disciplines: immunology, biochemistry and genetics to work jointly on a problem". He expressed the hope that "similar cooperation in other areas of research will be profitable for us all" (4).

ROCHELLE HIRSCHHORN

I have had a varied scientific and medical background, including genetics, biochemistry, cell biology and immunology as well as medicine and pediatrics. I believe this explains why I immediately recognized and accepted the significance of the report in *The Lancet* in November 1972 by Eloise Giblett et al. (as described by Hilaire Meuwissen) (1). I had no problem with invoking nucleotide involvement in possible mechanisms of the immunological disease, and I would like to outline briefly some relevant history, my own and that of Elo Giblett.

In 1952, my senior year at Barnard College, I decided to go into medicine. I had taken enough science courses almost to qualify as a chemistry major, but I needed to write a senior science thesis to allow me to apply to medical school. I found a report by a Professor Harry Harris dealing with inborn errors of metabolism, using electrophoresis and staining for enzymes. Garrod, whom Harris had studied, had hypothesized that enzyme defects caused by inborn errors of metabolism could be responsible for certain diseases. Dr Harris helped prove this theory. My choice of this subject for a senior thesis was followed by studies with Professor Harris in England, and later by work in immunology. This background was seminal for my involvement with studies of ADA deficiency and my search for further understanding of that disease.

I entered NYU Medical School in 1953, joining Kurt Hirschhorn, who was a medical student there and whom I had married six months previously. I participated in several student research projects (two were genetic), and graduated in 1957 with Honors (AOA). I then left with Kurt for Uppsala, Sweden where, with a newly awarded MS degree in genetics in addition to his MD, he had received a fellowship at the Genetics Institute. I was fortunate in obtaining a position at the institute working with a well known Italian geneticist, Marco Fracarro. Among other studies, we were trying to determine which compound in a mix reported in a recent paper was allowing peripheral blood lymphocytes to multiply. It turned out to be phytohemagglutinin (PHA), now a standard compound used to test the immunocompetence of lymphocytes. Marco and I did not discover this, although several years later I did publish findings as to why PHA functioned to activate lymphocytes.

After visiting many of the genetics centers in Europe, Kurt and I returned to the USA in 1958. I served as an intern in Medicine at NYU-Bellevue and obtained some pediatric training, as my children's pediatrician had convinced me to join her at the Bellevue Well Baby Clinic on a part-time basis. I had intended to return to genetics part time, and to join my husband in a group he was heading. However, I was told by my Chairman of Medicine, Lewis Thomas,

that studying genetics with my husband would be "nepotism" and to come up with a different plan. I surveyed the faculty for an imaginative, brilliant mentor who would accept me and ended up as a Rheumatology Fellow in the laboratory of Professor Gerald Weissmann,

I gave the lectures on gout to the NYU medical students, but did research on white blood cells. Between 1966 and 1969, I was a Fellow of the Arthritis Foundation. These were exciting times at NYU, and I watched several major biological discoveries by the NYU immunology/rheumatology group, including the work that resulted in the Nobel Prize for discovering the genes involved in the HLA system. In, for me, a landmark 1970 publication, Gerry Weissmann and I, for convoluted reasons, hypothesized and then demonstrated that a variety of adenine compounds (including adenosine, AMP, ADP, ATP and cyclic AMP) inhibited the synthesis of DNA by PHA stimulated lymphocytes. I was therefore not surprised to later learn that absence of the enzyme ADA, which should elevate the concentration of adenosine, would interfere with the mitosis of immunocompetent lymphocytes.

In 1971, Kurt and I (with three children) left for London to work with Harry Harris, now head of the famous Galton Laboratory at University College London. I was very interested in the mechanisms for activation of immunocompetent lymphocytes. Since molecular DNA and protein analysis was in its infancy, I proposed to look for changes in different forms of enzymes (isozymes) associated with stimulation or inhibition of immunocompetent cells. Harry had defined a whole series of isozymes that differed among people, and published a very clear manual and a book for biochemical genetics. These enzyme polymorphisms "complemented typing of blood groups" and, at about the same time (1969), Elo Giblett published a superb single author book dealing with blood groups and with known red cell enzyme differences. (In her excellent 2006 autobiographical essay (6), Elo described how she had spent six months in 1955 in London studying blood groups at Dr Patrick Mollison's Blood Transfusion Research Unit as part of a two year Hematology fellowship. She did this on the advice of Arno Motulsky, then a young, but well trained and wise geneticist. Seattle at the time had one of the leading centers for genetics in the USA, and Arno is still a major figure in Human Genetics.)

I was particularly interested in polymorphisms of ADA, which the Harris group defined in a series of publications from 1968 to 1971. In an addendum at the back of her book, Elo refers to this work on ADA. The two groups corresponded with each other for the next almost 20 years (1967–1985), as seen from the Harris archives at the University of Pennsylvania where he had become an All University Professor.

Before leaving for London in 1971, I had applied to the NIH for a research development award (RCDA). Upon returning from London in 1972, I found that I had been funded for "Activation of Immunocompetent Lymphocytes". I also read with great interest the paper by Giblett et al. in the November 1972 issue of *The Lancet* describing ADA deficiency in two patients with SCID, followed in December by a brief report in *The Lancet* of the disease in Denmark by Dissing and Knudsen (7). It was in this setting that I began collaborations with the Albany group, and with others who were also drawn to the questions raised by the discovery of ADA deficiency. I became embedded in a part of the immunology community, particularly with the Harvard group led by Fred Rosen and the European community, as well as with David Martin and Bill Kelley whose interest was diseases of purine metabolism.

Among other early studies, I measured ADA in red cells of 24 patients of Fred Rosen with hypogammaglobulinemia to confirm that none had ADA deficiency. Together with Hilaire Meuwissen and the Albany group, I demonstrated that ADA activity in red cells and non-red cells was controlled by a single genetic locus, and that tissue isozymes represented binding of ADA to

a non-ADA protein, which was later demonstrated by several groups to be CD26/DPP IV. We published three papers on this theme from 1973 to 1975.

The Albany group designed a screen and undertook newborn testing for ADA deficiency (8). At the time, newborn screening for inherited metabolic disorders was relatively new and mostly for phenylketonuria (PKU). Screening in NY City was difficult because of poverty, changing location, no telephones etc. Bellevue had a very experienced nurse and a pediatric physician/geneticist, and when a child from NYC was lost to follow-up, I would participate in confirming screening results with the help of this nurse (sometimes I ended up exploring the south Bronx for a family by myself). I had the good fortune of following the children with Dr William Borkowsky, who was not only a pediatrician but an immunologist and infectious disease specialist. Most children identified with low red cell ADA activity by the screen were not immunodeficient. We were examining one of these babies, when Bill said to me, "You want to see an immunodeficient? Come to the Hospital Wards and I will show you a baby with thrush pouring out of his mouth". He not only had thrush, but also cartilage abnormalities at the rib junctions (lateral to classic rachitic rosaries). In order to not be biased, I gave the blood sample to my technician, telling her it was from a normal child. She came back and said this is an ADA-deficient baby! This was confirmed, and standard immunological tests showed combined immunodeficiency. From the records, he had been tested as a newborn, but for unknown reasons had not been identified.

At one of the "clinical meetings", a group of us were discussing ADA deficiency and its possible pathophysiology. We were each following affected children and we decided to collaborate by exchanging samples and determining metabolites. The critical result came from Amos Cohen in David Martin's Lab, who found markedly elevated deoxyATP in red cells, reported in 1978 (9).

We encountered a family with a pregnant mother who had a previous child with SCID. She offered without our asking to have prenatal diagnosis but would deliver the child independent of the diagnosis. The fetus was found to be ADA deficient, confirmed at birth. There was no candidate for a bone marrow transplant. We were desperate. Together with Steven Polmar, and based on *in vitro* and *in vivo* studies, we proposed to give the baby partial exchange transfusions with frozen irradiated blood (10). Several patients of Drs Arye Rubinstein, George Papageorgiu and others received this pseudo-enzyme replacement therapy with initially promising results. However, the treatment was complicated by high iron overload, rare red blood cell types, incompatibility and unknown exposure to chicken pox.

In 1980, I was Attending Physician on the prison ward of Bellevue. I developed a febrile illness that lasted over three months, which fortunately was not the about-to-be discovered illness caused by HIV. I slowly recovered over the next five years to be able to do research. During this period, I was approached vis à vis testing PEG-ADA but had to refuse due to this illness. Fortunately, Michael Hershfield, who had been actively studying the pathogenesis of ADA deficiency, undertook PEG-ADA therapy (as described below).

In 1985, I began mutation analysis of ADA during a sabbatical with Stuart Orkin at Harvard (work I pursued for many years afterward, as did Michael Hershfield). Around this time, I spoke at the Institute of Medicine, predicting that ADA deficiency would be the first of the successful gene therapies.

MICHAEL S. HERSHFIELD

Searching for Mechanisms

In 1972, after a fellowship at the NIH studying bacteriophage DNA replication, I began a second postdoc with J.E. Seegmiller at UC San Diego (UCSD). A few years earlier, Jay's lab (then at the NIH) had identified a deficiency of the purine salvage enzyme HPRT in the Lesch–Nyhan syndrome (uric acid nephropathy, a severe movement disorder, and compulsive self-mutilation). That discovery was reflected in the name of his division at UCSD: Rheumatic and Genetic Diseases. Within a decade, such divisions were renamed Rheumatology and Immunology, but at the time, "inborn errors research" was attracting young PhDs and NIH-trained MDs hoping to discover the biochemical mechanisms responsible for the intriguing phenotypes associated with inherited enzyme deficiencies. This was the challenge posed by the recent report that another purine salvage enzyme, ADA, was absent in two patients with severely defective cell-mediated immunity (1): the clinical consequences of HPRT and ADA deficiency could not have been more different.

At UCSD, the Seegmiller lab had begun to use human lymphoblastoid cell lines (LCL) and somatic cell genetic techniques to study HPRT deficiency and potentially other inborn errors. Without appreciating where it would lead, I decided to work out a new method for studying *de novo* purine nucleotide synthesis in intact LCL, including variants lacking salvage pathway enzymes. In one of these experiments, I found that, when ADA was inhibited, adenosine was toxic to LCL lacking adenosine kinase, which I had isolated. This result was puzzling, as nucleoside toxicity was thought to require conversion to nucleotides, and it seemed inconsistent with the then-prevailing view that adenosine-induced ATP pool expansion was responsible for the lymphopenia associated with ADA deficiency. We proposed that a novel "nucleotide-independent" mechanism might contribute, possibly via a receptor or enzyme with high affinity for adenosine (11).

In mid-1976, I joined the Rheumatology division at Duke, and decided to work on the pathogenesis of ADA deficiency. My lab was located in research space that had been set up on Rankin

Ward, the Duke Clinical Research Unit (CRU), by Jim Wyngaarden, the chair of Medicine, who had a strong interest in purine disorders. Fortuitously, Rebecca Buckley had a clinical research unit (CRU) protocol to study bone marrow transplantation in patients with SCID, for whom a special isolation facility had been constructed on Rankin Ward. I began to test SCID patients for ADA and PNP deficiency, and identified several over the next few years. From time to time, I joined Rebecca and her team when they rounded on these patients, which was my introduction to Pediatric Immunology.

In January 1977, while writing my first NIH grant and RCDA applications ("A Lymphoblast Model for Diseases of Purine Metabolism"), I found a report of a cytoplasmic adenosine binding protein of unknown function. I set up a ^{3}H-adenosine binding assay, found this activity in LCL extracts, and proposed that it might mediate nucleotide-independent adenosine toxicity. After the applications were submitted, I began to purify an adenosine-binding protein from human placenta. Serendipity soon revealed its function, and a role in ADA deficiency, when Nicholas Kredich, another Duke Rheumatologist, returned from a 1976–77 sabbatical in the lab of his former Duke classmate, David Martin, at UC San Francisco, where he had also worked on adenosine toxicity.

Nick's main interest was sulfur amino acid metabolism. He realized that ADA facilitated the hydrolysis of S-adenosylhomocysteine (AdoHcy), a product and inhibitor of all S-adenosylmethionine (AdoMet)-dependent transmethylation reactions. At UCSF, Nick obtained evidence that adenosine impaired AdoHcy breakdown to inhibit DNA methylation in ADA-inhibited mouse T lymphoma cells, findings soon to appear in *Cell* (12). As Nick was interested in AdoHcy hydrolase, I suggested that he assay for the enzyme in column fractions I had generated in isolating placental adenosine-binding protein. To our joint surprise, AdoHcyase and adenosine-binding

activities co-purified. We quickly reported this finding (13), and went on to show that AdoHcy-mediated inhibition of methylation could account for nucleotide-independent adenosine toxicity to ADA-inhibited human LCL (14).

A twist... the Role of Deoxyadenosine (With AdoHcyase as One of its Targets)

In early 1978, Amos Cohen et al. and Mary Sue Coleman et al. independently made the very important discovery that dATP, not ATP, accumulated in red cells of ADA-deficient patients (9, 15). Other reports from the labs of Anne Simmonds, Dennis Carson (with whom I had overlapped in the Seegmiller lab) and David Martin showed that deoxyadenosine was more toxic than adenosine to ADA-inhibited PBL and LCL (16–19). Mary Sue Coleman suggested that dATP accumulation might alter the product of terminal deoxynucleotidyl transferase (TDT) in lymphoid cells to limit immunological diversity. The Martin and Carson groups suggested that high deoxynucleoside kinase activity would favor dATP accumulation in ADA-deficient T-cells, resulting in inhibition of ribonucleotide reductase and a block in DNA replication; an analogous mechanism involving accumulation of dGTP from 2'-deoxyguanosine was suggested as the cause of T lymphopenia in PNP deficiency. Dennis Carson and Buddy Ullman (working in the Martin lab) went on to use mutant human B and T LCL to define the roles of specific nucleoside kinases in activating deoxyadenosine and deoxyguanosine, and I collaborated in this work (20, 21).

In the spring of 1978, I discovered an unexpected action of deoxyadenosine. Although it was not a substrate and only a weak competitive inhibitor of AdoHcyase, I found that ^{3}H-deoxyadenosine formed a complex with the enzyme. This binding led to the loss of catalytic activity, which was not restored by dialysis. The kinetics and irreversibility suggested a recently described mechanism termed "suicide inactivation" (22).

The potential clinical significance became apparent when we found that red cells from three ADA-deficient SCID patients had <2% of normal AdoHcyase activity (23). AdoHcyase inactivation and inhibition of methylation accounted for the ability of deoxyadenosine to inhibit the growth of kinase-deficient ADA-inhibited LCLs that were unable to convert deoxyadenosine to dATP.

A Pharmacological Model of ADA Deficiency

Relating findings in cultured LCL to the *in vivo* situation was not straightforward. The *in vitro* experiments used arbitrary concentrations of either adenosine or deoxyadenosine, whereas both nucleosides are present *in vivo*, and at levels determined by their individual rates of formation, excretion and metabolism. Measuring metabolites in lymphoid cells from patients with SCID was not possible. At the time, a potent and specific ADA inhibitor, 2'-deoxycoformycin, was being investigated as a treatment for acute T-lymphoblastic leukemia (T-ALL), and I thought this offered a relevant *in vivo* pharmacological model.

Between 1980 and 1984, we collaborated with Beverly Mitchell, Michael Grever and oncologists at Duke to monitor closely the biochemical consequences of infusing deoxycoformycin in patients with T-ALL. Inducing ADA deficiency resulted in the rapid inactivation of AdoHcyase in their circulating lymphoblasts, followed by an increase in AdoHcy to levels that inhibited RNA methylation; these changes coincided with a marked accumulation of dATP in these cells (24, 25). Besides documenting AdoHcyase inactivation and methylation inhibition in lymphoid cells *in vivo*, a dramatic and rapid change from a T lymphoblastoid to a myeloid leukemic phenotype in one patient raised the possibility, still not fully explored, that ADA deficiency may affect the program of differentiation of hematopoietic stem cells (25).

Introducing Enzyme Replacement Therapy With PEG-ADA

By 1983, much of what is known today about the pathogenesis of SCID due to ADA deficiency had been discovered, and was reviewed that year in a chapter that Nick Kredich and I wrote for the 5th edition of *The Metabolic Basis of Inherited Disease* (26). Missing from the chapter was the relationship of genotype to phenotype, as the first ADA cDNA sequences were only reported in 1983 (27, 28), and Dan Wiginton et al. did not publish the sequence of the entire 32kb *ADA* gene until 1986 (29). Partial exchange transfusion was reviewed, but the only effective therapy was still transplantation of bone marrow from an HLA-identical sibling, just as for SCID patients whose molecular defects remained unknown. By 1984, a proposal from the NIH to undertake gene therapy for ADA deficiency was beginning to attract wide attention, but few knew of efforts by a small New Jersey pharmaceutical company to test a novel form of enzyme replacement for ADA deficiency.

One afternoon in mid-1985, Rebecca Buckley stopped me in the parking lot we shared to ask if I would look over some material she had received from Enzon, Inc. The thin folder contained animal toxicology and pharmacokinetic data for a purified bovine ADA conjugated to polyethylene glycol. The company's founders, Frank Davis, a biochemist at Rutgers University, and his student Abe Abuchowski, held the patent on PEGylation; they had been trying for several years to organize a trial of PEG-ADA for ADA deficiency. At the time, no PEGylated protein, and no form of enzyme replacement therapy, had been approved by the Food and Drug Administration (FDA) for any disease.

Though not a proof of principle (an ADA knockout mouse would not be available for ten more years), the Enzon data showed that PEG-ADA was not toxic to rats and mice, and it had a much longer circulating life in plasma and was less immunogenic in mice than unmodified bovine ADA. Tissue uptake was not examined, but published studies suggested that PEG-ADA was unlikely to be taken up by lymphocytes. I was not concerned by this, as ADA substrates could rapidly equilibrate between cells and plasma, and partial exchange transfusion markedly reduced deoxyadenosine excretion and the level of red cell dATP in ADA-deficient patients (9). It seemed possible that PEG-ADA could provide much more ADA activity than exchange transfusion, and it might be far safer for chronic therapy.

I agreed with Rebecca that a test of PEG-ADA would be worthwhile for a three year old patient we had diagnosed in 1983, who had not engrafted after two haploidentical transplants, and whose immune function had not improved while receiving biweekly red cell transfusions for more than a year. I thought we could easily tell if PEG-ADA was able to correct the metabolic effects of ADA deficiency, and Rebecca's lab could evaluate restoration of immune function. On April 1, 1986, a representative from Enzon flew to Durham to hand deliver several vials of PEG-ADA. The next morning, Rebecca and I met in a conference room on Rankin Ward, where we were interviewed by CBS news about the prospects for gene therapy for ADA deficiency. Once that was over, we walked down the hall to witness the first IM injection of PEG-ADA.

By the next morning, it was clear that the enzyme was circulating in the patient's plasma and her red cells already showed a decrease in dATP and an increase in AdoHcyase activity. After four months of slow dose escalation, these parameters had normalized and her lymphocyte counts had begun to increase. We began to treat a second patient, and reported these results in 1987 (30). In March 1990, PEG-ADA (Adagen®) received FDA approval. At about the same time, two of the 12 children receiving PEG-ADA became the first two human patients to enter a trial of gene therapy at the NIH (the topic of gene therapy is reviewed in Chapter 26).

References

1. Giblett ER, Anderson JE, Cohen F, Pollara B, Meuwissen HJ. Adenosine deaminase deficiency in two patients with severely impaired cellular immunity. *Lancet* 1972;2:1067–9.
2. Giblett ER, Ammann AJ, Wara DW, Sandman R, Diamond LK. Nucleoside-phosphorylase deficiency in a child with severely defective T-cell immunity and normal B-cell immunity. *Lancet* 1975;1:1010–3.
3. Gatti RA, Meuwissen HJ, Allen HD, Hong R, Good RA. Immunological reconstitution of sex-linked lymphopenic immunological deficiency. *Lancet* 1968;2:1366–9.
4. Meuwissen HJ, Pickering RJ, Pollara B, Porter IH, editors. *Combined Immunodeficiency Disease and Adenosine Deaminase Deficiency. A Molecular Defect.* New York: Academic Press; 1975.
5. Meuwissen HJ, Pollara B, Pickering RJ. Combined immunodeficiency disease associated with adenosine deaminase deficiency. Report on a workshop held in Albany, New York, October 1, 1973. *J Pediatr* 1975;86:169–81.
6. Giblett ER. Back to the beginnings: an autobiography. *Transfus Med Rev* 2006;20:318–21.
7. Dissing J, Knudsen B. Adenosine-deaminase deficiency and combined immunodeficiency syndrome. *Lancet* 1972;2:1316.
8. Moore EC, Meuwissen HJ. Screening for ADA deficiency. *J Pediatr* 1974;85:802–4.
9. Cohen A, Hirschhorn R, Horowitz SD, et al. Deoxyadenosine triphosphate as a potentially toxic metabolite in adenosine deaminase deficiency. *Proc Natl Acad Sci USA* 1978;75:472–5.
10. Polmar SH, Stern RC, Schwartz AL, Wetzler EM, Chase PA, Hirschhorn R. Enzyme replacement therapy for adenosine deaminase deficiency and severe combined immunodeficiency. *N Engl J Med* 1976;295:1337–43.
11. Hershfield MS, Snyder FF, Seegmiller JE. Adenine and adenosine are toxic to human lymphoblast mutants defective in purine salvage enzymes. *Science* 1977;197:1284–7.
12. Kredich NM, Martin Jr DW. Role of S-adenosylhomocysteine in adenosine- mediated toxicity in cultured moust T-lymphoma cells. *Cell* 1977;12:931–8.
13. Hershfield MS, Kredich NM. S-adenosylhomocysteine hydrolase is an adenosine-binding protein: a target for adenosine toxicity. *Science* 1978;202:757–60.
14. Kredich NM, Hershfield MS. S-adenosylhomocysteine toxicity in normal and adenosine kinase-deficient lymphoblasts of human origin. *Proc Natl Acad Sci USA* 1979;76:2450–4.
15. Coleman MS, Donofrio J, Hutton JJ, et al. Identification and quantitation of adenine deoxynucleotides in erythrocytes of a patient with adenosine deaminase deficiency and severe combined immunodeficiency. *J Biol Chem* 1978;253:1619–26.

16. Carson DA, Kaye J, Seegmiller JE. Lymphospecific toxicity in adenosine deaminase deficiency and purine nucleoside phosphorylase deficiency: Possible role of nucleoside kinase(s). *Proc Natl Acad Sci USA* 1977;**74**:5677.

17. Carson DA, Kaye J, Seegmiller JE. Differential sensitivity of human leukemic T-cell lines and B-cell lines to growth inhibition by deoxyadenosine. *J Immunol* 1978;**121**:1726–31.

18. Simmonds HA, Panayi GS, Corrigall V. A role for purine metabolism in the immune response: Adenosine-deaminase activity and deoxyadenosine catabolism. *Lancet* 1978;**1**:60–3.

19. Ullman B, Gudas LJ, Cohen A, Martin Jr DW. Deoxyadenosine metabolism and cytotoxicity in cultured mouse T lymphoma cells: a model for immunodeficiency disease. *Cell* 1978;**14**:365–75.

20. Hershfield MS, Fetter JE, Small WC, et al. Effects of mutational loss of adenosine kinase and deoxycytidine kinase on deoxyATP accumulation and deoxyadenosine toxicity in cultured CEM human T-lymphoblastoid cells. *J Biol Chem* 1982;**257**:6380–6.

21. Ullman B, Levinson BB, Hershfield MS, Martin Jr DW. A biochemical genetic study of the role of specific nucleoside kinases in deoxyadenosine phosphorylation by cultured human cells. *J Biol Chem* 1981;**256**:848–52.

22. Hershfield MS. Apparent suicide inactivation of human lymphoblast S-adenosylhomocysteine hydrolase by 2'-deoxyadenosine and adenine arabinoside. A basis for direct toxic effects of analogs of adenosine. *J Biol Chem* 1979;**254**:22–5.

23. Hershfield MS, Kredich NM, Ownby DR, Ownby H, Buckley R. In vivo inactivation of erythrocyte S-adenosylhomocysteine hydrolase by 2'-deoxyadenosine in adenosine deaminase-deficient patients. *J Clin Invest* 1979;**63**:807–11.

24. Hershfield MS, Kredich N.M., Koller CA, et al. S-adenosylhomocysteine catabolism and basis for acquired resistance during treatment of T-cell acute lymphoblastic leukemia with 2'-deoxycoformycin alone and in combination with 9-beta-D-arabinofuranosyladenine. *Cancer Res* 1983;**43**:3451–8.

25. Hershfield MS, Kurtzberg J, Harden E, Moore JO, Whang-Peng J, Haynes BF. Conversion of a stem cell leukemia from a T-lymphoid to a myeloid phenotype induced by the adenosine deaminase inhibitor 2'-deoxycoformycin. *Proc Natl Acad Sci USA* 1984;**81**:253–7.

26. Kredich NM, Hershfield MS. Immunodeficiency diseases caused by adenosine deaminase deficiency and purine nucleoside phosphorylase deficiency. In: Stanbury JB, Wyngaarden JB, Fredrickson DS, Goldstein JL, Brown MS, editors. *The Metabolic Basis of Inherited Disease*. 5th edn. New York: McGraw-Hill; 1983. p. 1157–83.

27. Valerio D, Duyvesteyn MGC, Meera Kahn P, van Kessel AG, de Waard A, van der Eb A. Isolation of cDNA clones for human adenosine deaminase. *Gene* 1983;**25**:231–40.

28. Wiginton DA, Adrian GS, Friedman D, Suttle DP, Hutton JJ. Cloning of cDNA sequences of human adenosine deaminase. *Proc Natl Acad Sci USA* 1983;**80**:7481–5.

29. Wiginton DA, Kaplan DJ, States JC, et al. Complete sequence and structure of the gene for human adenosine deaminase. *Biochemistry* 1986;**25**:8234–44.

30. Hershfield MS, Buckley RH, Greenberg ML, et al. Treatment of adenosine deaminase deficiency with polyethylene glycol-modified adenosine deaminase. *N Engl J Med* 1987;**316**:589–96.

21

The Leukocyte Adhesion Deficiency Story

Amos Etzioni

Meyer Children Hospital, Rambam Medical Campus, Rappaport Faculty of Medicine, Technion, Haifa, Israel

OUTLINE

INTRODUCTION

The phenomena of leukocyte adhesion and migration were observed more than 150 years ago, but their physiological significance remained elusive for many years. Rudolf Virchow, for instance, believed that leukocyte emigration served a function in the nourishment of tissues (1). Only towards the end of the 19th century did Elie Metchnikoff reveal the link between the leukocyte and the host defense mechanism.

In 1888, Theudore Leber demonstrated for the first time migration of leukocytes toward chemical stimuli and noted "the property of the leukocyte to topic migration by substances foreign to the organism is of the greatest importance in making possible an extensive counteraction of the organism against external factors, since only in this way is the accumulation of a large number of leukocytes at the site of noxa assured" (2).

It took another half century to understand the clinical relevance of chemotaxis with the recognition of pathological disorders in functions of human leukocytes. The first disease affecting neutrophils, chronic granulomotous disease (CGD) (see Chapter 13) was described in the early 1960s. Some 20 years later, a defect in leukocyte adhesion was reported (3–6). As in many other clinical conditions, the description of the clinical symptoms preceded the finding of the molecular defects causing the disease. While leukocyte adhesion deficiency (LAD) I and II were "experiments of nature" and affected patients were the driving force to reveal the importance of the integrins and selectins, for LAD III "knockout" mouse models were required to provide the final answer.

Primary Immunodeficiency Disorders: A Historic and Scientific Perspective

The LAD history is an excellent example to show how clinical and basic research can work hand in hand to increase our understanding of fundamental host defense processes.

Evaluation of human leukocyte mobilization *in vivo* is limited to the histopathological assessment of "skin window" techniques initially described by Rebuck (7). However, difficulties encountered in creating uniform skin lesions, in addition to the pain and the possibility of infectious complications, limited the suitability of this technique as a standard clinical test.

Thus, leukocyte adhesion and migration studies are most commonly performed *in vitro*. For many years, the micropore filter method, originally described by Boyden (8), and the under agarose assay, were the main *in vitro* methods for the study of leukocyte adhesion and migration. More recently, intravital microscopy, which uses transparent tissues (for example rat or rabbit mesentery) became available, allowing the direct observation of leukocyte rolling, adhesion and transmigration. The use of flow chambers, another new technique, can define shear stress equivalent to the *in vivo* situation. Such systems use purified layers of the various adhesion molecules and can define precisely the different phases of the adhesion process (1).

The first gene involved in leukocyte adhesion was cloned in 1986 (9), several years after the encoded protein was found to be absent in children with a leukocyte adhesion defect. This molecule was called integrin to denote "its role as an integral membrane complex involved in the transmembrane association between the extracellular matrix and the cytoskeleton" (9). Later it was found that different integrins are involved in the extra matrix, platelet aggregation and leukocyte adhesion to the endothelium (10).

Several years later, another family of adhesion molecules was discovered. They were called selectins to indicate that they select the involvement of carbohydrate recognition in their function (11). This family is constituted of three different proteins: the E (endothelial), the P (platelet) and the L (lymphocyte) selectins. The main ligand for E-selectin, Sialyl Lewis X, was found a year later (12).

In 1991, Butcher (13) was the first to propose the idea that the adhesion cascade is an active process requiring at least three sequential events. The first, which is mediated by the selectins, causes the leukocyte to roll on the endothelial wall of the blood vessels. Mediators (later found to be chemokines) will then turn on the leukocyte (by activating the integrin molecules). During the third phase of the adhesion cascade, a firm adhesion of the leukocyte occurs, mediated by activated integrin on the cell surface which binds to its ligand, the intracellular adhesion molecule (ICAM 1) (13), which is located on the endothelial cell.

Defects in each of the three phases have been described in humans. In LAD I, the firm leukocyte adhesion is defective due to mutations in the β-submit (CD18) of the integrin molecule (14). In LAD II, a specific defect in fucose metabolism leads to the absence of fucosylate glycoprotein from the cell surface including Sialyl Lewis X, the ligand for the selectins (15). This will interfere with the ability of the leukocyte to roll. The second phase, the activation of integrin, is defective in LAD III due to mutations in Kindlin3, a crucial molecule involved in activating integrin (16).

LAD I

During the 1970s, several reports of patients with features compatible with the leukocyte adhesion deficiency syndromes appeared in the literature, but they were incomplete, both in the clinical description as well as the laboratory findings. For example, Smith and Goldman had observed a patient with recurrent infections and severe periodontitis with defective neutrophil migration as demonstrated by time-lapse photography of random neuthrophil migration. The results of Rebuck skin window tests suggested a marked decrease in neutrophil chemotaxis (17). The full clinical description of LAD I was reported in *The Lancet* in 1979 by Hayward et al., who presented an infant with delayed separation of the umbilical cord, widespread infections and defective neutrophil mobility (3). A year later, Bowen (18), presented a complete description of a neutrophil adhesion defect with its clinical characteristics. However, the major event in revealing the mystery behind this condition occurred a short time later, when Crowely et al. (4) reported in *The New England Journal of Medicine* a similar patient in whom a neutrophil protein of 110 000 kD was missing. Subsequently, Arnaout et al. (5) and Bowen et al. (6) reported patients with deficiencies of neutrophil surface glycoproteins of 150 000 kD and 180 000 kD, respectively.

This range in molecular weight was consistent with the α subunit of Mac-1, which was described as an αβ-complex (CD18) on leukocyte surfaces (19). The availabilities of monoclonal antibodies against the two subunits of the complex provided an opportunity to define the molecular defect precisely (20). Based on the fact that all three α subunits and the common β subunit were absent, Springer et al. (20) proposed that the primary defect was in the β subunit and that this subunit was necessary for cell surface expression of the α subunits. Several names were given to the syndrome (Mo-1 deficiency, LFA-1 deficiency) and, thus, in the interest of brevity and comprehensiveness, in 1987 Anderson and Springer coined the name leukocyte adhesion deficiency (LAD) (19).

With the accumulation of more patients, it became possible to divide the cases into two groups, one of which being the severe phenotype, with less than 2% expression of CD18 on the surface of the leukocyte. These patients suffer from life-threatening infection and usually die early in life without curative therapy. In the "moderate" type, there is between 2 and 15% CD18 expression, and affected individuals suffer from infectious episodes that can be controlled by antibiotics (21). Because consanguinity has often been

observed and the number of male and female patients was equal, autosomal recessive inheritance was strongly suggested (19). Kishimoto et al., in 1987, were the first to report the causative gene (*ITGB2*), which is located on chromosome 21, and encodes for the β subunit of CD18 (14). As of today, more than 100 unique mutations across the entire gene, *ITGB2*, have been reported without hot spots or apparent phenotype/genotype correlation (22).

Several specific clinical features were recognized to be unique to LAD I: delayed separation of the umbilical cord, persistent leukocytosis (above 20 000 mm^3), recurrent infections predominantly affecting the skin, impaired wound healing and, later in life, chronic periodontitis (23).

Because of the high mortality in patients with the severe form of LAD, Alain Fischer's group (24) performed successful bone morrow transplantation in two patients as early as 1983. This is the only curative measure for severe LAD I with an overall survival rate of 75% (25). The idea that LAD may be a good candidate for gene therapy was considered more than 20 years ago. The proof of concept was provided in an *in vitro* study by Wilson et al. (26), who corrected CD18 deficient lymphocytes by retrovirus-mediated gene transfer. It was subsequently shown that gene therapy can be successful in the canine model of LAD (27) but, unfortunately, success in humans has not yet been reported.

LAD II

The second LAD syndrome was recognized almost ten years after LAD I. We observed a child of consanguinous parents of Arabic ethnicity with recurrent infections, severe psychomotor retardation, peculiar facial features and marked leukocytosis. As our hospital was unable to type his blood, we sent a sample to the Israeli national blood bank, where this patient was recognized as the first ever Israeli with the

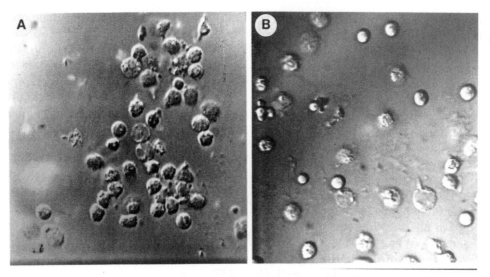

FIGURE 21.1   Neutrophil spreading and aggregation on glass. (A) Control, (B) LAD II patient.

rare Bombay blood group, lacking the H antigen on his erythrocytes.

Searching the literature, we were unable to find any report of a similar case and thus were not sure if it was indeed a new syndrome found just in one person. Preliminary studies showed that patient neutrophils failed to spread on glass and to aggregate, unlike control neutrophils (Fig. 21.1). Furthermore, in the under agarose assay, his leukocytes did not migrate towards a chemotactic factor and random migration was markedly defective.

Several months later, a pediatrician from Tel-Aviv called and told me that he also had a patient with the Bombay phenotype and, answering my question, confirmed that indeed his patient had marked leukocytosis. When his patient (again with Arabic ethnicity and consanguinity) came to see us, we were astonished how alike the two patients looked, although there was no blood relation between the two families.

At that point it was clear that we were dealing with a new syndrome and we called it "Rambam–Hasharon syndrome" in reference to the two hospitals were the patients were cared for (28). The report of this syndrome provided a

clinical description, since at that time we did not understand the causative molecular defect. Although a leukocyte adhesion and migration defect was observed, the characteristic of the defect was different from that of the known leukocyte adhesion deficiency, LAD I.

At about the same time, the first member of the selectin family (ELAM-1, later called E-selectin) had been cloned (29) and one year later, in 1990, Jim Paulson's group described Sialyl Lewis X as the ligand for ELAM-1 (12). We asked Jim Paulson for his anti Sialyl-Lewis X antibody and discovered that Sialyl Lewis X (CD15a) was not expressed in the leukocytes of our two patients. As both Sialyl Lewis X and the H antigen incorporate fucose into their protein complex, we postulated a general defect in fucose metabolism as the primary defect (15). At this point, we were convinced that the adhesion defect in these two patients was due to the inability of leukocyte to bind to selectin, and had nothing to do with the integrins. Thus, we suggested calling our new adhesion defect LAD II and to refer to the adhesion defect caused by mutations in CD18 as LAD I (15). In a subsequent study using the intravital microscopy assay, we could demonstrate that the defect in LAD II affects the rolling phase, while in LAD I the defect resides in the firm adhesion phase (30).

At that time, little was known about fucose metabolism except for the fact that two enzymes take part in the conversion of mannose to fucose, accounting for more than 90% of human fucose. Together with Michela Tonneti from Genova, we were able to demonstrate that the activity of one of these enzymes, GDP-d-mannose 4,6-dehydratase, was markedly decreased in patient leukocytes (31). However, no mutation was found in the cDNA of this enzyme and it was present in amounts comparable to normal controls (31). It was thus clear that the reduced enzymatic activity must be related to an alteration of some other aspect of the fucose metabolism and not to a genetic defect directly involving this enzyme.

Then, unexpectedly, a third patient of Turkish origin with symptoms resembling LAD II was reported (32). This patient was found to have a defect in a specific transporter of fucose from the cytoplasm to the Golgi apparatus where fucosylation into glycoprotein occurs (33). We checked our two patients and found the same defect (34).

Using a functional cloning approach to isolate a cDNA that complemented the phenotype of the patient fibroblasts, we were able to identify the specific GDP-fucose transporter and to determine the specific mutation in the gene (35). To date, fewer than ten LAD II patients have been described and several different mutations identified (36). We speculated, based on genetic testing, that in our Arab population LAD II is a relative new disease and may date back only about 100 years (37).

The infections observed in LAD II are much milder than in LAD I and, later in life, present as chronic severe periodontitis, the main infectious problem after childhood (38). While in some cases fucose supplementation may improve the neutrophil defect (39), this was not observed in others (40) and improvement in the mental and growth defects was never achieved.

While the adhesion defect is clearly related to the absence of Sialyl Lewis X, the pathological mechanisms causing mental and growth abnormalities are not understood. An attractive possibility was a defect in Notch, an important protein in brain development, which is also fucosylated. However, we found that O-fucosylation of Notch was virtually unaffected in our LAD II patients (41). This is consistent with a subsequent finding that O-fucosylation does not occur in the Golgi bodies but rather in the endoplasmatic reticulum (42). Several animal studies point to a Notch-independent involvement of fucosylation in neurodevelopment (36) which may be relevant for LAD II. We are now working with an induced pluripotent stem cell technique, and have generated neurons from fibroblasts from a LAD II patient; we hope to gain new insight into the mental abnormalities observed in this syndrome.

LAD III

In 1997, Kuijpers et al. described a patient who presented as a mild LAD I phenotype, with a defect in the activation of β_2 integrin without a mutation in the gene encoding for CD18 (43). They designated this new entity as LAD I/variant. In this report, β_1 activation was normal. Several years later, another similar case was reported in which both β_1 and β_2 activation was defective while β_3 activation was not studied (44).

In 2003, a third patient was discovered, in whom activation of all three β subunits was defective (45). In the same year, our group reported a child with the same phenotype of recurrent infections, profound leukocytosis and a bleeding tendency. While CD18 expression was normal and sequencing the gene failed to reveal a mutation, we observed a defect in the activation of all integrins (46). As no defect in the structure of CD18 was observed, we proposed to name this syndrome LAD III. While in LAD I the defect is in the firm adhesion phase, here the defect is in the activation phase of the adhesion cascade and thus the term LAD III is more appropriate than LAD I/variant (47).

The primary genetic defect remained elusive until recently, when integrin activation defects were studied in several knockout mice models, finally revealing the molecular basis of LAD III. A critical component of integrin activation by G-protein coupled receptor (GPCR) signals is the GTPase, Rap-1. Although Rap-1 expression occurs widely in all tissues, "inside-out" activation of integrins triggered by GPCR signals was recently linked to a novel Rap-1 guanine exchange factor (GEF), Cal-DAG-GEF-1, which is expressed mainly in platelets and leukocytes but also in the brain. Cal-DAG-GEF-1 knockout mice exhibited defects in β_1, β_2 and β_3 integrin activation in both platelets and leukocytes, similarly to LAD III patients, suggesting that this mouse model is representative of leukocyte adhesion deficiency type III (48). Indeed, a large group of LAD III patients (all of Turkish origin) exhibit a splice junction mutation in their *Cal-DAG-GEF-1* (official symbol RASGRP2) gene with nearly complete loss of messenger RNA and protein expression both in patient platelets, neutrophils and primary T lymphocytes (48). However, LAD III patients of non-Turkish origin were found to express Cal-DAG-GEF-1 normally.

A breakthrough in the search for the molecular defect underlying LAD III occurred when a family of integrin cytoplasmic tail binding adaptors, termed Kindlins, was recognized to be crucial for maximal integrin activation. One of these three family members, Kindlin 3, is almost exclusively expressed on hematopoietic lineage cells, and integrin activation was diminished in Kindlin 3 knockout mice (50) leading to total loss of β_2 integrin adhesiveness both *in vivo* and *in vitro* (51). Thus, Kindlin 3 knockout mice represented another genetic model for the LAD III-like phenotype. Since, unlike talin, Kindlin 3 expression is restricted to platelets, leukocytes and erythrocytes, the global deficiency of both

β_1 β_2 and β_3 integrin expression in these knock-out mice suggested that Kindlin 3 deficiency, on its own, could be the cause of LAD III in patients expressing intact Cal-DAG-GEF-1.

Indeed, mutational analysis of Kindlin 3 (official symbol FERMT3) revealed homozygous stop codon mutations in all LAD III patients regardless of their origin, including p.R513X in three Turkish patients who also carried the Cal-DAG-GEF-1 mutation (16).

Kindlin 3 is encoded in the same locus as Cal-DAG-GEF-1 (11q13), although the two genes are 500 000 bp apart, suggesting that both mutations were inherited as a common allele in these patients. Further studies (52, 53) described several distinct mutations in Kindlin 3 in non-Turkish LAD III patients, all from the Eastern Mediterranean area. Transfection of patient-derived lymphoblastoid cell lines with wild-type Kindlin 3 rescued the LAD III defect, restoring integrin-mediated adhesion, spreading and motility (53), even in those derived from patients in whom mutations in Cal-DAG-GEF-1 were found. The role of Cal-DAG-GEF-1 in human leukocytes will only be revealed when patients with an isolated defect in Cal-DAG-GEF-1 are identified.

To date less than 20 patients with LAD III have been identified, all requiring blood transfusions and prophylactic antibiotics. The only curative treatment, stem cell transplantation, appears less favorable compared than in LAD I (43). This may be due to the fact that most patients with LAD III were found to have increased bone density (54).

The LAD story has two important repercussions. First, through the study of the three different forms of the disease, the various *in vivo* steps of the adhesion cascade were confirmed. The important role of selectins, and their ligand, in the rolling phase was authenticated in LAD II, while the crucial function of integrin β2 was validated in LAD I and III.

The second facet highlights the different paths used to unravel the molecular structure underlying the diseases, prior to the whole-genomic sequence (WGS) era. In LAD I, the defective protein was recognized thereby allowing gene identification. Indeed, LAD I was among the first primary immunodeficiency syndromes to be deciphered at the molecular level. In LAD II, as the essence of the defect was unknown, DNA complementary assays were employed to identify the fault in the specific transporter of fructose to the Golgi apparatus. Ultimately, a mouse knockout model of Kindlin 3, mimicking LAD III phenotype, showed the involvement of Kindlin 3 as the primary defect underlying LAD III.

References

1. Ley K. The selectins as rolling receptor. In: Vestweber D, editor. *The selectins*. Amsterdam: Hartwood Academic; 1997. p. 63–104.

2. Leber T. Ueber die Entstehung der Entzundung die Wirkung der entzundungse-regedon Schadlichkeiten. *Fortschr Med* 1888;**6**:460.

3. Hayward AR, Leonard J, Wood C, et al. Delayed separation of the umbilical cord, widespread infections and defective neutrophil mobility. *Lancet* 1979;**1**:1011–99.

4. Crowley CA, Curnutte JT, Rosin RE, et al. An inherited abnormality of neutrophil adhesion: its genetic transmission and its association with a missing protein. *N Engl J Med* 1980;**302**:1163–8.

5. Arnaout MA, Pitt J, Cohen HJ, et al. Deficiency of granulocyte-membrane glycoprotein (gp150) in a boy with recurrent bacterial infections. *N Engl J Med* 1982;**306**:693–9.

6. Bowen TJ, Ochs HD, Altman LC, et al. Severe recurrent bacterial infections associated with defective adherence and chemotaxis in two patients with neutrophil deficient in a cell-associated glycoprotein. *J Pediatr* 1982;**101**:932–40.

7. Rebuck JW, Crowley JH. A method of studying leukocyte function in vivo. *Ann NY Acad Sci* 1955;**59**:1955.

8. Boyden S. The chemotactic effect of mixtures of antibody and antigen on polymorphonuclear leukocytes. *J Exp Med* 1962;**115**:453.

9. Tamkun JW, DeSimone DW, Fonda D, et al. Structure of Integrin, a glycoprotein involved in the transmembrane linkage between fibronectin and actin. *Cell* 1986;**46**:271–82.

10. Hynes RO. A family of cell surface receptors. *Cell* 1987;**48**:549–54.

11. Bevilacqua M, Butcher E, Furle B, et al. Selectins: A family of adhesion receptors. *Cell* 1991;**67**:233.

12. Phillips ML, Nudelman ED, Gaeta FCA, et al. ELAM-1 mediates cell adhesion by recognition of a carbohydrate ligand, sialyl-Lex. *Science* 1990;**250**:1130–2.

13. Butcher EC. Leukocyte-endothelial cell recognition: Three (or more) steps to specificity and diversity. *Cell* 1991;**67**:1033–6.

14. Kishimoto TK, Hollander N, Roberts TM, et al. Heterogeneous mutations in the β subunit common to LFA-1, Mac-1, and p150,95 glycoproteins cause leukocyte adhesion deficiency. *Cell* 1987;**50**:193–202.

15. Etzioni A, Frydman M, Pollack S, et al. Severe recurrent infections due to a novel adhesion molecule defect. *N Engl J Med* 1992;**327**:1789–92.

16. Mory A, Feigelson SW, Yareli N, et al. Kindlin-3: a new gene involved in the pathogenesis of LAD III. *Blood* 2009;**114**:2541.

17. Schmalstieg FC. Discovery of the leukocyte adherence defect – A historical perspective. *Sem Hematol* 1993;**30**:66–71.

18. Bowen TJ, Ochs HD, Altman LC, et al. A cellular, chemotactic, and adherence defect associated with recurrent bacterial infections, abnormal umbilical cord separation, and severe periodontitis. *Pediatr Res* 1980;**14**:544A (abstract).

19. Anderson DC, Springer TA. Leukocyte adhesion deficiency: or inherited defect in the Mac-1, LFA-1 and p150,95 glycoproteins. *Annu Rev Med* 1987;**38**:175–94.

20. Springer TA, Thompson WS, Miller LJ, et al. Inherited deficiency of the Mac-1, LFA-1, p150,95 glycoprotein family and its molecular basis. *J Exp Med* 1984;**160**:1901–18.

21. Anderson DC, Schmalstieg FC, Finegold MJ, et al. The severe and moderate phenotypes of heritable Mac-1, LFA-1 deficiency: Their quantitative definition and relation to leukocyte dysfunction and clinical features. *J Infect Dis* 1985;**152**:668–89.

22. van de Vijver E, Maddalena A, Sanal O, et al. Hematologically important mutations: Leukocyte adhesion deficiency (first update). *Blood Cells Mol Dis* 2012;**48**:53–61.

23. Anderson DC, Smith CW, Springer TA. Leukocyte adhesion deficiency and other disorders of leukocyte motility. In: Scriber CR, editor. *The Metabolic Basis of Inherited Disease*. 6th edn. McGraw Hill; 1989. p. 2751–77.

24. Fischer A, Descamps-Latscha B, Gerota I, et al. Bone marrow transplantation for inborn error of phagocytic cells associated with defective adherence, chemotaxis, and oxidative response during opsonised particle phagocytosis. *Lancet* 1983;**2**:473–6.

25. Qasim W, Cavazzana-Calvo M, Davies EG, et al. Allogeneic hematopoietic stem-cell transplantation for leukocyte adhesion deficiency. *Pediatrics* 2009;**123**:836–40.

26. Wilson JM, Ping AJ, Krauss JC, et al. Correction of CD18 deficient lymphocytes by retrovirus-mediated gene transfer. *Science* 1990;**248**:1413–6.

27. Bauer Jr TR, Allen JM, Hai M, et al. Successful treatment of canine leukocyte adhesion deficiency by foamy virus vectors. *Nat Med* 2007;**14**:93–7.

28. Frydman M, Etzioni A, Eidlitz-Markus T, et al. Rambam-Hasharon syndrome of psychomotor retardation, short stature, defective neutrophil motility, and Bombay phenotype. *Am J Med Genet* 1992;**44**:297–302.

29. Bevilacqua MP, Stengelin S, Gimbrone Jr MA, et al. Endothelial leukocyte adhesion molecule 1: an inducible receptor for neutrophils related to complement regulatory proteins and lectins. *Science* 1989;**243**:1160–5.

30. von Andrian UH, Berger EM, Ramezani L, et al. In vivo behavior of neutrophils from two patients with distinct inherited leukocyte adhesion deficiency syndromes. *J Clin Invest* 1993;**91**:2893–7.

31. Sturla L, Etzioni A, Bisso A, et al. Defective intracellular activity of GDP-d-mannose-4.6-dehydratase in leukocyte adhesion deficiency type II syndrome. *FEBS Lett* 1998;**429**:274–8.

32. Marquardt T, Brune T, Luhn K, et al. Leukocyte adhesion deficiency II syndrome, a generalized defect in fucose metabolism. *J Pediatr* 1999;**134**:681–8.

33. Lubke T, Marquardt T, von Figura K, Korner C. A new type of carbohydrate deficiency glycoprotein syndrome due to a decreased import of GDP-fucose into the Golgi. *J Biol Chem* 1999;**274**:25986–9.

34. Sturla L, Puglielli L, Tonetti M, et al. Impairment of the Golgi GDP-L-fucose transport and unresponsiveness to fucose replacement therapy in LAD II patients. *Pediatr Res* 2001;**49**:537–42.

35. Lubke T, Marquardt T, Etzioni A, et al. Complementation cloning identities CGD-11c, a new type of congenital disorder of glycosylation, as a GDP-fucose transporter deficiency. *Nat Gen* 2001;**28**:73–6.

36. Juhn K, Wild UK. Human deficiencies of fucosylation and sialylation affecting selectin ligands. *Semin Immunopathol* 2012;**34**:383–99.

37. Etzioni A, Sturla L, Antonellis A, et al. LADII/CDG-IIc founder effect and genotype/phenotype correlation. *Am J Med Gen* 2002;**110**:131–5.

38. Etzioni A, Gershoni-Baruch R, Pollack S, Shehadeh N. Leukocyte adhesion deficiency type II: long term follow up. *J Allergy Clin Immunol* 1998;**102**:323–4.

39. Marquardt T, Juhn K, Greeze HH, Harns E, Vesteweber D. Correction of leukocyte adhesion deficiency type II with oral fucose. *Blood* 1999;**94**:3976–85.

40. Etzioni A, Tonetti M. Fucose supplementation in leukocyte adhesion type II. *Blood* 2000;**94**:3641–2.

41. Sturla L, Rampal R, Haltiwanger RS, et al. Differential terminal fucosylation of N-linked glycans versus protein O-fucosylation in leukocyte adhesion deficiency type II (CDG IIc). *J Biol Chem* 2003;**278**:26727–33.

42. Luo Y, Haltiwanger RS. O-fucosylation of notch occurs in the endoplasmic reticulum. *J Biol Chem* 2005;**280**:11289–94.

43. Kujpers TW, vanLier RAW, Hamain D, et al. Leukocyte adhesion deficiency type 1 (LAD-1)/variant. *J Clin Invest* 1977;**100**:1725–33.

44. Harris ES, Shigeoku AO, Li W, et al. A novel syndrome of variant leukocyte adhesion deficiency involving defects in adhesion mediated by β_1 and β_2 integrins. *Blood* 2001;**97**:767–72.

45. McDowall A, Inwald D, Leitinger B, et al. A novel form of integrin dysfunction involving β_1, β_2 and β_3 integrins. *J Clin Invest* 2003;**111**:51–60.

46. Alon R, Aker M, Feigelson S, et al. A novel genetic leukocyte adhesion deficiency in subsecond triggering of integrin avidity by endothelial chemokines results in impaired leukocyte arrest on vascular endothelium under shear flow. *Blood* 2003;**101**:4437–45.

47. Alon R, Etzioni A. LAD III a novel group of leukocyte integrin activation deficiencies. *Trends Immunol* 2003;**24**:561–6.

48. Bergmeier W, Goerge T, Wang HW, et al. Mice lacking the signaling molecule CalDAGGEF1 represent a model for leukocyte adhesion deficiency type III. *J Clin Invest* 2007;**117**:1699–707.

49. Pasvolsky R, Feigelson SW, Kilic SS, et al. A LAD III syndrome is associated with defective expression of the Rap 1 activator CalDAGGEF1 in lymphocytes, neutrophils and platelets. *J Exp Med* 2007;**204**:1571–82.

50. Moser M, Nieswandt B, Ussar S, et al. Kindlin 3 is essential for integrin activation and platelet aggregation. *Nat Med* 2008;**14**:325–30.

51. Moser M, Bauer M, Schmid S, et al. Kindlin 3 is required for β_2 integrin mediated leukocyte adhesion to endothelial cells. *Nat Med* 2009;**15**:300–5.

52. Molinin NL, Zhang L, Choi J, et al. A point mutation in Kindlin 3 ablated activation of three integrin subfamilies in humans. *Nat Med* 2009;**15**:313–8.

53. Svensson L, Howarth K, McDowall A, et al. Leukocyte adhesion deficiency III is caused by mutations in Kindlin 3 affects integrin activation. *Nat Med* 2009;**15**:306–12.

54. Kilic SS, Etzioni A. The clinical spectrum of leukocyte adhesion deficiency (LAD) III due to defective CalDAGGEF2. *J Clin Immunol* 2010;**45**:117–22.

How Common Variable Immune Deficiency has Changed Over Six Decades

Charlotte Cunningham-Rundles[1], Helen Chapel[2]

[1]Mount Sinai School of Medicine, New York, NY, USA
[2]Nuffield Department of Medicine, University of Oxford, Oxford University Hospitals, Oxford, UK

INTRODUCTION

Two years after the first publication of agammaglobulinemia in a male child (1), Janeway and Gitlin reported nine additional male children with similar findings, and suggested sex-linked inheritance for this immune defect (2). The first report that agammaglobulinemia was not a disease restricted to either males or children appeared in 1954 when Sanford et al. described a 39-year-old agammaglobulinemic woman, who had what we now recognize as the most frequent complications of common variable immune deficiency disorder (CVID), including chronic bronchitis, episodes of bacterial pneumonia, *Haemophilus influenzae* meningitis, chronic diarrhea with malabsorption leading to weight loss and severe hypocalcemia (3). Following this, a series of other reports of adults with recurrent infections suggested that agammaglobulinemia and hypogammaglobulinemia were more widely recognized, as the more common use of antibiotics permitted longer survival and subsequently clinical recognition (4–7). Most of the earlier reports described the development of severe bronchiectasis and respiratory failure, which was the commonest cause of death.

Primary Immunodeficiency Disorders: A Historic and Scientific Perspective

As for the original case of Bruton, the loss of gamma globulin in the serum was determined by the newly-introduced method of Tiselius zone electrophoresis; the proteins moving in the gamma region, absent in patients with recurrent infections, have been known as immunoglobulins since the early 1960s when their antibody nature was recognized. Laboratory studies confirmed the lack of isohemagglutinins in most cases, but other analyses included the positive Schick and Dick tests, the standard means of confirming susceptibility to diphtheria or scarlet fever by demonstrating lack of specific antibodies. Although, as in the first case, these patients were treated with intramuscular gamma globulin concentrates that had been recently purified from human plasma, premature death was common. Autopsy reports from these early cases showed lymph nodes with poorly-defined cortices, no follicle formation and absence of plasma cells in lymph nodes and bone marrow (7). While it was suggested that plasma cell development was somehow impaired, the significance of this was uncertain, as the origin of gamma globulin was not yet established (6–8). Some evidence pointed to the plasma cell as producers of gamma globulin, as the lymph nodes of children with agammaglobulinemia given immunizations showed no germinal-center formation, no plasma or pre-plasma cells and absent antibodies. Furthermore, lymph nodes of patients with low serum immunoglobulins did not respond with plasmacytosis after antigen stimulation (9). Advancing this connection, Good et al. also reported that loss of plasma cells was characteristic of males with agammaglobulinemia, further pointing to the origin of gammaglobulin production (10, 11). Excess loss of gamma globulin had already been excluded by labeling globulins with I^{131} and noting that the disappearance of tracer was no different from that of normal subjects (7). While early authors were not clear on the cellular origin of gamma-globulin, Seltzer et al. found that while their hypogammaglobulinemic patient could not be

successfully immunized, he still had a positive histoplasmin skin test, suggesting the lack of correlation of "delayed tissue hypersensitivity and humoral immunity", or two types of immunity, a novel concept at that time (4). The formal division of cellular and humoral immunity was not demonstrated until a decade later when Cooper and Good, based on work in the chicken, suggested that there was a functional division of labor in immune functions (12).

As for physicians seeing patients with CVID at that time, the authors of early reports were struck by the observation that patients commonly

> present a history indicating that they have been well and without inordinate trouble from infection up to a certain point in their lives. Then suddenly, just as with congenital agammaglobulinemia, life for them becomes a continuous round of severe, life-threatening bacterial infections (13).

Because of the apparent onset in adult life, the term "acquired hypogammaglobulinemia" was applied to distinguish these subjects from patients with congenital agammaglobulinemia or transient hypogammaglobulinemia of infancy. In contrast to males with congenital agammaglobulinemia, these older subjects, examined by more sensitive tests, tended to have somewhat more measurable gamma globulin in their blood, suggesting that hypogammaglobulinemia was a more appropriate term.

In 1971, the WHO initiated an informal discussion to organize the emerging immune deficiency diseases into a framework. This resulted in a publication that same year; here the syndrome relating to patients with late onset immunoglobulin failure were first called *variable immune deficiency (common, largely unclassified)* and was segregated from other, more distinct forms of immunodeficiency, namely the infantile forms of agammaglobulinemia, transient forms, IgA deficiency, DiGeorge syndrome, Wiskott–Aldrich syndrome, immune deficiency with short-limbed dwarfism, severe combined immune deficiency, ataxia telangiectasia, immune defects with thymoma and hyper IgM syndrome, rather similar to the current day classification scheme (14–16). However, most early reports still considered the immunoglobulin failure syndromes together; for example, the UK Medical Research Council in 1971 reported that of 176 subjects with hypogammaglobulinemia, 87.6% of patients had upper and lower respiratory infections, pyoderma, diarrhea and repeated episodes of meningitis, septicemia and pyelonephritis (17). There were 51 deaths in this cohort, usually due to respiratory failure, but with diarrhea, infections and malignant disease being other common causes. Although not ascribed a precise diagnosis, the cases are so well described that even today a reader can determine the diagnosis in most patients, whether XLA, CD40L deficiency or CVID, e.g. the deaths include five infants who died of *Pneumocystis jirovecii* pneumonia, suggesting other (combined) immune defects in these children.

CLINICAL CONDITIONS

With the increased recognition of the CVID syndrome in Europe and the USA, various centers gathered sufficient numbers of late onset patients to describe the general features. Males and females were diagnosed in about equal numbers, and more often in the third decade of life, but with a very wide range of ages of onset of symptoms. It was also clear in these early reports that the family members of these subjects, while not commonly overtly immune deficient, were more likely to be IgA deficient, and/or have various forms of autoimmunity (18, 19). The incidence of a positive family history has gradually been reduced, as many more patients have been recognized and diagnosed. The incidence of familial CVID is now well below 5%, though that of autoimmunity or selective IgA deficiency remains higher.

After these original reports, the conditions and complications that we now recognize as part of the CVID syndrome were described in single cases and in small series. Sinopulmonary infections were predominantly caused by *Haemophilus influenzae, Diplococcus pneumoniae* (as *Strep. pneumoniae* was then), *Streptococcus pyogenes* and *Staphylococcus aureus*. Bronchiectasis occurred in 28% of cases (20). Other common manifestations cited included gastrointestinal giardia infections, lymphoid hyperplasia, autoimmune or infectious conditions of joints, thrombocytopenia, hemolytic anemia, achlorhydria, pernicious anemia, regional enteritis and granulomatous disease suggesting sarcoidosis (20–22). Clinically, lymphadenopathy was common and splenomegaly noted in 28% of cases. The coincidence of agammaglobulinemia and malignancy was first noted in 1954 by Arends, who described a woman with a history of severe respiratory and skin infections who developed malignant lymphoma but had no detectable serum gamma globulin and no plasma cells in her bone marrow (8). By 1971, a literature review estimated the incidence of cancer in primary immune deficiency at 10%, and highlighted the mostly lymphoid types, including leukemia, several different kinds of lymphoma, but also included adenocarcinoma of the stomach along with other tumors (23). It is important to note though that, over the years, this incidence has fallen as more patients have been diagnosed. By 1985 the rate was reduced (24) and more recently, was lower still (25).

TREATMENT

Although the treatment of subjects with intramuscularly injected immune globulin concentrates, starting in the 1950s, clearly reduced the number of bacterial infections, the amount of immunoglobulin that could be delivered easily by intramuscular injection was limited. The dose of 100 mg/kg/month was the accepted amount, although the British Medical Research

Council had found that 50 mg/kg/week dose was more beneficial; however, the discomfort related to the larger injection was prohibitive, especially for adult subjects who required volumes of up to 10–25 mL each week. Furthermore, serious adverse reactions were common. Thus acute bacterial respiratory infections continued to be the main clinical concern in CVID until the introduction and widespread use of intravenous immune globulins in the middle 1980s (26). After this improvement, allowing increased dosing, subjects with CVID had fewer acute infections and longer life spans, allowing studies of the natural history of this syndrome, and investigation into the further causes of morbidity and mortality. Unfortunately, chronic sinopulmonary infections continued, along with other common problems, including chronic diarrhea, joint disease and various forms of hepatitis (27). The joint disease was occasionally associated with ureaplasma and other mycoplasma species, which were sometimes also found in patients with cystitis (28). Among malignancies, lymphoma and stomach cancer appeared clearly increased (24, 29). Aside from *Giardia*, the first gastrointestinal infections in CVID reported were *Cryptosporidia*, *Campylobacter* and *Salmonella*. In 1999, the main categories of clinical illness in CVID patients were infections, autoimmune disease, hepatitis, granulomatous infiltrations, gastrointestinal or pulmonary disease, lymphoma and other cancers. For this cohort, survival 20 years after diagnosis was 64% for males and 67% for females, compared to the expected 92% population survival for males and 94% for females (30).

As satisfactorily large doses of immunoglobulin became standard therapy by the 2000s, more and more patients had achieved trough IgG levels within the normal range for the majority of their lives. But the non-infectious conditions, noted all along, began to emerge as the most difficult aspects of patient management. With increased survival, what also emerged was that, for still unclear reasons, patients tended to have rather stereotypical sets of complications, or patterns of inflammation/autoimmunity, that appeared stable over many years and that were not adequately addressed by immune globulin replacement. These complications were clearly associated with worse survival, and even today present the most complex therapeutic challenges (31–33). The good news is that for uncomplicated patients, life expectancy now is almost normal (34).

PATHOGENESIS

Many investigators puzzled over the intriguing question of why the B-cells of CVID subjects apparently lose function over time and, indeed, this remains a continuing question. The term "acquired", as due to an event after birth, was used simply because of the late onset. However, soon after the first description, Bram and Morton wrote that they found it difficult to envision any acquired process that would selectively "assault" globulin production, without affecting other closely related proteins or functions; on this basis, they suggested that single gene defects would be more likely (35). Taking another tactic, however, Waldman and others described experiments that seemed to show that some groups of circulating white blood cells were capable of suppressing immunoglobulin production or B-cell differentiation *in vitro* (36). These observations fitted quite well with the seemingly adult onset of CVID. With the identification of lymphocyte subsets, further dissection of CVID T- and B-cell phenotypes emerged (37). T-cell proliferative functions and reconstitution of B-cell function in a number of *in vitro* studies remained a prime topic of investigation, identifying catalogs of various proliferative and functional defects (38–42).

As CVID is so heterogeneous, many groups then attempted to classify these subjects into clinically relevant groups, based on laboratory parameters. As *in vitro* cell culture methods became much more sophisticated, CVID B-cells could be examined in detail. Using polyclonal activators to stimulate immunoglobulin production, subjects could be divided into those who had very few, if any, B-cells, those for whom B-cells failed to respond with any stimulus, those whose B-cells responded to an extent and in which limited immunoglobulin synthesis could be detected and, into a final group, who appeared to have normal or almost normal immunoglobulin production (43–45). As techniques for measuring cytokines became available, studies on CVID showed that again, numerous defects were documented; the first of these was the relative lack of interleukin 2 (IL-2) production (46, 47) followed by the description of a number of other cytokine deficiencies, of IL-4, IL-12, IL-10 and interferon γ (INF-γ) in some cases (48–51); but excess production of IL-6, IL-12, and INF-γ was observed in others (48, 52–54). These experiments seemed to confirm that there were many potential failed mechanisms that might be involved in different patients, mechanisms that were not necessarily mutually exclusive. Recently, investigations have focused on B-cell phenotypes, as dissected by flow cytometry. Several studies have examined the memory B-cell phenotypes, concluding that patients with the most severe disease had the lowest proportion of isotype switched memory B-cells (55–58).

GENES AND CVID

While study of monogenic primary immune defects has yielded a treasure trove of genes essential for normal immune function, CVID, always understood as a complex syndrome, has been harder to unravel. Major efforts in collecting numbers of families with several affected members have been undertaken, with clues predominantly pointing to the MHC region (59, 60). To approach CVID using genome-wide association studies, single nucleotide polymorphism arrays have been used to examine genomic

DNA. Here again, a number of gene associations, including the MHC region, were revealed (61). The most interesting finding, through the use of a Support Vector Machine (SVM) algorithm, was that >90% of CVID patients have a huge excess of copy number variations, suggesting polygenic etiologies. Furthermore, the unique signature could be used successfully for diagnosis of CVID with an overall accuracy of 91% and a positive predictive value of 100% (62).

However, as predicted by Bram and Morton (35), in a minority of familial cases, it could be shown that autosomal recessive genetic mutations can lead to the CVID syndrome, including inducible co-stimulatory (*ICOS*) gene (63), CD19 (64, 65), CD20 (66), CD21 (67) and CD81 (68). While both heterozygous and homozygous mutations in the gene for the B-cell receptor transmembrane activator and CAML interactor (*TACI*) can be found in 8–10% of subjects with CVID, some of the same mutations can be found in healthy controls and phenotypically "normal" relatives (69–73), suggesting susceptibility mutations rather than disease-causing changes in this apparently non-essential gene. Similarly, mutations in the gene for B-cell activating factor (BAFF) receptor are likely to be related to susceptibility too (74) if, as with many other late onset conditions such as SLE, CVID is confirmed as mostly polygenic.

CONCLUSION

Over the past 60 years since its first description, CVID has remained a syndrome characterized by hypogammaglobulinemia and loss of functional antibody. Among the primary immune defects, CVID is the most clinically important, based on the numbers of medical encounters over many years. While some subjects are almost completely agammaglobulinemic, others have a relative loss of several immune globulin isotypes. The clinical syndrome varies widely; some have serious and organ-damaging infections or inflammatory complications while others, on now adequate Ig replacement, appear to do very well over years of follow-up care with little or no disability.

References

1. Bruton OC. Agammaglobulinemia. *Pediatrics* 1952;9: 722–8.
2. Janeway CA, Apt L, Gitlin D. Agammaglobulinemia. *Trans Assoc Am Physicians* 1953;66:200–2.
3. Sanford JP, Favour CB, Tribeman MS. Absence of serum gamma globulins in an adult. *N Engl J Med* 1954;250: 1027–9.
4. Seltzer G, Baron S, Toporek M. Idiopathic hypogammaglobulinemia and agammaglobulinemia; review of the literature and report of a case. *N Engl J Med* 1955;252: 252–5.
5. Young II, Wolfson WQ, Cohn C. Studies in serum proteins; agammaglobulinemia in the adult. *Am J Med* 1955;19: 222–30.
6. Collins HD, Dudley HR. Agammaglobulinemia and bronchiectasis; a report of two cases in adults, with autopsy findings. *N Engl J Med* 1955;252:255–9.
7. Greenhouse AH. Gamma globulin deficiency; report of a case and survey of the literature. *J Kans Med Soc* 1956;57:611–8.
8. Arends T, Coonrad EV, Rundles RW. Serum proteins in Hodgkin's disease and malignant lymphoma. *Am J Med* 1954;16:833–41.
9. Keuning FJ, van der Slikke LB. The role of immature plasma cells, lymphoblasts, and lymphocytes in the formation of antibodies, as established in tissue culture experiments. *J Lab Clin Med* 1950;36:162–82.
10. Good RA. Studies on agammaglobulinemia. II. Failure of plasma cell formation in the bone marrow and lymph nodes of patients with agammaglobulinemia. *J Lab Clin Med* 1955;46:167–81.
11. Good RA, Varco RL. A clinical and experimental study of agammaglobulinemia. *J Lancet* 1955;75:245–71.
12. Cooper MD, Peterson RD, Good RA. Delineation of the thymic and bursal lymphoid systems in the chicken. *Nature* 1965;205:143–6.
13. Good RA, Varco RL, Aust JB, Zak SJ. Transplantation studies in patients with agammaglobulinemia. *Ann NY Acad Sci* 1957;64:882–924 discussion 928.
14. Fudenberg HH, Good RA, Hitzig W, et al. Classification of the primary immune deficiencies: WHO recommendation. *N Engl J Med* 1970;283:656–7.
15. Fudenberg H, Good RA, Goodman HC, et al. Primary immunodeficiencies. Report of a World Health Organization Committee. *Pediatrics* 1971;47:927–46.
16. Al-Herz W, Bousfiha A, Casanova J-L, et al. Primary immunodeficiency diseases: an update on the Classification from the International Union of Immunological Societies Expert Committee for Primary Immunodeficiency. *Front Immunol* 2011;2:1–26.
17. Hill LE, Mollison PL. Hypogammaglobulinaemia in the United Kingdom. *Spec Rep Ser Med Res Counc* 1971;310: 4–8.
18. Friedman JM, Fialkow PJ, Davis SD, Ochs HD, Wedgwood RJ. Autoimmunity in the relatives of patients with immunodeficiency diseases. *Clin Exp Immunol* 1977;28:375–88.
19. Yount WJ, Seligmann M, Hong R, Good R, Kunkel HG. Imbalances of gamma globulin subgroups and gene defects in patients with primary hypogammaglobulinemia. *J Clin Invest* 1970;49:1957–66.
20. Hermans PE, Diaz-Buxo JA, Stobo JD. Idiopathic late-onset immunoglobulin deficiency. Clinical observations in 50 patients. *Am J Med* 1976;61:221–37.
21. Webster AD, Platts-Mills TA, Jannossy G, Morgan M, Asherson GL. Autoimmune blood dyscrasias in five patients with hypogammaglobulinemia: response of neutropenia to vincristine. *J Clin Immunol* 1981;1:113–8.
22. Webster AD, Slavin G, Shiner M, Platts-Mills TA, Asherson GL. Coeliac disease with severe hypogammaglobulinaemia. *Gut* 1981;22:153–7.
23. Gatti RA, Good RA. Occurrence of malignancy in immunodeficiency diseases. A literature review. *Cancer* 1971;28: 89–98.
24. Kinlen LJ, Webster AD, Bird AG, et al. Prospective study of cancer in patients with hypogammaglobulinaemia. *Lancet* 1985;1:263–6.
25. Dhalla F, da Silva SP, Lucas M, Travis S, Chapel H. Review of gastric cancer risk factors in patients with common variable immunodeficiency disorders, resulting in a proposal for a surveillance programme. *Clin Exp Immunol* 2011;165:1–7.
26. Ochs HD, Wedgwood RJ, et al. Comparison of high-dose and low-dose intravenous immunoglobulin therapy in patients with primary immunodeficiency diseases. *Am J Med* 1984;76:78–82.
27. Hermaszewski RA, Webster AD. Primary hypogammaglobulinaemia: a survey of clinical manifestations and complications. *Q J Med* 1993;86:31–42.
28. Johnston CL, Webster AD, Taylor-Robinson D, Rapaport G, Hughes GR. Primary late-onset hypogammaglobulinaemia associated with inflammatory polyarthritis and septic arthritis due to Mycoplasma pneumoniae. *Ann Rheum Dis* 1983;42:108–10.
29. Cunningham-Rundles C, Siegal FP, Cunningham-Rundles S, Lieberman P. Incidence of cancer in 98 patients with common varied immunodeficiency. *J Clin Immunol* 1987;7:294–9.

30. Cunningham-Rundles C, Bodian C. Common variable immunodeficiency: clinical and immunological features of 248 patients. *Clin Immunol* 1999;**92**:34–48.

31. Chapel H, Lucas M, Lee M, et al. Common variable immunodeficiency disorders: division into distinct clinical phenotypes. *Blood* 2008;**112**:277–86.

32. Chapel H, Lucas M, Patel S, et al. Confirmation and improvement of criteria for clinical phenotyping in common variable immunodeficiency disorders in replicate cohorts. *J Allergy Clin Immunol* 2012;**130**:1197–8.

33. Resnick ES, Cunningham-Rundles C. The many faces of the clinical picture of common variable immune deficiency. *Curr Opin Allergy Clin Immunol* 2012;**12**:595–601.

34. Chapel HaC-R. Update in understanding common variable immunodeficiency disorders (CVIDs) and the management of patients with these conditions. *Br J Haematol* 2009;**145**:709–27.

35. Bram TH, Morton MD. Defective serum gamma globulin formation. *Ann Intern Med* 1955;**43**:465–70.

36. Siegal FP, Siegal M, Good RA. Suppression of B-cell differentiation by leukocytes from hypogammaglobulinemic patients. *J Clin Invest* 1976;**58**:109–22.

37. Farrant J, Spickett G, Matamoros N, et al. Study of B and T-cell phenotypes in blood from patients with common variable immunodeficiency (CVID). *Immunodeficiency* 1994;**5**:159–69.

38. Jaffe JS, Eisenstein E, Sneller MC, Strober W. T-cell abnormalities in common variable immunodeficiency. *Pediatr Res* 1993;**33**(1 Suppl):S24–27 discussion S7-8.

39. North ME, Webster AD, Farrant J. Primary defect in CD8+ lymphocytes in the antibody deficiency disease (common variable immunodeficiency): abnormalities in intracellular production of interferon-gamma (IFN-gamma) in CD28+ ('cytotoxic') and CD28- ('suppressor') CD8+ subsets. *Clin Exp Immunol* 1998;**111**:70–5.

40. Kondratenko I, Amlot PL, Webster AD, Farrant J. Lack of specific antibody response in common variable immunodeficiency (CVID) associated with failure in production of antigen-specific memory T-cells. MRC Immunodeficiency Group. *Clin Exp Immunol* 1997;**108**:9–13.

41. Funauchi M, Farrant J, Moreno C, Webster AD. Defects in antigen-driven lymphocyte responses in common variable immunodeficiency (CVID) are due to a reduction in the number of antigen-specific CD4+ T-cells. *Clin Exp Immunol* 1995;**101**:82–8.

42. Stagg AJ, Funauchi M, Knight SC, Webster AD, Farrant J. Failure in antigen responses by T-cells from patients with common variable immunodeficiency (CVID). *Clin Exp Immunol* 1994;**96**:48–53.

43. Eisenstein EM, Chua K, Strober W. B-cell differentiation defects in common variable immunodeficiency are ameliorated after stimulation with anti-CD40 antibody and IL-10. *J Immunol* 1994;**152**:5957–68.

44. Saiki O, Ralph P, Cunningham-Rundles C, Good RA. Three distinct stages of B-cell defects in common varied immunodeficiency. *Proc Natl Acad Sci USA* 1982;**79**:6008–12.

45. Bryant A, Calver NC, Toubi E, Webster AD, Farrant J. Classification of patients with common variable immunodeficiency by B-cell secretion of IgM and IgG in response to anti-IgM and interleukin-2. *Clin Immunol Immunopathol* 1990;**56**:239–48.

46. Eisenstein EM, Jaffe JS, Strober W. Reduced interleukin-2 (IL-2) production in common variable immunodeficiency is due to a primary abnormality of CD4+ T-cell differentiation. *J Clin Immunol* 1993;**13**:247–58.

47. Kruger G, Welte K, Ciobanu N, et al. Interleukin-2 correction of defective in vitro T-cell mitogenesis in patients with common varied immunodeficiency. *J Clin Immunol* 1984;**4**:295–303.

48. Cambronero R, Sewell WA, North ME, Webster AD, Farrant J. Up-regulation of IL-12 in monocytes: a fundamental defect in common variable immunodeficiency. *J Immunol* 2000;**164**:488–94.

49. Sewell WA, North ME, Cambronero R, Webster AD, Farrant J. *In vivo* modulation of cytokine synthesis by intravenous immunoglobulin. *Clin Exp Immunol* 1999;**116**:509–15.

50. Fritsch A, Junker U, Vogelsang H, Jager L. On interleukins 4, 6 and 10 and their interrelationship with immunoglobulins G and M in common variable immunodeficiency. *Cell Biol Int* 1994;**18**:1067–75.

51. Zhou Z, Huang R, Danon M, Mayer L, Cunningham-Rundles C. IL-10 production in common variable immunodeficiency. *Clin Immunol Immunopathol* 1998;**86**:298–304.

52. Agarwal S, Smereka P, Harpaz N, Cunningham-Rundles C, Mayer L. Characterization of immunologic defects in patients with common variable immunodeficiency (CVID) with intestinal disease. *Inflamm Bowel Dis* 2011;**17**:251–9.

53. Mannon PJ, Fuss IJ, Dill S, et al. Excess IL-12 but not IL-23 accompanies the inflammatory bowel disease associated with common variable immunodeficiency. *Gastroenterology* 2006;**131**:748–56.

54. Adelman DC, Matsuda T, Hirano T, Kishimoto T, Saxon A. Elevated serum interleukin-6 associated with a failure in B-cell differentiation in common variable immunodeficiency. *J Allergy Clin Immunol* 1990;**86**:512–21.

55. Warnatz K, Denz A, Drager R, et al. Severe deficiency of switched memory B-cells (CD27(+)IgM(-)IgD(-)) in subgroups of patients with common variable immunodeficiency: a new approach to classify a heterogeneous disease. *Blood* 2002;**99**:1544–51.

56. Sanchez-Ramon S, Radigan L, Yu JE, Bard S, Cunningham-Rundles C. Memory B-cells in common variable immunodeficiency: clinical associations and sex differences. *Clin Immunol* 2008;**128**:314–21.

57. Wehr C, Eibel H, Masilamani M, et al. A new CD21 low B-cell population in the peripheral blood of patients with SLE. *Clin Immunol* 2004;**113**:161–71.

58. Piqueras B, Lavenu-Bombled C, Galicier L, et al. Common variable immunodeficiency patient classification based on impaired B-cell memory differentiation correlates with clinical aspects. *J Clin Immunol* 2003;**23**: 385–400.

59. Vorechovsky I, Cullen M, Carrington M, Hammarstrom L, Webster AD. Fine mapping of IGAD1 in IgA deficiency and common variable immunodeficiency: identification and characterization of haplotypes shared by affected members of 101 multiple-case families. *J Immunol* 2000; **164**:4408–16.

60. Offer SM, Pan-Hammarstrom Q, Hammarstrom L, Harris RS. Unique DNA repair gene variations and potential associations with the primary antibody deficiency syndromes IgAD and CVID. *PLoS One* 2010;**5**:e12260.

61. Orange JS, Glessner JT, Resnick E, et al. Genome-wide association identifies diverse causes of common variable immunodeficiency. *J Allergy Clin Immunol* 2011;**127**: 1360–7.

62. Keller MD, Jyonouchi S. Chipping away at a mountain: Genomic studies in common variable immunodeficiency. *Autoimmun Rev* 2013;**12**:687–9.

63. Grimbacher B, Hutloff A, Schlesier M, et al. Homozygous loss of ICOS is associated with adult-onset common variable immunodeficiency. *Nat Immunol* 2003;**4**:261–8.

64. van Zelm MC, Reisli I, van der Burg M, et al. An antibody-deficiency syndrome due to mutations in the CD19 gene. *N Engl J Med* 2006;**354**:1901–12.

65. Kanegane H, Agematsu K, Futatani T, et al. Novel mutations in a Japanese patient with CD19 deficiency. *Genes Immun* 2007;**8**:663–70.

66. Kuijpers TW, Bende RJ, Baars PA, et al. CD20 deficiency in humans results in impaired T-cell-independent antibody responses. *J Clin Invest* 2010;**120**:214–22.

67. Thiel J, Kimmig L, Salzer U, et al. Genetic CD21 deficiency is associated with hypogammaglobulinemia. *J Allergy Clin Immunol* 2012;**129**:801–10.

68. van Zelm MC, Smet J, Adams B, et al. CD81 gene defect in humans disrupts CD19 complex formation and leads to antibody deficiency. *J Clin Invest* 2010;**120**:1265–74.

69. Salzer U, Chapel HM, Webster AD, et al. Mutations in TNFRSF13B encoding TACI are associated with common variable immunodeficiency in humans. *Nat Genet* 2005;**37**:820–8.

70. Castigli E, Wilson SA, Garibyan L, et al. TACI is mutant in common variable immunodeficiency and IgA deficiency. *Nat Genet* 2005;**37**:829–34.

71. Pan-Hammarstrom Q, Salzer U, Du L, et al. Reexamining the role of TACI coding variants in common variable immunodeficiency and selective IgA deficiency. *Nat Genet* 2007;**39**:429–30.

72. Zhang L, Radigan L, Salzer U, et al. Transmembrane activator and calcium-modulating cyclophilin ligand interactor mutations in common variable immunodeficiency: clinical and immunologic outcomes in heterozygotes. *J Allergy Clin Immunol* 2007;**120**:1178–85.

73. Salzer U, Bacchelli C, Buckridge S, et al. Relevance of biallelic versus monoallelic TNFRSF13B mutations in distinguishing disease-causing from risk-increasing TNFRSF13B variants in antibody deficiency syndromes. *Blood* 2009;**113**:1967–76.

74. Warnatz K, Salzer U, Rizzi M, et al. B-cell activating factor receptor deficiency is associated with an adult-onset antibody deficiency syndrome in humans. *Proc Natl Acad Sci USA* 2009;**106**:13945–50.

From Subcutaneous to Intravenous Immunoglobulin and Back

Melvin Berger[1], E. Richard Stiehm[2]

[1]CSL Behring LLC, King of Prussia, PA, and Adjunct Professor of Pediatrics and Pathology, Case Western Reserve University, Cleveland, OH, USA
[2]Department of Pediatrics, David Geffen School of Medicine at UCLA, Los Angeles CA, USA

INTRODUCTION

Although primary antibody deficiency disorders (PIDD) were not recognized or treated with immunoglobulin replacement therapy until Bruton's report in 1952 (1), the passive administration of antibodies for treatment and/or prevention of infectious diseases was well established long before that time. The use of immune animal sera for the treatment of diphtheria was established in the late 1800s, as recognized by the award of the first Nobel Prize in Medicine or Physiology to Emil von Behring in 1901 (2). In the 1930s, convalescent human and/or hyperimmune animal serum came into wide use for pneumococcal pneumonia (3), and large trials were carried out using immunoglobulin-containing placental extracts for modifying and preventing measles in children (4).

Human immune serum globulin (ISG) became available as a by-product of Cohn cold ethanol fractionation to produce albumin for resuscitating wounded soldiers during World War II. These ISG preparations were used for the prevention of hepatitis among deployed soldiers (5) and for the prevention and amelioration of measles in children in the Northeastern USA (6).

ISG replacement therapy for PIDD began in 1952 following the description of agammaglobulinemia by Ogden Bruton (1). Intravenous immunoglobulin (IVIG) was first licensed in the USA in 1981, and its use by slow subcutaneous infusion was licensed several years later. IVIG was also found to be of value in the treatment of autoimmune and inflammatory diseases as well as immune-mediated neurological diseases. High-titered hyperimmune globulins for specific infections, and monoclonal antibodies for the diagnosis and treatment of a variety of illnesses are more recent but equally important advances.

DEVELOPMENT OF HUMAN IMMUNOGLOBULIN

Anticipating the entry of the USA into World War II, the National Research Council asked protein chemists Edwin Cohn and J.L. Oncley of Harvard University to develop a stable blood substitute to resuscitate wounded soldiers going into shock on the battlefield. In 1941, Cohn was given $10 000 to develop large-scale production of albumin from American Red Cross plasma donations. He recognized that blood plasma contained multiple different proteins of unique therapeutic value and had written a quotation from Goethe's *Faust* (1806) on his office blackboard: "Blut ist ein ganz besonderer Saft (blood is a most special juice)". Cohn had considerable expertise in fractionating proteins with different concentrations of salts such as ammonium sulfate, but for the large-scale fractionation of plasma, he used ethanol. This had the advantage that, as a liquid, it could be mixed with aqueous solutions at low temperatures, thus minimizing the risks of bacterial contamination (7).

The stepwise precipitation of different plasma proteins with increasing percentages of cold ethanol yielded a preparation of albumin termed "Fraction V" which could be lyophilized and stored as a dry powder or held as a 25% liquid. The globulins containing most of the antibody activity of serum were found in Fractions II and III, which also contained clotting factors and lipoproteins. Oncley modified the procedure by manipulating the pH, ionic strength and ethanol concentration, and succeeded in recovering most of the 7S gamma globulins in Fraction II, with the clotting factors and faster sedimenting globulins in fraction III (7, 8).

As already noted, Fraction II ISG was used in the early 1940s for the prevention of hepatitis and measles (5, 6). In these studies, the ISG was given intramuscularly (IM), as had been done previously with placental immune globulin extracts in the 1930s (4). Janeway et al. in Boston (9), and Barandun in Switzerland (10) attempted to give ISG intravenously, but their efforts were abandoned because of severe immediate reactions including hypotension, chills and fever (see below for more details of these classic experiments).

Early Use of Immune Serum Globulin in PIDD

Several developments came together to allow the recognition of PIDD in the early 1950s: the availability of sulfonamides, then penicillin and other antibiotics to keep patients with recurrent infections alive; advances in protein chemistry such as serum protein electrophoresis allowed the diagnosis of serum protein abnormalities; and alcohol fractionation allowed industrial-scale production of ISG.

In 1952, Ogden Bruton, a pediatrician at the Walter Reed Army Hospital, showed that the γ-globulin band was missing from the serum protein electrophoresis pattern of an eight-year-old boy who had suffered 19 episodes of sepsis in the preceding years, most with typeable pneumococci, to which he did not develop specific antibodies (1). Additional cases of agammaglobulinemia in young boys were soon reported by Janeway et al. (11) and in a young woman by Sanford et al. (12). Bruton treated his patient with subcutaneous (SC) injections of ISG; this resulted in clinical improvement as well as detectable γ-globulin upon electrophoresis of his serum. Janeway and colleagues continued with IM injections using monthly doses of 0, 100 mg/kg (0.6 mL of 16% ISG/kg).

A country-wide study in the UK in the 1960s sponsored by the Medical Research Council (MRC) enrolled 176 subjects with antibody deficiency and compared two doses of IM ISG: 25 mg/kg/week or 50 mg/kg/wk (13). The higher dose resulted in fewer febrile episodes, fewer episodes of otitis media and pneumonia, lower levels of C-reactive protein and fewer deaths. However, the MRC concluded that the results did not justify routine use of the higher dose (13). The serum IgG levels achieved in that study were mostly <250 mg/dL in the low-dose group and 300–400 mg/dL in the high-dose group (13). Even the lower dose of 25 mg/kg/wk (0.15 mL/kg/wk) had to be divided between multiple injection sites as it carried a risk of sciatic nerve injury, particularly in children who were underweight and poorly nourished due to chronic gastrointestinal disease.

Early on it was realized that the intramuscular injection of ISG had a number of additional disadvantages: the procedure was painful, the amount that could be given was limited, and if some of the material was inadvertently injected into a vein, it could cause severe anaphylactoid reactions. To overcome the limitations of intramuscular injection of ISG, Silvio Barandun and co-investigators performed a series of seminal *in vivo* and *in vitro* experiments at the Tiefenauspital and the Central Laboratory of the Swiss Red Cross in Bern, Switzerland (10). When they intravenously injected a diluted ISG preparation (final concentration 1.48% gammaglobulin) made from standard commercial 16% gammaglobulin (Cohn fraction II), these investigators observed that only seven of 55 normal controls (12.7%) developed "intolerance" (flushing, lumbar pain, chills, fever), while 14 of 15 patients with antibody deficiency (93.3%) developed "intolerance" (including severe anaphylactic shock). Unexpectedly, a substantial decrease in serum complement was detected in all patients developing intolerance. It was further noticed that when the symptoms following the first infusion had disappeared, the patients became "refractory" to reactions to subsequent gammaglobulin infusions for four to five days.

This phenomenon was explained by activation of the complement by IgG aggregates in the unmodified ISG preparation and/or by the formation of complexes between antigens which had accumulated in the patients' blood because of their immune deficiencies and the infused IgG molecules. By manipulating the standard gammaglobulin (16%: Fraction II), to remove the aggregates, for instance by ultracentrifugation, exposure to pH4, or treatment with pepsin (or reducing agents), the researchers obtained immunoglobulin preparations that, when given intravenously, were tolerated well by PIDD patients. This modified preparation no longer exhibited complement-activating properties and had lost the induction of a refractory phase (10). These experiments, performed in the early 1960s, eventually led to the first widely used IVIG preparations some 20 years later.

One alternative to the IM injections of ISG was the use of unfractionated plasma. Stiehm et al. described treatment with 20 mL of plasma per kg per month (14), which contains approximately 200 mg of IgG, similar to the higher dose regimen studied by the MRC. Family members were used as donors to decrease the risk of hepatitis transmission, and many PIDD patients were maintained on regular infusions of plasma from "buddy donors" throughout the 1970s and 1980s.

Intravenous (IVIG) and Subcutaneous Immune Globulin (SCIG) Treatment of Antibody Deficiencies

Roughly 40 years passed between the initial use of Cohn Fraction II as prophylaxis or treatment of infections and the licensing of the first intravenous immune globulin (IVIG) preparations in the USA in 1981 (15, 16). However, since the mid-1980s, rapid progress in improving the safety and tolerability of IVIG preparations has led to a gradual increase in the doses used for replacement therapy, leading to a decreased frequency of serious infections. More recently, convenience and quality of life have been the focus of efforts to improve the use of ISG for replacement therapy in PIDD.

With the availability of reasonably well-tolerated IVIG preparations in the 1980s, doses began to increase. Eibl et al. adjusted the frequency of IV infusions to maintain serum IgG levels above a minimum of 300 mg/dL (16), and Barandun et al. experimented with doses as high as 1000 mg/kg/month (17). Roifman et al. carried out a randomized cross-over study of 12 patients and showed that patients had fewer infections and improved pulmonary function when they maintained serum IgG trough levels above 500 mg/dL on a dose of 600 mg/kg/month, as compared to increased infections and deteriorating pulmonary function when their serum IgG was less than 500 mg/dL on a lower IVIG dose (200 mg/kg/month) (18). Eijkhout et al. in a randomized, cross-over study of 41 subjects demonstrated that "high dose" IVIG therapy (600 mg/kg/month in adults and 800 mg/kg/month in children) was associated with significantly fewer episodes of infection and days of illness than "standard" doses (300 mg/kg/month in adults and 400 mg/kg/month in children with PIDD) (19). Recent guidelines continue to recommend higher doses for PIDD patients with chronic lung and/or sinus disease.

Slow subcutaneous infusion of immunoglobulin (SCIG) for antibody deficiency using a small pump was introduced by Berger et al. in 1980 for administering a 16% preparation (Gammastan) to three CVID patients who had reactions to IMIG (20), and then used by Ugazio et al. and Roord et al. in 1982 (21, 22). The subcutaneous route was first implemented regularly in the late 1980s by Gardulf et al. in Sweden (23). Many PIDD patients now receive their immunoglobulin by slow subcutaneous infusion using 10, 16 or 20% products, the latter developed especially for subcutaneous injection. The subcutaneous route has the advantages of not needing intravenous access, steadier IgG levels, higher IgG trough levels with comparable doses and, most importantly, considerably fewer systemic side effects than IVIG infusions (23–25).

The usual dose for SCIG is 300–500 mg/kg/month, similar to the IVIG dose (26, 27). Infusions are usually given weekly at 25% of the monthly dose. Thus a weekly dose of 100 mg/kg requires a subcutaneous infusion of 1.0 mL/kg of a 10% IG product or 0.5 mL/kg of a 20% IG product. This is usually given over 1–2 hours using a portable infusion pump. Multiple sites on the abdominal wall, thighs or arms can be used so that no more than 5 to 30 mL is given at a single site, depending on the patient's weight. Because of the lessened risk of systemic side effects, SCIG can be given at home, by self-infusion, and without premedication (22–27).

ADVERSE REACTIONS TO IVIG AND SCIG

Immediate Reactions

Mild reactions to presently available IVIG infusions are common, occurring in 15–20% of infusions and in 50% of PIDD patients at one time or another (28–32). These adverse events (AEs) are mostly uncomfortable and/or unpleasant but rarely serious. Symptoms may include headache, nausea, musculoskeletal pain, and flushing and tachycardia. In most cases, symptoms can be easily managed by slowing or temporarily stopping the infusion until the symptoms subside, and/or by treatment with acetaminophen, non-steroidal anti-inflammatory drugs (NSAIDS) and/or antihistamines (33). Some patients may require corticosteroids; and many patients are given NSAIDs or steroids prophylactically. Most AEs are related to the rate of infusion and can be avoided by beginning the infusion slowly (1 mg/kg/min) and gradually increasing the rate step-wise as tolerated. Decreasing the interval between IV infusions may also help to eliminate AEs.

Patients who are naïve to IVIG replacement, have had interruptions in their therapy and/or who are actively or chronically infected have an increased risk of infusion-related AEs. This may be related, in part, to the accumulation of circulating antigen and the formation of antigen–antibody complexes as the IgG is being given, and/or the rapid release of lipopolysaccharide or other components of pathogens already present in the recipient. The risk of these reactions can be reduced by making sure patients are afebrile and that those with active infections are on antibiotics before giving an IVIG dose. The incidence of reactions may increase when patients already on therapy are given a different brand of IVIG (34), so whenever this occurs, it is prudent to begin the infusion slowly and/or to premedicate the patient.

When severe, IVIG infusion reactions may resemble anaphylaxis, but usually do not involve IgE, and should be termed "anaphylactoid" (27, 35–38). A key difference between these anaphylactoid reactions that accompany IVIG infusions and true IgE-mediated anaphylaxis is that the former are usually associated with hypertension, rather than hypotension. Furthermore, anaphylactoid IVIG reactions often are less severe with subsequent infusions rather than more severe as would be expected with true allergy.

True anaphylaxis may occur in patients receiving IVIG (see below), particularly in those who are deficient in IgA but still have the capacity to produce IgE (36, 37). This occurs very rarely, but may be life threatening, and can be avoided by slow administration of low-IgA products (36, 37) and/or by administering the IgG subcutaneously (39).

Systemic reactions to SCIG infusions are rare. Gardulf et al. reported only 30 systemic reactions in 25 immunodeficient patients given 3232 infusions (0.93%) (23). A subsequent review by Ann Gardulf which included additional clinical trials totaling over 40 000 infusions, showed that only one study reported a rate of systemic AEs >1% (38). Although nearly 75% of patients may have some local discomfort associated with the swelling and redness at the site of the infusions (38–40), the swelling and local symptoms usually subside within 24 to 48 hours, and rarely deter patients from continuing with their SCIG regimen. Because of the infrequency of systemic AEs with SCIG, premedication is not necessary nor is close monitoring required during the infusion. SCIG has thus emerged as an ideal route for home use for many PIDD patients (39, 40).

Product Improvements to Minimize Immediate Reactions

Studies through the 1950s and 1960s suggested that the vasomotor reactions and other immediate AEs that accompanied the early trials of IVIG were attributable to complement activation by aggregates of the 7S globulins. Barandun et al. showed that these aggregates could be removed by ultracentrifugation or dissociated by treatment with low concentrations of pepsin, incubation at pH 4 and/or reduction of disulfide bonds followed by alkylation of free thiols (10, 17). Other protein modifications such as S-sulfonylation and treatment with β-propiolactone were also studied (10, 16, 17, 40, 41). The first commercial IVIGs were pepsin- or plasmin-treated preparations which contained fragmented IgG molecules that sedimented more slowly than intact IgG (i.e. at 5–6.5S), and had shortened half-lives and decreased effector function compared to the intact IgG in the IM preparations (16, 41).

By the late 1970s, combined treatments including mild reduction and alkylation and reconstitution in 0.3 M glycine, and/or treatment at pH4 with low concentrations of pepsin, were tested to prevent aggregate formation. These two methods were eventually used in the first IVIG preparations licensed in the USA: Gamimune® and Sandoglobulin®, respectively. The most important contribution to the tolerability of these preparations, however, was arguably the use of high concentrations of sugars as stabilizers: 10% maltose in Gamimune® (41) and sucrose in Sandoglobulin® (42).

Despite minimization of complement-binding aggregates, early IVIG preparations often caused hypotension and/or other signs associated with vasodilatation and increased capillary permeability, which were shown to be associated with the presence of contaminating amounts of proteins such as prekallikrein activator and kallikrein itself (43). Contamination with Factor XIa was also found to be responsible for procoagulant activity in many ISG and early IVIG preparations (43). In addition, other studies showed that isolated Fc fragments, IgG aggregates and antigen–antibody complexes induced secretion of prostaglandin E_2 from monocytes, suggesting that this mediator might also be contributing to the adverse effects of early IVIG preparations (44).

In 1981, the World Health Organization published a set of "Desirable Characteristics of IVIG Preparations" (45):

- IVIG should be extracted from a pool of at least 1000 individual donors
- IVIG should contain as little IgA as possible
- The IgG should be modified biochemically as little as possible and possess opsonizing and complement-fixing activities
- IVIG should be free from preservatives or stabilizers that might accumulate *in vivo*.

Current IVIG Products

Additional methods to increase the yield of IgG per liter of plasma, to improve the convenience of infusing IVIG preparations and to minimize AEs have been developed in recent years. Many manufacturers utilize new purification procedures which employ only a single ethanol precipitation step and substitute precipitation with fatty acids such as caprylate or medium chain alcohols, and depth filtration for the serial Cohn–Oncley ethanol precipitations steps (46). Anion-exchange column chromatography was added to improve the purity by decreasing the concentrations of IgA and potentially vasoactive and/or thrombogenic protein contaminants, and some manufacturers also added cation exchange chromatography further to purify the IgG product.

To improve the convenience of preparing and administering IVIG, most manufacturers have IVIG preparations which are available as 10% liquids that can be kept at room temperature for at least part of their shelf life. As of 2013, three products: Gammagard® (Baxter), Gamunex® (Grifols) and Privigen® (CSL Behring) dominate the US market. Gammagard® and Gamunex® are stabilized with glycine, and Privigen® is stabilized with L-proline. All are available as 10% liquids, none contain any sugars, and the IgA content of all three is less than 50 μg/mL.

Other Serious IVIG Reactions

The increasing use of IVIG, particularly with large doses for inflammatory and autoimmune diseases, resulted in other serious adverse events, including transmission of hepatitis C, aseptic meningitis, renal failure, thromboembolism and hemolytic anemia (47–53). This has led to the requirement (in the USA) that all IVIG products contain a "Black Box" warning about acute renal dysfunction/failure (47), and "Warnings and Precautions" about the risk of thromboembolic events, hemolytic anemia, aseptic meningitis syndrome and transfusion-associated acute lung injury (TRALI) (48). Most reports of acute renal failure/dysfunction suggest that these were due to osmotic nephrosis related to the use of sucrose as a stabilizer in certain products (47). The risk of aseptic meningitis and thromboembolic events seems to be highest in patients receiving high dose IVIG for neurological diseases (49, 50).

Prevention of Pathogen Transmission

Although clotting factor concentrates prepared by Cohn fractionation were known to have transmitted what is now known as hepatitis B, Fraction II ISG had been given to hundreds of children for measles prophylaxis with only one case of apparent transmission of that virus (9). The subsequent widespread use of Fraction II ISG for prophylaxis of "endemic" (type A) hepatitis, measles and polio also provided confidence that that this product carried little risk of transmission of blood-borne infectious agents.

Unfortunately, that turned out to be false confidence. Even after the recognition of AIDS, complacency was still fostered by reports that HIV (then termed HTLV III) was inactivated and/or partitioned out of Fraction II by the ethanol precipitation procedure (51). In retrospect, the apparent safety of Fraction II ISG was, in part, due to the presence of antibodies in some plasma donations which neutralized undetected viruses in others, causing neutralization and/or facilitating removal of the potentially pathogenic viruses in immune complexes which formed upon pooling.

In the late 1980s, reports of "non-A, non-B" hepatitis, now termed hepatitis C virus (HCV), among IVIG recipients appeared. In 1994, the US Centers for Disease Control reported over 100 cases of HCV in PIDD patients, most of whom had received a single IVIG preparation (52). In the same year, a report from Scandinavia noted that 17 of 20 recipients of a different IVIG preparation used in that region also developed clinical hepatitis and became seropositive to the newly recognized HCV (53).

After it became apparent that the fractionation procedure was not efficient at inactivating or partitioning free viruses, most manufacturers added a process for disrupting the lipids of enveloped viruses, for instance by using the solvent tri-(n-butyl)-phosphate together with a detergent such as Triton X-100 or Tween. This process, known as solvent detergent (S/D) treatment, was developed by the New York Blood Center (54).

However, S/D treatment does not inactivate non-enveloped viruses such as Parvovirus B-19, nor does it inactivate the prions responsible for transmissible spongiform encephalopathies such as variant Creutzfeldt–Jakob (mad cow) disease. To reduce the risk that agents such as these might be in the plasma units from which products such as IVIG are made, government regulators and the plasma fractionators instituted stringent donor screening and record-keeping procedures and now use sensitive nucleic-acid amplification (PCR) testing. In addition, quarantine or "hold-back" programs have been instituted in which any given unit of plasma is held separately and may not be added to a pool until the same donor gives a subsequent unit which also tests negative. This insures that no units are used that were obtained during the "window" period when viremia is clinically inapparent.

Besides these methods to prevent tainted plasma from getting into the starting material, all US manufacturers have incorporated multiple viral inactivation procedures such as pH4 incubation, treatment with fatty acids such as caprylate (octanoic acid) and/or prolonged treatment at 60°C (pasteurization). These procedures are more effective at eliminating the pathogenicity of enveloped than non-enveloped viruses, so depth filtration in the presence of particulate adsorbents ("filter-aids") and nanofiltration through membranes which exclude viruses larger than 20 nm is also used with most preparations.

The efficacy of these steps is verified by adding model animal viruses and prions to pilot-size plasma pools fractioned in the same way as used for actual production to verify the absence of these infectious agents in the final preparations (55).

The use of multiple, dedicated viral safety steps in the manufacturing processes has led to new generations of IVIG products with intact, unaltered IgG molecules which are purer, safer and easier to administer than those used in the 1980s. Thus, three decades after the World Health Organization published its "Desirable Characteristics", and more than 70 years after Cohn pioneered fractionation of plasma, the IVIG preparations now available meet those criteria.

Thromboembolic Events

Determining the true rates of thromboembolic events (TEE) and hemolytic incidents is difficult because the Food and Drug Administration (FDA) and manufacturers' pharmacovigilance efforts rely on voluntary efforts of patients/providers and because there are few data with which to formulate a denominator in terms of doses given or patients treated. High-dose IVIG can cause hyperviscosity and slow blood flow in critical vascular beds in some patients (50); and endothelial cell and/or platelet activation may also contribute to TEEs. There are multiple reports of myocardial infarction, transient cerebral ischemic attacks and strokes related to high-dose IVIG therapy (56). Best estimates suggest a baseline incidence of 0.16 to 0.6 TEEs for every 1 million grams of IVIG used (56–58). If one uses 50 grams as a median adult dose, these rates represent roughly 0.8 to 3 cases per 100 000 doses. In a recent well-investigated episode, TEEs were reported in nine patients involving seven different lots of a single IVIG product (56–58).

Subtle changes in the production procedure of that product, including the use of resins to isolate certain clotting system proteins, apparently increased contact activation of Factor XI which had co-purified with the IgG during the initial ethanol precipitation (58). Because the total Factor XI plus XIa was still within acceptable limits by the assays in use at that time, the affected product passed all criteria for safety. Only recently developed thrombin-generation assays have sufficient sensitivity to detect activated FXIa (56, 58). Factor XI and kallikrein are difficult to separate from IgG because they have isoelectric points similar to IgG and co-precipitate with it during ethanol precipitation (42).

On the one hand, the results suggest that contamination of only one or two individual IVIG lots was not responsible for the increase in factor XIa, since many lots were affected (56, 58). On the other hand, even with the affected lots, TEEs were extremely rare, suggesting that multiple risk factors in the affected patients also contributed.

Better chromatographic purification methods and the use of specific immunoadsorbents to remove FXI/FXIa from products which are made by multiple precipitations without ion-exchange chromatography should decrease the risk of Factor XIa-related AEs (58). Furthermore, the use of new thrombin-generation assays should ensure the absence of procoagulant activity in current and future IgG products.

Hemolytic Reactions

An analogous problem is the presence of antibodies to erythrocytes (isoagglutinins), resulting in positive Coombs' tests, occasional cases of clinically significant hemolytic anemia and extremely rare episodes of acute severe intravascular hemolysis (59, 60). In the original Cohn–Oncley fractionation scheme, isoagglutinins, which have higher isoelectric points than other IgGs, were greatly reduced by removing Fraction III and continuing with Fraction II alone (8). Several recent production schemes use the combined Fraction II and III precipitates as the starting material for IgG purification and many manufacturers have substituted a precipitation step with caprylic acid for the ethanol steps that removed more of the isoagglutinins, resulting in two- to fourfold increases in isoagglutinin titers in the final products.

As with thromboembolic events, true rates of Coombs' positivity, hemolytic anemia and acute severe hemolysis due to IVIG therapy are difficult to estimate. However, risk factors which have been identified include non-O blood groups, underlying associated inflammatory state and high cumulative doses of IVIG over several days (48). Some studies of IVIG therapy of ITP or neurological diseases report that as many as 20% of patients may convert to Coombs' positivity shortly after high dose infusions, but the incidence of clinically significant anemia or acute severe hemolysis is significantly lower than that (59, 60). Analysis of recent US and Canadian series reported clinically significant hemolysis after IVIG in a combined total of 37 patients, including 23 patients with blood type A, nine with type B, four with type AB, and only one with type O (59, 60).

Pre-screening of donors to avoid using plasma units with high isoagglutinin titers in the pools from which the IVIG is prepared, and the use of specific immunoadsorbents to lower the titers of anti-A and anti-B are steps now being introduced to decrease this problem (61).

Events that Affect Treatment in PIDD Patients With IVIG

Immunodeficient patients on maintenance ISG therapy may experience accelerated immunoglobulin catabolism during febrile illnesses, or may have increased gastrointestinal loss of immunoglobulin with diarrheal diseases. Extra doses of IVIG may be considered if the patient has prolonged fever or other signs of slow or delayed recovery. Refractory pulmonary or staphylococcal infections and chronic lung disease associated with bronchiectasis may also

benefit from high-dose immunoglobulin therapy (62, 63).

Refractory diarrhea in immunodeficiency has been treated successfully with immunoglobulin. In some situations, oral immunoglobulin (100–150 mg/kg/day) has been used to control gastrointestinal infections or to interrupt viral excretion (64). Oral immunoglobulin has been used to limit viral shedding in chronic rotavirus or polio infections. Chronic cryptosporidial infection in patients with secondary immunodeficiency has been successfully treated with high-dose IVIG (65).

The monoclonal antibody palivizumab (Synagis®) is used to prevent respiratory syncytial virus (RSV) in premature and high-risk infants aged under two years during the RSV season (66). It is also indicated for immunodeficient infants, including those on regular IgG therapy, since standard IVIG or SCIG preparations lack sufficient antibodies against RSV to be effective. The earlier use of an IVIG preparation to prevent RSV provides interesting insights into the use of standard versus hyperimmune or monoclonal globulins: to get a monthly protective dose of anti-RSV antibody requires 3750 mg/kg of standard IVIG, 750 mg/kg of the RSV hyperimmune RespiGam® (67), or only 15 mg/kg of the monoclonal Synagis® (68).

CNS infection with coxsackie, echovirus and poliovirus are known complications of patients with antibody deficiency, particularly those with X-linked agammaglobulinemia on low-dose IMIG (69). High-dose IVIG, using a preparation with a high titer to the virus, has been used successfully in clearing the CNS virus. Occasionally, the IVIG is given intrathecally (62).

HYPERIMMUNE IMMUNOGLOBULINS

In parallel with developing intramuscular and intravenous immune globulins was the recognition that these products could prevent certain specific infections, sometimes in conjunction with administration of the necessary vaccine in an unimmunized person. However, the titers of antibodies to specific pathogens, e.g. hepatitis A or B, varicella or rabies in different donor pools were variable, so that the standard replacement dose was inadequate in many situations. In contrast, in other situations, the antibody content was too large, and the response to a concomitantly given vaccine was inhibited. Thus, the need for standardization of antibody content in regular ISG preparations and the development of hyperimmune globulins with known titers of antibody to specific pathogens became evident (70).

To develop a hyperimmune globulin, plasma donors are selected from immunized, convalescing or screened normal subjects with high titers to the desired microbe or antigen. Plasma is processed as for regular ISG or IVIG and the final product is tested to ensure an adequate antibody titer to the microbe or other antigens. Examples include hyperimmune globulin to cytomegalovirus (CytoGam®), and hyperimmune globulin to the Rh (D) red cell antigen (Rh immunoglobulin) or ovine hyperimmune globulin to digitalis. If rapid clearance of an antigen is desired, the immunoglobin is cleaved to the Fab_2 fragment (Digibind®).

The first hyperimmune products were derived from animal sera, including tetanus and diphtheria antitoxins. Throughout the 1930s, convalescent human sera were used for treatment of infections, particularly when there were large outbreaks of pneumococcal pneumonia. Several animal antisera are still used for some conditions (diphtheria antitoxin, snake and spider envenomation) when a high-titered human immunoglobulin is unavailable.

The first hyperimmune human immunoglobulin was a 16% IM product with a high titer of hepatitis B antibodies, which was used to prevent hepatitis in high-risk institutionalized children, prevention of maternal–infant transmission, or following accidental exposure to hepatitis B (62, 63). This is still used (in conjunction with vaccine) for infants of hepatitis B-exposed mothers, and following liver transplantation into HBsAg-positive recipients, to prevent the new liver from becoming infected.

Other hyperimmunes are listed in Table 23.1. Several are now available for intravenous use. As noted above, immunodeficient subjects including those on SCIG or IVIG may require these products in case of specific exposure, since standard immunoglobulin products may lack sufficient protective antibodies.

DEVELOPMENT OF THERAPEUTIC MONOCLONAL ANTIBODIES

Production of therapeutic monoclonal antibodies followed the 1965 work of Milstein and Kohler (for which they were awarded the Nobel prize in 1984), who demonstrated the feasibility of fusing a B cell making a single antibody with a malignant myeloma cell that had lost its ability to secrete antibody. The resultant cell line secretes a single isotype of antibody with the specificity of the B cell and the immortality of the tumor cell, thus providing a continuous source of antibody to a single antigen. The first monoclonal antibodies were derived from murine cells, but new techniques have allowed the development of chimeric human–murine antibodies or of completely "humanized" antibodies to limit their antigenicity (70).

Such monoclonal antibodies were first used as reagents to assess specific human cell or microbial antigens for diagnostic purposes but, since 1984, monoclonal antibodies have also been used as therapeutic agents. Over 30 monoclonals to cells, tumor cell antigens, inflammatory cytokines, immunoglobulins, complement components and respiratory syncytial virus are now available for the prevention and treatment of a variety of inflammatory, immunological and malignant disorders.

TABLE 23.1 Antibody preparations available for passive immunity in the USA

1. POLYCLONAL HUMAN IMMUNOGLOBULINS

Immune globulin, intravenous (5%, 10%) (IGIV, IVIG)	Treatment of antibody deficiency Immune thrombocytopenic purpura Kawasaki syndrome Other immunoregulatory diseases
Immune globulin, intramuscular (16%) (IGIM, ISG)	(Treatment of antibody deficiency) Prevention of measles, hepatitis
Immune globulin, subcutaneous (10%, 20%) (IGSC, SCIG)	Treatment of antibody deficiency

2. HYPERIMMUNE HUMAN IMMUNOGLOBULINS FOR INTRAMUSCULAR USE

Hepatitis B immune globulin (HBIG)	Prevention of hepatitis B
Varicella-zoster immune globulin (VZIG)	Prevention or modification of chickenpox
Rabies immune globulin (RIG)	Prevention of rabies
Tetanus immune globulin (TIG)	Prevention or treatment of tetanus
Vaccinia immune globulin (VIG)	Prevention or treatment of vaccinia
Rh(D) immune globulin (RhoGAM, Rhophylac)	Prevention of Rh hemolytic disease

3. HYPERIMMUNE HUMAN IMMUNE GLOBULINS FOR INTRAVENOUS USE

Cytomegalovirus immune globulin (CytoGam, CMV-IVIG, CMVIG)	Prevention or treatment of cytomegalovirus infection
Hepatitis B immune globulin (HepaGam B)	Prevention of hepatitis B (including liver transplantation)
Vaccinia immune globulin intravenous (VIG-IVIG)	Prevention or treatment of vaccinia, prevention of smallpox
Rho(D) immune globulin, intravenous (WinRho SDF)	Treatment of immune thrombocytopenic purpura
Botulism immune globulin intravenous (BIG, Baby BIG)	Treatment of infantile botulism

4. ANIMAL SERUMS AND GLOBULINS

Tetanus antitoxin (equine) TAT	Prevention or treatment of tetanus (when TIG unavailable)
Diphtheria antitoxin (equine) DAT	Treatment of diphtheria
Botulinum antitoxins (equine)	Treatment of botulism
Latrodectus mactans antivenin (equine)	Treatment of black widow spider bites
Crotalidae polyvalent antivenin (equine)	Treatment of most snake bites
Crotalidae polyvalent immune Fab (ovine)	Treatment of most snake bites
Micrurus fulvius antivenin (equine)	Treatment of coral snake bites
Digoxin immune Fab Digibind, (ovine) DigiFab®	Treatment of digoxin or digitoxin overdose
Anti-lymphocyte/thymocyte immune globulin (equine) ATG,	Immunosuppression
Anti-lymphocyte/thymocyte immune globulin (rabbit) ATG, thymoglobulin	Immunosuppression

A LOOK TO THE FUTURE

Expanding Markets

Immunoglobulin use is expanding 10–13% annually and is estimated to become a 9 billion dollar/year industry by 2018. Countries in which IVIG is readily available also experience economic growth, an expanding and aging population, improved health care and new indications for high-dose IVIG, factors that are likely to continue to fuel this expansion. Early diagnosis of immunodeficiency by newborn screening and the recognition that doses above the standard regimen of 300–500 mg/kg/month are indicated for some immunodeficient subjects will also add to the overall use of IgG products. The possibility of a worldwide shortage is real and would be devastating for patients depending on Ig replacement therapy.

Immunoglobulin use will also expand in less-developed countries, driven in part by increased recognition of PIDDs, aided by worldwide efforts of the Jeffrey Modell Foundation, including wide distribution of "The 10 warning signs of primary immunodeficiency", which has been translated into more than 60 languages, physician education, international immunodeficiency registries, and improved local and regional laboratory facilities. One measure of this globalization is the fact that many countries now manufacture their own IVIG from local donor plasma, rather than purchasing product from US or European manufacturers.

New Indications

The use of immunoglobulin to prevent rejection of transplanted kidneys and other solid organs is expanding. Treatment of neurological disorders with high-dose immunoglobulin is continuing to increase, as disorders such as PANDAS, regional pain syndromes, peripheral neuropathies and possibly Alzheimer's disease have been added to the illnesses in which immunoglobulin may be beneficial (71, 72). A review of clinicaltrials.gov shows the broad range of clinical trials now underway.

New Products

We have seen the introduction of 20% IgG products to facilitate subcutaneous therapy. The anticipated licensure of an IgG product containing recombinant human hyaluronidase, which facilitates more rapid absorption of immunoglobulin from subcutaneous sites, suggests that SCIG can be given more easily for PIDD patients and that the subcutaneous route may be used for high dose immunomodulatory therapy.

Considerable work is underway to identify the mechanisms by which immunoglobulin exerts its therapeutic benefits other than simple replacement of passive antibodies. Preparations enriched for anti-inflammatory activity may be forthcoming. Of particular interest are the observations by Kaneko et al. suggesting that a small fraction of IgG molecules (<5%) containing fully sialylated F_c domains may be responsible for much of the anti-inflammatory activity of IVIG (73). Efforts to enrich this subpopulation of IgG might result in products with enhanced immunomodulatory activity at much lower doses of IVIG (74).

Other IVIG preparations, enriched for anti-idiotypic antibodies, are also being studied in autoimmune conditions. Antibodies to specific β-amyloid peptides for Alzheimer's disease, antibodies for refractory and/or toxin-producing staphylococcal infections, broadly neutralizing antibodies to HIV and antibodies which neutralize multiple strains of influenza may also be forthcoming for the prevention and/or treatment of these disorders. IgM and IgA-enriched products are also under development, and may prove useful by providing a unique spectrum of specificities and for topical uses, such as gastrointestinal disorders (75).

In the more distant future, engineered antibodies, antibodies produced by plants and/or in eggs, and additional monoclonal antibodies to pathogens, cellular antigens and regulatory molecules, complement components and cytokines may offer new therapeutic options to prevent infection, modulate inflammation and reverse degenerative diseases of the brain and other systems may be forthcoming.

SUMMARY

The recognition of neutralizing, opsonizing, and autoaggressive antibodies in the serum, the development of vaccines to induce their production and of techniques to isolate and enrich these proteins and formulate safe preparations for treatment of PIDD patients has taken a bit more than one hundred years. While immunoglobulin replacement therapy for primary and secondary immunodeficiencies is now well established, it is also used for the prevention and treatment of established infections and for immune modulation of many autoimmune, inflammatory and neurological disorders. New indications and improved products will result in its greatly expanded use both in developed and developing countries, possibly leading to shortages and the need for stricter allocations.

References

1. Bruton O. Agammaglobulinemia. *Pediatrics* 1952;**9**: 722–8.
2. von Behring E. Serum therapy in therapeutics and medical science. Nobel Lecture, December 12, 1901. Available at: http://www.nobelprize.org/nobel_prizes/medicine/laureates/1901/behring-lecture.html.
3. Podolsky SH. *Pneumonia Before Antibiotics*. Baltimore: The Johns Hopkins University Press; 2006.
4. Robinson ES, McKhann CF. Immunological application of placental extracts. *Am J Publ Hlth Nations Hlth* 1935;**25**:1353–8.
5. Gellis SS, Stokes J, Brother GM, Hall WM, Gilmore HR, Morrissey RA. The use of human immune serum globulin (Gamma Globulin) in infectious (epidemic) hepatitis in the Mediterranean theater of operations. *J Am Med Assoc* 1945;**128**:1062–3.
6. Stokes J, Maris EP, Gellis SS. Chemical, clinical and immunological studies on the products of human plasma fractionation. XI. The use of concentrated normal human serum gamma globulin (human immune serum globulin) in the prophylaxis and treatment of measles. *J Clin Invest* 1944;**23**:531–40.
7. Cohn EJ. The history of plasma fractionation. In: Andrus EC, Bronk DW, Carden GA Jr, editors. *Advances in Military Medicine*, Vol 1. Boston: Little, Brown; 1948. p. 364–443.
8. Oncley JL, Melin M, Richert DA, et al. The separation of the antibodies, isoagglutinins, prothrombin, plasminogen and β-1 lipoprotein into subfractions of human plasma. *J Am Chem Soc* 1949;**71**:41–550.
9. Janeway CA. The development of clinical uses of immunoglobulins: a review. In: Merler E, editor. Immunoglobulins. National Academy of Sciences: Washington, DC, 1970, p. 3–14.
10. Barandun S, Kistler P, Jeunet F, Isliker H. Intravenous administration of human g-globulin. *Vox Sang* 1962;**7**: 157–74.
11. Janeway CA, Apt L, Gitlin D. Agammaglobulinemia. *Trans Assoc Am Physicians* 1953;**66**:200–2.
12. Sanford JP, Favour CB, Tribeman MS. Absence of serum gamma globulins in an adult. *N Engl J Med* 1954;**250**: 1027–9.
13. MRC Working Party on Hypogammaglobulinaemia, Hypogammaglobulinaemia in the United Kingdom. Her Majesty's Stationery Office: London, 1971.
14. Stiehm ER, Vaerman J-P, Fudenberg H. Plasma infusions in immunologic deficiency states: metabolic and therapeutic studies. *Blood* 1966;**28**:918–37.
15. Ammann AJ, Ashman RF, Buckley RH, et al. Use of intravenous gamma globulin in antibody immunodeficiency. Results of a multicenter controlled trial. *Clin Immunol Immunopathol* 1982;**22**:60–7.
16. Eibl MM. History of immunoglobulin replacement. *Immunol Allergy Clin N Am* 2008;**28**:737–64.
17. Barandun S, Morell A, Skvaril F. Clinical experiences with immunoglobulin for intravenous use. In: Alving BM, Finlayson JS, editors. *Immunoglobulins: Characteristics and uses of intravenous preparations*. Washington, DC: US Department of Health and Human Services; 1980. p. 31–5.
18. Roifman CM, Levison H, Gelfand EW. High-dose versus low-dose intravenous immunoglobulinin hypogammaglobulinaemia and chronic lung disease. *Lancet* 1987;**1**: 1075–7.
19. Eijkhout HW, van Der Meer JW, Kallenberg CG, et al. The effect of two different dosages ofintravenous immunoglobulin on the incidence of recurrent infections in patients with primary hypogammaglobulinaemia: a randomized, double-blinded, multicenter crossover trial. *Ann Intern Med* 2001;**135**:165–74.

20. Berger M, Cupps TR, Fauci AS. Immunoglobulin replacement by slow subcutaneousinfusion. *Ann Intern Med* 1980;**98**:55–6.

21. Ugazio AG, Duse M, Re R. Subcutaneous infusions of gammaglobulin in management of agammaglobulinemia. *Lancet* 1982;**1**:226.

22. Roord JJ, Van der Meer JWM, Kuis M, et al. Home treatment in patients with antibodydeficiency by slow subcutaneous infusion of gammaglobulin. *Lancet* 1982;**I**:689–90.

23. Gardulf A, Hammarstrom L, Smith CIE. Home treatment of hypogammaglobulinaemia with subcutaneous gammaglobulin by rapid infusion. *Lancet* 1991;**338**:162–6.

24. Berger M. Subcutaneous immunoglobulin replacement in primary immunodeficiencies. *Clin Immunol* 2004;**112**:1–7.

25. Gardulf A, Nicolay U, Math D, et al. Children and adults with primary antibody deficiencies gain quality of life by subcutaneous IgG self-infusions at home. *J Allergy Clin Immunol* 2004;**114**:936–42.

26. Huang F, Feuille E, Cunningham-Rundles R. Home care use of intravenous and subcutaneous immunoglobulin for primary immunodeficiency in the United States. *J Clin Immunol* 2013;**33**:49–54.

27. Berger M, Broyles R, Bullinger A, Rodden L. US immunologists prescribe the same doses of SCIG and IVIG for the treatment of PIDD. *J Clin immunol* 2013; (in press).

28. Stiehm ER. Adverse effect of human immunoglobulin therapy. *Transfus Med Rev* 2013;**27**:171–8.

29. Nydegger UE, Sturzenegger M. Adverse effects of intravenous immunoglobulin therapy. *Drug Safety* 1999;**3**:171–8.

30. Pierce LR, Jain N. Risks associated with the use of intravenousimmunoglobulin. *Transfus Med Rev* 2003;**17**:241–51.

31. Orbach H, Katz U, Sherer Y, Shoenfeld Y. Intravenous immunoglobulin. Adverse effects and safe administration. *Clin Rev Allergy Immunol* 2006;**29**:173–84.

32. Bonilla FA. Intravenous immunoglobulin: Adverse reactions and management. *J Allergy Clin Immunol* 2008;**122**:1238–9.

33. Ballow M. Safety of IGIV therapy and infusion-related adverse events. *Immunol Res* 2007;**38**:122–32.

34. Ameratunga R, Sinclair J, Kolbe J. Increased risk of adverse events when changing intravenous immunoglobulin preparations. *Clin Exp Immunol* 2004;**136**:111–3.

35. Burks AW, Sampson HA, Buckley RH. Anaphylactic reactions after gamma globulin administration in patients with hypogammaglobulinemia. Detection of IgE antibodies to IgA. *N Engl J Med* 1986;**314**:560–4.

36. Rachid R, Bonilla FA. The role of anti-IgA antibodies in causing adverse reactions to gamma globulin infusion in immunodeficient patients: a comprehensive review of the literature. *J Allergy Clin Immunol* 2012;**129**:628–34.

37. Sundin U, Nava S, Hammarström L. Induction of unresponsiveness against IgA in IgA-deficient patients on subcutaneous immunoglobulin infusion therapy. *Clin Exp Immunol* 1998;**112**:341–6.

38. Gardulf A. Immunoglobulin treatment for primary antibody deficiencies: Advantages of the subcutaneous route. *Biodrugs* 2007;**21**:106–16.

39. Abolhassani H, Sadaghiani MS, Aghamohammadi A, Ochs HD, Rezaei N. Home-based subcutaneous immunoglobulin versus hospital-based intravenous immunoglobulin in treatment of primary antibody deficiencies: Systematic review and meta analysis. *J Clin Immunol* 2012;**6**:1180–92.

40. Wasserman RL. Progress in gammaglobulin therapy for immunodeficiency: From subcutaneous to intravenous infusions and back again. *J Clin Immunol* 2012;**6**:1153–64.

41. Berger M. A history of immune globulin therapy, from the Harvard crash program to monoclonal antibodies. *Curr Allergy Asthma Rep* 2002;**2**:368–78.

42. Ochs HD, Buckley RH, Pirofsky B, et al. Safety and patient acceptability of intravenous immune globulin in 10% maltose. *Lancet* 1980;**2**:1158–9.

43. Alving BM, Tankersley DL, Mason BL, Rossi F, Aronson DL, Finlayson JS. Contact-activated factors: contaminants of immunoglobulins preparations with coagulant and vasoactive properties. *J Lab Clin Med* 1980;**96**:334–46.

44. Passwell J, Rosen FS, Merler E. The effect of Fc fragments of IgG on human mononuclear cell responses. *Cell Immunol* 1980;**52**:395–403.

45. World Health Organization. Appropriate uses of human immunoglobulin in clinical practice: Memorandum from an IUIS/WHO meeting. *Bull World Health Org* 1982;**60**:43–7.

46. Lebing W, Remington KM, Schreiner C, Paul HI. Properties of a new intravenous immunoglobulin (IGIV-C, 10%) produced by virus inactivation with caprylate and column chromatography. *Vox Sang* 2003;**84**:193–201.

47. Centers for Disease Control Prevention. Renal insufficiency and failure associated with immune globulin intravenous therapy. United States, 1985–1998. *Morb Mortal Wkly Rep* 1999;**48**:518–21.

48. FDA Safety Communication. Updated information on the risks of thrombosis and hemolysis potentially related to administration of intravenous, subcutaneous and intramuscular human immune globulin product 11-13-2012. Available at: http://www.fda.gov/Biologics BloodVaccines/SafetyAvailability/ucm327934.htm. Accessed 5/6/2013.

49. Sekul EA, Cupler EJ, Dalakas MC. Aseptic meningitis associated with high-dose intravenous immunoglobulin therapy: Frequency and risk factors. *Ann Intern Med* 1992;**15**:259–62.

50. Dalakas MC. High-dose intravenous immunoglobulin and serum viscosity: risk of precipitating thromboembolic events. *Neurology* 1994;**44**:223–6.

51. Wells MA, Wittek AE, Epstein JS, et al. Inactivation and partition of human T-cell lymphotrophic virus, type III, during ethanol fractionation of plasma. *Transfusion* 1986; **26**:210–3.

52. Centers for Disease Control and Prevention (CDC). Outbreak of hepatitis C associated with intravenous immunoglobulin administration – United States, October 1993–June 1994. *Morb Mortal Wkly Rep* 1994;**43**:505–9.

53. Bjoro K, Froland SS, Yun Z, Samdal HH, Haaland T. Hepatitis C infection in patients with primary hypogammaglobulinemia after treatment with contaminated immune globulin. *N Engl J Med* 1994;**331**:1607–11.

54. Horowitz B, Wiebe ME, Lippin A, et al. Inactivation of viruses in liable blood derivatives. I. Disruption of lipid-enveloped viruses by tri(n-butyl)phosphate detergent combinations. *Transfusion* 1985;**25**:516–22.

55. Biesert L. Virus validation studies of immunoglobulin preparations. *Clin Exp Rheumatol* 1996;**14**:S47–52.

56. FDA public workshop. Risk mitigation strategies to address procoagulant activity in immune globulin products. May 17, 2011. Transcript available at: http://www.fda.gov/downloads/BiologicsBloodVaccines/NewsEvents/WorkshopsMeetingsConferences/UCM258022.pdf.

57. Funk MB, Gross N, Gross S, et al. Thromboembolic events associated with immunoglobulin treatment. *Vox Sang* 2013;doi: 10.1111/vox.12025 [Epub ahead of print].

58. Roemsch JR. Identification of activated factor XI as the major biochemical root cause in IVIg batches associated with thromboembolic events. Analytical and experimental approaches resulting in corrective and preventive measures implemented into the Octagam® manufacturing process. http://www.webmedcentral.com/article_view/2002.

59. Daw Z, Padmore R, Neurath D, et al. Hemolytic transfusion reactions after administration of intravenous immune (gamma) globulin: a case series analysis. *Transfusion* 2008;**28**:1598–601.

60. Morgan S, Sorensen P, Vercelotti G, Zantek ND. Haemolysis after treatment with IVIg due to anti-A. *Transfus Med* 2011;**21**:267–70.

61. Dhainaut F, Guillaumat P-O, Dib H, et al. In vitro and in vivo properties differ among liquid intravenous immunoglobulin preparations. *Vox Sang* 2013;**104**:115–26.

62. Stiehm ER, Keller MA. Passive Immunization. In: Plotkin SA, Orenstein WA, Offit PA, editors. *Vaccines*. 6th edn. Elsevier/Saunders: Philadelphia, PA; 2013. p. 80–7.

63. Barry J, Lacroix-Desmazes S, Kazatchkine MD, et al. Intravenous immunoglobulin for infectious diseases: back to the pre-antibiotic and passive prophylaxis era? *Trends Pharmacol Sci* 2004;**25**:306–8.

64. Losonsky GA, Johnson JP, Winkelstein JA, Yolken RH. Oral administration of human serum immunoglobulin in immunodeficient patients with viral gastroenteritis. A pharmacokinetic and functional analysis. *J Clin Invest* 1985;**76**:2362–7.

65. Borowitz SM, Saulsbury FT. Treatment of chronic cryptosporidial infection with orally administered human serum immune globulin. *J Pediatr* 1991;**119**:593–5.

66. Impact-RSV Study Group. Palivizumab, a humanized respiratory syncytial virus monoclonal antibody, reduces hospitalization from respiratory syncytial virus infection in high-risk infants. *Pediatrics* 1998;**102**:531–7.

67. RespiGam® Prescribing Information. Available at: www.drugs.com/mmx/respigam.html.

68. Synagis® Prescribing Information. Available at: http://www.medimmune.com/pdf/products/synagis_pi.pdf.

69. Wilfert CM, Buckley RH, Mohanakumar T, et al. Persistent and fatal central-nervous-system ECHOvirus infections in patients with agammaglobulinemia. *N Engl J Med* 1977;**296**:1485–9.

70. Stiehm ER, Keller MA, Vyas GN. Preparation and use of therapeutic antibodies primarily of human origin. *Biologicals* 2008;**36**:363–74.

71. Gelfand EW. Intavenous immune globulin in autoimmune and inflammatory diseases. *N Engl J Med* 2012;**367**: 2015–25.

72. Miescher SM, Käsermann F. The future of immunoglobulin therapy: an overview of the 2nd international workshop on natural antibodies in health and disease. *Autoimmun Rev* 2013;**6**:639–42.

73. Kaneko Y, Nimmerjahn F, Ravetch JV. Anti-inflammatory activity of immunoglobulin G resulting from Fc sialylation. *Science* 2006;**313**:670–3.

74. Anthony RM, Wermeling F, Ravetch JV. Novel roles for the IgG Fc glycan. *Ann NY Acad Sci* 2012;**1253**:170–80.

75. Longet S, Miled S, Lötscher M, Miescher SM, Zuercher AW, Corthésy B. Human plasma-derived polymeric IgA and IgM antibodies associate with secretory component to yield biologically active secretory-like antibodies. *J Biol Chem* 2013;**6**:4085–94.

History of Hematopoietic Stem Cell Transplantation

Richard A. Gatti[1], Rainer Storb[2]

[1]Departments of Pathology and Laboratory Medicine, and Human Genetics,
David Geffen School of Medicine at UCLA, Los Angeles, CA, USA
[2]Fred Hutchinson Cancer Research Center, University of Washington,
School of Medicine, Seattle, WA, USA

INTRODUCTION

Research involving human hematopoietic stem cells has progressed at an amazing pace, culminating in their clinical use for the treatment of malignant and non-malignant diseases. Hematopoietic stem cells have served to correct genetically defective hematopoiesis, for example, immunodeficiency diseases or hemoglobinopathies, and have been used to restore acquired loss of marrow function such as in aplastic anemia. Additionally, they have allowed increases in the magnitude of chemoradiation therapy used for eradicating malignant diseases such as leukemias and lymphomas whereby the infused stem cells serve to rescue the patient from therapy-induced marrow aplasia. These studies also discovered the concept of adoptive immunotherapy, a powerful immunological effect of donor lymphocytes contained in the graft which kill recipient tumor cells, a phenomenon which is called graft-versus-tumor effect.

The excitement in the field of hematopoietic cell transplantation (HCT) renders this review of its history a timely effort. To appreciate fully the enthusiasm of researchers, clinicians and patients, it is important to reflect upon how the field has developed and to document the enormous efforts by numerous investigators throughout the world who have worked to bring the experimental field of HCT to clinical reality. This chapter will highlight the seminal discoveries that have brought this field from one that was declared dead in the 1960s to the amazing clinical results obtained today. The chapter will conclude with a look to future research in the field of allogeneic HCT.

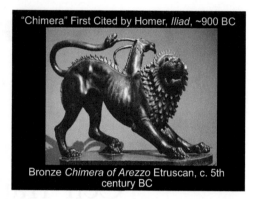

"Chimera" First Cited by Homer, *Iliad*, ~900 BC

Bronze *Chimera of Arezzo* Etruscan, c. 5th century BC

FIGURE 24.1   Idea of a Chimera in ancient times. This figure is reproduced in color in the color section.

THE EARLY YEARS

Major research efforts on how to repair or avert radiation effects were set in motion by the observations of radiation damage in various organs which were made in survivors of the first atomic bomb explosions in Japan. In the course of these studies, Leon Jacobson and colleagues made the important observation that mice could be protected from the fatal effects of X-irradiation on bone marrow by shielding their spleens or femurs with lead (1). Two years later, Lorenz and colleagues reported that protection against irradiation in both mice and guinea pigs could also be achieved by intravenous infusions of marrow cells (2). Initially, many investigators, including Jacobson, thought that the protection was accomplished by some humoral factor(s) from the spleen or marrow that led to endogenous marrow recovery. However, by the mid-1950s, the "humoral hypothesis" was firmly rejected and several laboratories, using various genetic markers in blood cells, demonstrated that the radiation protection was due to colonization of the recipient marrow by donor cells (3–5).

This exciting discovery was greeted with enthusiasm by hematologists, immunologists and radiation biologists because of the implications not only for cell biology but also for the treatment of patients with life-threatening hematological disorders. The principle of clinical bone marrow transplantation was simple: high doses of chemoradiation therapy would be used to destroy the diseased marrow and to suppress the patient's immune cells so that a marrow graft from a different donor with a competent immune system could be accepted, thus creating a bone marrow chimera (Fig. 24.1). The marrow graft, in turn, replaced the diseased marrow. Within one year of the publications of the pivotal studies in rodents (3–5), Don Thomas and colleagues, in 1957, published an article in the *New England Journal of Medicine* showing that large amounts of marrow could be infused intravenously into leukemia patients with safety and would engraft at least temporarily, even though, in the end, the patients were not cured of their leukemia (6). In the following year, Mathé's group attempted the dramatic rescue, by marrow transplantation, of six nuclear reactor workers who had become victims of an irradiation accident (7). Four of the six patients survived, although persistence of donor cells was only partial and transitory. In 1965, Mathé and colleagues were the first to describe the survival of an allogeneic marrow graft in a leukemia patient (8). The patient was treated by irradiation and then given marrow infusions from six relatives. The marrow from one of the relatives engrafted. The physicians were able to show chimerism and anti-leukemic effects. Unfortunately, while the leukemia remained in remission, the patient eventually succumbed to a complication, graft-versus-host disease (GVHD). Mathé also coined the term graft-versus-leukemia effect. In 1970, Bortin summarized the results of all human marrow grafts reported between 1957 and 1967 (9). Of the 200 patients included in his review, 73 were transplanted for aplastic anemia, 115 for hematological malignancies and 12 for immunodeficiency diseases. None of these patients survived the procedure: 125 died of graft failure, 47 who engrafted died of GVHD, while others died of infections or recurrence of their underlying malignancies.

All of these transplants were performed before the discovery of histocompatibility matching. They were based directly on work in inbred mice, for which histocompatibility matching is not an absolute requirement. In 1967, the prominent experimental hematologist Dirk van Bekkum said, "These failures have occurred mainly because the clinical applications were undertaken too soon, most of them before even the minimum basic knowledge required to bridge the gap between mouse and patient had been obtained"(10). Clinical allogeneic HCT was declared a complete failure and prominent immunologists were of the opinion that the allogeneic barrier between individuals could never be crossed. van Bekkum and de Vries in their book "Radiation Chimeras" (10) noted that "these factors have caused many investigators to abandon the idea that bone marrow transplantation can ever become a valuable asset to clinical medicine".

RENEWED EFFORTS: FOCUS ON ANIMAL EXPERIMENTS

While most investigators left the field, pronouncing it a dead end, a small number of laboratories around the world persevered and continued efforts in experimental animals to overcome the obstacles encountered in humans undergoing allogeneic HCT. van Bekkum's group at the Rijswijk Radiobiological Institute in Holland chose a primate model, George Santos at Johns Hopkins chose rats and the Seattle group at the University of Washington chose random-bred dogs as experimental animals. The reason behind the use of dogs as study animals was that, besides humans, only the dog has that rare combination of unusual genetic diversity due to centuries of canine breeding, and a widespread, well mixed gene pool. Additionally, dogs share spontaneous hematological diseases with humans, such as non-Hodgkin lymphoma, severe hemolytic anemia from pyruvate kinase deficiency and, of particular interest here, X-linked severe combine immunodeficiency (SCID) due to a spontaneous mutation in the interleukin 2 (IL-2) receptor γ-chain (11). In addition to determining the most appropriate ways to administer total body irradiation (TBI), new drugs were introduced that had myeloablative or immunosuppressive qualities, such as cyclophosphamide and busulfan. These compounds improved levels of graft acceptance and provided tumor cell killing in a manner that was equivalent to TBI. Based on the mouse histocompatibility system defined ten years earlier, an *in vitro* histocompatibility typing system for dogs was developed and the results were reported in 1968 (12, 13). One publication showed that dogs immunosuppressed by high-dose TBI and given grafts from dog leukocyte antigen (DLA)-matched littermates survived significantly longer than their DLA-mismatched counterparts (12). Another publication showed similar findings with unrelated grafts even though the histocompatibility typing techniques at that time were very primitive (13). The basis for this typing proved to be a system similar to that described in humans and termed HLA, even though the complexity of the genetic region coding for major histocompatibility antigens was as yet not fully understood. While serious GVHD was first described in H–2 mismatched mice and in unrelated randomly selected monkeys, the canine studies first drew attention to the fact that fatal GVHD could also occur after transplantation across minor histocompatibility barriers (14). These pivotal observations triggered the search for effective drug regimens delivered after transplantation to contain the detrimental effects of GVHD. The drug with the most promise in this regard was the folic acid antagonist methotrexate (15). Further studies in canines showed that transfusion-induced sensitization to minor histocompatibility antigens caused rejection of DLA-identical grafts (16). Continued studies in dogs eventually led to ways of understanding, preventing and overcoming transfusion-induced sensitization. These methods resulted in significant improvements in the outcome of transplantation for human patients with severe aplastic anemia (17, reviewed in 18). Next, the mechanisms of graft–host tolerance without the need for lifelong immunosuppressive drugs were investigated. It turned out that immunosuppression could generally be discontinued after 3–6 months (19), and donor-derived T lymphocytes were identified that downregulated immune reactions of other donor T-cells against host antigen (20). Immune reconstitution in canine marrow recipients receiving lethal total body irradiation or a single dose of cyclophosphamide (100 mg/kg) was found to be complete in long-term chimeras (>200 days post-grafting): antibody responses, tuberculin conversion and skin graft rejection were similar to normal controls and the transplanted animals were able to live drug-free in an unprotected environment without increased susceptibility to infections or other diseases (19) (Fig. 24.2). Techniques for isolating transplantable stem cells from peripheral blood were refined in dog and primate transplantation models (21–23). Importantly, studies in dogs with spontaneous non-Hodgkin lymphoma showed that these tumors could be cured in part due to graft-versus-tumor effects (24).

FIGURE 24.2 Long-term canine chimeras, fully immune reconstituted, living in an unprotected environment without immunosuppression or antimicrobial prophylaxis, generated in the early 1970s. This figure is reproduced in color in the color section.

THE RETURN TO CLINICAL TRANSPLANTATION: 1968–1980

The second half of the 1960s saw the development of high-intensity conditioning regimens. These included maximally tolerated doses of TBI and/or chemotherapeutic agents such as cyclophosphamide or busulfan. These regimens served to eliminate the underlying diseases of the patients, such as leukemia, and they suppressed the hosts' immune systems so that marrow grafts would be accepted. Based on the canine studies, histocompatibility matching of donor recipient pairs was recognized to be of utmost importance to reduce the risks of both graft rejection and GVHD. It was recognized that even when donor and recipient were well matched, GVHD remained a problem unless post-grafting immunosuppression was given with the antimetabolite drug, methotrexate, which reduced the tempo of donor lymphocyte replication in response to encountering recipient antigens (15). Rapid advances in the understanding of the major human histocompatibility complex – human leukocyte antigens (HLA) – helped to set the stage for better histocompatibility matching between human donor recipient pairs (25).

By 1968, the stage was set for clinical trials to resume. The first successful transplants were reported for patients with primary immune deficiency disorders (PIDD) (26, 27), even though during the first seven or eight years most clinical studies were carried out in patients with advanced and refractory hematological malignancies who were in poor general condition, and in patients with severe aplastic anemia (28, 29). These patients presented tremendous challenges in supportive care. Given their underlying condition and the profound destruction of marrow function through the conditioning regimens, patients required transfusions of blood products and therapy to prevent or treat bacterial, fungal and viral infections. Therefore, in addition to discoveries made in bone marrow transplantation, these early clinical trials stimulated progress in infectious disease and transfusion research.

The results of the initial clinical studies showed that GVHD still occurred in approximately half of the patients, even though donors were HLA matched and patients received methotrexate after transplantation in hopes of controlling GVHD. This finding was consistent with the earlier observations in dogs, and stimulated further research into pre-clinical model systems. Major improvements in GVHD prevention and patient survival were made by combining methotrexate with calcineurin inhibitors such as cyclosporine or tacrolimus (30). Combinations of these drugs remain the most widely used method for GVHD prevention to date.

Early results in patients with severe aplastic anemia given marrow grafts from HLA-identical siblings following conditioning with cyclophosphamide showed 45% long-term survival (17, 18). The major impediment to greater success was graft rejection, as was predicted from canine studies on transfusion-induced sensitization to minor histocompatibility antigens. Acute and chronic GVHD also remained serious problems. Subsequent studies in the laboratory identified dendritic cells in the transfusion product as the key element in transfusion-induced sensitization (31). Depleting red blood cell and platelet transfusions of white cells, therefore, reduced the risk of graft rejection. Further canine studies resulted in the development of a conditioning regimen for human patients consisting of alternating cyclophosphamide and antithymocyte globulin, which greatly reduced the risk of graft rejection (32). Finally, irradiation of blood products with 2000 cGy *in vitro* almost completely averted sensitization to minor histocompatibility antigens (33). Therefore, the problem of graft rejection in transplantation for aplastic anemia has largely been overcome, and current survivals with HLA-identical sibling grafts approach 100% (34).

Some transplant centers have focused on removing T-cells from the bone marrow as a means of preventing GVHD. Initial studies showed high incidences of graft rejection, relapse of underlying malignancies and infections (35). More recent studies showed that, while disease relapse after T-cell depletion has remained a problem in patients with chronic myeloid leukemia, it seemed a lesser problem in patients with acute leukemia (36). Other centers have used T-cell depletion with subsequent careful monitoring for minimal residual disease, and treating patients in case of disease recurrence with donor leukocyte infusions in hopes of initiating graft-versus-leukemia responses without causing GVHD (37).

The 1990s saw the first use of granulocyte colony stimulating factor (G-CSF)-mobilized peripheral blood stem cells (38). Randomized prospective studies demonstrated that marrow and peripheral blood stem cells were equivalent as far as engraftment and overall survival were concerned. However, there seemed to be an increase in the incidence of chronic GVHD following the use of peripheral blood stem cells (39). For patients with non-malignant diseases, such as aplastic anemia, marrow has remained the preferred source of stem cells in order to keep the rate of chronic GVHD relatively low.

Due to the fact that only approximately 35% of patients have HLA-identical siblings, the use of alternative donors has been explored, to include extended family members who share one HLA haplotype and to unrelated individuals who are HLA-matched. Following up on canine studies published in 1977 (40), the first successful transplant for a patient with acute leukemia from an unrelated HLA-matched donor was reported in 1980 (41). In order to expand the donor pool, national donor registries were established, and currently more than 20 million HLA-typed volunteers are included in these registries. The likelihood of finding suitably HLA-matched unrelated stem cell graft donors is approximately 80% for Caucasian patients, but this percentage declines dramatically for patients from minority groups. Recent studies have focused on the use of megadose CD34+ cells from HLA-haploidentical donors for treating leukemia (42). In that setting, the potential anti-leukemic effect of natural killer cells from the donor has also been identified. Most recently, a minimal-intensity regimen for HLA-haploidentical HCT has been described (43).

BONE MARROW TRANSPLANTATION IN SCID AND WISKOTT–ALDRICH SYNDROME

After more than 20 years of experiments in rodents, dogs and non-human primates, and a multitude of clinical trials that ended disappointingly in failure (9), the ball was picked up by clinical immunologists who cared for patients with potentially lethal congenital immune defects, e.g. SCID (26) and the Wiskott–Aldrich syndrome (WAS) (27).

SCID (see Chapters 14 and 15) was recognized in 1958 by two groups of Swiss pediatricians, one from Zürich (Walter Hitzig and colleagues) (44) and one from Bern (45). Together, they reported four infants, three male and one female, who died of severe fungal and bacterial infections. The early onset of symptoms, the apparent autosomal recessive inheritance, the lymphopenia and universal fatality differentiated these infants from Bruton's agammaglobulinemia. Because both the plasma cells and the lymphocytes were absent, the Swiss investigators proposed a combined humoral and cellular immune defect. The syndrome became known as "Swiss-type agammaglobulinemia" and, in 1970, it was designated as SCID by a WHO expert-committee.

Because SCID patients were found to lack immunologic competence – a set up for severe bacterial, viral and fungal infections, causing early death – they were considered candidates for "the transplantation of histocompatible lymphoid cells from an immunologic competent donor" as suggested by Fred Rosen and his colleagues (46).

This hypothesis was tested as early as 1966, when the Boston group intravenously injected maternal bone marrow into a six-month old boy with classic X-linked SCID. The boy, whose older brother had died at age 5.5 months of recurrent infections and moniliasis, received intravenously approximately 50×10^6 nucleated bone marrow cells from his mother. In the days post-transplant, the patient temporarily improved, his lymphocyte count increased tenfold, but he died 13 days later, having developed severe diarrhea, edema, icterus and bleeding resembling acute GVHD (46). Two years later, the prediction of the Boston team was fulfilled in Minneapolis, when a five-month old boy was referred to the University of Minnesota. The infant, diagnosed in Boston, as having "thymic alymphoplasia and agammaglobulinemia", was most likely affected with X-linked SCID with a three-generation family history of 11 male deaths from infections in early infancy (Fig. 24.3). Because of low serum IgG, IgM and IgA, the patient was being treated with immunoglobulin injections and antibiotics for persistent pneumonia. A chest radiograph revealed the absence of the thymus. No specific antibodies were detected in his blood, he was lymphopenic, and proliferative responses

Family Pedigree of X-SCID

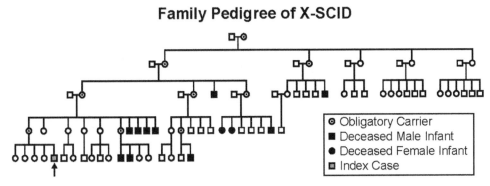

FIGURE 24.3 Pedigree of the X-SCID family with multiple affected males, including the index case who was the first successful marrow transplant recipient from his HLA-matched "sister No. 3" (26, 47, 48). *(Modified from de la Morena and Gatti, 2011 (49).)*

to mitogen stimulation were negative. Tonsils, adenoids and peripheral lymph nodes could not be detected (26). Luckily, the boy had four full sisters, increasing the chances of finding a matched sibling for marrow transplantation. By 1968, two techniques of histocompatibility testing had been developed: serological typing for HLA (Class 1 only) and cellular typing by mixed leukocyte culture (MLC). HLA typing of the patient and the four sisters, performed by Terasaki at the University of California, Los Angeles, suggested one of the four (sister three) to be the best match, but not perfect as she lacked one antigen that was present in the patient. This somewhat imperfect match was a subject of debate, since the importance of histocompatibility was recognized at that time to be a *conditio sine qua non*. In addition, sister three was also ABO incompatible. Several years later, with improved HLA typing techniques, was it possible to recognize a crossover between the two HLA loci in the donor cells, explaining the serological typing obtained in 1968 (50). Driven by the almost certain fatal outcome of the disease without treatment, a decision was made to attempt the bone marrow transplant using the eight-year old sister three as the donor (26). Both peripheral blood (to provide hematopoietic stem cells) and bone marrow (1.35×10^9 nucleated cells obtained from the iliac crest and the tibia), were collected from the donor and infused intraperitoneally, primarily to avoid having to filter out bone spicules and thereby reduce the number of cells available for engraftment. It was estimated that in today's terms approximately 1.25×10^6 CD34+ cells per kilogram were infused (49).

One week after the transplant, GVHD symptoms appeared, involving the skin, gut and liver. This was accompanied by a hemolytic anemia thought to be caused by donor/host ABO incompatibility. No immunosuppressive therapy was given because of concern that immunosuppressive drugs would impede stem cell engraftment, and because of experimental evidence that a mild GVHD would spontaneously subside. One week later, the high fever and rash disappeared, and engraftment ensued. Proliferative responses to mitogens normalized, delayed-type hypersensitivity reaction could be demonstrated for the first time, and bone marrow aspirate showed that 25% of bone marrow cells were female – and of donor origin – by karyotyping. However, approximately 120 days after the transplant, the patient developed a severe aplastic anemia, possibly due to GVHD reaction involving the marrow. Donor-specific cytotoxic antibodies were demonstrated. A second bone marrow infusion was given intraperitoneally

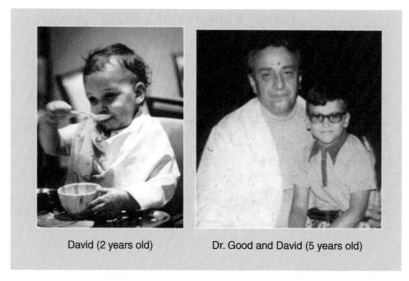

David (2 years old)   Dr. Good and David (5 years old)

FIGURE 24.4   The first successful bone marrow transplantation performed in an infant with X-linked SCID, photographed at 2 and 5 years of age, together with Robert Good, who led the BM transplantation team at the University of Minnesota. This figure is reproduced in color in the color section.

and into the iliac crest, again without conditioning and from the same donor (47). Within two weeks, the leukocyte counts improved, GVHD subsided and the proportion of erythrocytes with the host's Group A type had begun to decline, shifting instead to the donor's Group O blood type which has persisted to date. The patient has remained cured from his SCID diagnosis, demonstrating (1) that both SCID and aplastic anemia can be cured by bone marrow transplantation; (2) that histocompatibility matching plays a crucial role in protection from graft-versus-host disease; and (3) confirming the hypothesis that the central defect in SCID patients is the lack of pluripotent stem cells and not a defect of thymic microenvironment. A two-year post-transplant evaluation demonstrated not only stable immunological reconstitution, but also the transfer of T-cell memory, as evidenced by a positive skin test to mumps, which occurred despite the fact that the recipient never had mumps, while the donor had developed mumps shortly before the cells were harvested for transplant (Fig. 24.4) (47, 49).

In 1968, Fritz Bach and colleagues from the University of Wisconsin reported their experience with bone marrow transplantation in a 22-month-old boy with the diagnosis of Wiskott–Aldrich syndrome (27). In contrast to patients with SCID, those with WAS have evidence of considerable immunological function: they can make antibodies to at least some antigens and their lymphocytes proliferate when cultured in the presence of mitogens. To overcome the potential for rejection, pre-transplant conditioning is required. For this reason, it was decided that the WAS patient from Wisconsin should receive a mild immunosuppressive therapy with azathioprine (5 mg/kg), and prednisone (2 mg/kg) two days before the transplant. Marrow (6.5×10^9 nucleated cells) from his HLA/MLC matched sister was infused intravenously through a femoral line. No clinical improvement was noticed and engraftment was considered a failure. A second marrow transplant from the same donor was given 36 days later, after the recipient received four daily doses (a total of 200 mg/kg) of cyclophosphamide, a more intensive conditioning regimen. Thirty-six hours after the last cyclophosphamide dose, freshly aspirated bone marrow (2.7×10^9 nucleated cells) was given intravenously. When the story was published in December of 1968, the post- (second) transplant observation was only six weeks. At that time, the patient was gaining weight, lymphocyte and neutrophil counts had normalized, but he was still thrombocytopenic, although he had become "chimeric", as evidenced by the presence of 100% female cells in his peripheral blood. Fifteen years later, the patient was re-evaluated and noted to have full T-cell and partial B-cell chimerism, but no evidence of hematopoietic engraftment; he remained thrombocytopenic, although no major bleeding had occurred (48).

MOVING AHEAD: THE 1990s AND BEYOND

By now, more than one million HCTs have been carried out in more than 500 transplantation centers worldwide. Unfortunately, conventional HCT following high-intensity, marrow ablative approaches is risky and must be carried out in specialized intensive care wards. This restricts the therapy to relatively young patients in good medical condition. To allow the inclusion of older patients, and to consider the need for a less toxic conditioning regimen for PIDD patients, a number of transplant centers have developed less intensive conditioning programs (51). In the case of malignancies, these rely less on chemoradiation therapy and shift the burden of tumor cell kill towards graft-versus-tumor effects. This strategy combines pre-transplant fludarabine, an inhibitor of DNA synthesis, with small doses of TBI and a post-transplant regimen of immunosuppression that includes an inhibitor of purine synthesis (mycophenolatemofetil) and a calcineurin inhibitor (cyclosporine A or tacrolimus). This approach serves to prevent

allograft rejection and to control GVHD. This reduced-intensity regimen has in part been directly translated from pre-clinical models (52, 53) into the clinic. The reduced risk of chemoradiation-related toxicities has allowed the extension of allogeneic HCT to include both elderly and medically infirm patients who were not eligible for high-intensity conditioning regimens (54) and, for the same reason, is more acceptable for patients with T-cell-deficient PIDD (55–57).

A publication in 2013 highlighted the results of the first 1092 patients with advanced hematological malignancies who were treated with this approach and given either HLA-matched related or unrelated grafts (54). The median patient age was 56 years, half had co-morbidity scores of ≥ 3 and the median follow-up time was five years. Depending on the relapse risk of the underlying malignancies, on co-morbidities, and on GVHD, lasting remissions were seen in 45–75% of patients, and five-year survivals ranged from 25% to 60%. The five-year overall non-relapse mortality rate was 24% (mostly as a result of GVHD), and the overall relapse mortality rate was 34.5%. While encouraging, the results could be further improved by controlling the progression of malignancy during the early post-transplantation period and effectively preventing GVHD.

In 2008, a series of 113 patients with different types of PIDD who had undergone non-myeloablative conditioning between 1998 and 2006, reported an overall survival of 82%. There was no significant difference between HLA-matched and -mismatched donors (57).

In the future, reduced-intensity or minimal-intensity HCT might become an even more powerful therapeutic tool. Better understanding of tissue-specific polymorphic minor histocompatibility antigens might result in the development of vaccines that could be used to direct donor immune cells toward hematopoietic targets, rather than inducing general GVHD.

When CD34 was recognized as a glycoprotein that helped to identify hematopoietic progenitor cells, their isolation from peripheral blood provided another stem cell source. These peripheral blood stem cells (PBSCs) are capable of forming colonies of granulocytes/macrophages, erythrocytes and other multipotential or immature progenitors (58). The introduction of growth factors such as granulocyte colony-stimulating factor (filgrastim) and other agents (59) has contributed to the successful mobilization of CD34+ cells into the peripheral circulation and thus their dose salvage therapy to patients with refractory malignancies, resulting in prolonged survival and impeding tumor progression (60). Umbilical cord blood (UCB) represents another alternative source of HSCs (61) and this has become a standard option for both children and adults with hematopoietic disorders and malignancies. The advantages include: (1) a low rate of viral contamination; (2) lower rates of GVHD; and (3) readily available units (62). Two single-center experiences with UCB have been reported (63, 64); in both series, immunological reconstitution was achieved.

CURRENT AND FUTURE STATUS OF HCT FOR PIDD

After a stormy start and many failures, HCT has, through careful animal studies and clinical trials, become a highly effective and often curative procedure, in the treatment of both malignant and non-malignant diseases of the blood. Clinical immunologists treating patients with severely compromised immune systems had only to fear GVHD, which could be controlled by careful donor selection through increasingly sophisticated histocompatibility matching, but were not restricted by the need for chemotherapy and the concern of relapse of malignancy. Based on the spectacular success of the first stem cell transplantations for SCID and WAS, HCT has been rapidly accepted for the treatment of

immunodeficiency disorders predicted to result in premature death. Centers devoted to HCT of PIDD patients have been established throughout the developed and developing world. New and innovative protocols were developed by European, North American and Asian consortiums. SCID, classic WAS and other PIDDs are now considered to represent the type of patients for which HCT has become the standard of care. Survival for SCID following stem cell transplantation ranges from 81% to 92% when an HLA matched sibling donor is available, and 52% to 54% when HLA-haploidentical or HLA mismatched donors are used (65, 66). The lack of donor availability, variable evidence of long-term immunological reconstitution and other limitations have led to the extensive use of HLA-matched unrelated donors (MUD) as the source of stem cells for SCID. The outcomes have improved over recent years, with survival of those receiving MUD transplants ranging from 63% to 80% (67, 68). Buckley recently reported the long-term outcome of 166 consecutive SCID infants who had received non-conditioned related donor bone marrow transplants at Duke University over 28 years (69). Age at diagnosis ranged from newborn to 21 months. All grafts of HLA-haploidentical transplants were T-cell depleted. No pre-transplant conditioning or post-transplant GVHD prophylaxis was given. The overall survival rate was 76%. Importantly, the survival of the 48 infants transplanted during the first 3.5 months of life was significantly higher (94%) than the survival of the patients who were transplanted after that age (69%).

The largest group of non-SCID patients with primary immunodeficiency for which HCT has been consistently successful includes WAS (see Chapter 9) and chronic granulomatous disease (CGD) (see Chapter 13). A collaborative study of the international bone marrow transplant registry analyzed 170 transplants performed for WAS between 1968 and 1996. The overall five-year probability of survival was 70%. Best outcomes were noted for patients receiving transplants from HLA identical siblings (87% survival) compared with 52% for those receiving from other related donors. Matched unrelated donor transplants resulted in a survival rate of 71%. Of interest, if boys receiving HLA-matched unrelated donor marrow were transplanted before the age of five years, the outcome was similar to HLA-matched sibling transplants (70). In a subsequent retrospective collaborative study, the long-term outcome and donor cell engraftment in 194 patients with WAS who received HCT in the period 1980 through 2009 were reviewed (71). Overall survival was 84%, and five-year survival was even higher (89.1%) for those who received HCT after the year 2000, reflecting recent improvements in outcome after transplantation from mismatched family donors and for patients who received HCT from an unrelated donor older than five years. Retrospective analysis of lineage-specific donor cell engraftment showed that stable full donor chimerism was achieved in 72.3% of patients who survive for at least one year after HCT. Mixed chimerism was associated with an increased risk of autoimmunity, and with persistent thrombocytopenia (71). These observations indicate continuous improvement of outcome after HCT for WAS, and may have important implications for the development of novel protocols aiming to obtain full correction of the disease and to reduce post-HCT complications.

Allogenetic HCT for CGD is becoming more common and reflects the realization of a poor long-term outcome if treated conservatively and increased overall success of HCT. Survival has increased from approximately 85% before 2000 to 90–95% on recently reported outcomes (72–74). Other PIDDs for which HCT has been successfully applied include X-linked lymphoproliferative syndrome due to mutations in *SH2D1A* (XLP1) (75), familial hemophagocytic lymphohistiocytosis (76) (see Chapter 12), leukocyte adhesion deficiency (LAD) (77) (see Chapter 21), CD40 ligand deficiency (78) (see Chapter 16) and immune dysregulation, polyendocrinopathy,

endocrinopathy, X-linked (IPEX) (79). However, a recently completed international survey of HCT for patients with XLP caused by mutations in *XIAP* (XLP2) reports poor outcomes, especially if busulfan-containing myeloablative conditioning regimens were used, with only one out of seven patients surviving the transplant, or if the patients had active hemophagocytosis (80).

SUMMARY

The last 40 years, starting with carefully designed animal experiments, has seen the emergence of HCT as a therapeutic modality for potentially fatal blood diseases and as a curative option for individuals born with inherited disorders that carry limited life expectancy and poor quality of life. Despite the rarity and heterogeneity of PIDDs, these disorders have led the way toward innovative curative therapies and also provide insights into mechanisms of immunological reconstitution applicable to all HCT transplants. Critical analysis of outcomes and prospective multicenter clinical trials will be necessary to further advance our understanding of the best therapeutic approaches for patients with PIDD and other blood diseases.

References

1. Jacobson LO, Marks EK, Robson MJ, Gaston EO, Zirkle RE. Effect of spleen protection on mortality following X-irradiation. *J Lab Clin Med* 1949;**34**:1538–43.
2. Lorenz E, Uphoff D, Reid TR, Shelton E. Modification of irradiation injury in mice and guinea pigs by bone marrow injections. *J Natl Cancer Inst* 1951;**12**:197–201.
3. Main JM, Prehn RT. Successful skin homografts after the administration of high dosage X radiation and homologous bone marrow. *J Natl Cancer Inst* 1955;**15**:1023–9.
4. Ford CE, Hamerton JL, Barnes DW, Loutit JF. Cytological identification of radiation-chimaeras. *Nature* 1956;**177**:452–4.
5. Trentin JJ. Mortality and skin transplantability in x-irradiated mice receiving isologous, homologous or heterologous bone marrow. *Proc Soc Exp Biol Med* 1956;**92**:688–93.

6. Thomas ED, Lochte Jr HL, Lu WC, Ferrebee JW. Intravenous infusion of bone marrow in patients receiving radiation and chemotherapy. *N Engl J Med* 1957;**257**: 491–6.

7. Mathé G, Jammet H, Pendic B, et al. [Transfusions and grafts of homologous bone marrow in humans after accidental high dosage irradiation]. *Rev Fr Etud Clin Biol* 1959;**4**:226–38.

8. Mathé G, Amiel JL, Schwarzenberg L, Cattan A, Schneider M. Adoptive immunotherapy of acute leukemia: experimental and clinical results. *Cancer Res* 1965;**25**: 1525–31.

9. Bortin MM. A compendium of reported human bone marrow transplants. *Transplantation* 1970;**9**:571–87.

10. van Bekkum DW, de Vries MJ. *Radiation Chimaeras*. London: Logos Press; 1967.

11. Felsburg PJ, Hartnett BJ, Henthorn PS, Moore PF, Krakowka S, Ochs HD. Canine X-linked severe combined immunodeficiency. *Vet Immunol Immunopathol* 1999;**69**: 127–35.

12. Epstein RB, Storb R, Ragde H, Thomas ED. Cytotoxic typing antisera for marrow grafting in littermate dogs. *Transplantation* 1968;**6**:45–58.

13. Storb R, Epstein RB, Bryant J, Ragde H, Thomas ED. Marrow grafts by combined marrow and leukocyte infusions in unrelated dogs selected by histocompatibility typing. *Transplantation* 1968;**6**:587–93.

14. Storb R, Rudolph RH, Thomas ED. Marrow grafts between canine siblings matched by serotyping and mixed leukocyte culture. *J Clin Invest* 1971;**50**:1272–5.

15. Storb R, Epstein RB, Graham TC, Thomas ED. Methotrexate regimens for control of graft-versus-host disease in dogs with allogeneic marrow grafts. *Transplantation* 1970;**9**:240–6.

16. Storb R, Epstein RB, Rudolph RH, Thomas ED. The effect of prior transfusion on marrow grafts between histocompatible canine siblings. *J Immunol* 1970;**105**:627–33.

17. Thomas ED, Storb R, Fefer A, et al. Aplastic anaemia treated by marrow transplantation. *Lancet* 1972;**1**: 284–9.

18. Georges GE, Storb R. Hematopoeitic cell transplantation for aplastic anemia. In: Appelbaum FR, Forman SJ, Negrin RS, Blume KG, editors. *Thomas' Hematopoietic Cell Transplantation*. Oxford: Wiley-Blackwell; 2009. p. 707–26.

19. Ochs HD, Storb R, Thomas ED, et al. Immunologic reactivity in canine marrow graft recipients. *J Immunol* 1974;**113**:1039–57.

20. Storb R, Thomas ED. Graft-versus-host disease in dog and man: the Seattle experience. *Immunol Rev* 1985;**88**:215–38.

21. de Revel T, Appelbaum FR, Storb R, et al. Effects of granulocyte colony-stimulating factor and stem cell factor, alone and in combination, on the mobilization of peripheral blood cells that engraft lethally irradiated dogs. *Blood* 1994;**83**:3795–9.

22. Ageyama N, Hanazono Y, Shibata H, et al. Safe and efficient collection of cytokine-mobilized peripheral blood cells from cynomolgus monkeys (*Macaca fascicularis*) with human newborn-equivalent body weights. *Exp Anim* 2005;**54**:421–8.

23. Storb R, Epstein RB, Ragde H, Bryant J, Thomas ED. Marrow engraftment by allogeneic leukocytes in lethally irradiated dogs. *Blood* 1967;**30**:805–11.

24. Weiden PL, Storb R, Deeg HJ, Graham TC, Thomas ED. Prolonged disease-free survival in dogs with lymphoma after total-body irradiation and autologous marrow transplantation consolidation of combination-chemotherapy-induced remissions. *Blood* 1979;**54**: 1039–49.

25. Amos DB. Genetic and antigenetic aspects of human histocompatibility systems. *Adv Immunol* 1969;**10**: 251–97.

26. Gatti RA, Meuwissen HJ, Allen HD, Hong R, Good RA. Immunological reconstitution of sex-linked lymphopenic immunological deficiency. *Lancet* 1968;**2**:1366–9.

27. Bach FH, Albertini RJ, Joo P, Anderson JL, Bortin MM. Bone-marrow transplantation in a patient with the Wiskott-Aldrich syndrome. *Lancet* 1968;**2**:1364–6.

28. Thomas E, Storb R, Clift RA, et al. Bone-marrow transplantation (first of two parts). *N Engl J Med* 1975;**292**: 832–43.

29. Thomas ED, Storb R, Clift RA, et al. Bone-marrow transplantation (second of two parts). *N Engl J Med.* 1975;**292**:895–902.

30. Storb R, Deeg HJ, Whitehead J, et al. Methotrexate and cyclosporine compared with cyclosporine alone for prophylaxis of acute graft versus host disease after marrow transplantation for leukemia. *N Engl J Med* 1986;**314**:729–35.

31. Deeg HJ, Aprile J, Storb R, et al. Functional dendritic cells are required for transfusion-induced sensitization in canine marrow graft recipients. *Blood* 1988;**71**: 1138–40.

32. Storb R, Etzioni R, Anasetti C, et al. Cyclophosphamide combined with antithymocyte globulin in preparation for allogeneic marrow transplants in patients with aplastic anemia. *Blood* 1994;**84**:941–9.

33. Bean MA, Storb R, Graham T, et al. Prevention of transfusion-induced sensitization to minor histocompatibility antigens on DLA-identical canine marrow grafts by gamma irradiation of marrow donor blood. *Transplantation* 1991;**52**:956–60.

34. Burroughs LM, Woolfrey AE, et al. Success of allogeneic marrow transplantation for children with severe aplastic anaemia. *Br J Haematol* 2012;**158**:120–8.

35. Maraninchi D, Gluckman E, Blaise D, et al. Impact of T-cell depletion on outcome of allogeneic bone-marrow transplantation for standard-risk leukaemias. *Lancet* 1987;**2**:175–8.

36. Devine SM, Carter S, Soiffer RJ, et al. Low risk of chronic graft-versus-host disease and relapse associated with T-cell-depleted peripheral blood stem cell transplantation for acute myelogenous leukemia in first remission: results of the blood and marrow transplant clinical trials network protocol 0303. *Biol Blood Marrow Transplant* 2011;**17**:1343–51.

37. Mackinnon S, Papadopoulos EB, Carabasi MH, et al. Adoptive immunotherapy evaluating escalating doses of donor leukocytes for relapse of chronic myeloid leukemia after bone marrow transplantation: separation of graft-versus-leukemia responses from graft-versus-host disease. *Blood* 1995;**86**:1261–8.

38. Bensinger WI, Weaver CH, Appelbaum FR, et al. Transplantation of allogeneic peripheral blood stem cells mobilized by recombinant human granulocyte colony-stimulating factor. *Blood* 1995;**85**:1655–8.

39. Anasetti C, Logan BR, Lee SJ, et al. Peripheral-blood stem cells versus bone marrow from unrelated donors. *N Engl J Med* 2012;**367**:1487–96.

40. Storb R, Weiden PL, Graham TC, Lerner KG, Thomas ED. Marrow grafts between DLA-identical and homozygous unrelated dogs: evidence for an additional locus involved in graft-versus-host disease. *Transplantation* 1977;**24**:165–74.

41. Hansen JA, Clift RA, Thomas ED, Buckner CD, Storb R, Giblett ER. Transplantation of marrow from an unrelated donor to a patient with acute leukemia. *N Engl J Med* 1980;**303**:565–7.

42. Aversa F, Tabilio A, Velardi A, et al. Treatment of high-risk acute leukemia with T-cell-depleted stem cells from related donors with one fully mismatched HLA haplotype. *N Engl J Med* 1998;**339**:1186–93.

43. Luznik L, O'Donnell PV, Symons HJ, et al. HLA-haploidentical bone marrow transplantation for hematologic malignancies using nonmyeloablative conditioning and high-dose, posttransplantation cyclophosphamide. *Biol Blood Marrow Transplant* 2008;**14**:641–50.

44. Hitzig WH, Biro Z, Bosch H, Huser HJ. [Agammaglobulinemia & alymphocytosis with atrophy of lymphatic tissue]. *Helv Paediatr Acta* 1958;**13**:551–85.

45. Tobler R, Cottier H. [Familial lymphopenia with agammaglobulinemia & severe moniliasis: the essential lymphocytophthisis as a special form of early childhood agammaglobulinemia]. *Helv Paediatr Acta* 1958;**13**:313–38.

46. Rosen FS, Gotoff SP, Craig JM, Ritchie J, Janeway CA. Further observations on the Swiss type of agammaglobulinemia (alymphocytosis). The effect of syngeneic bone-marrow cells. *N Engl J Med* 1966;**274**:18–21.

47. Meuwissen HJ, Gatti RA, Terasaki PI, Hong R, Good RA. Treatment of lymphopenic hypogammaglobulinemia and bone-marrow aplasia by transplantation of allogeneic marrow. Crucial role of histocompatiility matching. *N Engl J Med* 1969;**281**:691–7.

48. Meuwissen HJ, Bortin MM, Bach FH, et al. Long-term survival after bone marrow transplantation: a 15-year follow-up report of a patient with Wiskott-Aldrich syndrome. *J Pediatr* 1984;**105**:365–9.

49. de la Morena MT, Gatti RA. A history of bone marrow transplantation. *Hematol Oncol Clin North Am* 2011;**25**:1–15.

50. Gatti RA, Meuwissen HJ, Terasaki PI, Good RA. Recombination within the HL-A locus. *Tissue Antigens* 1971;**1**:239–41.

51. Sandmaier BM, Storb R. Reduced-intensity conditioning followed by hematopoietic cell transplantation for hematologic malignancies. In: Appelbaum FR, Forman SJ, Negrin RS, Blume KG, editors. *Thomas' Hematopoietic Cell Transplantation*. Oxford: Wiley-Blackwell; 2009. p. 1043–58.

52. Storb R, Yu C, Wagner JL, et al. Stable mixed hematopoietic chimerism in DLA-identical littermate dogs given sublethal total body irradiation before and pharmacological immunosuppression after marrow transplantation. *Blood* 1997;**89**:3048–54.

53. Storb R, Yu C, Barnett T, et al. Stable mixed hematopoietic chimerism in dog leukocyte antigen-identical littermate dogs given lymph node irradiation before and pharmacologic immunosuppression after marrow transplantation. *Blood* 1999;**94**:1131–6.

54. Storb R, Gyurkocza B, Storer BE, et al. Graft-versus-host disease and graft-versus-tumor effects after allogeneic hematopoietic cell transplantation. *J Clin Oncol* 2013;**31**:1530–8.

55. Gaspar HB, Amrolia P, Hassan A, et al. Non-myeloablative stem cell transplantation for congenital immunodeficiencies. *Recent Results Cancer Res* 2002;**159**:134–42.

56. Burroughs LM, Storb R, Leisenring WM, et al. Intensive postgrafting immune suppression combined with nonmyeloablative conditioning for transplantation of HLA-identical hematopoietic cell grafts: results of a pilot study for treatment of primary immunodeficiency disorders. *Bone Marrow Transplant* 2007;**40**:633–42.

57. Satwani P, Cooper N, Rao K, Veys P, Amrolia P. Reduced intensity conditioning and allogeneic stem cell transplantation in childhood malignant and nonmalignant diseases. *Bone Marrow Transplant* 2008;**41**:173–82.

58. Krause DS, Fackler MJ, Civin CI, May WS. CD34: structure, biology, and clinical utility. *Blood* 1996;**87**:1–13.

59. Greinix HT, Worel N. New agents for mobilizing peripheral blood stem cells. *Transfus Apher Sci* 2009;**41**:67–71.

60. Kessinger A, Armitage JO, Smith DM, Landmark JD, Bierman PJ, Weisenburger DD. High-dose therapy and autologous peripheral blood stem cell transplantation for patients with lymphoma. *Blood* 1989;**74**:1260–5.

61. Gluckman E, Rocha V, Boyer-Chammard A, et al. Outcome of cord-blood transplantation from related and unrelated donors. Eurocord Transplant Group and the European Blood and Marrow Transplantation Group. *N Engl J Med* 1997;**337**:373–81.

62. Gluckman E, Rocha V. Cord blood transplantation: state of the art. *Haematologica* 2009;**94**:451–4.

63. Knutsen AP, Wall DA. Umbilical cord blood transplantation in severe T-cell immunodeficiency disorders: two-year experience. *J Clin Immunol* 2000;**20**:466–76.

64. Bhattacharya A, Slatter MA, Chapman CE, et al. Single centre experience of umbilical cord stem cell transplantation for primary immunodeficiency. *Bone Marrow Transplant* 2005;**36**:295–9.

65. Antoine C, Muller S, Cant A, et al. Long-term survival and transplantation of haemopoietic stem cells for immunodeficiencies: report of the European experience 1968–99. *Lancet* 2003;**361**:553–60.

66. Grunebaum E, Mazzolari E, Porta F, et al. Bone marrow transplantation for severe combined immune deficiency. *J Am Med Assoc* 2006;**295**:508–18.

67. Rao K, Amrolia PJ, Jones A, et al. Improved survival after unrelated donor bone marrow transplantation in children with primary immunodeficiency using a reduced-intensity conditioning regimen. *Blood* 2005;**105**:879–85.

68. Roifman CM, Grunebaum E, Dalal I, Notarangelo L. Matched unrelated bone marrow transplant for severe combined immunodeficiency. *Immunol Res* 2007;**38**:191–200.

69. Buckley RH. Transplantation of hematopoietic stem cells in human severe combined immunodeficiency: longterm outcomes. *Immunol Res* 2011;**49**:25–43.

70. Filipovich AH, Stone JV, Tomany SC, et al. Impact of donor type on outcome of bone marrow transplantation for Wiskott-Aldrich syndrome: collaborative study of the International Bone Marrow Transplant Registry and the National Marrow Donor Program. *Blood* 2001;**97**:1598–603.

71. Moratto D, Giliani S, Bonfim C, et al. Long-term outcome and lineage-specific chimerism in 194 patients with Wiskott-Aldrich syndrome treated by hematopoietic cell transplantation in the period 1980–2009: an international collaborative study. *Blood* 2011;**118**:1675–84.

72. Seger RA, Gungor T, Belohradsky BH, et al. Treatment of chronic granulomatous disease with myeloablative conditioning and an unmodified hemopoietic allograft: a survey of the European experience, 1985–2000. *Blood* 2002;**100**:4344–5430.

73. Soncini E, Slatter MA, Jones LB, et al. Unrelated donor and HLA-identical sibling haematopoietic stem cell transplantation cure chronic granulomatous disease with good long-term outcome and growth. *Br J Haematol* 2009;**145**:73–83.

74. Kang EM, Marciano BE, DeRavin S, Zarember KA, Holland SM, Malech HL. Chronic granulomatous disease: overview and hematopoietic stem cell transplantation. *J Allergy Clin Immunol* 2011;**127**:1319–26 quiz 1327-1328.

75. Hoffmann T, Heilmann C, Madsen HO, Vindelov L, Schmiegelow K. Matched unrelated allogeneic bone marrow transplantation for recurrent malignant lymphoma in a patient with X-linked lymphoproliferative disease (XLP). *Bone Marrow Transplant* 1998;**22**:603–4.

76. Marsh RA, Jordan MB, Filipovich AH. Reduced-intensity conditioning haematopoietic cell transplantation for haemophagocytic lymphohistiocytosis: an important step forward. *Br J Haematol* 2011;**154**:556–63.

77. Qasim W, Cavazzana-Calvo M, Davies EG, et al. Allogeneic hematopoietic stem-cell transplantation for leukocyte adhesion deficiency. *Pediatrics* 2009;**123**:836–40.

78. Duplantier JE, Seyama K, Day NK, et al. Immunologic reconstitution following bone marrow transplantation for X-linked hyper IgM syndrome. *Clin Immunol* 2001;**98**:313–8.

79. Burroughs LM, Torgerson TR, Storb R, et al. Stable hematopoietic cell engraftment after low-intensity nonmyeloablative conditioning in patients with immune dysregulation, polyendocrinopathy, enteropathy, X-linked syndrome. *J Allergy Clin Immunol* 2010;**126**:1000–5.

80. Marsh RA, Rao K, Satwani P, Lehmberg K, et al. Allogeneic hematopoietic cell transplantation for XIAP deficiency: an international survey reveals poor outcomes. *Blood* 2013;**121**:877–83.

David's Story

William T. Shearer[1,2], Carol Ann Demaret[1,2]

[1]The David Center, Texas Children's Hospital, Baylor College of Medicine,
Houston, Texas, USA

[2]The Department of Pediatrics, Immunology, Allergy, and Rheumatology,
Baylor College of Medicine, Houston, Texas, USA

INTRODUCTION

David Vetter was born when the world of immunology was exploding with new discoveries of T-cells, B-cells, antibody replacement therapy, and restoration of severe immune deficiencies with bone marrow containing hematopoietic stem cells. David's life within the sterile bubble taught the world about the power of the immune system by demonstrating what precautions a child with severe combined immunodeficiency (SCID) had to endure to avoid life-threatening infections. Investigators of primary immune deficiency disease and secondary immune deficiency (e.g. HIV infection) benefited from the study of David's immune system and his unfortunate demise subsequent to reactivation of Epstein–Barr (EBV)-containing donor lymphocytes that released infectious virus. Had David survived, he would be 43 years of age. Here is his story.

BIRTH AND EARLY YEARS

David was the second son with SCID born in 1971 to young parents in suburban Houston. His brother had passed away within a few months of his birth, due to uncertainty of the cause of his infections and delayed diagnosis. Soon after the brother's diagnosis of SCID was made, he expired from *Pnueumocystis jirovecii* pneumonia. With the subsequent pregnancy, prenatal testing of David's mother revealed that her fetus was a male, and preparations were made to deliver the baby by Caesarean section and to place him immediately in a sterile plastic isolator (originally designed by NASA engineers for the isolation of moon rocks and subsequently modified for David). Just after David entered that permanent isolation, he was baptized with sterile holy water by one of his doctors, Raphael Wilson, who was also a catholic brother of the Holy Cross Order. To the dismay of all, David's lymphocyte testing showed a lack of E-rosetting cells (i.e. T-cells; see Chapter 5) and absence of stimulation of these cells with the universal lymphocyte stimulant, phytohemagglutinin (PHA). Also disappointing was the non-matching of David's and his only sibling's (Katherine) HLA antigens, thus preventing a regular bone marrow transplant. David's doctors at that time, Mary Ann South, MD, Jack R. Montgomery, MD, MBA, and Raphael Wilson, PhD, were frustrated at the inability to restore David's immune system because HLA-mismatched hematopoietic stem cell transplants were not available at that time. A decision was made by the parents and physicians to keep David in his sterile environment until further developments might provide immune reconstitution: either David's immune system would develop in a germ-free environment, or technology would be designed to cross the HLA barrier. With that decision, the hospital provided an expanding suite of rooms to accommodate a growing child, (larger and larger bubbles). All items of life into and out of the sterile system had to be passed through a sterilizing chamber. An impressive list of consultants gave advice on every aspect of David's life, including neurodevelopment and neurocognition, so that David would have as close to a normal life as possible, given the unique nature of his existence.

RESEARCH STUDIES ON DAVID

Scientific publications on David were reviewed for preparation of this chapter that chronicles his life from birth in 1971 to his death in 1984. These publications include research on his family history (1–3), infection control (4, 5), immune system and responses (6–10), hematology (11), nutrition (12), mental and psychosocial development (13), speech and language development (14), psychiatric condition (15), bone marrow transplantation (16), development of cancer (17), detection of genetic mutation for SCID (17) and post-mortem B-cell studies (18,19).

IMMUNOLOGICAL STUDIES

Many of these publications concerned David's life up to four years of age and are contained in a monograph authored by the medical professionals who were responsible for his medical care, such as his nutritional status, neurological and language development, and psychiatric condition (1). These experts continued to follow David after his early years as the need arose. Aside from his immunological deficits, other studies revealed a normally developing boy without major problems, due to the intense efforts of parents and medical personnel to provide him with a normal young life.

A longitudinal assessment of David's few lymphocytes demonstrates that his serum immunoglobulin values and his T- and B-cell numbers changed little over his 12 year life span (Tables 25.1–25.4). At birth, his absolute lymphocyte count was 300–440 cells/μL (normal 3400–7600) and his lymphocytes showed no response to phytohemagglutinin (PHA) and only a weak response to allogeneic antigens in mixed lymphocyte culture. The serum IgM was present, but low; IgA was absent and the serum IgG value showed a decline of maternal IgG over the ensuing months eventually becoming undetectable (7) (Table 25.1). There was no antibody response to keyhole limpet hemocyanin;

TABLE 25.1 David's serum immunoglobulin values over the first 4 years

	Immunoglobulins (mg/100 mL)		
Age	IgM	IgG	IgA
Birth	16	560	0
1 wk	22	560	0
3 wk	30	ND*	ND
7 wk	10	305	0
3 mo	32	130	0
4 mo	29	135	0
6 mo	22	120	0
7 mo	17	105	0
10 mo	11	37	0
11 mo	8	27	0
12 mo	11	28	ND
13 mo	7	0	0
28 mo	6	0	0
39 mo	15	0	1.2
42 mo	17	0	3.7
43 mo	12	0	6.7
46 mo	10	0	<5.8

*ND: not done.
Taken with permission from South MA, Montgomery JR, Richie E, et al. Immunologic studies. Pediatr Res 1977;**11**:71-78.

TABLE 25.2 Mitogenic response of David's lymphocytes in tests using isolated leukocyte technique (birth to 27 months)

	Simulation index*: patient (control)		
Age, months (+ days)	MLC†	PHA	PWM
(Birth)		<1.0 (8.0)	1.5 (12.0)
0 (1)		0.6 (8.6)	1.5 (1.5)
1		1.0 (5.0)	1.3 (15.0)
2 (22)	<1.0 (25.9)		
	(5.9)		
3 (16)	<1.0 (10.0)		
4 (5)		1.4 (30.0)	1.3 (9.0)
5 (3)		1.2 (5.0)	0.9 (3.5)
5 (20)	1.5 (15.8)		
	1.2 (18.5)		
6 (30)			<1.0 (83.0)
11 (23)		<1.0 (453.0)	1.0 (77.0)
13 (20)		0.7 (483.0)	1.1 (188.0)
15 (20)		0.9 (127.0)	
15 (23)		0.9 (101.0)	
16 (12)	2.6 (31.0)	0.6 (130.0)	0.4 (114.0)
	2.4 (22.7)		
18 (8)		0.8 (250.0)	1.2 (122.0)
18 (23)	1.8 (11.0)	0.7 (24.0	1.0 (27.0)
27 (26)		0.7 (15.0)	2.0 (15.0)

*The stimulation index is the ratio of counts per min in stimulated lymphocytes to counts per min in unstimulated lymphocytes in radioactive assay. †MLC: mixed lymphocyte culture; PHA: phytohemagglutinin; PWM: pokeweed mitogen. Values for control subject are within parentheses.
Taken with permission from South MA, Montgomery JR, Richie E, et al. Immunologic studies. Pediatr Res 1977;**11**:71-78

David's chest X-ray showed no thymic shadow; phagocytic function was normal by the Rebuck skin window technique; and serum complement values were normal. T-cell function was absent as measured by the blastogenic response to PHA and there was no change in the results over several years (7,10) (Tables 25.2, 25.4). Also, in assessing his T-cell responses, the child failed to reject an allogeneic skin patch (from Dr South) placed on his arm (7). Injections of thymosin did not change his T-cell function (7). Lymphocyte markers revealed an average of 50–100% for B-cells (surface Ig+) and 3–12 percent for T-cells (E-rosette forming lymphocytes). There was no evidence of a delayed ontogeny of the T-cell subsets taking place over the years.

As David grew older, additional immunological studies confirmed all of the above findings with upgraded laboratory technology (10).

TABLE 25.3 David's immunoglobulin data 8–12 years of age

Date	Age	Serum Ig levels (mg/dL)			Unstimulated immunoglobulin-secreting cells/10^6 Mononuclear cells (mg/dL)			Fluorescent anti-human Ig staining (%)				
		IgG	IgM	IgA	IgG	IgM	IgA	IgM	IgD	IgG	IgA	Polyvalent
10/24/79	8 yr 1 mo	5	40	20				13	18			15
12/9/80	9 yr 3 mo	8	49	16	2	44	93					
6/7/82	10 yr 9 mo	8	37	13								
6/1/83	11 yr 9 mo	8	42	12				30	42	1	6	41
Normal range*		631–1298	56–258	70–312	102–2337	61–1384	3–358	2.3–16.3	1.6–14.4	0.0–8.0	0.0–4.7	6.1–17.5

*95% confidence limits (normal values for serum Ig levels were 9–11-year-old control subjects).
Taken with permission from Paschall VL, Brown LA, Lawrence EC, et al. Immunoregulation in an isolated 12-year-old boy with congenital severe combined immunodeficiency. Pediatr Res 1984;18: 723–728.

TABLE 25.4 David's mitogen specific antigen, and mixed leukocyte culture responses 8–12 years of age

Date	Age	Unstimulated response (cpm $\times 10^3$)	PHA (10 µg/mL) net response (cpm $\times 10^3$)	Con A (10 µg/mL) net response (cpm $\times 10^3$)	PWM (1:640) net response (cpm $\times 10^3$)	Unstimulated response (cpm $\times 10^3$)	Candida (1:100) net response (cpm $\times 10^3$)	SK-SD (1:100) net response (cpm $\times 10^3$)	Tetanus (1:100) net response (cpm $\times 10^3$)	MLC response (cpm $\times 10^3$)
5/24/79	7 yr 8 mo	1.2 (109)†	23.2 (8)†	15.4 (7)†	17.2 (19)†					2.2 (0.8)†
7/18/79	7 yr 10 mo					0.2 (1)†	0.1 (4)†	0.2 (4)†		
8/13/79	7 yr 11 mo	0.02 (10)	2.4 (2)	0.9 (1)	4.5 (4)					
3/12/80	8 yr 6 mo					0.02 (2)	0.192	0.02 (2)	0.5 (7)†	
3/27/81	9 yr 6 mo	0.2 (28)	6.6 (4)	0.3 (.6)	5.4 (9)	0.4 (45)	0.1 (4)	0.2 (7)	0.1 (5)	
6/7/82	10 yr 9 mo	0.8 (26)	13.5 (9)	8.4 (6)	9.4 (7)	0.2 (3)	0.2 (2)	0	0.2 (5)	
6/7/83	11 yr 9 mo	0.2 (18)	9.8 (5)	7.3 (4)	5.0 (3)	0.3 (1)	0	0	0	
Normal range*	0–3	21–235	6–125	34–226	0–6	0–13	0–17	0–9	>2.0	3.0

Stimulation index †Percentage of control of the same day ‡Monocyte depleted §95% confidence limits. Only the concentration of mitogens or antigens showing the largest responses of the patient are presented.
Taken with permission from Paschall VL, Brown LA, Lawrence EC, et al. Immunoregulation in an isolated 12-year-old boy with congenital severe combined immunodeficiency. Pediatr Res 1984;18: 723–728.

Table 25.3 gives results of testing serum Ig levels and secreting B-cells in culture from age eight to 12 years of age. David's B-cells secreted some IgM, IgD and IgG, but virtually no IgG. His *in vitro* T-cell responses to several mitogens and antigens were essentially zero, confirming his complete inability to activate his T-cells; however, his lymphocytes did possess weak responses to allogeneic antigens (see Tables 25.2, 25.4) (7, 10). With the arrival of the flow cytometer, it became possible to detect surface markers on blood mononuclear cells. These markers for T-cells and B-cells failed to demonstrate a movement to normal values with time. Table 25.5 compares the older E-rosetting technique with the new technique of staining surface markers (e.g. T3 equivalent to CD3) and the results generally confirm each other. Also shown is the high percentage of B-cells (B1 equivalent to CD20, the mature B-cell marker). Based on *in vitro* experiments of David's B-cells cultured in the presence of normal monocytes,

his immunoglobulin production was only 12% of normal values. In addition to these T- and B-cell defects, David's natural killer (NK) cell function was grossly abnormal. The results showed that his NK cells could bind to target tumor cells but could not kill them (10).

Aside from the abnormal immunological system that David inherited, his life was as normal as could be expected, growing from infancy to early adolescence in a "sterile" world. Despite his weak T- and B-cell immune response, David possessed some measure of immune protection, presumably due to his normal neutrophil function, since after years of isolation, it was determined that his bubble system actually contained a biofilm of low virulence bacteria (5). David's immune system from the beginning to the end was characterized as lymphopenia with low percentages and absolute counts of CD4+ T and CD8+ T-cells, and his T-cells did not respond to the usual extracellular signals, such as mitogens and antigens.

TABLE 25.5 Characteristics of David's mononuclear cell subpopulations, 9–12 years of age

A. Date	Age	E rosettes (%)
7/18/79	7 yr 10 mo	5.0
8/13/79	7 yr 11 mo	6.9
3/27/81	9 yr 6 mo	15.8
6/7/82	10 yr 9 mo	4.0
6/1/83	11 yr 9 mo	15.1
Normal range		43.4–81.0

B. Date	Age	%T11	%T3	%T4	%T8	%T6	%T10	T4:T8	%Ia1	%B1	%M1
5/20/81	9 yr 8 mo		30.9	19.3	15.3			1.3	47.7		0
7/21/81	9 yr 10 mo		40.4	21.1	17.7			1.2	24.9	17.7	13.4
10/26/81	10 yr 1 mo	22.7	25.0	14.3	3.7			3.9	49.3	42.2	6.2
12/16/81	10 yr 3 mo		12.9	5.5	1.0	1.1		5.5	60.1	32.6	25.0
6/7/83	11 yr 9 mo	11.6	11.0	12.7	3.8	0	13.1	3.3	68.1	45.4	29.7
Normal* range		54–97	60–85	45–60	15–25	0–2	3–5	1.5–3.5	5–20	5–10	10–25

Equivalence of surface markers: T11 (total T-cells); T3, (CD3), T8 (CD8), T6 (immature thymocytes), T10 (Immature activated T-Cells), Ia1 (DR), B1 (CD20), and M1 (CD14).

95% confidence limits.

Taken with permission from Paschall VL, Brown LA, Lawrence EC, et al. Immunoregulation in an isolated 12-year-old boy with congenital severe combined immunodeficiency. Pediatr Res 1984;18(8):723-728.

HAPLOIDENTICAL T-CELL DEPLETED BONE MARROW TRANSPLANT

At 12 years of age, David began verbally to express his concern that he might never get out of his bubble. He was able to do that to a limited degree in the sterile transport isolator that he used for going to and from his suburban home, where another large isolator system maintained him. He had daily contact with grade school friends who came by with his homework assignments, and David was connected to his classroom through a speaker phone system. Despite this participation in home and school activities, David longed to be free from his isolator system. At this time, developments in transplantation were taking place in the research laboratories and hospitals whereby it would be possible to cross the HLA barrier and avoid fatal graft-versus-host disease (20, 21). This new approach greatly reduced the number of mature T-cells contained in bone marrow given to patients. These mature T-cells were removed from donor bone marrow by lysis using, for instance, anti-CD6 monoclonal antibody and complement (20); thus eliminating most attacking T-cells and allowing the stem cells in the donor bone marrow to repopulate the patient with donor stem cells that develop into normal T-cells. With approval of the Baylor College of Medicine Institutional Review Board and with David's assent and his parent's written informed consent, the transplantation of David's sister's bone marrow (7.87×10^7 cells/kg of T-cell depleted) was performed on October 21, 1983 without conditioning and without prophylaxis for graft-versus-host disease (16). Resident T-cells in the donor bone marrow were reduced from 21 to 1% after the third lytic process. David had been brought from his home isolator via the transplant isolator to his hospital isolator at Texas Children's Hospital for the procedure. He was monitored for chimerism, immune reconstitution and graft-versus-host disease.

David went home for the Christmas holiday, but when he returned to the hospital, he developed high-spiking fever, thrombocytopenia, erratic lymphocytosis and hyper IgM globulinemia (Fig. 25.1). At day 109 of the transplant, David had severe abdominal pain and bloody stools. David was removed from his isolator system with his parents' consent and was noted

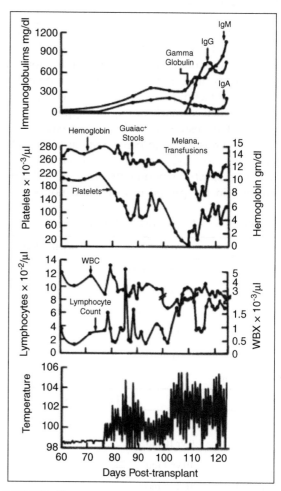

FIGURE 25.1 Clinical course of David after bone marrow transplantation. WBC denotes white blood cell count. *Taken with permission from Shearer WT, Ritz J, Finegold MJ, et al. Epstein–Barr virus-associated B-cell proliferations of diverse clonal origins after bone marrow transplantation in a 12-year-old patient with severe combined immunodeficiency. N Engl J Med 1985;**312**:1151-1159.*

to be dehydrated with an enlarged liver. Intravenous IgG (400 mg/kg), and packed, washed, and irradiated blood and platelets were infused. Solumedrol (3 mg/kg) was given on the presumption of graft-versus-host disease on the basis of focal crypt epithelial necrosis seen at intestinal and rectal biopsy. Radiolabeled erythrocytes and nuclear scanning revealed multiple small bleeding sites in the terminal ileum, cecum, and colon. He expired on transplant day 124.

POST-MORTEM DISCOVERIES

The principal post-mortem finding was the presence of numerous small tumor nodules scattered throughout the patient's intestines and infiltrating his liver, gall bladder, spleen, thymus, parathyroids and stomach. Light microscopy revealed highly pleomorphic infiltrates of plasma cells and large and small lymphoid cells. Many mitotic figures were observed in these infiltrating cells. In addition, immunoblasts with vesicular nuclei were prominently seen in the jejunum. A large retroperitoneal mass of lymphoproliferative cells was seen. Hassall's corpuscles were absent in the apparent thymus, but mostly infiltrating cells of a malignant nature were found.

None of the tumor tissue contained the sister's HLA-B35 haplotype (not shared by David). Several tests of David's pre-mortem mononuclear blood cells did not contain this donor antigen, suggesting that no engraftment of donor cells had taken place. Instead Epstein–Barr virus was found in transformed B-cells in tumor masses and in peripheral blood B-cells from day 117 of the transplant (Fig. 25.2). The source of the virus was likely the donor B-cells contained in

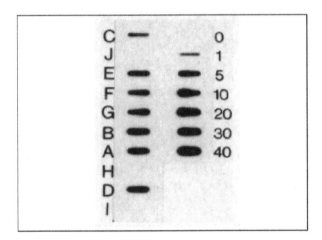

FIGURE 25.2 Autoradiograms of slot-blot DNA hybridizations for EBV genomes in multiple tumor specimens. DNA specimens were hybridized with a radiolabeled EBV-specific DNA probe after application of tissue DNA to nylon membranes. Tissue DNA was applied to the membrane with a slot-blot apparatus that deposits the DNA in a uniform area on the membrane. Specimen A was derived from a thymic lesion, B from a lung lesion, C from a liver lesion, D from a spleen lesion, E from a jejunal lesion, F from an appendiceal lesion, G from a retroperitoneal lesion, H from a grossly and microscopically normal kidney, and I from a peripheral-blood leukocyte fraction from the allograft donor. J was derived from grossly and microscopically normal heart tissue removed at autopsy. For each tissue sample the intensity of the dark hybridization signal correlates with the number of EBV genomes in the tissue. For use as standards, varying amounts of EBV probe DNA were added to EBV-negative tonsil DNA in mixtures totaling 5 µg of DNA, before application to the membrane. The numbers of genome equivalents per cellular DNA equivalent in these standards are indicated at the right. *Taken with permission from Shearer WT, Ritz J, Finegold MJ, et al. Epstein–Barr virus-associated B-cell proliferations of diverse clonal origins after bone marrow transplantation in a 12-year-old patient with severe combined immunodeficiency. N Engl J Med 1985;312:1151-1159.*

the transplant. The EBV serology of the donor was positive for past infection (antibody to viral capsid antigens): IgM titer 1:5 and IgG titer 1:320; antibody to EBV nuclear antigen 1:8, and antibody to early antigen (R or D) <1:10. Studies of immunoglobulin gene rearrangement in multiple tumor specimens indicated that the tumors arose from monoclonal and oligoclonal B-cell proliferation (14).

In retrospect, the events of David' attempted bone marrow transplant indicated that the donor lymphocytes never engrafted (Table 25.6).

TABLE 25.6   Results of the pre- and post haploidentical T-cell-depleted transplant analysis of peripheral-blood lymphocytes

	Before Transplant	After Transplant					Normal Range (±2 S.D.)
		10 days	30 days	60 days	90 days	120 days	
Leukocytes (per microliter; ×10⁻³)	5.1	5.5	4.7	3.4	3.3	3.0	4.5–14.5
Lymphocytes							
Per cent	4	7	7	6	8	27	28–48
Per microliter	204	385	329	204	264	810	1260–6900
Surface markers (%)							
E rosettes (T-cells)	15	6	1	2	ND	ND	43–74
SIg cells (B-cells)	41	38	47	63	61	55	3–19
Antigens on cells (%)							
T3	11	7	8	3	2	1	38–86
T4	13	5	5	2	2	2	23–58
T8	4	2	7	2	4	2	13–33
T6	1	3	0	0	1	0	0–4
T10	13	5	15	12	11	58	0–26
T11	12	11	12	13	21	7	51–100
NKH1	ND	4	15	15	20	40	5–33
B1	45	17	15	13	25	49	2–14
Mo2	ND	12	27	19	14	14	1–30
Ia	68	18	38	25	21	79	11–40
Mitogen reactivity (×10⁻³ cpm)							
Phytohemagglutinin	10	7	4	1	0	0	64–297
Concanavalin A	7	2	1	0	0	0	69–283
Pokeweed mitogen	5	5	2	1	0	0	34–226

SIg denotes suface immunoglobulin, and ND not done.
See Table 25.5 for identification of surface markers. NK H1: natural killer cells, Mo2: monocytes.
Taken with permission Shearer WT, Ritz J, Finegold MJ, et al. Epstein-Barr virus-associated B-cell proliferations of diverse clonal origins after bone marrow transplantation in a 12-year-old patient with severe combined immunodeficiency. N Engl J Med 1985;312:1151-1159.

Suggestive evidence is that the weak 2-to-3-fold response to alloantigen in mixed lymphocyte cultures prevented engraftment (10). Also, there must have been some degree of immune resistance in David since he lived 12 years in a "germ-free" environment that contained small amounts of several bacteria. Malignant B-cells, all containing the EBV, caused David's death. None of these tumors appeared on the surface of David's body (as they have in other cases of post-transplant B-cell lymphoproliferative disease) rendering definitive diagnosis only at autopsy.

DAVID'S IMMORTALIZED B-CELL LINE: CONTRIBUTIONS TO SCIENCE

In the early 1990s, intense efforts were initiated to establish the genetic causes of primary immunodeficiencies, particularly the genetic defect that caused David's disease, namely X-linked SCID (13). Noguchi et al. (17) were the first to report that X-linked SCID was caused by mutations in the gamma chain of the interleukin-2 receptor. This landmark discovery was made in the lymphocytes of three male unrelated children who carried the phenotypic diagnosis of X-linked SCID. One of these patient specimens was that of David, his EBV-transformed B-cell line (18,19) (Fig. 25.3). Each of the children's lymphocytes carried a different mutation in the Xg13 region of the human X chromosome. The concordance of the genetic defect observed in David's B-cell line with that of two other non-related children firmly established the nature and chromosome locations of the gene responsible for X-linked SCID.

PERSONAL REFLECTIONS

Memories of David the Bubble Boy

On my first day as a new recruit to Baylor College of Medicine, Department of Pediatrics, and Texas Children's Hospital, I knew from David's firm handshake through plastic gloves, his direct eye contact and strong personality that he was a special child, far advanced in mental age and very much aware of his extraordinary life. On that first day at Texas Children's Hospital on September 1, 1978 (Fig. 25.4), I had no idea what this young boy would teach the world about human courage, immune resistance to infection and cancer, and about the gene discovery of the cause of X-linked SCID. David taught the world about T-cells, graft-versus-host disease, inheritance patterns of immunodeficiency, escape of virus from carrier B-cells and infection of target cells, rapid growth of virus-transformed tumors and how protective isolation prevents infection.

I became part of David's extended family, and grew to know his mother, Carol Ann Demaret, his father, David Vetter and his sister, Katherine Vetter, then just entering her teenage years. My team of faculty and fellows saw David almost every day and we were privileged to witness his bravery and his brilliance in understanding who he was to the world – a child who lived his entire life in a highly technological world never before and never since attempted. His questions to me were always sharp and well thought out, and many times caught me off-guard.

David would have been 40 years of age on September 21, 2011, but this celebration in his honor with his parents and family, friends, doctors, nurses and hospital staff is testimony to his immortal spirit, alive with perpetual youth and undiminished hope. Far more than any professor, book, or lecture, David's life was a lesson that will forever continue to teach us about the intricate yet mysterious mechanisms that permit life in a hostile microbial world.

Today, I speak to David's family and to all the families who have entrusted the care of their children to us. Thank you. Your children have taught us so much, not only about human immunity and its power to preserve life, but also about their will to succeed against great obstacles and emerge victorious. We are grateful for

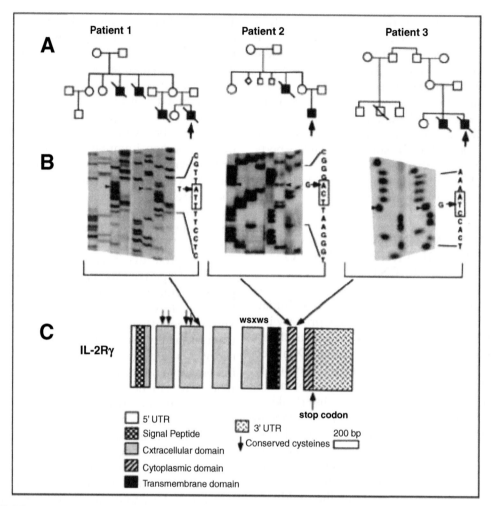

FIGURE 25.3 XSCID patients have mutations in the IL-2Rγ gene. (A) Pedigrees of the XSCID patients' studies. Circles, female; squares, male; closed squares, males with SCID; squares with slashes, deceased males. Small squares and diamond in the pedigree of patient 2 indicate miscarriages (male and of undetermined sex, respectively). David is patient 3. (B) Sequencing of the IL-2Rγ gene. Shown is the sequence of DNA from a normal donor (left panel in each pair) and of DNAs from patients 1, 2, and 3 (right panel in each pair). Patient 1 has an AAA (Lys) to TAA (stop codon) transversion in exon 3, resulting in a truncation of the carboxy-terminal 251 amino acids; patient 2 has a CGA (Arg) to TGA (stop codon) transition in exon 7, resulting in the truncation of 81 amino acids; and patient 3 has a TCG (Ser) to TAG (stop codon) transversion in exon 7, resulting in the truncation of 62 amino acids. The location of each mutation is indicated in the sequence to the right of each set of panels. The sequence shown is complementary to the coding strand. The boxed nucleotides for patients 1, 2, and 3 are complementary to TAA, TGA, and TAG stop codons, respectively. (C) Schematic showing locations of the artificial stop codons (diagonal arrows) present in the XSCID patients. Patient 1 has a premature stop codon in exon 3, and patients 2 and 3 have premature stop codons in exon 7. UTR: Untranslated region. *Taken with permission from Noguchi M, Yi H, Rosenblatt HM, et al. Interleukin-2 receptor gamma chain mutation results in X-linked severe combined immunodeficiency in humans. Cell 1993;***73***:147-157.*

FIGURE 25.4 The first time I met David, the bubble boy. Dr William T. Shearer meets David on September 1, 1978 on his first day at Texas Children's Hospital and shakes hands with his new patient. Dr Buford Nicols who directed David's nutritional program looks on. David is almost seven years of age in this picture.

the legacy David and all the special children after him have given to the world.

Delivered in David's memory upon the 40th anniversary of his birth, September 21, 2011, The David Center Texas Children's Hospital Houston, Texas.

A MOTHER'S RECOLLECTION: CAROL ANN DEMARET

Now and then I dream of David and we laugh and I feel touched and good when I awaken. In many ways, David's life at home was "normal". I could care for him, change his diapers as a baby and clothed him and fed him the foods that were packaged in sterile, air-locked space age "supple cylinders" that were replenished weekly. I couldn't touch him, but I could feel his warmth, as I embraced and held him close through the plastic and I believed he could feel mine, both as an infant and a boy. I cut his hair, and bathed him and changed his clothes when he was small and did most of the child care that all mothers do. As he grew, his bubbles grew in size and numbers – at home and at the hospital. He was happy to be at home with his family and he was happy to be at the hospital with his doctors and nurses and all the attention he received. However, he did not like having his blood drawn which was not an easy task.

As he grew, he had sterilized toys and games, was educated by way of at-home teachers, (sometimes accompanied by classmates). He received Holy Communion, after the host had been sterilized and consecrated. He had

FIGURE 25.5 The wall of plastic that separated David and his mother, Carol Ann Demaret. David is on the inside of his sterile isolator while his mother, Carol Ann, is on the outside. Among her many memories of David, she recalls with pleasure cutting his hair with sterile scissors. David is 11 years of age in this picture.

religious training, and family and friends who visited often. His sister slept on the floor next to him and they would talk and plot and giggle. He watched television and enjoyed favorite programs. He talked on the telephone, even had pets (outside the bubble, of course). He had a parakeet that he tried to teach to talk. There was a large window in his room, both at home and at the hospital, where he could view his surroundings.

David Phillip grew with fine health, dark and handsome, highly intelligent and astonishingly perceptive, with black locks and penetrating eyes that some said seemed to see into a far beyond (Fig. 25.5).

When a possible new medical technique that showed great promise finally became available, David himself decided that he wanted to participate. It was a complex treatment that involved a cellular transplant. At first it appeared to be working. But then it failed. All attempts to reverse the process were unsuccessful, and my gallant son died on February 22, 1984, age 12.

We were devastated. We wept. And we felt a nation wept with us.

David not only lives on in my soul, but in other ways, too.

Within a few days of David's passing, his family was approached by Texas Children's Hospital with an idea in tribute to David's memory to create "The David Center" where SCID babies, and all children born with compromised or faulty immune systems whether genetic or acquired, can be successfully treated, and with continued research, immune systems could be stirred into more vigorous action, and where important advances have been made in discovering the genetic defect for SCID.

A school was named after him, "The David Elementary" in the Woodlands, Texas, at one time said to be the only school in the country named after a child. One of David's unrealized dreams was to be able "to walk barefoot in the grass". With this in mind and spirit, each year the school sponsors a "David Dream Run" and raises contributions to "The David Center". There is a David Memorial Drive in the neighborhood where he lived.

In December of 2012, it was time to celebrate David's legacy as it marked the beginning of

SCID New born Screening in Texas. Texas joins 12 other states that are currently screening all newborns for SCID.

I now believe my prayer has been fully answered, and "that the bubble has burst" for all time.

For David's family and those who shared in his life, other lessons gleaned from David are important, too. He taught us that handicap of body is not necessary handicap of mind and spirit, that handicap cannot just be endured, but nobly endured. He has reminded the world of humble marvels, such as grass under our feet.

All children leave behind memories and a sense of yearning and loss, but David left behind some things for humankind, too, and it helps. "He never touched the world, but the world was touched by him", are the words inscribed on his gravestone.

I believe the promise for the very near future is not just to preserve some portion of precious life, but to enhance it at the core.

I carry in my heart sweet reminders of the past and great expectations for the future for all generations that follow.

Presented at the 3rd Pediatric Immunodeficiency Treatment Consortium Third Annual Workshop, May 2, 2013 The David Center Texas Children's Hospital Houston, Texas.

Acknowledgment

We thank Janice Hopkins and Janelle Allen for assistance in manuscript preparation, and for reviewing the memorabilia of David's life saved at the David Center, Texas Children's Hospital.

References

1. Williamson AP, Montgomery JR, South MA, et al. A special report: four-year study of a boy with combined immune deficiency maintained in strict reverse isolation from birth. *Pediatr Res* 1977;**11**:63–4.
2. Blattner RJ. Introduction. *Pediatr Res* 1977;**11**:65.
3. Montgomery JR, South MA, Wilson R, et al. Family background, early history, and diagnosis. *Pediatr Res* 1977;**11**:65–6.
4. Wilson R, Taylor GR, Kropp K, et al. Gnotobiotic care and infectious disease prevention. *Pediatr Res* 1977;**11**: 67–71.
5. Guerra IC, Shearer WT. Environmental control in management of immunodeficient patients: experience with "David". *Clin Immunol Immunopathol* 1986;**40**:128–35.
6. Mukhopadhyay N, Richie E, Montgomery J, et al. Peripheral blood T and B-cell characteristics in a patient with severe combined immune deficiency (SCID) maintained in a gnotobiotic environment. *Exp Hematol* 1976;**4**: 1–9.
7. South MA, Montgomery JR, Richie E, et al. Immunologic studies. *Pediatr Res* 1977;**11**:71–8.
8. Mukhopadhyay N, Richie E, Mackler BF, et al. A longitudinal study of T and B lymphocytes from a three-year-old patient with severe combined immunodeficiency (SCID) in 'gnotobiotic protection'. *Exp Hematol* 1978;**6**: 129–34.
9. Mackler BF, O'Neill PA. T-lymphocyte induction of non-T-cell-mediated nonspecific cytotoxicity. II. A clinical study of a severe combined immunodeficiency patient maintained in a gnotobiotic environment. *Clin Immunol Immunopathol* 1979;**12**:358–66.
10. Paschall VL, Brown LA, Lawrence EC, et al. Immunoregulation in an isolated 12-year-old boy with congenital severe combined immunodeficiency. *Pediatr Res* 1984;**18**: 723–8.
11. Fernbach DJ, Starling KA, Falletta JM. Hematology. *Pediatr Res* 1977;**11**:78–9.
12. Potts E, Huang CTL, Nichols BL. Nutritional care and related studies. *Pediatr Res* 1977;**11**:80–2.
13. Molish B, Murphy M, Desmond M. Mental, psychomotor, and psychosocial development. *Pediatr Res* 1977;**11**:82–3.
14. Musher KK. Speech and language development. *Pediatr Res* 1977;**11**:84–5.
15. Freedman DA. Psychiatric evaluation. *Pediatr Res* 1977;**11**:85–6.
16. Shearer WT, Ritz J, Finegold MJ, et al. Epstein-Barr virus-associated B-cell proliferations of diverse clonal origins after bone marrow transplantation in a 12-year-old patient with severe combined immunodeficiency. *N Engl J Med* 1985;**312**:1151–9.
17. Noguchi M, Yi H, Rosenblatt HM, et al. Interleukin-2 receptor gamma chain mutation results in X-linked severe combined immunodeficiency in humans. *Cell* 1993;**73**: 147–57.
18. Rosenblatt HM, Green CG, McClure JE, et al. Antibody to human lymphocyte actin regulates immunoglobulin secretion by an EBV-transformed human B-cell line. *Biochem Biophys Res Commun* 1986;**140**:399–405.
19. Cleary ML, Nalesnik MA, Shearer WT, et al. Clonal analysis of transplant-associated lymphoproliferations based on the structure of the genomic termini of the Epstein-Barr virus. *Blood* 1988;**72**:349–52.

20. Reinherz EL, Geha R, Rappeport JM, et al. Reconstitution after transplantation with T-lymphocyte-depleted HLA haplotype-mismatched bone marrow for severe combined immunodeficiency. *Proc Natl Acad Sci USA* 1982;**79**:6047–51.

21. Reisner Y, Kapoor N, Kirkpatrick D, et al. Transplantation for severe combined immunodeficiency with HLA-A, B, D, DR incompatible parental marrow cells fractionated by soybean agglutinin and sheep red blood cells. *Blood* 1983;**61**:341–8.

How Primary Immunodeficiencies Have Made Gene Therapy a Reality

Alain Fischer[1,2,3,4], *Salima Hacein-Bey-Abina*[1,2,5], *Marina Cavazzana-Calvo*[1,2,5]

[1]INSERM U1163, Paris, France
[2]Sorbonne Paris Cité, Université Paris Descartes, Imagine Institute, Paris, France
[3]Immunology and Pediatric Hematology Department, Necker Children's Hospital, Assistance Publique–Hôpitaux de Paris, Paris, France
[4]Collège de France, Paris, France
[5]Biotherapy Clinical Investigation Center, Groupe Hospitalier Universitaire Ouest, Assistance Publique–Hôpitaux de Paris, INSERM, Paris, France

INTRODUCTION

Gene therapy of primary immunodeficiencies (PIDs) has become a reality. At the time of writing, at least 49 patients with X-linked combined severe immunodeficiency (SCID)-X1 ($n = 17$) or adenosine deaminase (ADA) deficiency ($n = 32$) have gained clinical benefit from gene therapy over the last 14 years (1–7). Although these successes were mitigated by the advent of serious adverse events (leukemogenesis related to the use of first-generation vectors) in the SCID-X1

Primary Immunodeficiency Disorders: A Historic and Scientific Perspective

trial and other studies (8–13), they are paving the way towards extension of the indications of gene therapy to other PIDs (including Wiskott–Aldrich syndrome (WAS)), other forms of SCID, chronic granulomatous diseases (CGDs) and other serious conditions. Gene therapy for SCID-X1 due to mutations in the common γ-chain (γc) constituted the first successful application of gene therapy due to the reasons discussed below. In a sense, this is reminiscent of the fact that allogeneic hematopoietic stem cell transplantation (HSCT) was first successfully used as treatment for SCID (14) and WAS (15), prior to its extension to the much broader field of leukemia and other acquired and genetic disorders. We have come a long way since 1972, with publication of the first thoughts on gene transfer as treatment for inherited diseases in general and PIDs in particular (16, 17). A personal account of this story is given below (Table 26.1). A number of very detailed historical perspectives have been published (16–18). When considering gene therapy of PIDs, one has to take into account the diseases themselves, the target cells and the gene vectors.

PRIMARY IMMUNODEFICIENCIES

Since most PIDs are monogenic diseases, they are eligible for gene therapy based on either expression of the therapeutic gene, or the extinction of gene expression in cases where a disease is caused by a trans-dominant negative effect of the mutated gene. The very first genes to be causally related to a PID were *CYBB* (in X-linked CGD) and *ADA* (in ADA-deficient SCID) in 1984. Given that the expression rate of ADA is fairly constant and that ADA-deficient SCID has a poor prognosis, ADA was immediately considered as a prime target for gene therapy (17). Since then, the identification of over 190 PID-associated genes has potentially broadened the field of gene therapy to almost all life-threatening PIDs, including all forms of SCID,

some T-cell immunodeficiencies, innate immune deficiencies and regulatory disorders (such as hemophagocytic lymphohistiocytosis (HLH) and immune dysregulation, polyendocrinopathy, enteropathy, X-linked [IPEX] syndrome) (19).

TARGET CELLS

The vast majority of PIDs are intrinsic disorders of the hematopoietic system. The identification of murine bone marrow cells capable of giving rise to spleen-repopulating hematopoietic colonies and producing long-term hematological reconstitution upon transplantation by Till and McCulloch in 1961 (20) provided a strong scientific rationale for allogeneic HSCT. Along with the characterization of human leukocyte antigens in histocompatibility, this observation opened the way to the effective application of HSCT in PID diseases (14, 15). Since then, our understanding of hematopoietic stem cell (HSC) biology has progressed immensely (for an example, see (21) and Chapter 24). Understandably, the first attempts to transfer genes into HSCs were performed in murine models (22, 23), as discussed below, and this paved the way for HSC-based gene therapy of PIDs. However, for reasons that are also reviewed below, the first clinical trials were performed with peripheral T-cells from ADA-deficient patients receiving enzyme replacement therapy (ERT) (24).

VECTORS

Correction of a mammalian cell phenotype via the introduction of genetic material (DNA) was first performed chemically in a hypoxanthine guanine phosphoribosyltransferase (HPRT)-deficient cell line (25). Subsequently, calcium/phosphate precipitation was used to achieve the first transfer of a whole, stably expressed gene (coding for herpes simplex thymidine kinase) into mouse fibroblasts (26).

TABLE 26.1 Key steps in the gene therapy of PIDs

1911	Discovery of retroviruses
1961	Identification of spleen-repopulating hematopoietic colonies in mice
1968	First successful allogeneic bone marrow transplantation in the treatment of a PID patient (SCID)
1978	Expression of a foreign gene in a cell line
1983	First cell transduction based on retroviral vectors
1983	First *ex vivo* transduction of murine HSCs
1984	Cloning of two PID genes: *ADA* and *CYBB*
1986	Design of packaging cell lines capable of producing replication-free retroviral vectors
1989	The first clinical trial of gene therapy for ADA: *ex vivo* transduction of T lymphocytes
1991–1996	Improvements in the efficacy of transduction of human hematopoietic progenitor cells
1995	The Orkin–Motulsky report
1996	Design of a lentiviral (HIV-based) vector
1998	Design of a lentivirus-based self-inactivated (SIN) vector
2000	The first report of clinical success from the SCID-X1 trial
2002	The first report of clinical success from the ADA SCID trial
2003	Report of the occurrence of leukemia in the SCID-X1 trial
2007	Experimental use of engineered endonucleases to target the γc gene and enable gene correction through homologous recombination
2009	The first report of clinical success with a lentiviral vector (in the treatment of adrenoleukodystrophy)
2009	Initiation of new clinical trials for gene therapy (based on SIN retroviral and lentiviral vectors) of SCID-X1 and Wiskott–Aldrich syndrome

Although chemical methods, liposome fusion and electroporation have since been employed to introduce DNA into cells (17), none of these techniques has enabled sustained transgene expression upon cell division because the gene is not replicated. Since gene transfer within the hematopoietic system requires transmission of the transgene to daughter cells, a replication system is essential. The most obvious solution is to integrate the transgene into the cell's genome. In fact, gene therapy became a reality when it started to adopt retroviral vector technology. Retroviruses (RVs) had been discovered in chickens (27) and had since been characterized in terms of their structure, function and "life cycle" (reviewed in (28)). The RVs' genetic material is double-stranded RNA, which is reverse-transcribed into DNA within the host cell (29). This "provirus" then integrates into the target cell's genome. Many RVs are oncogenic because they can trigger the expression of endogenous oncogenes. As Varmus comments (28), "the incentive to use retroviruses as genetic vectors originated with the perception that retroviruses with viral oncogenes are naturally occurring genetic vectors". Mouse tumor viruses, Moloney viruses and mouse sarcoma viruses were the first to be considered as potential gene transfer vectors.

For use as a vector, the RV's structural genes were replaced by the transgene cDNA. These retroviral vectors could transduce cells but, nevertheless, contained replication-competent retroviral particles that were capable of propagation – as shown in the first *ex vivo* transductions of murine HSCs (22).

Further progress came with the differentiation between (1) plasmids containing the therapeutic genetic material (i.e. long terminal repeats [LTRs] and the cis-acting packaging element Ψ for encapsidation of the RNA within the viral particle) and (2) plasmids providing structural genes for forming retroviral particles in the transfected cell line (the so-called "packaging cell line") (30, 31). However, these retroviral vectors were still capable of recombination and thus could give rise to replication-competent viral particles. Nevertheless, these experiments led to sustained expression of the transgene (dihydrofolate reductase and HPRT) in murine hematopoietic cells and thus demonstrated for the first time the approach's feasibility and efficacy. Further refinement in the design of packaging cell lines (e.g. separation of the retroviral structural genes on distinct plasmids) led to the production of replication-free viral particles (32). This progress prompted researchers to test gene therapy in large animal models. Anderson et al. performed RV-mediated ADA gene transfer experiments in primates. In the absence of myeloablation, it was observed that expression of the human ADA gene was low and transient, indicating that HSCs had not been transduced and/or could not find their niche in the bone marrow (33). These findings prompted the researchers to consider transferring the ADA gene into peripheral T-cells (see below). The first approved gene therapy trial (performed in 1989) subsequently proved the feasibility and safety of injecting gene-marked tumor-infiltrating T lymphocytes into humans (34).

A key step forward in gene vector technology was the advent of constructs based on lentiviruses (primarily HIV) (35). The main advantage of lentiviral vectors is that (unlike retroviral vectors) their provirus can cross the nuclear membrane and thus readily infect stem cells during the G1 phase. As discussed below, use of this type of vector now enables efficient gene transfer into HSCs. In modified HIV-based vectors, deletion of the 3'-LTR enhancer and promoter element (reverse-transcribed into the 5'-LTR) meant that the LTR was no longer used for transgene expression and could be replaced by an internal promoter (36). As described below, this third-generation of "self-inactivated" (SIN) lentiviral vectors are in clinical use today. Furthermore, another class of RV (foamy virus) has been developed for potentially safer and more efficient cell transduction (37).

THE FIRST CLINICAL TRIALS OF GENE THERAPY

As mentioned above, Anderson et al. decided to target T-cells from ADA-deficient patients because (1) gene transfer into non-human primate HSCs was inefficient, (2) life-threatening ADA deficiency was a good candidate disease and (3) ADA gene expression was known not to be strictly controlled. Although this rationale was well-founded, the concomitant use of ADA ERT by the patients who received the transduced T-cells led to limited, barely evaluable efficacy because the expected selective growth advantage conferred by transgene expression (17) was lost in this setting. Nevertheless, two girls with ADA deficiency received repeated (up to 11) injections of *ex vivo* transduced T-cells from 1990 onwards (24). In one patient, very few transduced T-cells were detected over time, whereas a significant number of transduced CD8 T-cells (up to 20% of the total) have persisted in the second subject for more than 12 years, thus demonstrating the procedure's feasibility and safety (Table 26.2) (38).

A second clinical trial quickly followed: six patients on ERT received both transduced

TABLE 26.2 History of clinical trials of gene therapy for PIDs

Year of publication ref.	Diseases	n	Vector	CR	Target cells	SAE	Outcome
1995: 24, 38	ADA(ERT+)	2	γ-RV	-	T-cells	-	Long-term persistence of transduced T-cells, continued ERT
1998: 39	ADA(ERT+)	1	γ-RV	-	T-cells	-	Long-term persistence of transduced T-cells, continued ERT
1995: 40, 41	ADA(ERT+)	6	γ−RV	-	T+ CD34	-	Long-term persistence of transduced T-cells, continued ERT
1995: 42, 43	ADA(ERT+)	3	γ-RV	-	Cord blood CD34	-	Long-term persistence of transduced T-cells, continued ERT
1996: 44	ADA(ERT+)	3	γ-RV	-	CD34	-	Minimal engraftment, continued ERT
2000: 45	JAK3	1	γ-RV	-	CD34	-	Minimal engraftment
2000: 1,2	SCID-X1	10	γ-RV	-	CD34	4*	Long-term efficacy (n = 7)
2004: 3, 46	SCID-X1	10	γ-RV	-	CD34	1	Long-term efficacy (n = 10)
2005: 47	SCID-X1	2 (>10y old)	γ-RV	-	CD34	-	Failure
2007: 48	SCID-X1	3 (>10y old)	γ-RV	-	CD34	-	Minimal level of T-cell reconstitution
2002: 4,5	ADA(ERT-)	17	γ-RV		CD34	-	15/17 ERT-free
2011: 6	ADA(ERT-)	6	γ-RV		CD34	-	4/6 ERT-free
2012: 7	ADA(ERT-)	4	γ-RV	-	CD34	-	Failure
2012: 7	ADA(ERT-)	6	γ-RV		CD34	-	3/6 ERT-free
2010 (and unpublished): 12	WAS	10	γ-RV		CD34	7	Efficacy 9/10
1997: 49	CGD-AR	5	γ-RV	-	CD34 (PBSC)	-	Transient detection of transduced neutrophils
2010: 50	CGD-XL	3	γ-RV	-	CD34	-	Low level of long-term marking
2006:10, 2010:11 & 2011: 51	CGD-XL	4	γ-RV		CD34	4**	Transient efficacy
2013: 52	SCID-X1	9	SIN γ-RV	-	CD34	-	Short follow up
2013: 53	WAS	6	SIN LV		CD34	-	Short follow up
2013: 54	WAS	3	SIN LV		CD34	-	Short follow up

*: 1 death from leukemia; **: 3 patients transplanted, also one death. CR: conditioning regimen; ERT: enzyme replacement therapy; PBSC: peripheral blood stem cells.

T-cells and bone marrow cells marked with distinguishable RVs (40). Here again, persistent transduction of T-cells could be detected several years later (41). Nevertheless, the patients still required ERT. Next, Kohn et al. choose to perform *ex vivo* ADA gene transfer into cord blood cells from three ADA-deficient infants placed on ERT shortly after birth (42). As in the trial mentioned just above, only a small proportion (1–10%) of T-cells were gene-marked in the long term (43). Another trial consisting of *ex vivo* ADA gene transfer into CD34(+) bone marrow cells in three patients on ERT also failed because of a low transduction rate (44). Hence, two conclusions could be drawn from these early trials: the transduction technology was not yet efficient enough, particularly in CD34 cells, and ERT was an obstacle to the *in vivo* expansion of transduced lymphocytes.

TECHNICAL PROGRESS

Key advances in the 1990s partially solved the above-mentioned issues. This progress originated from carefully performed, albeit empirical, experiments and consisted of (1) definition of a cytokine cocktail that included SCF, TPO and Flt3L for triggering the division of CD34 cells and thus increasing the transduction rate, (2) production of higher viral particle titers by the packaging cell lines, (3) addition of a fibrinogen fragment to increase the transduction rate by coating cells and viruses together (55) and (4) the development of new viral envelopes, such as those based on the gibbon ape leukemia virus (GALV), to increase the cell infection rate (56).

Given that none of the clinical gene therapy protocols, which were primarily evaluated in the field of cancer, had provided convincing evidence of clinical benefits, the Orkin–Motulsky panel assembled by the NIH reviewed the field and, in 1995, set out clear guidelines for firmly establishing a scientific, rigorous approach to gene therapy (57). These guidelines included

the development of better vector systems and the in-depth assessment of disease mechanisms, notably in animal models. This prompted the generation of animal models of PIDs and SCIDs, including mice variously lacking γc, JAK3 or ADA. Sorrentino et al. demonstrated the correction of JAK-3-deficient SCID mice by *ex vivo* RV-mediated JAK3 cDNA transfer into bone marrow cells (58). However, a follow-up clinical trial failed to achieve significant correction of HSCs in a JAK3-deficient SCID patient. Several groups have since shown that γc-deficient mice can be immunologically corrected by a similar technique (59–61).

THE EFFICACY OF GENE THERAPY FOR SCID-X1 AND ADA DEFICIENCY

The presence of a growth advantage conferred by transgene expression in a SCID context (17) was confirmed by reports that somatic revertants of both ADA and γc genes led to milder forms of these PIDs (62, 63). In a SCID-X1 patient with a somatic reversion, it was estimated that the revertant precursor T-cell divided at least 10 to 11 times prior to T-cell receptor (TCR) rearrangement and thus could generate up to 1% of the TCR-beta repertoire (64). This "natural gene therapy" gave strong impetus to clinical trials based on improved, RV-mediated gene transfer technology and the demonstration of effective *in vitro* correction of the deficiency in patients' cells (65, 66).

The first SCID-X1 trial was initiated in 1999 and was soon followed by an ADA deficiency trial, in which the enrolled patients did not receive ERT in order to preserve the selective advantage of the corrected cells. In a second SCID-X1 trial, the RV used had a GALV envelope instead of an amphotropic envelope. In order to achieve a greater blood cell transduction rate, the ADA-deficient patients underwent myeloablation with 4 mg/kg busulfan in view of the overall enzyme deficiency that characterizes this condition. Fourteen years later, the following observations can be made. Of the 20 treated SCID-X1 patients, 18 are alive and 17 of them have enough T-cell immunity to overcome opportunistic infections (1–3). Efficacious gene transfer of γc into CD34 cells was achieved in 19 patients. The transfer failures were related to poor transduction in one case, hypersplenism in a second case and genotoxicity in a third case (see below). It was possible to detect T-cells within three months and the resulting cell counts, including naïve polyclonal T-cells, were sustained for ten years and more after gene therapy. Membrane γc expression was found to be close to physiological levels, bearing in mind that the latter is regulated by the expression of subunits to which γc binds (IL7Rα, IL2Rα/β, etc.). Various T-cell effector functions were restored, as shown *in vitro,* and *in vivo* by the control of varicella zoster virus and Epstein–Barr virus infections. The sustained restoration of a close-to-normal T-cell pool contrasted with persistently low natural killer cell counts (1–3), as had also been observed following non-myeloablative HSCT in SCID-X1 patients (67). Furthermore, transduced B-cells, observed soon after gene therapy, could not be detected several years later. Despite the persisting partial B-cell deficiency caused by defective IL-4R and IL-21R signaling, it is nevertheless important to note that over half the patients no longer require immunoglobulin substitution (1–3).

The sequencing of the human genome and the development of tools to assess the retroviral vector's integration sites (such as inverse polymerase chain reaction (PCR), ligation-mediated PCR and linear amplification-mediated PCR (68–70)) have enabled researchers to characterize in detail the population of marked cells. These experiments showed that, on average, no more than 1000 different integrations were detected in the population of peripheral blood T-cells – confirming that γc expression conferred a selective growth advantage on T-cell precursors. Initially, the observation that T-cells and a few transduced myeloid cells had the same integration sites was interpreted as suggesting that at least some multipotent progenitors had been transduced (1–3, 71, 72). However, as was subsequently found, only transduced T-cells persisted – an indication that there must be persistent, self-renewing T-cells, possibly localized in the thymus (2, 72).

In the field of ADA deficiency, a gene therapy trial was initiated in Milan and extended to London; a second trial was started in the USA jointly by the NIH and a Los Angeles group. These trials were based on (1) a similar gene transfer technology and (2) the use of low-dose busulfan or melphalan myeloablation. A total of 42 ADA-deficient patients have been treated over the last 12 years. All are alive and 31 are now off ERT, which constitutes a critical efficacy criterion (4–7, 73). The speed and amplitude of correction of the T-cell deficiency appear to be respectively slower and smaller when compared with those observed in γc deficiency; this likely reflects an intrinsic difference in the mechanism of SCID disease, because ADA deficiency involves impairments of the thymic epithelial component. As a consequence of myeloablation, transduced myeloid cells can be constantly detected over time – providing further support of metabolic detoxification, as judged by dAXP nucleotide assays (4–7). Overall, correction of the immunodeficiency is sustained and good enough to enable patients to live normally. The results of an integration site analysis in the transduced cells were similar to those seen in SCID-X1 trials. There was also additional evidence to suggest that some HSCs had been transduced because the different blood lineages continued to display the same integration site over time (5, 74). Overall, gene therapy of SCID-X1 or ADA SCID with first-generation retroviral vectors was associated with sustained clinical benefit in 42 patients. These results provided proof of principle and efficacy for gene therapy. The success rate was higher than that of haploidentical or unrelated cord blood stem cell transplantation performed in the same setting (75).

GENOTOXICITY

Five patients from the SCID-X1 trials (four in Paris and one in London) developed a form of T-cell leukemia (8, 9) 2 to 5.5 years after gene therapy. The leukemia was fatal in one patient (76). These serious adverse events (SAEs) attracted much attention at the time (77) and prompted the suspension of the other trials. Subsequent collaborative research revealed the main causal mechanism: insertional mutagenesis related to RV integration near or within a proto-oncogene and transactivation of the latter (LMO-2, in four cases) by the potent enhancer within the viral LTR. Secondary genomic alterations (translocations, activating Notch 1 mutations, etc.) then led to overt, clinical leukemia. A systematic review of RV integration sites showed that up to 60% are within gene loci, with a distribution centered by the transcription site (78). After a more comprehensive determination of RV integration sites (based on next-generation sequencing, the use of multiple restriction enzymes and non-restrictive linear amplification-mediated PCR (79)), further research generated an epigenetic map of accessible sites that included many active genes in hematopoietic progenitor cells (80, 81). There was some debate about this topic since, strikingly, no SAEs occurred in the successful ADA trials; this difference in incidence was statistically significant. The ADA trial's safety record is particularly impressive, since the pattern of RV integration sites was the same as in the SCID-X1 trials and included sites within the LMO-2 locus (74). This prompted some researchers to propose that γc itself (via overexpression and/or the triggering of continuous cell activation) had a role in the T-cell leukemia (82). This turned out not to be the case (83). Meanwhile, similar genotoxic events were found to occur in the otherwise successful RV-based gene therapy of WAS where seven out of nine cases developed leukemia, all associated with LMO-2 transactivation (12, 84) and of CGD with three cases (10, 11). In the CGD trial, use of a potent LTR, derived from the spleen focus-forming virus, with strong enhancer activity in myeloid cells, led to transactivation of the EVI1/PRDM1 genes and hence to a myelodysplastic syndrome that was fatal in one case. In summary, these distinct clinical trials established the unacceptably high risk of genotoxicity associated with first-generation retroviral vectors. The absence of such events in ADA deficiency has yet to be explained but may be related to the unfavorable thymic environment caused by the accumulation of toxic metabolites (73).

IMPROVEMENTS IN VECTOR SAFETY

Much effort has been devoted to developing safer but nevertheless effective vectors. As mentioned above, the construction of SIN vectors was a major advance (36, 85) because the enhancer contained within the 3'-LTR had been deleted and a weaker promoter was used to drive transcription of the transgene. In theory, this strategy should prevent gene transactivation without changing the overall pattern of viral integration into the genome. Researchers have considered additional measures, such as the use of a suicide gene (the herpes simplex virus 1 thymidine kinase (HSV1 TK) gene, for example) that can be induced by a drug, ganciclovir in the case of HSV-1 TK, so that transformed cells can be eliminated *in vivo*. A perhaps more attractive strategy is the addition of insulators. These are boundary sequences that prevent enhancer activities. The first insulator to be proposed was derived from the chicken globin locus (86) and has been added to some vectors used in clinical trials.

Lastly, modification of the vector integration site is an attractive strategy for preventing oncogene transactivation. As mentioned above, the use of lentiviral vectors (35, 36) enables the transduction of non-dividing cells and is of potential interest because these vectors do not integrate within the promoter regions of genes (87). Nevertheless, they do integrate within coding sequences. The first clinical application of lentiviral vectors (in the treatment of adrenoleukodystrophy (ALD) in four patients) has produced reassuring safety data, with four to five years of follow-up data on lentiviral integration patterns in blood cells (88).

An important and much-debated question faced by all the gene therapy strategies tested to date is whether cell- or animal-based predictability assays are of value in assessing a vector's potential genotoxic risk prior to use in clinical trials. Baum et al. have developed an interesting *in vitro* clonogenicity assay in which SIN vectors appeared to be safe (89), although only myeloid cell clonogenicity has been tested. An assay of LMO-2 activation in a Jurkat cell line has also been developed (90). Several animal models have been designed; most of these are cancerprone mice for testing potential oncogenic effects (reviewed in 91). To date, there is no clearcut proof that these models can reliably be used as toxicity assays.

THE LATEST CLINICAL TRIALS

Based on the above-mentioned advances, several new trials in which a SIN RV is used to trigger γc expression under the weak elongation factor promoter have been initiated. The preliminary results appear to be promising in terms of both efficacy and safety (52), although more follow-up is required before firm conclusions can be drawn. A SIN lentivirus is being used in two clinical trials of WAS gene therapy, which is combined with partial myeloablation in order to achieve sufficient engraftment of the transduced stem cells. Here again, the preliminary data appear to be promising (53, 54). These trials were preceded by thorough assessment of the WAS gene vectors in *in vitro* experiments and in animal models. Interestingly, the endogenous WAS promoter is being used in this trial,

although other strategies are also being considered (92–94). Likewise, new LV vectors have been designed with a view to treating CGD (95). A chimeric promoter that contains binding sites for myeloid transcription factors achieved high expression of gp91 phox in myeloid cells and will probably soon be tested in clinical trials (95). One can therefore expect a combination of more efficient HSC transduction, increased transgene expression and the use of myeloablation (to free HSC niches) to provide safe, stable disease correction, even when a selective growth advantage is lacking, as is the case in CGD. This expectation is reasonable, given (1) the longterm (four to five year) transgene expression rate observed in leukocytes from patients in the ALD trial (96) and (2) recently reported results from a trial in metachromatic leukodystrophy (MLD). If confirmed, this is likely to open a new chapter in gene therapy of PIDs, since most lifethreatening PIDs (SCIDs, HLH, immune dysregulation syndromes, etc.) could be considered as potential indications. A remaining technological obstacle to broader usage of the LV vector is the current lack of stable producer cell lines for industrial-scale batch production.

CONCLUSION

The history of gene therapy for PIDs is short. Some solid rationales have been provided and careful extension of the indications for gene therapy can now legitimately be considered. Further steps along this pathway could include new technologies, such as homologous recombination mediated by engineered nucleases (meganucleases, zinc finger nucleases, transcription activator-like effector nucleases and the recently described CRISPR-associated nuclease) (98). These may further increase safety because they can target a "safe harbor" (99) in the genome and provide approaches for direct gene correction (100). Furthermore, gene-engineering strategies may also be considered for the inactivation of gene mutations with dominant negative effects in certain PIDs. However, a number of issues remain to be solved, such as the assessment of off-target events and the ability to correct genes in HSCs. Gene correction through the use of engineered pluripotent stem cells from PID patients might also become an option in the future (101). Opportunities and challenges thus lie ahead in our efforts to make gene therapy a widely used therapeutic option for PIDs.

References

1. Cavazzana-Calvo M, Hacein-Bey S, De Saint Basile G, et al. Gene therapy of human severe combined immunodeficiency (SCID)-X1 disease. *Science* 2000;**288**:669–72.
2. Hacein-Bey-Abina S, Hauer J, Lim A, et al. Efficacy of gene therapy for X-linked severe combined immunodeficiency. *N Engl J Med* 2010;**363**:355–64.
3. Gaspar HB, Cooray S, Gilmour KC, et al. Long-term persistence of a polyclonal T-cell repertoire after gene therapy for X-linked severe combined immunodeficiency. *Sci Transl Med* 2011;**3**:97.
4. Aiuti A, Slavin S, Aker M, et al. Correction of ADA-SCID by stem cell gene therapy combined with nonmyeloablative conditioning. *Science* 2002;**296**:2410–3.
5. Aiuti A, Cattaneo F, Galimberti S, et al. Gene therapy for immunodeficiency due to adenosine deaminase deficiency. *N Engl J Med* 2009;**360**:447–58.
6. Gaspar HB, Cooray S, Gilmour KC, et al. Hematopoietic stem cell gene therapy for adenosine deaminase-deficient severe combined immunodeficiency leads to long-term immunological recovery and metabolic correction. *Sci Transl Med* 2011;**3**:97-80.
7. Candotti F, Shaw KL, Muul L, et al. Gene therapy for adenosine deaminase-deficient severe combined immune deficiency: clinical comparison of retroviral vectors and treatment plans. *Blood* 2012;**120**:3635–46.
8. Howe SJ, Mansour MR, Schwarzwaelder K, et al. Insertional mutagenesis combined with acquired somatic mutations causes leukemogenesis following gene therapy of SCID-X1 patients. *J Clin Invest* 2008;**118**:3143–50.
9. Hacein-Bey-Abina S, Von Kalle C, Schmidt M, et al. LMO2-associated clonal T-cell proliferation in two patients after gene therapy for SCID-X1. *Science* 2003;**302**:415–9.
10. Ott MG, Schmidt M, Schwarzwaelder K, et al. Correction of X-linked chronic granulomatous disease by gene therapy, augmented by insertional activation of MDS1-EVI1, PRDM16 or SETBP1. *Nat Med* 2006;**12**:401–9.
11. Stein S, Ott MG, Schultze-Strasser S, et al. Genomic instability and myelodysplasia with monosomy 7 consequent to EVI1 activation after gene therapy for chronic granulomatous disease. *Nat Med* 2010;**16**:198–204.
12. Boztug K, Schmidt M, Schwarzer A, et al. Stem-cell gene therapy for the Wiskott-Aldrich syndrome. *N Engl J Med* 2010;**363**:1918–27.
13. Aiuti A, Bacchetta R, Seger R, Villa A, Cavazzana-Calvo M. Gene therapy for primary immunodeficiencies: Part 2. *Curr Opin Immunol* 2012;**24**:585–91.
14. Hong R, Cooper MD, Allan MJ, Kay HE, Meuwissen H, Good RA. Immunological restitution in lymphopenic immunological deficiency syndrome. *Lancet* 1968;**1**:503–6.
15. Bach FH, Albertini RJ, Joo P, Anderson JL, Bortin MM. Bone-marrow transplantation in a patient with the Wiskott-Aldrich syndrome. *Lancet* 1968;**2**:1364–6.
16. Friedmann T, Roblin R. Gene therapy for human genetic disease? *Science* 1972;**175**:949–55.
17. Anderson WF. Prospects for human gene therapy. *Science* 1984;**226**:401–9.
18. Nienhuis AW. Development of gene therapy for blood disorders. *Blood* 2008;**111**:4431–44.
19. Al-Herz W, Bousfiha A, Casanova JL, et al. Primary immunodeficiency diseases: an update on the classification from the international union of immunological societies expert committee for primary immunodeficiency. *Front Immunol* 2011;**2**:54.
20. Till JE, McCulloch CE. A direct measurement of the radiation sensitivity of normal mouse bone marrow cells. *Radiat Res* 1961;**14**:213–22.
21. Doulatov S, Notta F, Laurenti E, Dick JE. Hematopoiesis: a human perspective. *Cell Stem Cell* 2012;**10**:120–36.
22. Joyner A, Keller G, Phillips RA, Bernstein A. Retrovirus transfer of a bacterial gene into mouse haematopoietic progenitor cells. *Nature* 1983;**305**:556–8.
23. Mann R, Mulligan RC, Baltimore D. Construction of a retrovirus packaging mutant and its use to produce helper-free defective retrovirus. *Cell* 1983;**33**:153–9.
24. Blaese RM, Culver KW, Miller AD, et al. T lymphocyte-directed gene therapy for ADA- SCID: initial trial results after 4 years. *Science* 1995;**270**:475–80.
25. Szybalska EH, Szybalski W. Genetics of human cess line. IV. DNA-mediated heritable transformation of a biochemical trait. *Proc Natl Acad Sci USA* 1962;**48**:2026–34.
26. Pellicer A, Wigler M, Axel R, Silverstein S. The transfer and stable integration of the HSV thymidine kinase gene into mouse cells. *Cell* 1978;**14**:133–41.
27. Rous P. Transmission of a malignant new growth by means of a cell-free filtrate. *J Am Med Assoc* 1911;**56**:198.
28. Varmus H. Retroviruses. *Science* 1988;**240**:1427–35.
29. Temin HM, Mizutani S. RNA-dependent DNA polymerase in virions of Rous sarcoma virus. *Nature* 1970;**226**:1211–3.

30. Williams DA, Lemischka IR, Nathan DG, Mulligan RC. Introduction of new genetic material into pluripotent haematopoietic stem cells of the mouse. *Nature* 1984;**310**:476–80.

31. Miller AD, Eckner RJ, Jolly DJ, Friedmann T, Verma IM. Expression of a retrovirus encoding human HPRT in mice. *Science* 1984;**225**:630–2.

32. Miller AD, Buttimore C. Redesign of retrovirus packaging cell lines to avoid recombination leading to helper virus production. *Mol Cell Biol* 1986;**6**:2895–902.

33. Kantoff PW, Gillio AP, McLachlin JR, et al. Expression of human adenosine deaminase in nonhuman primates after retrovirus-mediated gene transfer. *J Exp Med* 1987;**166**:219–34.

34. Rosenberg SA, Aebersold P, Cornetta K, et al. Gene transfer into humans – immunotherapy of patients with advanced melanoma, using tumor-infiltrating lymphocytes modified by retroviral gene transduction. *N Engl J Med* 1990;**323**:570–8.

35. Naldini L, Blomer U, Gallay P, et al. *In vivo* gene delivery and stable transduction of nondividing cells by a lentiviral vector [see comments]. *Science* 1996;**272**:263–7.

36. Zufferey R, Dull T, Mandel RJ, et al. Self-inactivating lentivirus vector for safe and efficient *in vivo* gene delivery. *J Virol* 1998;**72**:9873–80.

37. Russell DW, Miller AD. Foamy virus vectors. *J Virol* 1996;**70**:217–22.

38. Muul LM, Tuschong LM, Soenen SL, et al. Persistence and expression of the adenosine deaminase gene for 12 years and immune reaction to gene transfer components: long-term results of the first clinical gene therapy trial. *Blood* 2003;**101**:2563–9.

39. Bordignon C, Notarangelo LD, et al. Gene therapy in peripheral blood lymphocytes and bone marrow for ADA- immunodeficient patients. *Science* 1995;**270**:470–5.

40. Aiuti A, Vai S, Mortellaro A, et al. Immune reconstitution in ADA-SCID after PBL gene therapy and discontinuation of enzyme replacement. *Nat Med* 2002;**8**:423–5.

41. Kohn DB, Weinberg KI, Nolta JA, et al. Engraftment of gene-modified umbilical cord blood cells in neonates with adenosine deaminase deficiency. *Nat Med* 1995;**1**:1017–23.

42. Kohn DB, Hershfield MS, Carbonaro D, et al. T lymphocytes with a normal ADA gene accumulate after transplantation of transduced autologous umbilical cord blood CD34+ cells in ADA-deficient SCID neonates. *Nat Med* 1998;**4**:775–80.

43. Hoogerbrugge PM, van Beusechem VW, Fischer A, et al. Bone marrow gene transfer in three patients with adenosine deaminase deficiency. *Gene Ther* 1996;**3**:179–83.

44. Onodera M, Ariga T, Kawamura N, et al. Successful peripheral T-lymphocyte-directed gene transfer for a patient with severe combined immune deficiency caused by adenosine deaminase deficiency. *Blood* 1998;**91**:30–6.

45. Bunting KD, Lu T, Kelly PF, Sorrentino BP. Self-selection by genetically modified committed lymphocyte precursors reverses the phenotype of JAK3-deficient mice without myeloablation. *Hum Gene Ther* 2000;**11**:2353–64.

46. Gaspar HB, Parsley KL, Howe S, et al. Gene therapy of X-linked severe combined immunodeficiency by use of a pseudotyped gammaretroviral vector. *Lancet* 2004;**364**:2181–7.

47. Thrasher AJ, Hacein-Bey-Abina S, Gaspar HB, et al. Failure of SCID-X1 gene therapy in older patients. *Blood* 2005;**105**:4255–7.

48. Chinen J, Davis J, De Ravin SS, et al. Gene therapy improves immune function in preadolescents with X-linked severe combined immunodeficiency. *Blood* 2007;**110**:67–73.

49. Malech HL, Maples PB, Whiting-Theobald N, et al. Prolonged production of NADPH oxidase-corrected granulocytes after gene therapy of chronic granulomatous disease. *Proc Natl Acad Sci USA* 1997;**94**:12133–8.

50. Kang EM, Choi U, Theobald N, et al. Retrovirus gene therapy for X-linked chronic granulomatous disease can achieve stable long-term correction of oxidase activity in peripheral blood neutrophils. *Blood* 2010;**115**:783–91.

51. Grez M, Reichenbach J, Schwable J, Seger R, Dinauer MC, Thrasher AJ. Gene therapy of chronic granulomatous disease: the engraftment dilemma. *Mol Ther* 2011;**19**:28–35.

52. Hacein-Bey Abina S, Caccavelli L, Malani N, et al. Comparison of integration site profiles between SCID-X1 gene therapy trials using gammaretroviral vectors with intact or deleted (SIN) LTRs. Abstract ASGCT no. 310. *Mol Ther* 2013;**21**(Suppl 1):S119.

53. Aiuti A, Biasco L, Scaramuzza S, et al. Lentiviral hematopoietic stem cell gene therapy in patients with Wiskott–Aldrich syndrome. *Science* 2013;**341**:1233151.

54. Hacein-Bey Abina S, Blondeau J, Caccavelli L, et al. Lentiviral vector-based gene therapy for Wiskott–Aldrich syndrome: preliminary results from the french center, Abstract ASGCT no. 306. *Mol Ther* 2013;**21** (Suppl 1):S117.

55. Hanenberg H, Xiao XL, Dilloo D, Hashino K, Kato I, Williams DA. Colocalization of retrovirus and target cells on specific fibronectin fragments increases genetic transduction of mammalian cells. *Nat Med* 1996;**2**:876–82.

56. Miller AD, Garcia JV, von Suhr N, Lynch CM, Wilson C, Eiden MV. Construction and properties of retrovirus packaging cells based on gibbon ape leukemia virus. *J Virol* 1991;**65**:2220–4.

57. Orkin SH, Motulsky AG. Report and recommendations of the panel to assess the NIH investment in research on gene therapy. *http://www.nih.gov/news/panelrep.html*, 1995.

58. Bunting KD, Sangster MY, Ihle JN, Sorrentino BP. Restoration of lymphocyte function in Janus kinase 3-deficient mice by retroviral-mediated gene transfer [see comments]. *Nat Med* 1998;**4**:58–64.

59. Lo M, Bloom ML, Imada K, et al. Restoration of lymphoid populations in a murine model of X-linked severe combined immunodeficiency by a gene-therapy approach. *Blood* 1999;**94**:3027–36.

60. Otsu M, Anderson SM, Bodine DM, Puck JM, O'Shea JJ, Candotti F. Lymphoid development and function in X-linked severe combined immunodeficiency mice after stem cell gene therapy. *Mol Ther* 2000;**1**:145–53.

61. Soudais C, Tusjino S, Sharara LI, et al. Stable and functional lymphoid reconstitution common cytokine receptor gamma chain deficient mice by retroviral-mediated gene transfer. *Blood* 2000;**95**:3071–7.

62. Hirschhorn R, Yang DR, Puck JM, Huie ML, Jiang CK, Kurlandsky LE. Spontaneous in vivo reversion to normal of an inherited mutation in a patient with adenosine deaminase deficiency [see comments]. *Nat Genet* 1996;**13**:290–5.

63. Stephan V, Wahn V, Le Deist F, et al. Atypical X-linked severe combined immunodeficiency due to possible spontaneous reversion of the genetic defect in T-cells. *N Engl J Med* 1996;**335**:1563–7.

64. Bousso P, Wahn V, Douagi I, et al. Diversity, functionality, and stability of the T-cell repertoire derived in vivo from a single human T-cell precursor. *Proc Natl Acad Sci USA* 2000;**97**:274–8.

65. Hacein-Bey S, Cavazzana-Calvo M, Le Deist F, et al. gamma-c gene transfer into SCID X1 patients' B-cell lines restores normal high-affinity interleukin-2 receptor expression and function. *Blood* 1996;**87**:3108–16.

66. Hacein-Bey S, Basile GD, Lemerle J, Fischer A, Cavazzana-Calvo M. gamma c gene transfer in the presence of stem cell factor, FLT-3L, interleukin-7 (IL-7), IL-1, and IL-15 cytokines restores T-cell differentiation from gammac(-) X-linked severe combined immunodeficiency hematopoietic progenitor cells in murine fetal thymic organ cultures. *Blood* 1998;**92**:4090–7.

67. Fischer A, Le Deist F, Hacein-Bey-Abina S, et al. Severe combined immunodeficiency. A model disease for molecular immunology and therapy. *Immunol Rev* 2005;**203**:98–109.

68. Mueller PR, Wold B. *In vivo* footprinting of a muscle specific enhancer by ligation mediated PCR. *Science* 1989;**246**:780–6.

69. Schmidt M, Schwarzwaelder K, Bartholomae C, et al. High-resolution insertion-site analysis by linear amplification-mediated PCR (LAM-PCR). *Nat Methods* 2007;**4**:1051–7.

70. Wang GP, Garrigue A, Ciuffi A, et al. DNA bar coding and pyrosequencing to analyze adverse events in therapeutic gene transfer. *Nucleic Acids Res* 2008;**36**:e49.

71. Deichmann A, Hacein-Bey-Abina S, Schmidt M, et al. Vector integration is nonrandom and clustered and influences the fate of lymphopoiesis in SCID-X1 gene therapy. *J Clin Invest* 2007;**117**:2225–32.

72. Wang GP, Berry CC, Malani N, et al. Dynamics of gene-modified progenitor cells analyzed by tracking retroviral integration sites in a human SCID-X1 gene therapy trial. *Blood* 2010;**115**:4356–66.

73. Gaspar HB. Gene therapy for ADA-SCID: defining the factors for successful outcome. *Blood* 2012;**120**:3628–9.

74. Aiuti A, Cassani B, Andolfi G, et al. Multilineage hematopoietic reconstitution without clonal selection in ADA-SCID patients treated with stem cell gene therapy. *J Clin Invest* 2007;**117**:2233–40.

75. Fernandes JF, Rocha V, Labopin M, et al. Transplantation in patients with SCID: mismatched related stem cells or unrelated cord blood? *Blood* 2012;**119**:2949–55.

76. Hacein-Bey-Abina S, Garrigue A, Wang GP, et al. Insertional oncogenesis in 4 patients after retrovirus-mediated gene therapy of SCID-X1. *J Clin Invest* 2008;**118**:3132–42.

77. Check E. Harmful potential of viral vectors fuels doubts over gene therapy. *Nature* 2003;**423**:573–4.

78. Wu X, Li Y, Crise B, Burgess SM. Transcription start regions in the human genome are favored targets for MLV integration. *Science* 2003;**300**:1749–51.

79. Paruzynski A, Arens A, Gabriel R, et al. Genome-wide high-throughput integrome analyses by nrLAM-PCR and next-generation sequencing. *Nat Protocols* 2010;**5**:1379–95.

80. Schwarzwaelder K, Howe SJ, Schmidt M, et al. Gammaretrovirus-mediated correction of SCID-X1 is associated with skewed vector integration site distribution in vivo. *J Clin Invest* 2007;**117**:2241–9.

81. Baum C. Parachuting in the epigenome: the biology of gene vector insertion profiles in the context of clinical trials. *EMBO Mol Med* 2011;**3**:75–7.

82. Woods NB, Bottero V, Schmidt M, von Kalle C, Verma IM. Gene therapy: therapeutic gene causing lymphoma. *Nature* 2006;**440**:1123.

83. Thrasher AJ, Gaspar HB, Baum C, et al. Gene therapy: X-SCID transgene leukaemogenicity. *Nature* 2006;**443**:E5–6 discussion E-7.

84. Cavazzana-Calvo M, Fischer A, Hacein-Bey-Abina S, Aiuti A. Gene therapy for primary immunodeficiencies: part 1. *Curr Opin Immunol* 2012;**24**:580–4.

85. Thornhill SI, Schambach A, Howe SJ, et al. Self-inactivating gammaretroviral vectors for gene therapy of X-linked severe combined immunodeficiency. *Mol Ther* 2008;**16**:590–8.

86. Gaszner M, Felsenfeld G. Insulators: exploiting transcriptional and epigenetic mechanisms. *Nat Rev Genet* 2006;**7**:703–13.

87. Schroder AR, Shinn P, Chen H, Berry C, Ecker JR, Bushman F. HIV-1 integration in the human genome favors active genes and local hotspots. *Cell* 2002;**110**:521–9.

88. Cartier N, Hacein-Bey-Abina S, Bartholomae CC, et al. Hematopoietic stem cell gene therapy with a lentiviral vector in X-linked adrenoleukodystrophy. *Science* 2009;**326**:818–23.

89. Modlich U, Navarro S, Zychlinski D, et al. Insertional transformation of hematopoietic cells by self-inactivating lentiviral and gammaretroviral vectors. *Mol Ther* 2009;**17**:1919–28.

90. Ryu BY, Evans-Galea MV, Gray JT, Bodine DM, Persons DA, Nienhuis AW. An experimental system for the evaluation of retroviral vector design to diminish the risk for proto-oncogene activation. *Blood* 2008;**111**:1866–75.

91. Corrigan-Curay J, Cohen-Haguenauer O, O'Reilly M, et al. Challenges in vector and trial design using retroviral vectors for long-term gene correction in hematopoietic stem cell gene therapy. *Mol Ther* 2012;**20**:1084–94.

92. Scaramuzza S, Biasco L, Ripamonti A, et al. Preclinical safety and efficacy of human CD34(+) cells transduced with lentiviral vector for the treatment of Wiskott–Aldrich syndrome. *Mol Ther* 2013;**21**:175–84.

93. Charrier S, Dupre L, Scaramuzza S, et al. Lentiviral vectors targeting WASp expression to hematopoietic cells, efficiently transduce and correct cells from WAS patients. *Gene Ther* 2007;**14**:415–28.

94. Astrakhan A, Sather BD, Ryu BY, et al. Ubiquitous high-level gene expression in hematopoietic lineages provides effective lentiviral gene therapy of murine Wiskott–Aldrich syndrome. *Blood* 2012;**119**:4395–407.

95. Santilli G, Almarza E, Brendel C, et al. Biochemical correction of X-CGD by a novel chimeric promoter regulating high levels of transgene expression in myeloid cells. *Mol Ther* 2011;**19**:122–32.

96. Bartholomae CC. European Society of Gene and Cell Therapy French Society of Cell and Gene Therapy Collaborative Congress 2012. October 25–29, 2012. *Methods Mol Biol* 2012;A1–A173.

97. Biffi A, Montini E, Lorioli L, et al. Lentiviral hematopoietic stem cell gene therapy benefits metachromatic leukodystrophy. *Science* 2013;**341** 1233158.

98. van der Oost J. Molecular biology. New tool for genome surgery. *Science* 2013;**339**:768–70.

99. Papapetrou EP, Lee G, Malani N, et al. Genomic safe harbors permit high beta-globin transgene expression in thalassemia induced pluripotent stem cells. *Nat Biotechnol* 2011;**29**:73–8.

100. Lombardo A, Genovese P, Beausejour CM, et al. Gene editing in human stem cells using zinc finger nucleases and integrase-defective lentiviral vector delivery. *Nat Biotechnol* 2007;**25**:1298–306.

101. Pessach IM, Ordovas-Montanes J, Zhang SY, et al. Induced pluripotent stem cells: a novel frontier in the study of human primary immunodeficiencies. *J Allergy Clin Immunol* 2011;**127**:1400–7.

Index

Color Plate

FIGURE F.2 Ralph Wedgwood: Connected immunodeficiency with autoimmunity.

FIGURE F.5 Max Cooper: Established the concept of B and T lymphocytes and adaptive immunity.

FIGURE F.3 Fred Rosen: Instrumental in creating the field of PIDD as a specialty in medicine.

FIGURE F.8 Maxime Seligman: Early pioneer of PID in Europe.

FIGURE 1.1 Lady Mary Wortley Montagu, wife of the British Ambassador to Istanbul, capital of the Ottoman Empire. She introduced variolation to the West. *(Source: Wikipedia, Public Domain.)*

Impaired Cellular Immunity

1. Lymphocyte depletion

2. Impaired allograft rejection,
   delayed type hypersensitivity
   and GVH potential

3. Many plasma cells, but
   impaired antibody responses

Impaired Humoral Immunity

1. No germinal centers
   or plasma cells

2. No antibodies

3. Normal cellular immunity

FIGURE 2.2 The immune system effects of thymectomy or bursectomy combined with near-lethal irradiation. GVH: graft versus host.

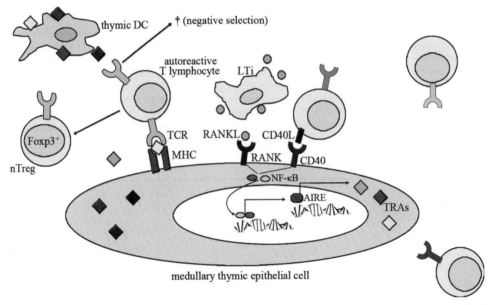

FIGURE 4.1 Mechanisms of T-cell tolerance in the thymus. Cross-talk between medullary thymic epithelial cells and lymphoid cells induces maturation of medullary thymic epithelial cells. In particular, secretion of RANK ligand by lymphoid tissue induced (LTi) cells secrete RANK ligand (RANKL) already at early stages during thymic development, and newly generated thymocytes express CD40 ligand (CD40L). These molecules bind to RANK and CD40, respectively, which are both expressed by medullary thymic epithelial cells, triggering activation of the NF-κB signaling pathway, ultimately inducing expression of the autoimmune regulator (*AIRE*) gene. The transcription factor AIRE drives expression of tissue-restricted antigens (TRAs), which are presented in association with major histocompatibility complex class II (MHC-II) molecules on the surface of medullary thymic epithelial cells or of thymic dendritic cells (DC). Recognition of TRA/MHC-II complex by self-reactive thymocytes that are generated in the thymus leads to clonal deletion or to conversion into natural regulatory T-cells (nTreg) expressing the FOXP3 transcription factor.

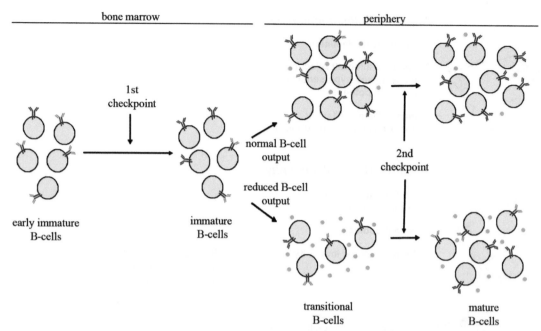

FIGURE 4.2 Checkpoints of B-cell tolerance in the bone marrow and in the periphery. Stochastic rearrangements of the V, D, and J elements of immunoglobulin genes in bone marrow B-cell progenitors allow generation of a diversified repertoire of early immature B lymphocytes, including a significant fraction of cells with self-reactive specificity (indicated by a blue trait on the immunoglobulin variable region). Prolonged expression of the RAG genes allows secondary V(D)J rearrangement that modifies the immunoglobulin specificity, thus leading to a decrease in the proportion of self-reactive cells among immature B-cells (first checkpoint). Upon egress to the periphery, B-cells respond to growth/survival factors, including BAFF (green solid circles). Anergic self-reactive B-cells express lower amounts of BAFF receptors. In the presence of normal B-cell output from the bone marrow, and of steady-state BAFF concentration in the serum, survival of non self-reactive mature B-cells is favored, resulting in a decrease in the proportion of mature self-reactive B-cells (secondary checkpoint). However, generation of a reduced number of transitional B-cells is associated with higher serum BAFF levels, thereby allowing rescue of self-reactive mature B-cells.

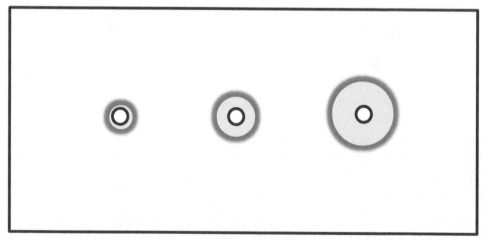

FIGURE 5.1 Radial immunodiffusion (RID) assay. In this example, anti-IgG is contained within the agar of this slide assay system and serum is placed in the pre-cut wells. IgG migrates radially from the well and precipitates when antigen–antibody equivalence has been reached. In this example of three serum samples, the precipitin ring diameter reflects the level of serum IgG. The actual concentration of IgG (or other immunoglobulin class, complement proteins, etc.) is determined by comparing the diameter of the precipitin ring from the patient sample to a series of standards.

A.

B.

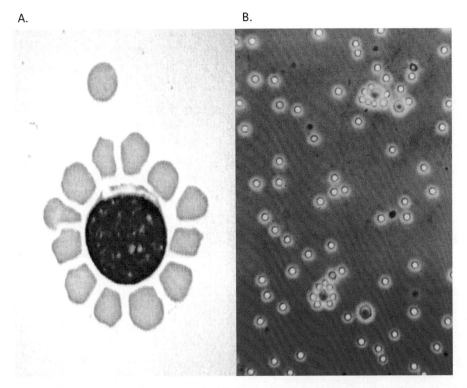

FIGURE 5.2 E-rosette. Panel A demonstrates a T-cell with sheep red cells forming a rosette highlighted by staining. Panel B demonstrates a direct preparation of peripheral blood mononuclear cells mixed with sheep red blood cells, demonstrating T-cell erythrocyte rosette formation; this represents the test system used to quantify T-cells.

A.

B.

C.

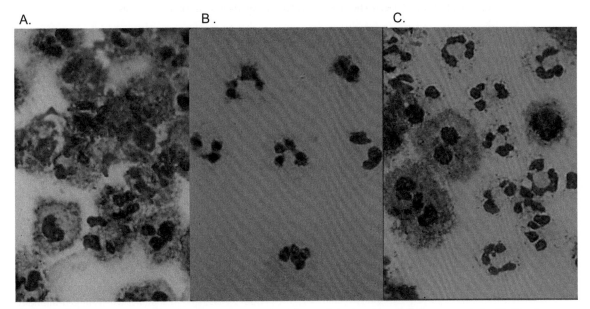

FIGURE 5.3 NBT test. Panel A demonstrates normal oxidase activity in neutrophils from a control with blue formazan dye clearly present. Panel B demonstrates absence of oxidase activity in the neutrophils of an X-linked CGD patient (no blue dye). Panel C demonstrates a combination of normal and abnormal neutrophils from an X-linked CGD carrier.

FIGURE 6.1 ASID Inaugural meeting in Casablanca, October 2008. Aziz Bousfiha (arrow) and, on his right Claude Griscelli, Luigi Notarangelo (Boston) and Bouchra Benhayoun (Hajar Association). Jean-Laurent Casanova (ESID president, star) in the middle of several African delegates but also several other experts from the five continents.

Vicki and Fred Modell with the picture of their son Jeffery.

FIGURE 8.3 Robert P. Sedgwick, neurologists from Children's Hospital, Los Angeles, who described the syndrome as ataxia-telangiectasia in 1957 (3).

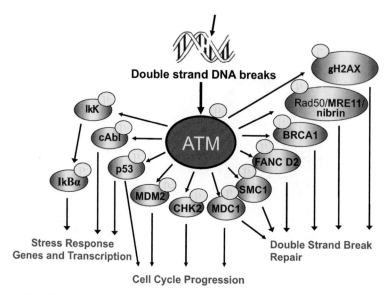

FIGURE 8.4 The multiple functions of ATM in response to double-stranded DNA breaks, including the rapid repair of DNA breaks, and its impact on cell cycle progression, stress response, genes and transcription.

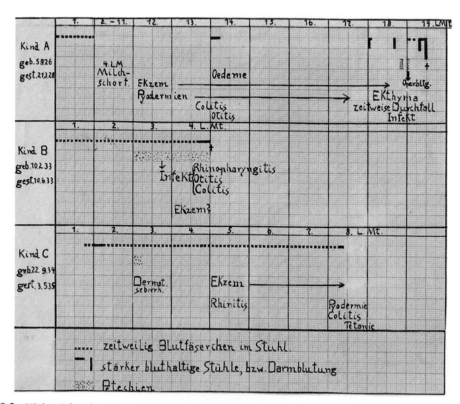

FIGURE 9.2 Wiskott's handwritten summary of the clinical course of the three brothers (A, B, C) presented in 1936 at the Pediatric meeting in Wuerzburg and published one year later in *Monatsschrift für Kinderheilkunde* (1). LM = Lebensmonat (month of life), ZeitweiseDurchfall = intermittent diarrhea, Zeitweilige Blutfäserchenim Stuhl = intermittent bloody strings in stool.

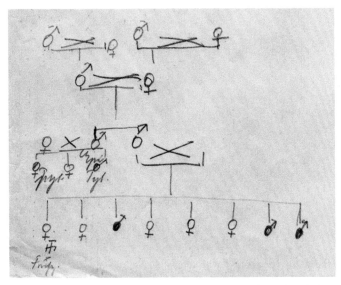

FIGURE 9.4 Pedigrees of the Bavarian family with three affected boys: Wiskott's hand drawn family tree; he seems to emphasize the father's side of the family. The eldest daughter is indicated to have died of prematurity. *(From Binder V. et al., New England Journal of Medicine, 2006, with permission; 7.)*

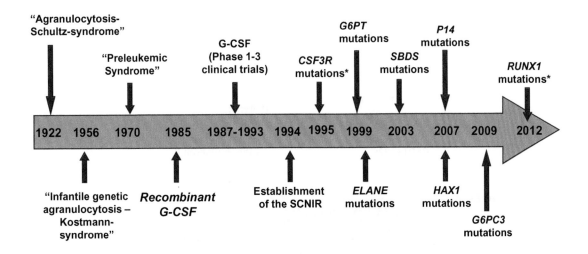

* = acquired mutations

FIGURE 10.1 Milestones in the history of congenital neutropenias.

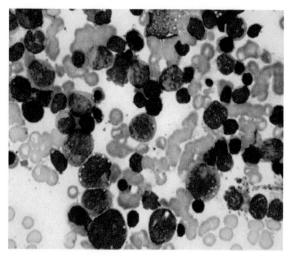

FIGURE 10.2   Typical bone marrow of a patient with congenital neutropenia.

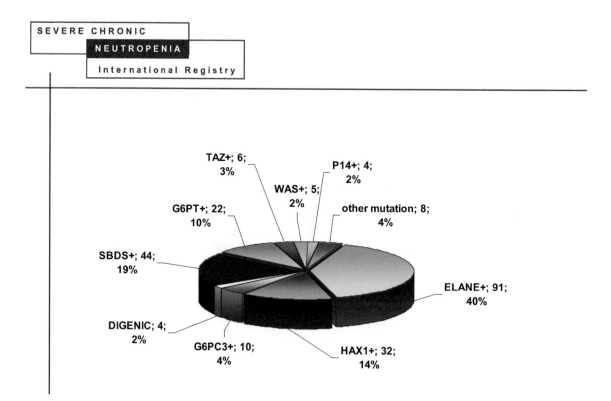

FIGURE 10.5   Distribution of gene mutations in European congenital neutropenia patients.

FIGURE 11.4 One of the discoverers of the genetic defect meets the discoverer of the disease in 1993. Dr Bruton is seated, his wife Kathryn D Bruton is standing to his left, Dr Jeffrey D Thomas is standing to his right and Dr CI Edvard Smith is standing behind him. *(Reproduced with permission from (37).)*

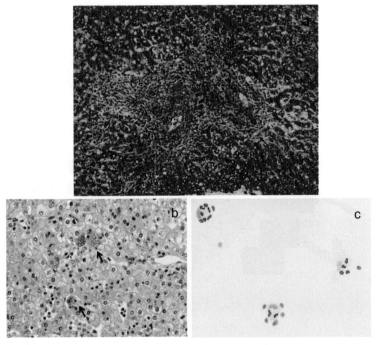

FIGURE 12.1 Cellular infiltration and hemophagocytosis: (A) H&E staining shows massive infiltration of portal tract and lobules of liver biopsy by mononuclear cells; (B) active phagocytosis of erythrocytes (arrow) by macrophages; (C) hemophagocytosis of blood elements by macrophages in the cerebrospinal fluid. *(Courtesy of Pr Nicole Brousse, Hôpital Necker, Paris, France.)*

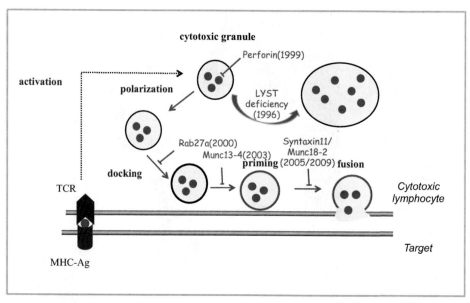

FIGURE 12.2 Inborn errors in the cytotoxic activity of lymphocytes causing HLH: the genetic defects causing hemophagocytic lymphohistiocytic syndrome (HLH) affect discrete steps of the cytotoxic machinery, i.e. granule biogenesis/morphology, granule content, docking, priming or fusion. In brackets, year of gene discovery.

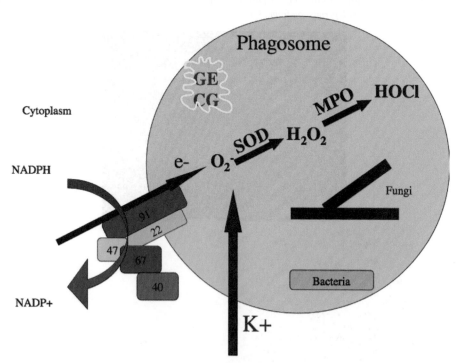

FIGURE 13.4 The assembled NADPH oxidase on the wall of the phagolysosome. The generation of superoxide, hydrogen peroxide, and hypohalous acid are shown. GE is granulocyte elastase and CG is cathepsin G, both enzymes that may be activated by the processes set in motion after superoxide generation.

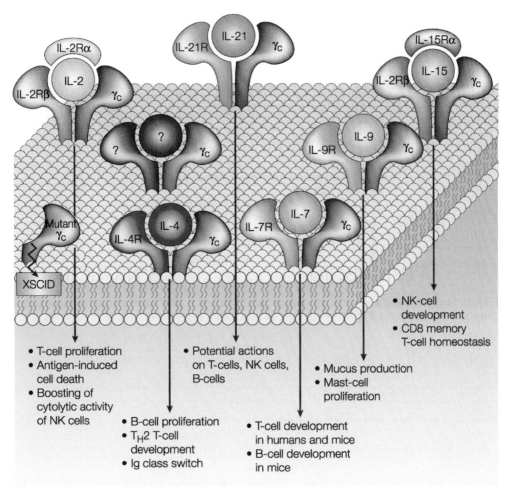

FIGURE 15.4 The γ_c family of cytokines. Originally discovered as IL-2Rγ, γ_c is now known to be an essential component of the receptors for IL-2, IL-4, IL-7, IL-9, IL-15 and IL-21. *(From Leonard et al. 2001) (36).*

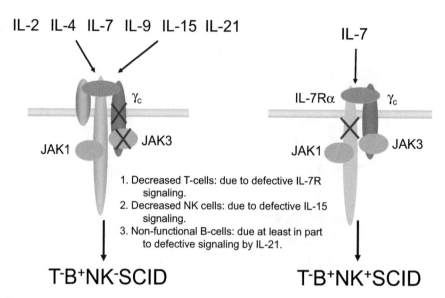

FIGURE 15.5 JAK1 associates with IL-2Rβ, IL-4Rα, IL-7Rα, IL-9R, and IL-21R, whereas JAK3 associates with γc. XSCID and JAK3-deficient SCID result in a T⁻B⁺NK⁻ phenotype whereas IL-7Rα-deficient SCID results in a T⁻B⁺NK⁺ phenotype.

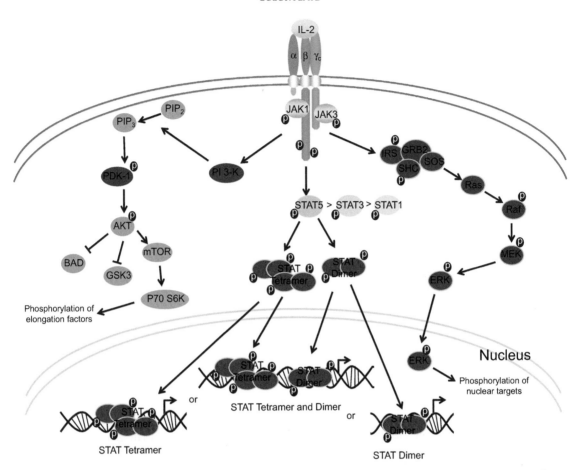

FIGURE 15.6 Signaling pathways utilized by IL-2. IL-2 signals via JAK-STAT, PI 3-kinase/Akt, and RAS-MAP kinase pathways. STAT5A and STAT5B are the principal STAT proteins activated by IL-2, with STAT1 and STAT3 activated to a lesser degree. *(Modified from a figure in (8).)*

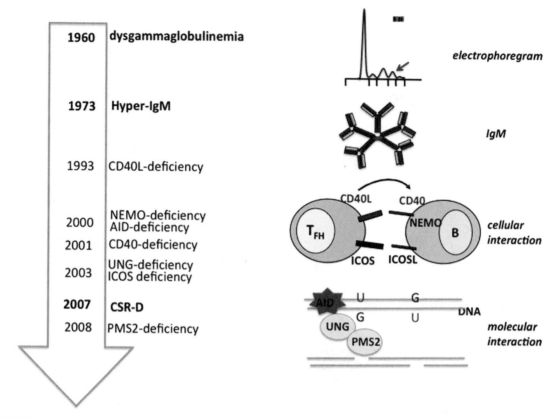

FIGURE 16.1 Milestones in the characterization of CSR-Ds. An electropherogram reveals dysgammaglobulinemia (elevated 19S globulins and a lack of γ-globulins). Pentameric IgM ("M" stands for "macroglobulinemia"). Cellular interactions: T follicular helper (T$_{FH}$) cells cross-talk with B-cells through interaction between CD40L/CD40 and – to a lesser extent – ICOS/ICOSL. CD40 signaling in B-cells involves the NF-κB pathway via activation of NF-κB essential modulator (NEMO). Molecular interactions: Activation induced cytidine deaminase (AID) introduces cytidine > uridine DNA lesions within S and V regions. The U:G base pairs that are misintegrated into the DNA are mostly processed by uracil-N glycosylase (UNG), leading to double-strand DNA breaks. PMS2 endonuclease activity contributes to DNA breakage downstream from UNG. The names of the disease entities described at the different time periods are indicated in bold letters.

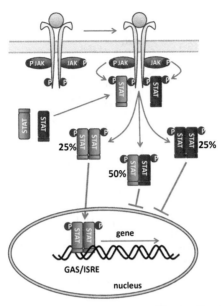

FIGURE 19.2 Dominant-negative effect of *STAT3* mutations on STAT3 signaling. Heterozygous *STAT3* mutations result in half of the STAT3 proteins being non-functional, allowing only ≈25% intact STAT3 homodimers to form. These are thought to be necessary for cell survival but seem to be insufficient for the proper function of some subsets of cells, such as leukocytes and cells that participate in bone formation. Mutated STAT3 is depicted red, wild-type STAT3 green. (*Figure courtesy of C. Woellner.*)

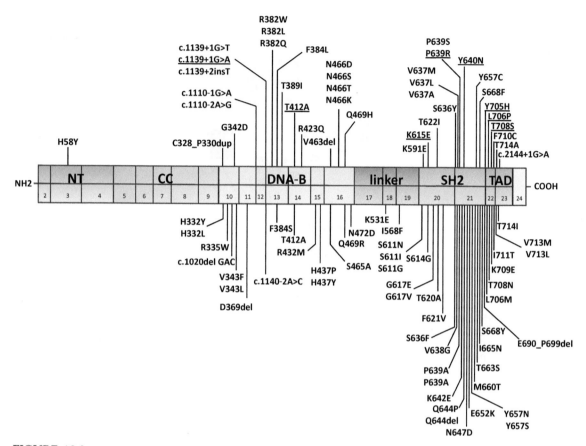

FIGURE 19.3 Schematic representation of *STAT3* showing mutational hotspots in the DNA-binding (DNA-B) and Src homology 2 (SH2) domain. Further mutations are located in the amino-terminal (NT), the linker and the transcriptional activation (TAD) domains. Mutations on the upper part of the cartoon were identified by Woellner et al., 2010. Underscored mutations have not yet been published in the literature. *(Figure modified after M.O. Chandesris et al., 2012 (59).)*

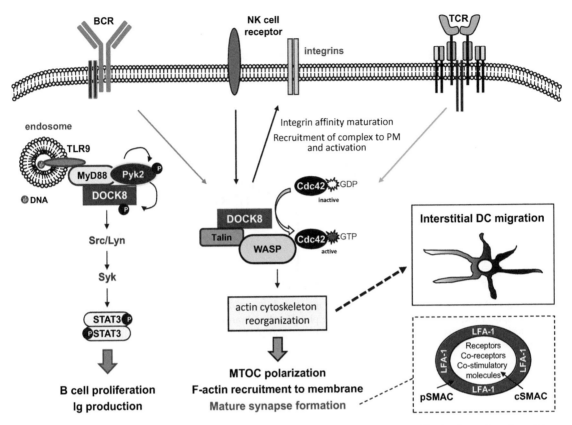

FIGURE 19.5 Role of DOCK8 in immune cells. In B-, T- and NK cells, DOCK8 is crucial for the formation of a stable immunological synapse with an integrin-containing pSMAC. It has been shown in NK cells that DOCK8 forms a complex with WASp and talin, leading to F-actin recruitment to the membrane, MTOC polarization and formation of the cytotoxic synapse. In this complex, DOCK8 might function as a GEF for Cdc42 which, in turn, activates WASp followed by actin cytoskeleton rearrangements. In DCs, DOCK8 plays a role in spatial activation of Cdc42 at the leading edge and interstitial DC migration. In B cells, DOCK8 might have an additional role as an adaptor protein in a complex with MyD88 and Pyk2 downstream of TLR9 and upstream of STAT3, being important for B-cell proliferation and immunoglobulin production. BCR: B-cell receptor; DC: dendritic cell; GEF: guanine nucleotide exchange factor; MTOC: microtubule organizing center; NK cell: natural killer cell; PM: plasma membrane; pSMAC: peripheral supramolecular activation complex; TCR: T-cell receptor; WASp: Wiskott–Aldrich syndrome protein.

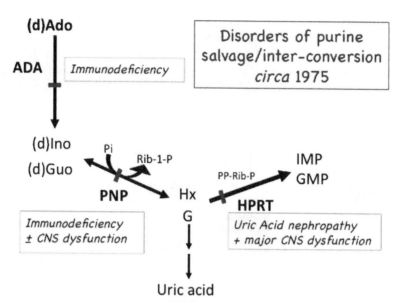

FIGURE 20.1 Major inherited disorders of purine metabolism *circa* 1975. ADA: adenosine deaminase; PNP: purine nucleoside phosphorylase; HPRT: hypoxanthine-guanine phosphoribosyltransferase; (d): deoxy; Ado: adenosine; Ino: inosine; Guo: guanosine; Hx: hypoxanthine; G: guanine; Pi: inorganic phosphate; Rib-1-P: ribose-1-phosphate; PP-rib-P: phosphoribosylpyrophosphate; CNS: central nervous system.

FIGURE 24.1 Idea of a Chimera in ancient times.

FIGURE 24.2 Long-term canine chimeras, fully immune reconstituted, living in an unprotected environment without immunosuppression or antimicrobial prophylaxis, generated in the early 1970s.

FIGURE 24.4 The first successful bone marrow transplantation performed in an infant with X-linked SCID, photographed at 5 years of age, together with Robert Good, who led the BM transplantation team at the University of Minnesota.

Additional Figures

FIGURE 1 Robert A. Good, MD, PhD, 1922–2003. This portrait is located at the Jeffrey Modell Immunology Center at Harvard Medical School.

FIGURE 2 Robert Good, Walter Hitzig, Richard Gatti and Norbibi Good, PIDD meeting in the 1990s.

FIGURE 3 Discussion of PIDD or politics, Ralph Wedgwood, Fred Rosen, Tom Waldmann, during the WHO-meeting, Hokkaido, Japan, 2001.

FIGURE 4 Claude Griscelli, who started the PID group at Hopital Necker, Tadamitsu Kishimoto from Osaka University and Ben Zeger, University of Utrecht, WHO meeting, Hokkaido, Japan, 2001.

FIGURE 5 Raif Geha, Erwin Gelfand, Fred Modell, Fred Rosen during a WHO meeting in Baden bei Wien, June 1998.

FIGURE 6 Philippe Schneider (center) and Silvio Barandun (R), both experts in IVIG, during one of the Interlaken meetings in Switzerland.

FIGURE 7 Robert Good (center), shortly before his death in 2008.

FIGURE 8 Vicki and Fred Modell, Fred Rosen, 2003 WHO meeting, Sintra, Portugal.

FIGURE 9 Past, present and future Presidents of ESID, photo taken by Hans Ochs during the WHO sponsored PID meeting in Bristol 1997: Alan Fischer (1st ESID president), Amos Etzioni (5th ESID president), Igor Resnik, Luigi Notarangelo (3rd ESID president), Edvard Smith (2nd ESID president), Owen Witte, Jean-Laurent Casanova (4th ESID president), Andrew Cant (6th ESID president).

FIGURE 10 A light moment during the 2003 AAAAI meeting in Denver, Colorado, Coors Brewery, in front: Elias Toubi (Palestinian) Amos Etzioni (Israeli), Hans Ochs (German).

FIGURE 11 IUIS expert committee for PID, 2011: (standing, left to right) Aziz Bousfiha, Chaim Roifman, Lennart Hammarström, Helen Chapel, Alan Fischer, Mary Ellen Conley, Reinhard Seger, Shigeaki Nonoyama, Jennifer Puck, Amos Etzioni, Charlotte Cunningham-Rundles, Mimi Tang; (front) Waleed Al-Herz, Jose Franco and Hans Ochs.

FIGURE 12 IUIS expert committee for PID, 2013: (left to right) Waleed Al-Herz. Amos Etzioni, Shigeaki Nonoyama, Kate Sullivan, Charlotte Cunningham-Rundles, Eric Oksenhendler, Luigi Notarangelo, Jose Franco, Bobby Gaspar, Jennifer Puck, Capucine Picard, Steve Holland, Talal Chatila, Mimi Tang, Mary Ellen Conley, Aziz Bousfiha and Hans Ochs.

FIGURE 13   The second European "Summer School" in Sintra, Portugal.

FIGURE 14   The first patient (second from left) successfully transplanted with bone marrow from his HLA-matched sister (second from right) at age 19. Richard Gatti at left.

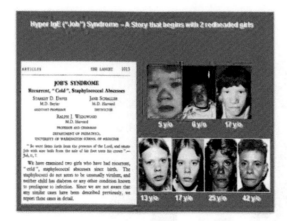

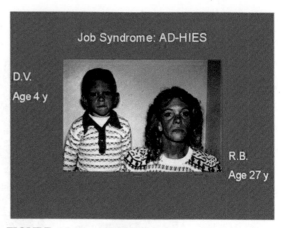

FIGURE 15 The 2 original "red headed girls" with Job syndrome, described by Davis, Schaller and Wedgwood in 1966, both succumbed to the disease, at age 18 and at age 53, respectively.

FIGURE 16 One of the original Job syndrome patients with her second affected son, demonstration autosomal dominant inheritance long before the causative gene, STAT3, was recognized as having a dominant negative effect if the mutation results in non-functional protein.

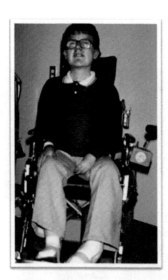

FIGURE 17 Patient BB at age 15, who had CGD, Duchenne's muscular dystrophy, retinitis pigmentosa and the McLeod blood phenotype. He volunteered to give blood many times and was instrumental in understanding the genetics of these X-linked diseases.

Dates of description and gene discovery of major PIDDs

Name of disease	Date of description	Date of genetic discovery
Ataxia–telangiectasia	1926	1995
Wiskott–Aldrich syndrome	1937	1994
Congenital neutropenia (Kostmann)	1950	2007
Agammaglobulinemia (X-linked)	1952	1993
Familial hemophagocytosis syndrome	1952	1999
Chronic granulomatous disease (X-linked)	1954	1987
Severe combined immunodeficiency (Swiss-type agammaglobulinemia)	1958	1993 (XSCID, cgamma)
Hyper IgM X-linked	1960	1993
Complement (C2)	1960	1992
DiGeorge syndrome	1965	1981
Hyper-IgE syndrome (Job syndrome)	1966	2007
ADA deficiency	1972	1985
Leukocyte adhesion deficiency 1 (LAD1)	1980	1987
Common variable immune deficiency	1954	

Printed and bound by CPI Group (UK) Ltd, Croydon, CR0 4YY

03/10/2024

01040321-0004